Unexplored by Whalemen.

RUSSIAN

AMERICA

N O R T H

A M E R I C A

KAMTCHATKA

ALEUTIAN ARCH.

ALTA CALIFORNIA

UPPER CANADA

CANADA

U N I T E D

S T A T E S

MEXICO

GULF OF MEXICO

CARIBBEAN SEA

Acapulco

Arequipa

BOLIVIA

PERU

BRASIL

S O U T H

Rio Janeiro

PATAGONIA

C. Blanco

all

od fishing may be had in these latitudes during winter (w) i. e. the Southern Summer.

OCEANOGRAPHY

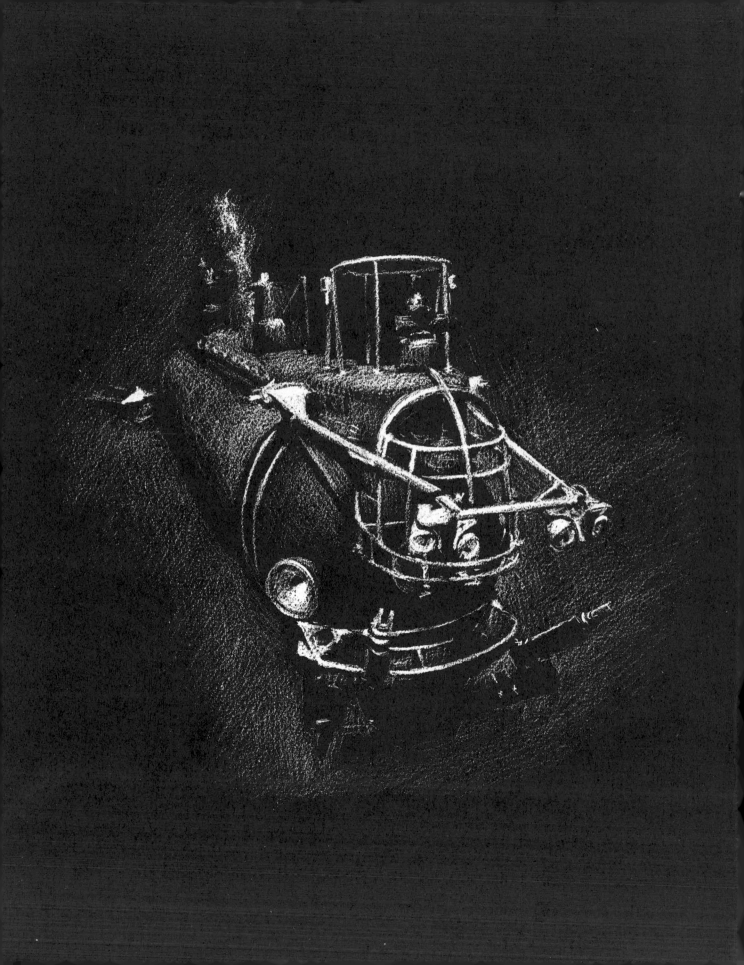

OCEANOGRAPHY
a view of the earth

M. GRANT GROSS

State University of New York
at Stony Brook

Illustrated by **Roger Hayward**

Prentice-Hall, Inc., Englewood Cliffs, New Jersey

To NANCY

The maps on pages 40, 53, 71, 126,
180, 181, 187, 214, 229, and 232
are from Goode's Base Map Series,
© University of Chicago.

10 9 8 7 6 5 4

ISBN: 0-13-629659-9

Library of Congress Catalog Card Number: 78-187899

Printed in the United States of America.

PRENTICE-HALL INTERNATIONAL, INC., London
PRENTICE-HALL OF AUSTRALIA PTY. LTD., Sydney
PRENTICE-HALL OF CANADA LTD., Toronto
PRENTICE-HALL OF INDIA PRIVATE LIMITED, New Delhi
PRENTICE-HALL OF JAPAN, Inc., Tokyo

CONTENTS

preface

one

OCEANOGRAPHY—THE SCIENCE   3

Oceanographers—those who study the ocean, 6   Physical oceanography, 7
Chemical oceanography, 9   Biological oceanography, 11
Geological oceanography and marine geophysics, 13   Ocean engineering, 15
Applications of oceanography, 17

two

LAND AND SEA   25

Distribution of land and ocean, 29   Pacific Ocean, 31   Atlantic Ocean, 34   Arctic Ocean, 35
Indian Ocean, 36   Sea level, 37   Ocean depths, 38   Continental margin, 41
Marginal-ocean basins, 45   Coastal ocean, 46

three

DEEP-OCEAN FLOOR   51

Oceanic rises, 52   Ocean basin, 59   Abyssal plains, 61   Submarine volcanoes, 62
Island arcs and trenches, 66   Coral reefs, 69

four

79   HISTORY OF OCEAN AND ATMOSPHERE

Crustal structure, *81*   Sea-floor spreading, *87*   Fracture zones and volcanic ridges, *90*
Magnetic stripes—the earth's growth lines, *91*   Evolution of ocean basins, *93*
Origin of ocean and atmosphere, *97*   Origin of life on earth, *99*

five

107   SEDIMENTS

Origin and classification of marine sediment, *108*   Sediment budget, *116*   Sediment transport, *119*
Accumulation of sediment in the ocean, *121*   Distribution of sediment deposits, *125*
Correlations and age determinations, *134*

six

141   SEAWATER

Molecular structure of water, *143*   Temperature effects on water, *146*   Density, *149*
Composition of sea salt, *150*   Salt composition and residence times, *154*   Salt in water, *155*
Dissolved gases, *159*   Physical properties of seawater, *161*

seven

169   TEMPERATURE, SALINITY, AND DENSITY

Light in the ocean, *172*   Heat budget and atmospheric circulation, *173*   Water budget, *179*
Density of seawater, *188*   Water masses, *192*   Sound in the sea, *198*   Sea-ice formation, *201*
Climatic regions, *206*

eight

211   OCEAN CIRCULATION

General surface circulation in the open ocean, *215*   Boundary currents, *218*   Langmuir circulation, *221*
Forces causing currents, *222*   Geostrophic currents, *227*   Thermohaline circulation, *230*
Atlantic Ocean circulation–a three-dimensional view, *233*

nine

239   WAVES

Ideal waves, *240*   Formation of sea and swell, *247*   Wave height and wave energy, *251*
Waves in shallow water, *254*   Internal and standing waves, *261*

ten

TIDES AND TIDAL CURRENTS   267

Tidal curves, *269*   Tide-generating forces and the equilibrium tide, *273*
Dynamical theory of the tide, *278*   Tidal currents, *283*   Tidal currents in coastal areas, *287*

eleven

ESTUARIES   295

Origin of estuaries, fjords, and lagoons, *296*   Estuarine circulation, *300*   Tides in estuaries, *304*
Biological productivity of estuaries, *305*   Chesapeake Bay, *306*   Pamlico Sound, *309*
Laguna Madre, *312*   Puget Sound, *315*   Long Island Sound, *319*

twelve

THE COASTAL OCEAN   327

Temperature and salinity in the coastal ocean, *329*   Coastal currents, *331*   Storm surges, *335*
Tide effects on river discharge, *337*   Oceanic effects on land climates, *340*   Coastal ocean regions, *341*
New York Bight, *343*   Northeast Pacific Coastal Ocean, *349*

thirteen

SHORELINES AND SHORELINE PROCESSES   357

Coastlines, *360*   Deltas, *362*   Beaches, *368*   Beach processes, *374*   Minor beach features, *378*
Salt Marshes, *381*   Modification of coastal regions, *383*

fourteen

THE MARINE ENVIRONMENT—AN ECOSYSTEM   391

Density, *394*   Light, *396*   Biogeochemical cycles, *398*   Dissolved and suspended organic matter, *404*
Salinity, *407*   Temperature effects on marine organisms, *410*

fifteen

MARINE PLANKTON   417

Phytoplankton, *418*   Zooplankton, *421*   Factors influencing zooplankton distribution, *433*
Zooplankton migration, *435*   Color and bioluminescence, *437*

sixteen

443   THE BENTHOS

Rocky beaches, *446*   Benthos of coastal sediments, *451*   Microorganisms in marine sediments, *456*
Reef communities, *459*   Deep-ocean benthos, *465*

seventeen

471   PRODUCTIVITY OF THE OCEAN

Primary Productivity, *473*   Geographical distribution of productivity, *476*
Effect of grazing by zooplankton, *478*   Organic growth factors, *481*   Food resources of the ocean, *481*
Estuaries and fish production, *485*   Waste disposal in the coastal ocean, *490*

appendix   1

507   EXPONENTIAL NOMENCLATURE, THE METRIC SYSTEM,
AND CONVERSION FACTORS

appendix   2

511   GRAPHS, CHARTS, AND MAPS

appendix   3

523   GLOSSARY

appendix   4

551   MARINE FISHES

560   INDEX

Oceanography—the scientific study of the ocean—presents a view of the earth that is new and useful to us. We have traditionally viewed the earth from the land and from a human frame of reference. But study of the ocean, the earth's most distinctive feature, shows us that continents are only large islands surrounded by a single body of water. Marine processes active in any single area are eventually felt throughout the world ocean. We see an earth with a finite surface area and a limited capacity for production or abuse, an earth whose weather and ocean currents are powered by energy from the sun and where we as humans have had little success in either controlling or modifying them.

This book examines the world ocean and those processes that control its major features and the life in it. It is intended for a general survey course, either for beginning science students or for students who may never take another course in science. The primary objective of the book is to investigate the major features of the ocean and to present some of the problems that have occupied oceanographers during the century since the beginnings of the science.

A second objective of the book is to illuminate the workings of modern science. Oceanography is still evolving at a rapid rate. Because of this rapid development, one must recognize that many of the "facts" given in this book are actually hypotheses—ideas on trial. Some will stand the test of careful examination, further observation, and more sophisticated analyses of available data. These ideas will survive. Many will not, however, and they are then discarded or replaced by other hypotheses that explain the observations more fully or make more accurate predictions. Hence, it should be remembered that our view of the earth represents a momentary glance at a rapidly changing picture.

A book of this nature necessarily omits or condenses many aspects of oceanography. In general the ocean areas and processes selected for discussion are those most likely to be seen by land dwellers or to affect our lives. For instance, those processes that affect the coastal ocean are emphasized—beach formation, coastal currents, and sea ice formation are just a few examples. Other interesting areas have been condensed, such as the discussion of the deep structure of the earth's crust or the early history of the earth and evolution of presently living organisms. References to more detailed treatments of these topics have been included at the end of the chapters, to aid in finding further material on specific subjects.

The view offered here of the coastal ocean and its limitations should be useful beyond the bounds of a college course. As citizens we are increasingly called on to make decisions or to evaluate recommendations about utilization of the coastal ocean and the coastline at its margins. Should a salt marsh be used for a housing development? A sanitary landfill? A marina? Or perhaps left in its natural state? Should waste disposal be permitted in a bay or on the continental shelf? Reasonable decisions in these areas require a knowledge of basic environmental processes. With an increased awareness and improved understanding of the processes affecting ocean and atmosphere, perhaps we can eventually break out of the cycle of buying time for today's problems by creating new problems for tomorrow. This is especially important, for the use of the ocean involves not just our own coastline but the entire planet.

PREFACE

OCEANOGRAPHY

OCEANOGRAPHY — THE SCIENCE

one

Challenger Expedition The first oceanographic expedition to circle the globe, sponsored by the British Navy and the Royal Society, was a scientific party headed by Sir Wyville Thomson aboard the *Challenger*, a full-rigged corvette with an auxiliary 1234-horsepower steam engine. She set out in December, 1872, to investigate "everything about the sea" and returned to England in May, 1876 with data that eventually filled fifty large volumes; working up the reports employed seventy-six authors over a 23-year period.

This immense project was inspired by the discovery in 1860, when the prevailing theory held that life below 600 meters was impossible, that strange and wonderful creatures existed at depths of 2 kilometers or more below sea level. The *Challenger* staff was asked to investigate physical and biological conditions in every ocean and to record everything that might influence the geographical distribution of marine species. This included taking water samples and temperature measurements of both bottom and surface waters, recording currents and barometric pressures, and collecting bottom samples in order to study sediments and attempt to find new species. Swimming animals were caught in nets dragged behind the ship.

Laboratory conditions and scientific equipment aboard the *Challenger* were the finest to be had, and first-class data were collected during the three-and-one-half-year expedition. Every ocean was sounded except the Arctic; 4717 new species of animals were collected and classified by specialists.

The *Challenger* Expedition was probably the most innovative single oceanographic research voyage ever made. It established the tradition of large-scale team efforts that characterize modern oceanographic research.

Oceanography is the scientific study of the world ocean, the surface feature that makes this planet unique among those in the solar system. Oceanographic research requires multidisciplinary team effort: a typical project employs scientists recruited from most of the major scientific disciplines, and involves problems that do not fit easily within the confines of a narrow, compartmentalized view of science.

The importance of oceanography to our understanding of the earth and to our everyday life has been underscored by two dramatic developments of the late 1960's and early 1970's—the beginnings of manned space exploration and increased concern about the state of the environment, including urban areas—especially those in coastal areas. Both have led to a new view of the earth in which the ocean occupies a central position. In this and the following chapters, we shall develop this ocean-oriented view of the earth—in sharp contrast with our traditional, land-oriented, man-dominated view.

Earth-orbiting artificial satellites have provided us with pictures of the earth, as in Fig. 1–1, showing it to be about 70 percent covered with ocean and about half obscured by clouds at any instant. To understand fully the earth and the processes acting on its surface, we must understand its characteristic feature—the ocean. This is more than just a subject for curious inquiry. Regardless of whether we live on the ocean coast or in the middle of the continent, the ocean plays a major role in our lives. Not only is it the key to weather and climate, the ocean is also our backyard, recreational area, highway, and dumping ground for wastes.

A large percentage of the United States' population (75.3 percent in 1970) live in 29 states bordering either the ocean or one of the Great Lakes. In the middle of the nineteenth century, the opening of the continent's interior reduced the proportion of people living in coastal counties from 37 percent in 1800 to 25 percent in 1850, but the percentage has been steadily rising since then. Our children will also most probably live near water in large cities on estuaries, harbors, or lakes. It has been estimated that by 1980, 7 out of every 10 Americans will live in metropolitan areas. Seven of the largest metropolitan areas in the U. S. are on the coast. Most of the projected "megalopoli" are also coastal—for example, the Boston–New York–Philadelphia–Washington complex, or the Vancouver–Seattle–Tacoma and Los Angeles–San Diego complexes.

In this century the shallow ocean bottom bordering the continents has become a source of raw materials and fuels for industry. In 1968, the production of petroleum from offshore oil fields supplied about 16 percent of the world's petroleum production. In 1978, this source is expected to provide about one-third of the world's demand. The production of natural gas from submerged fields is an important fuel source for countries around the North Sea.

For centuries the ocean provided a significant part of the world's protein supply. Fish provide about 3 percent of the direct human intake of protein, but probably account for about 10 percent of the total intake because of the large amounts of fish used for feeding animals such as chickens. Most experts agree that the take of fish from the ocean could be increased to twice their 1968 levels, and likely even more, if new equipment is developed, additional species of fish utilized, and new areas fished. The ocean is by no means an unlimited source of food and protein, but its resources can be better utilized than they have been in the past.

FIG. 1–1
A view of the earth taken from a space craft circling the moon. Much of the earth's surface is obscured by clouds. The cratered moon surface is seen in the bottom portion of the photograph. (Photograph courtesy NASA)

FIG. 1–2
Atlantis II, a modern vessel designed for oceanographic research, is operated by the Woods Hole Oceanographic Institution. (Photograph courtesy Woods Hole Oceanographic Institution)

FIG. 1–3a
FLIP (FLoating Instrument Platform) is a research platform 103 meters (355 ft.) long, shaped somewhat like a ship but lacking motive power, so it must be towed to location.

FIG. 1–3b
On location, one part of the hull is flooded causing it to rotate into a vertical position.

FIG. 1–3c
In the vertical position about 90 meters (300 ft.) is submerged. It forms an extremely stable platform from which to carry out scientific experiments. (Official U.S. Navy photographs).

The ocean also serves purposes that most of us take for granted. For the United States, both the Atlantic and Pacific Oceans are defense barriers; on the more pleasant side, the oceans serve as a major recreational area, from beaching to boating. With increased population and continued spoiling of open land spaces, the ocean will become increasingly valuable as a recreational resource.

Finally, the ocean is a major factor determining our weather, regardless of where we live on earth. The ocean is an important reservoir of the heat and water vapor that power atmospheric storms. Also the ocean stores heat in summer and releases it slowly in the fall and winter to modify climatic extremes; thus coastal areas do not experience the extremes of very cold winters and very hot summers characteristic of continental interiors.

OCEANOGRAPHERS—THOSE WHO STUDY THE OCEAN

The world ocean is too vast a subject for a single person to be an expert on all its aspects, even after a lifetime of study and work, if only because study of the ocean requires the application of a wide variety of scientific disciplines—primarily physics, chemistry, geology, and biology. Consequently, oceanographers usually specialize in one aspect or subdiscipline, although a broad understanding of related fields is required in order to work effectively with other scientists on mutually interesting problems.

As a profession, oceanography is a small field. In the mid-1960's there were probably only a few thousand persons in the United States who called themselves oceanographers, although the field grew rapidly during that decade. In 1960, there were about 1300 oceanographers employed at 24 universities and U.S. federal agencies. Their breakdown by subdiscipline showed:

biological oceanographers	48 percent
chemical oceanographers	7
geological oceanographers (including geophysicists)	11
physical oceanographers (including meteorologists)	34

Oceanographers being a diverse lot, the methods they use for studying the oceans are many and various. Certain technical constraints, however, are common to all aspects of oceanographic research. Most projects involve work aboard ship, whether it be a converted Army tug, an aged oyster boat for work in coastal and inshore waters, or a modern, fully equipped research ship 300 feet long, outfitted for spending months at sea. Even the largest research vessels are rather slow-moving (10–15 knots or about 18–27 km/hour) and provide an unsteady wave-tossed platform. Where stability is essential, other research platforms may be substituted, such as special-purpose ships like the FLIP (Floating Instrument Platform; shown in Fig. 1–3) or submersibles (small, special-purpose submarines). Helicopters, fixed-wing aircraft, or hydrofoils are used where speed is essential, but none is as versatile or as widely used as the research ship.

A characteristic feature of oceanographic research is the lowering of instruments or samplers on a wire line (Fig. 1–4). A winch aboard ship lets out a thin cable, and pulls it back aboard; a metered wheel indicates the amount of line let out, and by measuring the angle of the wire the instrument depth may be calculated. A small brass weight, called a *messenger*, runs down the wire to actuate the instrument. Limited to the use of such remotely controlled devices, an oceanographer can rarely observe his instruments at work or see in their natural state the water, sediment, or organisms he is trying to retrieve. The analogy has been drawn comparing the oceanographer who studies the ocean with a person studying the earth from a balloon that is perpetually above the clouds. Using ropes and grappling hooks, the latter might retrieve rocks, tree branches, and an occasional TV antenna, but he would never see the surface he is studying.

PHYSICAL OCEANOGRAPHY

Physical oceanographers study physical processes in the ocean, such as ocean currents and tides, or the interactions between the ocean and the atmosphere. One of their major contributions has been to map ocean surface currents. Most of the data for this project has been taken from ships' logs, which record the extent to which a ship has been deflected by currents in a particular part of the ocean.

Systematic mapping of currents, not yet complete, was begun in the mid-nineteenth century by the pioneer American oceanographer Matthew Fontaine Maury (1806–1873; see Vignette II at the end of this chapter). Recent long-range studies of such major currents as the Atlantic Gulf Stream and Pacific Kuroshio Current may finally answer questions first asked by eighteenth-century naturalists. Eventually it may be possible to predict changes in ocean-current patterns, including current strength and direction. Such predictions, combined with improved long-range weather and wave forecasts, should permit substantial savings in ship travel time.

Another task for physical oceanographers is mapping subsurface currents, both at intermediate depths and near the ocean bottom. These currents cannot be studied using information from ships' logs; instead elaborate apparatus is required, including current meters and floating devices combined with computer-assisted data reduction systems (Fig. 1–5). Such studies, while they require substantial investments of time and money, reveal much hitherto unavailable information about deep-ocean phenomena.

Where fresh water enters the ocean from river mouths, complex circulation patterns are set up as moving water masses meet

FIG. 1–4

A Nansen bottle attached to a wire line for lowering into the water. The messenger (a brass weight on the line above the bottle) releases the trigger permitting the bottle to invert and take a sample. Precision reversing thermometers in the two cases mounted on the bottle record the temperature at the time the sample is taken. (Photograph courtesy G M Manufacturing and Instrument Co.)

FIG. 1–5

An instrumented, general purpose buoy is used to support various current and atmospheric measuring devices. The buoy has a self contained power system to operate data recording and transmitting systems. Such buoys can be moored in deep ocean areas to measure ocean and atmospheric conditions eliminating the expense of maintaining ships at sea for long periods. (Photograph courtesy U. S. Office of Naval Research)

and flow past each other offshore and in estuaries and harbors. Seafaring peoples from earliest times have studied the daily and seasonal changes in the behavior of these waters in order to navigate them safely. Tides and the coastal currents associated with them were studied and at least partially understood by the ancient Phoenicians and Greeks. Men were making fairly accurate predictions about them by the Middle Ages. Today, one can go to any boating supply house and obtain tide tables and current tables for coastal waters all over the world (though they may not always be perfectly accurate).

Continuing studies of waves, tides, and storms are carried out in connection with coastal engineering projects such as polder reclamation and dike construction in the Netherlands. Increasingly sophisticated techniques for measuring tides and tidal-current strength are constantly being developed. Constructing mathematical models to predict tide-related phenomena constitutes a new and exciting branch of oceanography. In the United States, the effects of large marine construction activities are studied by the U.S. Army Corps of Engineers, using working hydraulic models to help predict the effect on bays and harbors of various structures (Fig. 1–6).

FIG. 1–6

A hydraulic model of New York harbor used to study processes causing siltation and shoaling (shallowing) in the lower Hudson River and to investigate the feasibility of various schemes to reduce shoaling or to improve dredging and disposing of the spoil. (Photograph courtesy U. S. Army Corps of Engineers, Waterways Experiment Station)

At sea, physical oceanographers are primarily engaged in making precise measurements of seawater temperature and *salinity* (the amount of salt in a given amount of water), frequently for the purpose of identifying the source and history of various water masses. In other words, they seek to determine that part of the world ocean in which a given water mass acquired its characteristic physical and chemical properties, and the subsequent conditions to which it has been exposed and which might have caused those properties to change. Water temperatures are measured by special *reversing thermometers* and recorded at the instant the water is sealed into a sampling bottle, to an accuracy of $\pm$ 0.02°C. Modern electronic sensors are also used to obtain accurate temperature readings; they have the additional advantage of providing continuous temperature records as opposed to the isolated single point temperatures provided by reversing thermometers.

Salinity is also accurately determined on the same samples. Knowing accurate temperatures and salinities, the physical oceanographer can identify the source of water masses or predict the speed and direction of surface currents.

Another, more costly method of studying surface current variables is by direct measurement with a *current meter*. Several of these are often strung together, supported by a single buoy. Data relayed back from them is fed into data processing and transmission equipment to be relayed to shore by radio. Such telemetering techniques have been used in the deep ocean between Bermuda and Cape Cod, and in inshore waters of Puget Sound and Long Island Sound.

Other physical oceanographers are more theoretically inclined. Their studies are often related to the atmospheric sciences or fluid dynamics, and call for relatively little work at sea.

CHEMICAL OCEANOGRAPHY

Ever since systematic oceanographic research began with the famous *Challenger* expedition of 1876, chemists have been constantly improving their methods for analyzing the components and properties of seawater. Concentrations of the major constituents were isolated and identified many years ago, and it has long been established that the chemical makeup of seawater anywhere in the world is nearly the same. More recent studies have shown, however, that biological and sedimentary processes cause some small changes in seawater chemistry.

Advances in scientific instruments and analytical techniques have permitted more refined analyses of the salts dissolved in seawater (sea salts), so that by the early 1970's, 74 elements had been found in the ocean and their concentrations determined. Detection of about 20 more elements—i.e., the remaining elements of the periodic table occurring in nature—may become possible with the development of still more sensitive analytical techniques.

The development of a new analytical technique or a new measuring instrument nearly always leads to new knowledge of the chemistry of the oceans. Thus this subdiscipline, and to a certain extent oceanography as a whole, is dependent on the instruments and techniques available. In keeping with the latest trends in analytical chemistry, these techniques tend increasingly to involve electronics, nuclear reactors, and mass spectrometers. The analyst can feed data obtained from chemical analyses directly into a computer, for example, get his results at once, and adjust any part of his procedure accordingly if necessary.

Chemical oceanographers also seek to determine the chemical form that elements assume in seawater. Knowledge of exactly which compounds are formed plays an important part in determining the chemical behavior of each element in the ocean. Such information is necessary in formulating a picture of the gigantic chemical system formed by the earth's ocean, atmosphere, and crust. Knowledge of these chemical processes also underlies predictions of the behavior and fate of the many materials that find their way into the ocean from the land.

Chemical oceanographers, in addition to working on their own problems, are often called upon to perform analyses for other oceanographers. They may work with biologists, for example, to help determine how concentrations of various substances dissolved in seawater affect the relative abundance of various types of marine life. Conversely, they may seek the aid of biological oceanographers to learn how the presence of certain marine organisms affects the concentration of nutrients and gases dissolved in the water: substances such as vitamin B-12 and minute quantities of various metals, for example, are known to influence the growth of organisms. Future efforts at domesticating or growing various marine organisms (*aquaculture*) de-

FIG. 1–7

Plankton are collected by towing a cone-shaped net of fine meshed cloth through seawater. Organisms unable to swim faster than the net are captured for study. (Photograph courtesy Woods Hole Oceanographic Institution)

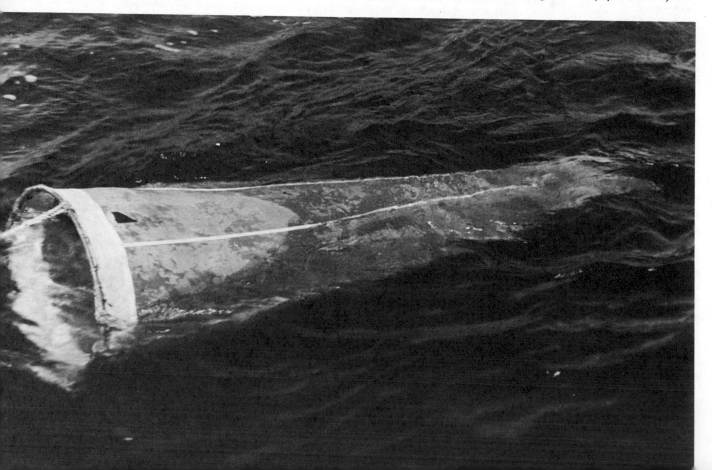

pend on reliable tests to determine whether seawater is suitable for the growth of a particular organism, just as soils must be tested to see if they are suited to the growth of a particular crop.

The past several years have witnessed a growing public awareness of man's effect on the environment, including the ocean. New substances, especially the by-products of various technological advances, are indiscriminately poured into coastal areas in vast quantities. Such disposal of petroleum, DDT, and sewage in various stages of treatment has received particularly widespread publicity. In many cases, the effects of these materials on ocean waters, marine organisms, and ocean-floor sediments cannot be anticipated, making difficult or impossible the prediction of undesirable side-effects. Clearly, chemical oceanography has an important contribution to make in this area.

For a chemical oceanographer, work begins aboard ship when sample bottles are washed with seawater from the sampling bottle, often a *Nansen bottle* (Fig. 1–4), in order to remove contaminants. Usually several samples are taken at each of the predetermined stations on the ship's course. Electronic instruments, often installed aboard ship, measure the conductivity of the seawater, permitting salinity to be determined to ± 0.003 parts per thousand—i.e., a precision of one part in 10,000. Analyses may also be made for special purposes, such as analyses of nutrients, dissolved substances needed for plant growth. Other samples may be frozen aboard ship to prevent chemical reactions from taking place in them before they can be analyzed in a lab on shore.

The study of radioactivity in the marine environment presents special technical problems. New methods of detecting radioactive elements provide opportunities to study both those naturally occurring and those introduced by testing of atomic weapons. Such methods involve measuring the energy given off in the decay of these elements, sometimes in the form of subatomic particles. Moreover, radioactive materials have proved to be excellent tracers for studying water movements, coincident with studies of the chemistry and biochemistry of the radioactive substances themselves.

BIOLOGICAL OCEANOGRAPHY

The ocean sustains an astonishing variety of living creatures, diverse in both physical form and life style. Botanists, zoologists, microbiologists, and fisheries experts have been attracted to oceanography as a result of their interest in plant and animal life in the ocean. Biological oceanographers study these organisms and their twofold relationship with the ocean environment —the effects of ocean conditions on the distribution and abundance of marine organisms (Fig. 1–7), and the effects of the metabolic processes of these organisms on the chemical properties of seawater (Fig. 1–8).

Marine ecology, a specialized field of interest for biological oceanographers, considers the influence of environmental factors on the success or failure to survive of populations (groups of

FIG. 1–8
Biological oceanographers use chemical techniques to estimate the abundance of photosynthetic organisms in seawater and to determine the rate at which they synthesize new organic matter. (Photograph courtesy Woods Hole Oceanographic Institution)

organisms). Marine ecologists also study the flow of energy through the marine food web, by which energy from the sun is transformed by photosynthesizing plants into food for plant-eaters, who are in their turn eaten by other animals in a chain which finally ends with unusable organic matter settling into the mud on the ocean floor. Theoretical marine ecologists create predictive models for application in energy-flow and population dynamics problems.

Still other biological oceanographers direct their efforts at systematic classification: describing, identifying, and classifying marine organisms according to their common morphological characteristics, and subjecting the mutual occurrence of various characteristics to statistical analysis. The consequent emergence of patterns of seemingly unrelated but mutually occurring physical features provides the base for a system that reduces the art of taxonomic classification to a statistically based observational science, a method called *numerical taxonomy*.

The most common application of biological oceanographic research is in the area of food production. Harvesting of known stocks on rich fishing grounds is only one method of exploiting the sea; another is finding previously unknown or unexploited food sources—the hunting stage of food resources management. Even more sophisticated techniques are applied to growing wild stocks, such as seeding oysters in one location, letting them fatten in another, and moving them to still a third before fully mature and ready to market. Also, new methods of food production have been developed or greatly expanded. For instance, periodically flooded rice paddies in Japan are also used for fish culture, the fish maturing before the fields must be drained for harvesting the rice.

There is great public interest in managing food resources in marine environments, and even hope that the vast reaches of the oceans may some day be cultivated to feed future generations. Oceanographers, however, are by no means confident that the seas can be harvested to support the unchecked growth of human populations.

For biological oceanographers, collecting samples at sea for laboratory study presents problems quite different from those faced by chemists or geologists. Although the average marine organism is about the size of a mosquito, organisms range in size from microscopic bacteria and other one-celled organisms to whales, the largest animal that apparently has ever lived on earth. Acquiring specimens thus requires an equally diverse array of techniques and equipment.

In studying the myriad of tiny floating or weakly swimming organisms known as *plankton*, biologists commonly use nets (basically cloth screens) towed horizontally behind a vessel which filter water and retain these organisms (Fig. 1–7). After towing the net for a predetermined time, it is brought aboard and the organisms recovered to be taken to a lab ashore where they are examined, identified, counted, and/or described.

Somewhat similar techniques are used to collect the *nekton*, those organisms which are strong swimmers. Larger nets are towed behind the ship or dragged over the sea bottom. Fish that cannot swim fast enough are captured by the net, while the

FIG. 1–9

A dredge is a large metal frame attached to a heavy chain bag, rugged enough to be dragged over the ocean bottom. Bottom-dwelling organisms and rocks are retained in the bag. (Photograph courtesy Woods Hole Oceanographic Institution)

fastest swimmers often escape. Variant methods include hauling the net vertically upward or dropping a special net from the surface.

Very large animals and fast swimmers are not easily collected. For example, our first-hand knowledge of the giant squid is primarily from specimens that have been washed ashore and from various collected fragments. Because it is often difficult or impossible to observe such animals in their native state, our knowledge of them remains extremely fragmentary.

Benthic animals (benthos) those living in or on the ocean bottom, are collected by basically the same techniques as are used to collect sediment. Dredges (as in Fig. 1–9) or grab samples are most commonly used, coring devices much less frequently.

A major problem for marine biologists is that marine organisms are not uniformly distributed, but like life on land, tend to occur in patches—many here, none there. Isolated samples may thus present a completely misleading picture of an organism's overall distribution and abundance. Furthermore, animals with good eyesight (or comparably effective sensory perception) often detect sampling devices and actively avoid them with the result that the sample is biased—that is, it does not accurately represent the assemblage of organisms on the bottom or in the water mass sampled.

GEOLOGICAL OCEANOGRAPHY AND MARINE GEOPHYSICS

Geological oceanographers study the ocean basin—particularly its origins, its structure, and its sediment cover—in order to reconstruct the history of the ocean and its basins and to understand the processes that shape the earth's surface.

Using echo-sounding techniques with equipment such as shown in Fig. 1–10, oceanographers measure ocean depths in order to construct maps of the mid-ocean ridges, incredibly flat abyssal plains, and great trenches in the ocean floor. Despite great technical advances in sounding techniques within the past half-century, by the early 1970's less than 3 percent of the ocean bottom had been mapped well enough to be useful for ship navigation. Most mapping has been done in shallow coastal areas.

Geological oceanographers are also interested in coastlines and their changes, studying beaches in particular to determine the effects of major storms, seasonal variations owing to changed wave conditions, and movement of sands by wave action.

Another concern is deep-ocean sediment (pictured in Fig. 1–11), a feature which has no counterpart on the continents. Oceanic sediments contain the only history of the ocean basin and its waters. Since the early 1960's, ships specially equipped for drilling in the deep ocean have obtained samples of loose sediment and sedimentary rocks as old as 150 million years. From studies of these samples and of the remains of minute

FIG. 1–10
A modern depth recorder provides a continuous record of water depths below the ship's track. Under certain circumstances it is possible to determine the depths of surficial sediment deposits because layers below the present ocean bottom also reflect sound pulses back to the ship.

FIG. 1–11
A thin coating of sediment has accumulated among the manganese nodules and rocks on the Blake Plateau on the lower continental slope off the southeastern United States. (Photograph courtesy Woods Hole Oceanographic Institution)

marine organisms contained in them, it is possible to piece together the historical record of the ocean basins. The preserved record of evolution of marine organisms provides a usable time scale; the types of fossils and their individual characteristics indicate the prevalent climate at the time they lived. Geological oceanographers attempt to provide a picture of ocean basin history just as other geologists work out the history of the continents.

Marine geophysics is the use of the techniques of physics to study ocean basins. Studies of such phenomena as the behavior of waves from earthquakes provide a three-dimensional picture of the ocean bottom to supplement the two-dimensional picture gained by mapping. Powerful sound sources (explosives, underwater electrical discharges, or gas explosions) generate sound waves that penetrate the oceanic crust to reveal the structures and sediment thicknesses of various parts of the oceanic crust and the relationships among them. Another technique involves the detailed analysis of the magnetic field over the ocean basins, which has provided important clues about the formation of crustal material at mid-ocean ridges and the associated movement of the continents.

Studies of ocean-basin history require collection of material from the ocean bottom. An early technique still used is *dredging*. A sturdy metal frame attached to a bag made of chain (as shown in Fig. 1–9) is connected to a strong wire and dragged across the ocean bottom, collecting the rocks, nodules or sediment dug up thereby. Because of the bag's open mesh, fine-grained sediment is washed out as the bag is brought to the surface, leaving only rocks and larger objects. In the deep ocean, dredging requires much costly ship time; also, many dredges are caught on the bottom and lost. Still, dredging has provided most of the rock fragments and manganese nodules available for study.

Sediment samples are widely used to study the ocean-bottom deposits and the record of ocean history recorded in the remains of marine organisms buried in the sediment. Various coring devices are used to penetrate the sediment–water interface. Basically, a *corer*, such as the one in Fig. 1–12, is a pipe which fills with sediment as it penetrates the ocean bottom. A valve-like device closes the bottom of the corer so that the sample is not lost when the corer is brought to the surface. Some corers collect samples tens of meters long.

The same techniques employed in drilling oil wells on land are also used by special ships to drill the ocean bottom, penetrating sediment and recovering the underlying rock. In the late 1960's and early 1970's drilling operations on the *Glomar Challenger*, shown in Fig. 1–13, recovered ancient sediment from the deep ocean floor, greatly expanding our knowledge of ocean-basin history.

Ocean basins are usually studied indirectly using *echo sounders*. These determine ocean depths by means of a transducer on the ship's hull transmitting a sound pulse, which is reflected from the bottom and then detected, as an echo, by the same transducer. The echo sounder measures the time elapsed between transmission and reception of the signal, and a profile of the ocean bottom under the ship's track is drawn by a recorder.

FIG. 1–12
Small free-fall gravity corer ready for sampling deep-ocean sediment. (Photograph courtesy Woods Hole Oceanographic Institution)

Geological oceanographers use such techniques to map the ocean bottom.

Sound waves from explosions or electric sparks are echoed back from buried layers beneath the ocean bottom. Mapping these buried surfaces provides information about sediment thickness and ocean-bottom structure. Marine geophysicists also use such strong signals to study variation of seismic-wave velocity in various layers beneath the ocean bottom. Usually the energy source and the receivers are located several kilometers apart, so that the seismic waves must travel through the subbottom strata. By timing the initial shot and the wave as detected by the receiver, it is possible to calculate wave velocity and to estimate thickness of the various layers. Variations in seismic-wave velocity provide important clues about the composition of various layers of the ocean bottom.

Variations in the earth's magnetic and gravitational fields provide other important clues about the material of the ocean bottom and indirectly about the forces affecting the oceanic crust. For example, the earth's magnetic field exhibits magnetic stripes, paralleling the mid-ocean ridges, which reflect changes in the field as ocean crust is generated at mid-ocean ridges. As another example the earth's gravitational field is markedly reduced over the deep trenches that border island arcs.

From all these bits of evidence has emerged a picture of a dynamic earth in which oceanic crust forms, moves, and is finally destroyed. In the process, the continents are carried along as passive flotsam. This picture contrasts sharply with the ancient cratered surfaces of the moon which lacks an active interior.

FIG. 1–13
A geophysicist aboard *Glomar Challenger* determines the velocity of sound in a sediment core recovered by deep-ocean drilling. Such information is necessary for more precise interpretations of geophysical observations on thickness and structures of rocks beneath the ocean bottom. (Photograph courtesy Deep-Sea Drilling Project, Scripps Institution of Oceanography under contract to National Science Foundation)

OCEAN ENGINEERING

Ocean engineering is the application of knowledge gained from all branches of science, including oceanography, to the practical problems of working in, on, and under the ocean. For many years, there was little communication in the U.S. between oceanographers, who usually worked in university research programs, and engineers, who worked either for the few private companies involved in marine enterprises or with federal agencies active in the ocean environment. Recent efforts to exploit the resources of the ocean and the ocean bottom have resulted in closer cooperation, with the result that several once-distinct fields are now lumped together under the heading ocean engineering.

Naval architecture—the design of ships and other floating structures—is probably the oldest branch of what we now call ocean engineering. Ships must be designed to withstand the forces exerted by winds, waves, and tides at the ocean surface. The generally excellent performance of ships built over the past twenty centuries or more is a tribute to the ingenuity of naval architects in solving the myriad problems that seafaring poses. The development of new vessels—such as Hovercraft, hydrofoils, and large submersibles—promises new approaches to old problems and provides exciting opportunities for future naval architects.

Designing and building docks, jetties, seawalls, and other structures to prevent or retard the loss of coastal areas is another well-established area of ocean engineering. Long before the development of computers and powerful earth-moving equipment, the Dutch were successfully reclaiming parts of their coastal areas for agriculture. Closing off a shallow arm of the sea, the Zuider Zee, was a major engineering triumph of the 1930's, and the damming of the many mouths of the Rhine River to prevent recurrence of the disastrous floods of 1953 is more recent evidence of the success of this branch of ocean engineering. On the other hand, frequent failures of seawalls or structures built in the shallow ocean show that much remains to be learned about the ocean and its effects on fixed structures of all kinds.

FIG.   1–14
Dock facilities in Upper New York Bay in various stages of disrepair. Such facilities are being replaced by new docks designed for modern freight handling procedures. (Photograph courtesy of The Port of New York Authority)

Changes in ship design and methods of cargo handling have had a major impact on the number and type of coastal structures needed. For example, because the drafts of many ships built after the second half of the 1960's, especially those of tankers, greatly exceeded the depths of the channels leading into major ports, extensive dredging was necessary to make the channels usable. Furthermore, the type and number of docks required to handle cargo has changed. The smaller ships of the past required many more docks for loading and unloading; newer ships, with cargo prepacked in containers, can be handled in much less time. Consequently, most major U.S. ports have found that many aged dock facilities, such as those in Fig. 1–14, are no longer needed for transportation, and have undertaken programs to fill in these areas to be used as parks, or as sites for new housing to relieve crowded urban areas.

Production of petroleum from offshore wells developed rapidly in the U.S. beginning in the late 1940's. Drilling platforms and structures for the production of petroleum taxed the ability of engineers to design installations that would withstand the storms, tidal currents, and high waves of coastal regions. Millions of dollars have been spent in many coastal areas to build these structures. Their high rate of failure has demonstrated that there is much to be learned about the forces acting at the ocean surface and about the characteristics and behavior of the ocean floor beneath.

The design of future structures for undersea habitation will be another concern of ocean engineers. Such structures must be designed not only to withstand forces acting at the ocean surface, but also to avoid problems imposed by high pressures, cold, and lack of light at depths of only a few hundred meters. Partial compensation for the difficulties of building and maintaining submarine structures is the fact that the forces which cause such problems at the ocean surface are greatly reduced just a few meters down; about one wavelength below the surface most of the forces associated with the passage of a wave are negligible.

APPLICATIONS OF OCEANOGRAPHY

Oceanography, like ocean engineering, is a practical business. Many—perhaps most—of the topics covered in this book can be traced to research initiated to solve practical problems. Some of these problems and their relation to oceanographic research are discussed below.

The practical application of oceanographic data has a long history in the United States. The colonial mathematician Nathaniel Bowditch (discussed at length in Vignette I) is known for his efforts at simplifying navigation for ships' officers. The pioneer American oceanographer Matthew Fontaine Maury (discussed in Vignette II) compiled observations of both currents and winds in the mid-nineteenth century, greatly aiding the sailing ships of that time. Then, as now, time meant money to ship operators. Furthermore, Maury compiled records on the abundance and distribution of whales to guide ships of the American whaling industry, then in its heyday.

Even though ships are no longer dependent on winds for their power, and deflection by currents no longer a major problem, mapping ocean currents is still necessary for modern shipping. Perhaps the most dramatic demonstration of this was the sinking on her maiden voyage (April 15, 1912) of the *Titanic*, the largest ship then afloat, when she collided with an iceberg in the North Atlantic. Of the 2224 persons aboard, 1513 were lost. Shortly afterward the International Ice Patrol was established to chart icebergs carried by the Labrador Current to the vicinity of North Atlantic shipping lanes in spring and early summer. The information gained from mapping currents has also been used to help locate survivors of shipwrecks. Largely because of the Ice Patrol, major disasters such as that of the *Titanic* have been avoided, despite the greatly increased amount of shipping in these lanes.

Detailed mapping of ocean-surface temperatures may permit long-range weather forecasts for large continental areas. Warming of surface waters in the North Pacific, for example, has been implicated as a cause of abnormally cold winters on the U.S. Atlantic coast. If such hypotheses are validated, then more accurate weather predictions may be possible over longer periods of time in the future.

The efficient operation of fisheries is a prime stimulus for oceanographic research. In a small country highly dependent on its fisheries such as Iceland, the success or failure of a fishery in any year may spell the difference between a healthy economy or one troubled by currency devaluations and governmental crises. Also, knowledge of the coastal ocean in many parts of the North Atlantic is an indirect result of studies on behavior of commercially important fishes.

Much of the equipment now used for oceanographic research was originally developed for military purposes. For instance, the echo sounder used to determine ocean depths is basically "sonar" —developed for detection of submarines and other vessels at sea. The same principles used for echo sounding and ship location can also be used to locate schools of fish and to guide fishing vessels to them.

Wartime research has otherwise substantially increased our knowledge of the ocean. Allied amphibious military operations in World War II required increased understanding of the waves and currents in nearshore areas. Our present understanding of coral reefs and atolls is the result of military operations in those areas of the Pacific Ocean. Such studies are not limited to wartime applications, however. Construction of coastal structures and preservation or restoration of beaches requires a similar detailed understanding of wave behavior and currents. The beach you visit this summer may have been built up by methods stemming from research carried out by oceanographers and engineers at several government and university laboratories.

In common with other sciences, oceanography seeks to develop the ability to predict the effects of disturbance on natural systems—as, for instance, changes in water quality resulting

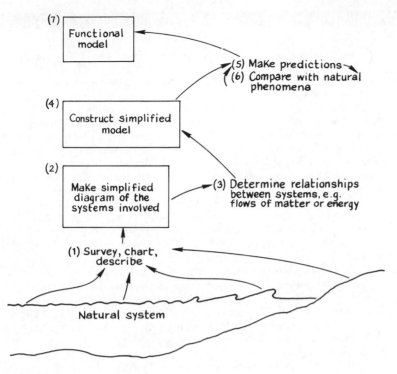

(7) Functional model

(5) Make predictions
(6) Compare with natural phenomena

(4) Construct simplified model

(2) Make simplified diagram of the systems involved

(3) Determine relationships between systems, e.g. flows of matter or energy

(1) Survey, chart, describe

Natural system

FIG. 1–15
Schematic representation of the development,
evaluation, and refinement steps leading to
the construction of a mathematical model for
predicting response to changes in a natural
system.

from waste-disposal operations. One way to accomplish this is
to construct mathematical models in order to isolate and analyze
some aspect of the ocean, such as that schematically represented
in Fig. 1–15. Such mathematical equations and computers serve
much the same functions that hydraulic models do for coastal
engineers.

Even simplified models help to organize and translate information for the purpose of determining relationships between
major parts of the system. Predictions thus made may then be
compared to the actual events being studied; usually a model
requires some degree of refinement before it can be used to
predict observable phenomena with acceptable accuracy. So far
mathematical models have been used, for example, in harbors
and coastal areas, to predict the amount of dissolved oxygen in
the water and the dilution and movement of wastes.

All these and many more problems occupy the attention
of oceanographers. Using the information obtained in earlier
studies, applying it to present problems, and then using and
further testing the knowledge so gained, scientists from diverse
fields clarify and refine our understanding of the ocean so that
we may use it more wisely and enjoy it more fully than we have
in the past.

The American navigator Nathaniel Bowditch (1773–1838) was a self-educated man. Poverty compelled him to leave school before his tenth birthday, and he was apprenticed at twelve to a ship's chandler in his home town of Salem, Massachusetts. When his gift for mathematics and desire for knowledge became known to the wealthy and learned citizens of his community, they opened their libraries to him. Before he was twenty-one, Bowditch had taught himself several languages and was better versed in mathematics and astronomy than anyone else in Salem. Following his apprenticeship, he signed as clerk on a cargo ship. At that time, determination of longitude was a highly difficult procedure, for chronometers were rare and expensive and contemporary mathematical methods too complex for most ship's officers. Bowditch developed a simple technique for determining longitude and proved its value by teaching illiterate seamen to calculate their position from the stars. He found so many errors in the standard British navigation manual, each a potential hazard to shipping, that he decided to recalculate the entire manual. The result, his New American Practical Navigator, is still in use today. Crammed with all sorts of navigational aids, including wind, tide, and current data, it was so written that every word would be intelligible to any sailor who could read.

An excellent astronomer, Bowditch also made many important contributions to that science. Furthermore, his feats of navigation by dead reckoning were related by seamen all over the colonies. Although in time he was awarded an honorary doctorate and a trusteeship at Harvard, as well as accorded membership in many learned socities, public life never interested him. For many years he was master of his own ship, and spent the last 35 years of his life as president of various insurance firms. (Photograph courtesy U. S. Naval Oceanographic Office)

Matthew Fontaine Maury (1806–1873) grew up on frontier farms in Virginia and Tennessee. He was appointed a midshipman at nineteen and by 1834 he had sailed around the world on U.S. Navy vessels.

In Maury's time, little information about ocean currents, prevailing winds, or storm tracks was available to young officers. There was then no Naval Academy, and only sketchy fragments of navigational science were taught aboard ship. Applying for an extended leave after ten years at sea, he wrote a text on navigation for junior officers, as well as a series of pseudononymous newspaper articles critical of Navy operations and training procedures.

After a stagecoach injury ended his seagoing career, Maury was named Superintendent of the depot of Charts and Instruments, later the Naval Observatory and Hydrographical Office, now the U.S. Naval Oceanographic Office. There he organized the vast amounts of wind, current, and seasonal weather information that he found in stored ships' logbooks. To supplement these data, he furnished blank charts to masters of Navy and merchant vessels, so that they could make daily records of weather and ocean conditions, showing the time and place of each instrument reading. The Wind and Current Charts he compiled from these data revolutionized the science of navigation and cut weeks off standard transoceanic runs. He also prepared a set of sailing directions for dangerous harbors, and described safe tracks for ships crossing the Atlantic.

In 1835, Maury organized the first international meteorological conference, in which ten nations participated and which led to international cooperation in collecting weather information at sea. His book The Physical Geography of the Sea, published in 1855, was the first major oceanographic work ever written in English, and he is generally credited with founding the science of physical oceanography. (Photograph courtesy U.S. Navy)

Major contributions to oceanographic research have often come from
versatile scientists with broadly ranging backgrounds and interests. No one
fits this pattern better than the Scottish naturalist John Murray (1841–1914).
As a young man he tried physics, then biology and geology, and finally
studied enough medicine to qualify as ship's doctor aboard an Arctic
whaler. Meeting Murray in an Edinburgh laboratory, Wyville Thomson
asked him to fill a vacancy among the naturalists aboard the Challenger
(see this chapter's Frontispiece); thus began one of the most distinguished
careers in the history of oceanography.

Murray quickly rose to the position of Thomson's first assistant, and
succeeded the latter as director of the Challenger Expedition Commission,
a government department set up in Edinburgh to analyze and publish
the results. Murray's studies of marine sediments were among the earliest,
and he published important works on plankton and coral reefs as well.
After persuading the British government to annex uninhabited Christmas
Island in the Indian Ocean, Murray made a fortune there in phosphate
mining enterprises. This wealth he used in part to promote oceanographic research; for example, he paid the entire scientific expenses of a
Norwegian expedition in the North Atlantic aboard the Michael Sars
in 1910. In collaboration with the director of that expedition, Dr. Johan
Hjort, Murray wrote his most important general work, The Depths of the
Ocean, which was for many years the most widely read and authoritative
oceanography text available in English. Later he published a smaller,
equally successful introduction to oceanography for the general reader.

The Norwegian scientist and adventurer Fridtjof Nansen (1861–1930) is best known for polar exploration, but his earliest achievements were as a zoologist and museum curator. On a pioneer expedition across the Greenland icecap (1888–89), he studied Eskimo life, and later published a book on the subject. Convinced that winds and current drift cause pack ice to flow between the North Pole and Franz Josef Land, from the Siberian Arctic to eastern Greenland, Nansen persuaded his government and the British Royal Geographical Society to help him test this novel theory; a special ship, the Fram, was built to carry a scientific party eastward from the Barents Sea toward the Pole, frozen in an ice floe. This expedition was a brilliant success. From September, 1893, to August, 1896, the Fram drifted with the ice, from 78°50' to an extreme north latitude of 85°57', closer to the Pole than any ship has been before or since. The crew then broke her free during the warm weather and sailed home. Nansen himself, with a companion, had left the ship at 84°N and set out toward the Pole with dogsleds. The two men spent fourteen months alone in the Arctic, but made only 86°14' before movements of the ice pack forced them to turn back toward Franz Josef Land. The remainder of Nansen's career was marked solid achievements in science and humanitarian enterprises. He published several books on the oceanography of the North Sea, designed the standard water sampling bottle which bears his name, and participated in many important scientific cruises. From 1905 onward, he was active in politics, and was awarded the Nobel Peace Prize in 1923 for contributions to the relief and repatriation of World War I refugees; he later became an expert on Soviet agricultural economics. Nansen was a lifelong athlete and a famous skier; he did much to popularize winter sports in Switzerland and elsewhere outside Scandinavia.

22

Representatives of many scientific disciplines cooperate to apply scientific method in oceanographic research

Major population centers are in coastal areas

Oceanographers are drawn from four major scientific disciplines; common problem is difficulty of access to area being studied

> Physical oceanography—research on currents, tides, waves; temperature and salinity data are an important key to oceanographic processes
>
> Chemical oceanography—study of chemistry of seawater and materials in contact with it
>
> Biological oceanography—interaction of marine organisms with their environment, individually and in terms of populations; research on food resources of the ocean
>
> Geological oceanography and marine geophysics—study of the ocean basins and their sediment cover, also coastlines and how they are changed by physical processes; techniques include echo sounding and deep drilling; picture of a dynamic earth has emerged from these studies.

Ocean engineering includes naval architecture, design of coastal structures, offshore drilling platforms, submarine installations

Applications of oceanography

> Most oceanographic research has been based on need to solve practical problems regarding shipping, weather, condition of fisheries, military operations
>
> Modern oceanography seeks to develop predictive capacity to understand effect of changes on natural systems; mathematical models used to analyze systems

These books cover the entire field of oceanography and provide information pertinent to most chapters in the book.

DIETRICH, GÜNTER. 1963. *General Oceanography; an Introduction.* Interscience, London, 588 pp. A comprehensive treatise, emphasizing physical oceanography; technical.

DUXBURY, A. C. 1971. *The Earth and Its Ocean.* Addison-Wesley Publishing Company, Inc., Reading, Massachusetts. 381 pp. Intermediate difficulty.

GROEN, P. 1967. *The Waters of the Sea.* Van Nostrand, London. 328 pp. Waves, tides, and currents, with a good section on ice; authoritative and readable.

PICKARD, GEORGE L. 1964. *Descriptive Physical Oceanography; an Introduction.* Pergamon Press, Elmsford, N.Y. 199 pp. An undergraduate level text; emphasis on water and heat budgets, currents, and water masses.

STUDY OF CRITICAL ENVIRONMENTAL PROBLEMS. 1970. Man's Impact on the Global Environment: Assessment and Recommendations for Action. The M.I.T. Press, Cambridge, Massachusetts. 319 pp. Authoritative analysis of environmental problems of worldwide significance, including those affecting atmosphere and ocean.

SVERDRUP, H. U., MARTIN W. JOHNSON, AND RICHARD H. FLEMING. 1942. *The Oceans; their Physics, Chemistry and General Biology.* Prentice-Hall, Englewood Cliffs, N.J. 1087 pp. Probably the best-known of all general oceanography texts; detailed and scholarly.

WEYL, PETER K. 1970. *Oceanography; an Introduction to the Marine Environment.* John Wiley, New York. 535 pp. A contemporary interpretation of oceanographic principles and their relationship with the earth and life sciences, emphasizes physical processes.

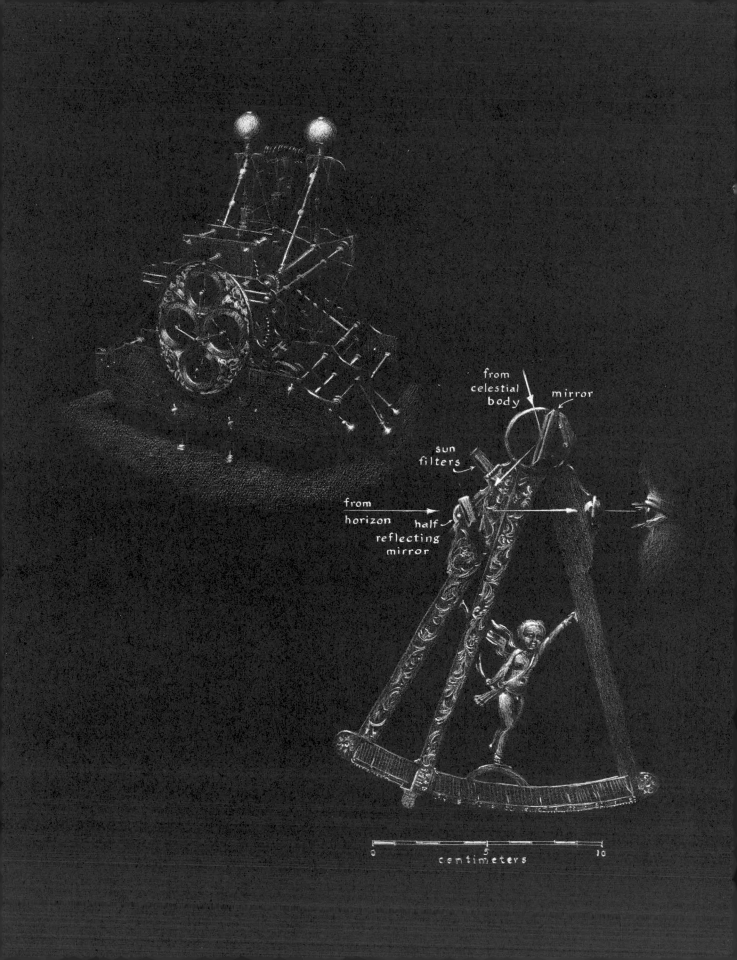

from
celestial
body

mirror

sun
filters

from
horizon

half
reflecting
mirror

centimeters

LAND AND SEA

Eighteenth-century navigational instruments: Octant (right) and Marine chronometer (left)   By means of the two mirrors on an octant, one half-silvered and one mounted on a movable arm, the image of a celestial body (sun, moon, or star) can be seen superimposed on an image of the horizon. The 45° of motion of the movable arm covers the angles of altitude from horizon to zenith. The diagonal lines of the calibration suggest that the required angles could be measured to within about 10 minutes of arc, equivalent to about 10 nautical miles (18.53 kilometers). This example, made of gilt bronze, is in the Altonaer Museum, Hamburg–Altona.

Modern instruments, with a small telescope in place of the simple sight, and worm wheel motions for the mirror arm, permit accuracy of about 1 second of arc. The modern sextant has a 60° arc calibrated to read 120°, permitting measurement to approximately 30° beyond the zenith, a convenience when the horizon is obscured by land.

The chronometer shown here, property of the National Maritime Museum of Greenwich, England, was built by John Harrison in 1736. It is one of several clocks he invented in competition for a prize offered by Parliament to the scientist who could design a mechanism whereby a ship could determine its longitude within specified limits after six weeks at sea. The chronometer is set for Greenwich time and longitude is calculated from the precise difference between Greenwich noon and local noon. The device illustrated here features self-correcting, counter-rotating spring pendulums. Any angular acceleration causing one of them to move faster also retards the other by the same amount. Harrison won the prize with a watch-sized chronometer that achieved a precision of 5 seconds during a voyage from England to Jamaica.

A satellite's view of the Atlantic Ocean from more than 37,000 kilometers (22,300 miles) above Brazil. Four continents can be seen: South America (center), North America (upper left quadrant), Europe (upper right), and Africa (center right). Antarctica (bottom of photograph) is obscured by clouds, the white fluffy masses often forming swirls as part of the large storms in the lower atmosphere. Note the large fraction of the earth's surface obscured by clouds. (Photograph courtesy NASA).

Water is the earth's most characteristic surface feature, covering more than two-thirds of the planet (as can be partially seen in Fig. 2–1), yet water represents only a small fraction of the earth's total mass as Table 2–1 indicates.

To begin our study of the water planet we should first establish its basic structure. To do this we divide the earth into concentric spheres, each having a distinctive composition, as illustrated in Fig. 2–2.

Core—the deep interior of the earth; made of very-high-density material with metallic properties; constitutes 39.1 percent of the earth's mass.

Mantle—the rock layer of density intermediate between crust and core; makes up 68.1 percent of the earth.

Crust—the outer shell of the solid portion of the earth; includes rocks of the continents and ocean bottom; constitutes 0.4 percent of the earth's mass.

Hydrosphere—all the water of the earth's exterior, including water in rocks; accounts for only 0.027 percent of total mass of the earth.

Atmosphere—the low-density layer of gases, primarily nitrogen, oxygen, and inert gases, mixed with smaller and variable quantities of water vapor and carbon dioxide.

The ocean is the largest part of the hydrosphere: 86.5 percent of the earth's free water is in it, as indicated in Fig. 2–3. Once it has been incorporated into sediments or sedimentary rocks,

SOME USEFUL DATA ABOUT THE EARTH AND THE OCEANS*

Size and Shape of the Earth

	Miles	Kilometers
Equatorial radius	3,963	6,378
Polar radius	3,950	6,357
Average radius	3,956	6,371
Equatorial circumference	24,902	40,077

Areas of the Earth, Land, and Ocean

	Millions of	
	Sq Miles	Sq km
Land (29.22 percent)	57.5	149
Ice sheets and glaciers	6	15.6
Oceans and seas (70.78 percent)	139.4	361
Land plus continental shelf	68.5	177.4
Oceans and seas minus continental shelf	128.4	332.6
Total area of the earth	196.9	510.0

Volume, Density, and Mass of the Earth and Its Parts

	Average thickness or radius (km)	Volume ($\times 10^6$ km^3)	Mean Density (g/cm^3)	Mass ($\times 10^{24}$g)	Relative Abundance (%)
Atmosphere	—	—	—	0.005	0.00008
Oceans and seas	3.8	1,370	1.03	1.41	0.023
Ice sheets and glaciers	1.6	25	0.90	0.023	0.0004
Continental crust[†]	35	6,210	2.8	17.39	0.29
Oceanic crust[‡]	8	2,660	2.9	7.71	0.13
Mantle	2,881	898,000	4.53	4,068	68.1
Core	3,473	175,500	10.72	1,881	31.5
Whole Earth	6,371	1,083,230	5.517	5,976	

Heights and Depths of the Earth's Surface

Land			Oceans and Seas		
	Feet	Meters		Feet	Meters
Greatest known height:			Greatest known depth:		
Mt. Everest	29,028	8,848	Marianas Trench	36,204	11,035
Average height	2,757	840	Average depth	12,460	3,808

Some Useful Conversion Factors

1 mile = 1.609 km	1 km = 0.621 mile
1 foot = 0.3048 meter (m)	1 m = 3.281 feet
1 sq mile = 2.59 km²	1 km² = 0.386 sq mile
1 cubic mile = 4.17 km³	1 km³ = 0.24 cubic mile
1 cubic mile = 4300 million metric tons of seawater	1 km³ = 1030 million metric tons of seawater

*From Holmes, 1965.
†Including continental shelves.
‡Excluding continental shelves.

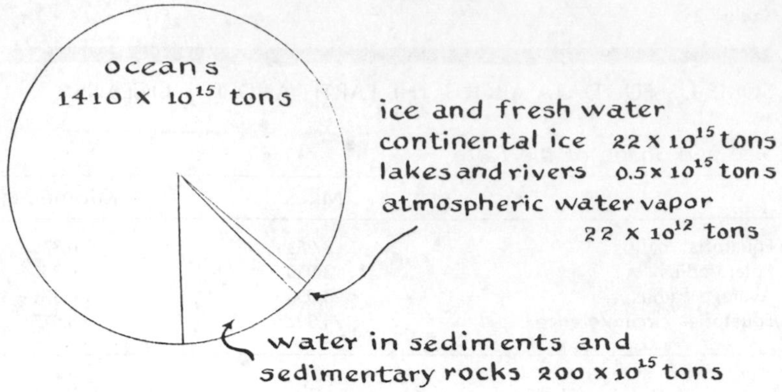

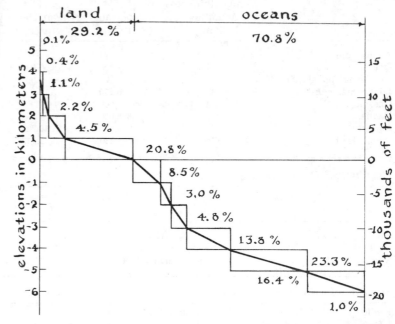

oceans
1410 × 10^15 tons

ice and fresh water
continental ice   22 × 10^15 tons
lakes and rivers   0.5 × 10^15 tons
atmospheric water vapor
22 × 10^12 tons

water in sediments and
sedimentary rocks 200 × 10^15 tons

FIG. 2–3
Mass of water (in tons) on the earth's surface.
Note the minute fraction of the total that
is in the form of fresh water at any instant.

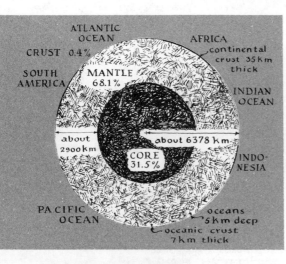

ATLANTIC
OCEAN

CRUST 0.4%

SOUTH
AMERICA

AFRICA
continental
crust 35 km
thick

MANTLE
68.1%

INDIAN
OCEAN

about
2900 km

about 6378 km

CORE
31.5%

INDO-
NESIA

PACIFIC
OCEAN

oceans
5 km deep

oceanic crust
7 km thick

FIG. 2–2
Cross-section through the earth's equator
showing the distribution of ocean basins and
continental blocks (exaggerated), otherwise
drawn approximately to scale, (From Holmes,
1965)

water is removed from the ocean for perhaps millions of years,
but eventually it too is returned to the ocean where it may remain for many thousands or millions of years.

Data on oceanic depths and elevations of the land are conveniently presented as a *hypsographic curve* (shown in Fig. 2–4),
in which the earth's surface is divided into two distinctly dif-

land   oceans

29.2 %   70.8 %

9.1 %
0.4 %
1.1 %
2.2 %
4.5 %
20.8 %
8.5 %
3.0 %
4.8 %
13.8 %
23.3 %
16.4 %
1.0 %

elevations in kilometers

thousands of feet

FIG. 2–4
The hypsographic curve shows the relative
proportion of the earth's surface at various
elevations above and below sea level.
The length of each bar shows the relative
proportion of the earth's surface within each
altitude zone. (From Arthur N. Strahler,
Physical geography, 2nd ed., John Wiley,
London, 1960)

ferent areas—continental blocks and ocean basins—each having its own distribution of depths or elevations. Continental blocks are partially exposed as land areas, mainly between sea level and an elevation of 1 kilometer (about 3000 feet) above sea level. Ocean basins are depressed, ranging approximately 3 to 5 kilometers (2 to 3 miles) deep, and lie between the continental blocks. Continental margins, ranging in depth from sea level to about 2 kilometers, are overlain by the coastal ocean.

Comparing the volume of continental blocks with that of ocean basins makes it apparent that the latter is much greater than the volume of continental material above sea level. In fact, if the earth's surface were planed smooth by removing material from the continents and filling in the ocean basins, it would be completely covered by ocean 2430 meters deep (about 8000 feet). On the average, land projects about 850 meters above sea level; the depths of ocean basins average 3730 meters below sea level.

Ignoring for the moment the problem of how these differences come about, note that, in terms of the scale in Fig. 2–2, they are smaller than the width of a pencil line on a cross-section through the earth's equator. Despite the great depth of the oceans and the enormous height of the mountains when compared to the height of a man, differences in elevation or depth for the earth as a whole are truly insignificant. Thus we can regard the ocean as a thin film of water on a nearly smooth globe.

DISTRIBUTION OF LAND AND OCEAN

Land and water are unequally distributed over the earth's surface: the ocean covers about 70.8 percent of the earth, while land makes up the remaining 29.2 percent. Furthermore, as Table 2–2 indicates, land is itself unequally distributed on the

Table   2–2

DISTRIBUTION OF LAND AND WATER ON THE EARTH'S SURFACE

	Land (%)	Ocean (%)
Northern Hemisphere	39.3	60.7
Southern Hemisphere	19.1	80.9

earth. Continental masses and ocean basins tend to be antipodal; that is, they occur opposite to each other on the earth. Antarctica, for instance, is antipodal to the Arctic Ocean, and Eurasia and Africa are opposite the Pacific Ocean. Only about 4 percent of land has other land antipodal to it.

About 70 percent of all land is in the Northern Hemisphere, while the Southern Hemisphere is almost entirely water (see Table 2–2). Furthermore, by designating a point near South Island, New Zealand, and its antipode on the western coast of France as "poles," we can make a Northern Hemisphere with

most of the land, and a Southern Hemisphere consisting of about 90 percent water, as Fig. 2–5 shows.

The continents may be seen as grouped into four great isolated masses—in order of size, Eurasia–Africa, the Americas, Antarctica, and Australia—in the midst of one vast ocean. Even though continents may be joined by land bridges, the connections tend to be narrow and often broken in the event of a relatively small rise in sea level. For example, a land bridge once connected the North American and Eurasian continents during the last Ice Age, more than 20,000 years ago, but was broken when a rise in sea level resulting from melting of the continental glaciers flooded the area.

Continents exhibit certain gross similarities. For example, as Fig. 2–6 illustrates, they all tend to be triangular-shaped with the sharpest points directed southward, most obvious in North America, South America, and Africa. As a result of their approximate north–south alignment, they block ocean currents moving east or west across ocean basins, particularly in the Northern Hemisphere. In the Southern Hemisphere, between latitudes 50° and 65°S, the ocean basins are connected and the ocean completely dominates. Surface and subsurface currents there move water between the major ocean basins.

FIG. 2–5

Two views of the earth from space, showing the uneven distribution of land and water. Most of the land occurs in the Northern Hemisphere; the Southern Hemisphere may properly be termed the *water hemisphere*.

FIG. 2–6

Map of the earth's surface showing the arbitrary boundaries drawn for the major oceans. Areas of the continents whose rivers drain into each of the ocean basins are shown as well as some of the major rivers.

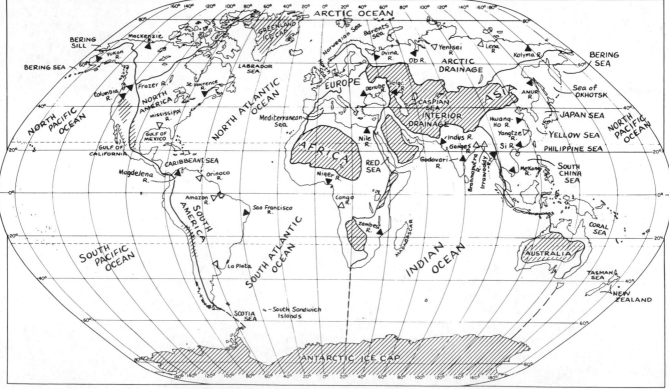

▽ RIVERS DISCHARGING > 15,000 m³/sec. (525,000 ft³/sec.)

▼ RIVERS DISCHARGING MORE THAN 3,000 m³/sec. (10⁵ ft³/sec.)

▨ NO RUNOFF TO OCEAN

Viewed simply, the ocean basins are essentially three "gulfs" trending northward from the ocean-dominated area around Antarctica. These "gulfs" are the Atlantic, Pacific, and Indian Oceans. Sometimes the waters around Antarctica are considered a fourth ocean, designated the Antarctic, or Southern Ocean, whose northern limit is set near 40°S latitude, where opposing currents cause a downward flow of water, here called the *subtropical convergence*.

Distribution of land and ocean at various latitudes is shown in Fig. 2–7. The relative amounts of land and water in the two hemispheres are quite obvious from this graph, which also shows that most of the ocean is located near the equator (about 52 percent between 30°N and 30°S).

Certain large islands, such as Greenland, Ceylon, and Madagascar, lie near continents to which they bear a close geological resemblance, making it seem quite likely that these islands and continents at one time constituted a single land mass. Certain large, shallow areas of the ocean floor called *microcontinents*—especially common in the Indian Ocean—more closely resemble continents than they do ocean basins.

PACIFIC OCEAN

Stretching about 17,000 kilometers (10,000 miles) from Ecuador to Indonesia, the Pacific Basin occupies more than one-third of

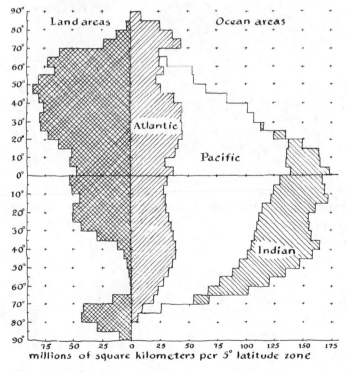

FIG. 2–7
Distribution of land and water in each 5° latitude belt. (Redrawn from M. Grant Gross, *Oceanography*, 2nd ed., Charles E. Merrill, Columbus, Ohio, 1971)

the earth's surface (see Fig. 2–1) and contains more than one-half of the earth's free water (indicated in Fig. 2–8). Active mountain building is a dominant feature of Pacific shorelines, the mountains creating a barrier between inland areas and ocean. Along the Pacific's entire eastern margin (the western coasts of the Americas from Alaska to Peru) rugged young mountain ranges parallel the coast.

All these mountains preclude direct access of rivers to the Pacific, and consequently relatively little fresh water or sediment is carried to it from the continents. Continental margins are narrow because the steep slope of the mountains tends to continue below sea level. As a result, the central Pacific is little affected by the continents that border it.

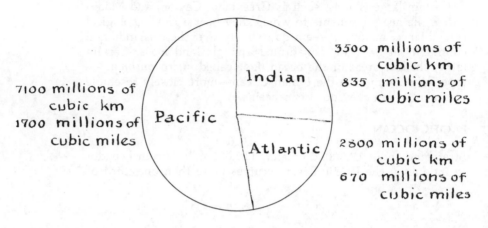

FIG.   2–8
Volumes of ocean basins.

The Pacific has the deepest, the coldest, and by far the largest of the ocean basins (see Table 2–3). Paradoxically, it is also the least salty (see Table 2–4), even though it receives the least river water of any major ocean (see Table 2–5).

Table   2–3

SURFACE AND DRAINAGE AREAS OF OCEAN BASINS AND THEIR AVERAGE DEPTHS*

Oceans[†]	Ocean Area (10^6 km²)	Land area drained[‡] (10^6 km²)	Ocean area / Drainage area	Average depth[‡] (m)
Pacific	180	19	11	3940
Atlantic	107	69	1.5	3310
Indian	74	13	5.7	3840
World ocean	361	101	3.6	3730

*From H. W. Menard and S. M. Smith, "Hypsometry of Ocean Basin Provinces," *Journal of Geophysical Research*, 71 (1966), 4305.

[†]Includes adjacent seas. Arctic, Mediterranean, and Black Seas included in the Atlantic Ocean.

[‡]Excludes Antarctica and continental areas with no exterior drainage.

Table 2-4

PACIFIC OCEAN   33

AVERAGE TEMPERATURE AND SALINITY OF WORLD OCEAN*

	Temperature (°C)	Salinity (⁰/₀₀)
Pacific	3.36	34.62
Indian	3.72	34.76
Atlantic	3.73	34.90
World	3.52	34.72

*From R. B. Montgomery, "Water Characteristics of Atlantic Ocean and of World Ocean," *Deep-Sea Research*, 5 (1958), 146.

Table 2-5

WATER SOURCES FOR THE MAJOR OCEAN BASINS (CENTIMETERS PER YEAR)*

Ocean	Precipitation	Runoff from adjoining land areas	Evaporation	Water exchange with other oceans
Atlantic	78	20	104	−6
Arctic	24	23	12	35
Indian	101	7	138	−30
Pacific	121	6	114	13

*From M. I. Budyko, *The Heat Balance of the Earth's Surface*, trans. N. A. Stepanova, Office of Technical Services, Dept. of Commerce. Washington, 1958.

There are more islands (over 25,000), submerged volcanoes, and underwater mountains in the Pacific Ocean basin than in all the rest of the world ocean. These tend to occur in long chains, some curved and others, such as the Hawaiian Islands, in straight lines. Four types of islands may be differentiated: continental, volcanic, low-carbonate, and high-carbonate islands. Their combined land area totals more than 2.6 million square kilometers (1 million square miles).

The largest islands, occurring along the western margin of the basin, exhibit a *continental* structure. They stretch from New Zealand in the south to Japan in the north, the largest being New Guinea. Consisting of continental-type rocks, these large islands are generally richer in industrial minerals and other natural resources than the smaller volcanic or carbonate islands.

Volcanic islands are the tops of volcanoes that rise from oceanic depths to penetrate the sea surface. These are among the largest volcanoes on earth, although the islands formed by their tops may be quite small. Such islands are generally mountainous, with large, deep valleys. Actively eroded by waves, the islands are often cut down to sea level within a few million years after volcanic eruptions have ceased.

Low-carbonate islands, which are quite small, are widely scattered throughout the Pacific, and conform to the usual image of

FIG. 2-9
Satellite view of parts of the Indian and Pacific Oceans taken from an altitude of about 10,000 kilometers (6568 miles). Much of Asia (upper left) is obscured by clouds except for portions of the China coast. Australia is clearly visible in the lower portion of the photograph, a consequence of dry atmospheric conditions prevailing there. (Photograph courtesy NASA)

a "South Sea island." Among them are 310 *atolls*—mere ribbon reefs encircling a lagoon. They are especially common in the southwestern Pacific.

High-carbonate islands are generally also quite small. They consist of reef rock which has been elevated above sea level. Such islands are scattered throughout the central and south Pacific, as Fig. 2-9 makes clear.

ATLANTIC OCEAN

The Atlantic Ocean Basin was the first ocean basin to be explored and has been studied more thoroughly than any other; thus more is known about it. Seemingly the youngest of the ocean basins, the Atlantic was apparently formed when North and South America split away from Africa and Eurasia less than 200 million years ago.

The Atlantic is relatively narrow, about 5000 kilometers (3000 miles) wide. The shorelines on either side of the basin are remarkably parallel, forming an S-shaped basin which is the major connection between the two polar regions of the world ocean (see Fig. 2-6).

In defining individual ocean basins, the delineation of boundaries must necessarily be somewhat arbitrary. This is especially the case in that part of the Southern Hemisphere where there are few land barriers. The boundary between the Atlantic and Indian Oceans is drawn from the Cape of Good Hope at the southern tip of Africa extending southward along longitude 20°E to Antarctica. The boundary between the Atlantic and Pacific Oceans is drawn between Cape Horn at the lower tip of South America to the northern end of the Palmer Peninsula of Antarctica.

On the north, the boundary also presents a problem. When the Arctic Ocean is included as part of the Atlantic, the northern boundaries of both the Atlantic and the Pacific are made to meet at the Bering Strait between Alaska and Siberia—a shallow area known there as the Bering Sill forms an easily recognizable boundary between the two oceans. (Although oceanographers draw such boundaries with ease, the ocean pays little attention to them. There is substantial movement of water across all these artificial boundaries, including the Bering Sill.) Defined this way, the Atlantic Ocean extends from the shore of Antarctica northward across the North Pole to about 65°N on the other side of the Earth. Neither of the other two major ocean basins extends so far in a north–south direction (Fig. 2-7).

The Atlantic is a relatively shallow ocean, having an average depth of only 3310 meters (10,800 feet), the result of an abundance of continental shelf in the ocean and shallow marginal seas (particularly the Arctic Ocean). The Atlantic also receives a relatively large amount of sediment from numerous rivers, which covers the ocean floor and thus reduces its depth.

There are relatively few islands in the Atlantic Ocean. Greenland, the world's largest island, is actually a part of the North American continent, isolated by a relatively narrow channel consisting of Davis Strait and Baffin Bay. Most other Atlantic islands are tops of volcanoes located on the great submarine ridge that runs north–south down the center of the Atlantic Basin. Iceland, for instance, is an exposed portion of this ridge. Other volcanic islands, such as the Azores or Tristan da Cunha, are smaller, generally being the exposed summits of large volcanoes; for instance, Bermuda is the top of an extinct volcano which is now covered by a limestone cap.

The Atlantic Ocean receives a large amount of fresh water from river discharge. The world's two largest rivers, the Amazon and the Congo (see Fig. 2–6), flow directly into the Atlantic equatorial region, together discharging about one-quarter of the world's runoff of fresh water to the ocean. Many other large rivers flow into the marginal seas that are a common feature of the North Atlantic. For example the Mississippi River, draining much of the North American continent, discharges into the Gulf of Mexico, and the St. Lawrence flows into a small gulf. Also, fresh water running from several large rivers into the Arctic Ocean eventually enters the Atlantic.

Despite its connections to both polar regions and the large amount of fresh water discharged into it, the Atlantic is the warmest (average temperature 3.73°C) and saltiest (salinity 34.90 parts per thousand—usually written as 34.90$^0/_{00}$) of the world ocean basins (average ocean temperature is 3.52°C, average salinity 34.72$^0/_{00}$). This apparent paradox is probably a result of the large number of marginal seas adjoining the Atlantic. The Mediterranean Sea in particular exerts a strong influence on the entire Atlantic. Waters flowing into the Mediterranean are warmed and exposed to dry atmospheric conditions. Substantial amounts of water are lost from the sea surface, so that the sea water becomes more saline. This warm, salty water then flows out into the Atlantic and is detectable at mid-depths over a wide area extending many thousands of kilometers from the Strait of Gibralter.

ARCTIC OCEAN

For most purposes we may consider the Arctic Ocean to be a landlocked arm of the Atlantic—this includes all waters north of Eurasia and the North American continent. Among oceanographers, debate continues as to whether the distinctive characteristics of this polar area justify its designation as a separate ocean. In any case, it is unique in several respects: continental shelf forms one-third of its floor, it is almost completely surrounded by land, and—especially characteristic—much of it is covered by pack ice during part or all of the year.

The Arctic as a whole is quite shallow, largely because of its wide continental shelves. On North America, between Alaska and Greenland, the shelf is only 100 to 200 kilometers (60 to 100 miles) wide, but offshore from western Alaska and Siberia it is much wider, 500 to 1700 kilometers (300 to 1000 miles) across, as shown in the map of Fig. 2–10. During much of the year it is unnavigable because its surface is covered by *pack ice* (sea ice in thick chunks that freeze together, break apart, and freeze again to form a solid surface) to a depth of 3 to 4 meters (10 feet or more). In the central Arctic this cover is permanent, though it may melt to a minimum thickness of about 2 meters by the end of August.

Because it is surrounded by continents, the Arctic Ocean is greatly influenced by river discharge. The large amounts of fresh water dilute sea salts and cause low salinity. Large amounts of sediment on the bottom of Arctic basins testify to extensive erosion of surrounding continents by glaciers during the past several million years. Much of this material was deposited as glaciers melted and lost the sediment they had accumulated while advancing over the land. Rivers also discharge considerable sediment into the Arctic region.

INDIAN OCEAN

The Indian Ocean is primarily a Southern Hemisphere ocean, extending from Antarctica to the Asian mainland at about 20°N, as shown in Fig. 2–6. We have already discussed the boundary

FIG. 2–10
Map of the Arctic Ocean showing submarine topography.

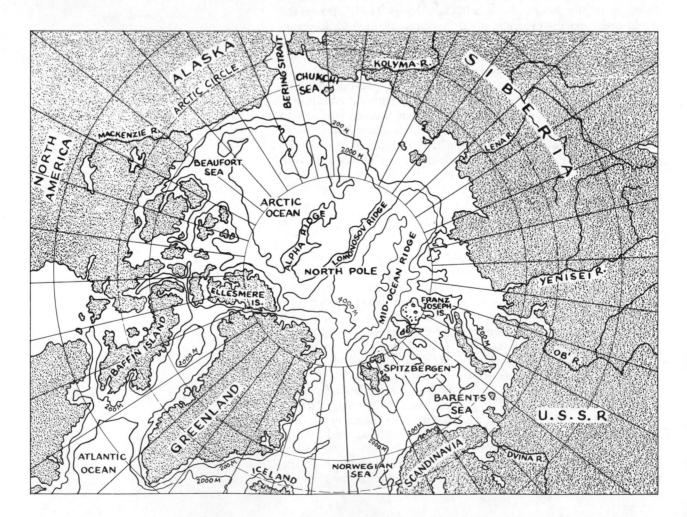

between the Indian and Atlantic Oceans running south from Cape Good Hope. The other boundary between the Pacific and the Indian Ocean runs through the Indonesian Islands and extends from Australia, through Tasmania and southward to Antarctica along longitude 150°E.

As delimited, the Indian Ocean is the smallest of the three major ocean basins. It is triangular in shape with a maximum width of about 15,000 kilometers (6000 miles) between South Africa and New Zealand. The continental shelves around the ocean are relatively narrow but a large amount of sediment has been deposited in the northern part of the basin, making it intermediate in average depth.

Three of the world's largest rivers (Ganges, Indus, Brahmaputra) discharge into the Indian Ocean along its northern margin, also the location of its only marginal seas. Of these the Red Sea and Persian Gulf have the greatest effect on the Indian Ocean as a whole, because they are areas of extensive evaporation. Most of the water discharged into the Indian Ocean drains from the Indian subcontinent (shown in Fig. 2–11); little comes from Africa.

There are relatively few islands in the Indian Ocean, Madagascar being the only large one. It is a microcontinental block, as are several shallow banks in the ocean basin—all apparently fragments of an ancient continental block. There are a few volcanic islands and several groups of low-carbonate islands, including atolls.

FIG. 2–11
The subcontinent of India and the island of Ceylon as seen from Gemini XI in September, 1966, from an altitude of 815 kilometers (507 miles). A well-developed sea breeze circulation has suppressed clouds along the coast, although clouds obscure much of the land. (Photograph courtesy NASA)

SEA LEVEL

While discussing distribution of land and water on the earth remember that we have used sea level to mark this boundary. In other words, the *shoreline*, the boundary between land and water, is the line where sea level intersects the continents. The position of the shoreline changes if the land rises (or falls).

Changes in sea level can also result from alteration of an ocean basin. The uplifting of large areas of ocean bottom (thereby decreasing the ocean-basin volume) will cause a rise in sea level. Such changes do not involve the volume of water in the oceans, but only its height relative to the land masses. Records of changes in ocean basins are perceived only in the flooding of large areas of low-lying continental blocks. In other words, we see the effect, but evidence of the cause usually disappears.

Another cause of sea level change is the formation and disappearance of glaciers, continent-sized ice sheets that permanently cover most of the Antarctic continent and some high mountains. The earth is now in a rather warm phase of an ice age that has lasted for several million years. During glacial periods, water is removed from the ocean and piled up on continents to form massive sheets of ice that may locally exceed a mile in thickness. Removal of this much water from the ocean causes a drop in sea level.

Over the past 20,000 years, the earth passed through a period

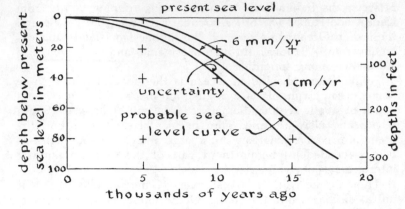

FIG. 2–12
Change in sea level over the past 20,000 years as dated by carbon-14 techniques. (Redrawn from Francis P. Shepard, *Submarine Geology*, 2nd ed., Harper & Row, New York, 1964)

of extensive glaciation when the sea stood perhaps 130 meters (440 feet) below its present level. As the glaciers melted and returned their water to the oceans, the sea level rose as indicated in Fig. 2–12, reaching its present level about 3000 years ago. Since that time there have been rather few, relatively minor changes in sea level.

The recent rise has occurred over a short period of geologic time (about 15,000 years), flooding large areas that earlier were low-lying dry land. Remains of large, presently extinct mammals, and of typically coastal shellfish, are found far out at sea on the shallow continental shelves that extend present shore-lines. These remains testify to the presence of dry land there thousands of years ago. Further rise of sea level would flood our present coastal plains, the site of many of the world's largest population centers.

OCEAN DEPTHS

Mapping the shorelines and land surface has proceeded over hundreds of years. Although many land areas are still poorly mapped, mapping them has been a relatively simple operation—after all, one can see the land that is being mapped.

Charting the ocean bottom is more difficult. Even the clearest ocean water permits only shallow ocean bottom to be observed directly. Therefore, it has been necessary to devise indirect means of surveying oceanic depths. The earliest depth determinations, called *soundings*, were made by lowering a weighted rope from a ship. Depths were determined by measuring the amount of line let out when the weight hit bottom, a costly and often inaccurate process. For great depths, it was difficult to determine when the weight hit bottom so that the depths charted were often wrong. Once an even greater problem was that of making accurate determinations of position at sea.

Since the 1920's, a different means of measuring oceanic depths has been developed, utilizing the fact that ocean water is nearly transparent to sound waves. The technique, known as *echo sounding*, depends on accurately timing the interval between transmission of a strong sound pulse by the ship and the detection of the echo from the bottom (Fig. 2–13). Making corrections for variations in sound velocity (affected by water temperature, salinity, and pressure), it is possible to make accurate depth determinations and to draw profiles of the ocean bottom while the ship is underway. Accurate location at sea remains a problem, although electronic devices using satellites now provide reliable information on position in all the world ocean.

On the ocean floor, three distinct provinces may be identified: *continental margins*, *oceanic rises*, and *ocean basins*, as shown in Fig. 2–14. Continents and their submerged margins account for about 43.7 percent of the earth's surface. About two-thirds of the continental blocks are exposed as land; the remaining one-third forms the continental margin. Oceanic rises cover about 23.1 percent of the earth's surface, and the ocean basins and other smaller oceanic provinces cover about 33.2 percent.

Shorelines do not mark major boundaries in the earth's crust. Instead, the 2-kilometer (1.2-mile) depth contour, as mapped in Fig. 2–15, is considered more useful to separate oceanic basin from continental block. Fundamental structural differences aid us in making a distinction between continental and oceanic portions of the earth's crust. Continental blocks are made of a thick mass (usually about 35 kilometers thick) of granitic rocks, usually light-colored and containing relatively high proportions of silica (SiO_2) and alumina (Al_2O_3). The upper surfaces, exposed on land, are commonly covered by a relatively thin layer of sedimentary rock that was originally deposited in the shallow ocean when parts of the continental blocks were submerged to form shallow seas. Rocks in the continental blocks are as old as 3.3 billion years.

In contrast, the oceanic basins are typically about 4 to 6 kilometers below sea level and underlain by a thin oceanic crust, typically about 6 kilometers thick. The rocks of the oceanic crust are normally basalts, dark-colored rocks that are slightly more dense than granite and contain far more magnesium and iron than granite does. Rocks older than about 200 million years are not known to occur in deep-ocean basins.

The transitional zone between these units is several tens of kilometers wide, typically averaging about 50 kilometers. Although we use the 2-kilometer depth contour for our purposes to mark this boundary, it is often slightly seaward of the continental slope and frequently covered by the continental rise, as indicated in Fig. 2–16. Shallow-ocean areas (less than 2 kilometers deep) tend to have a thick crust with properties more closely resembling continents than ocean basins. Areas deeper than 2 kilometers tend to have a thin crust, typical of ocean basins. On this basis, Greenland, Madagascar, and other large islands, as well as some shallow banks in the Indian Ocean and elsewhere, are more appropriately considered small continental blocks than oceanic-basin features.

FIG. 2–13

An echo sounder emits a sound pulse for a ship-mounted transmitter. The time elapsed between signal transmission and reception of the echo from the bottom provide means of measuring water depth.

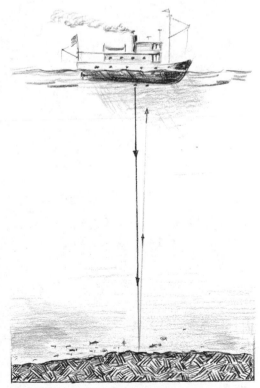

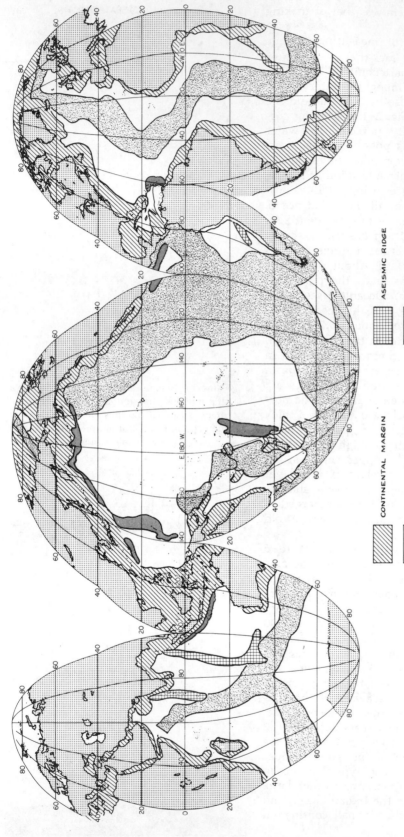

ASEISMIC RIDGE

ISLAND ARC & TRENCH SYSTEMS

CONTINENTAL MARGIN

OCEAN BASIN

OCEANIC RISE

FIG. 2-14

Ocean-bottom provinces showing relationships
between submerged continental margins,
mid-oceanic rises, ocean basins, island arcs,
and volcanic (aseismic) ridges. Note that
the trenches, the deepest parts of the ocean
bottom, occur along the basin margins.

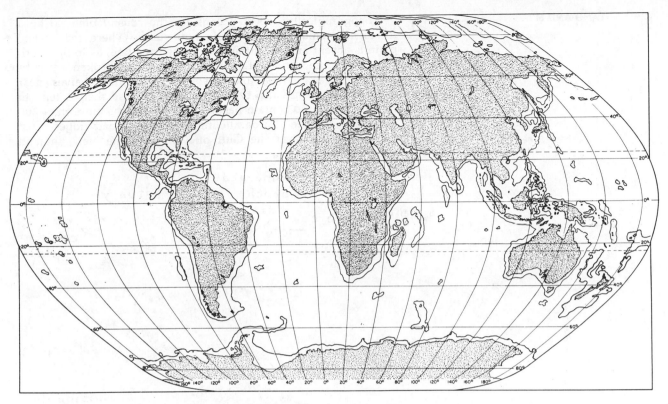

FIG. 2–15
Map showing relationship of continental land
masses to the 2000-meter depth contour
which marks the approximate outer limits of
the continental blocks. (From J. E. Williams
Prentice Hall World Atlas, 2nd ed., Prentice-
Hall, Inc., Englewood Cliffs, N.J., 1963)

CONTINENTAL MARGIN

The continental margin consists of the *continental shelf, slope,*
and *rise* (see Fig. 2–16). The continental shelf is the submerged
top of a continent's outer edge, and it begins at the shoreline.
This is the part of the continent that has been shaped by the
ocean, except where mountains occur at the edge of the continent. There the shaping force has been the processes that
formed the mountains. From the shoreline the shelf slopes gently
toward the *continental-shelf break* typically at a depth of about
130 meters (450 feet), except off Antarctica, where the shelf
break is more than 350 meters (1000 feet) deep. At the shelf
break the gently sloping shelf gives rise to the much steeper
continental slope.

On the average the continental shelf is about 70 kilometers
wide (40 miles). On much of the Pacific Coast it is relatively
narrow, a few tens of kilometers across. The widest shelves
occur in the Arctic Ocean. A broad continental shelf is the
sculptured build-up of sediment shed by a continent, particularly
where there has been no major mountain building at the margin over hundreds of millions of years.

Topographically, we may consider the *coastal plain* to be a terrestrial analog of the continental shelf. Where the coastal plain is absent and mountains form the coastline, as in Southern California, the continental margin is rugged, broken by submarine ridges, banks, and basins. Continental shelves near glaciated coasts are also typically rugged. These shelves tend to be quite wide and contain basins or deep troughs more than 200 meters deep, generally paralleling the coast. Broad ridges often divide these basins. The Gulf of Maine off the New England coast is such an area.

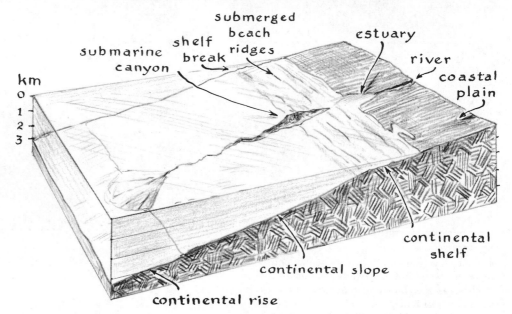

FIG. 2–16
Schematic representation of the continental margin showing the continental shelf and its relationship to the coastal plain, the continental slope, and continental rise. Note that the submarine canyon cuts across the shelf break and extends to the base of the continental slope, where it joins the continental rise.

Some troughs cut across the shelf and connect with flooded, deep indentations known as *fjords*. Such troughs can be quite large, such as the trough at the entrance to the Strait of Juan de Fuca connecting the Pacific Ocean with Puget Sound and the Strait of Georgia. Topography of these shelf areas was formed by glaciers during periods of lowered sea level, and have not been modified greatly by presently active marine processes.

On continental shelves not affected by glaciers, the bottom topography tends to be much smoother, commonly gently rolling and subdued. Elongate low sand ridges (sometimes containing gravel) generally parallel the shore. The ridges are only a few meters above the surrounding ocean bottom. Often between these ridges there are broad basins or troughs which are a few meters deeper than the adjacent ocean bottom. These low ridges and shallow basins are likely submerged shoreline features that were formed as the shoreline moved across the continental shelf.

In tropical oceans there are often coral reefs on the continental shelf. These commonly occur on the outer portion of the shelf and are separated from the land by a navigable channel paralleling the coast.

Continental shelves receive sediment brought to the ocean by rivers draining the continents. Sediment brought by rivers is usually deposited in the estuary. Relatively few rivers carry enough sediment to fill their estuaries and form extensive deposits on the adjacent continental shelf. Consequently about 70 percent of the world's continental shelf is covered by sediments deposited more than 20,000 years ago. These deposits are called *relict sediments*, since they were not deposited by modern oceanic processes.

The continental-shelf break is remarkably uniform in depth, averaging about 130 meters over most of the world ocean. The shelf break is thought to have formed when sea level stood at its lowest during Pleistocene glacial times. At this time the shoreline was at the edge of the present continental shelf, which was then a low-lying coastal plain. Estuaries and lagoons occurred along the edge, the estuaries occupying the sites where present submarine canyons indent the shelf break. Traces of ancient lagoons are not readily discernable although they must have been common in many areas.

Continental slopes are the relatively narrow, steeply inclined submerged edges of continental blocks. On an ocean-free earth these would be the most conspicuous boundaries on the earth's surface. Although their total relief is substantial, ranging from 1 to 10 kilometers (0.6 to 6 miles high), continental slopes are not precipitous. They average about 4°, ranging from 1° to 10° of slope, either quite straight or gently curving. The most spectacular slopes occur where a continental block borders a deep trench. On the western coast of South America, the peaks of the Andes Mountains rise to elevations of about 7 kilometers, while the adjacent Peru–Chile Trench is 8 kilometers deep. This adds up to a total relief of 15 kilometers (9 miles) within a few hundred kilometers. Continental slopes with trenches at their base are common in the Pacific and account for about half the world's continental slopes.

The origins of these slopes are not completely understood. In the Pacific Ocean, continental slopes appear to be formed by the same complex processes that form coastal mountain ranges and the offshore trenches and island arcs characteristic of Pacific margins. On the stable margins of continents, continental slopes appear to result from *faulting*, large-scale fracturing in the earth's crust. Rock is usually exposed on these faulted continental slopes. The faults themselves have usually been greatly altered by *slumping*, the sliding downslope of rock or sediment.

Faults on the slope often form small basins that trap sediment, changing the local topography from a complex of steep-walled basins to small, flat-floored basins that gradually become filled and smoothed over, eventually disappearing. With accumulation of a sufficient thickness of sediment, the original outlines of the continental slope are completely buried. Margins of many continental blocks have numerous buried ridges and ancient coral reefs, now covered by sediment deposits.

Thick accumulations of sediment cover many of the world's continental slopes, except along the Pacific Ocean. When sediment deposits become too thick, and the slopes too steep, the deposits may slump. This leaves large, spoon-shaped scars on the edge of the slope, and a mass of deformed sediment at the base of the slump. This probably is fairly common, and may account for many of the hills on the lower parts of continental rises.

Submarine canyons are large features that occur commonly on continental slopes (see Fig. 2–16). These canyons were found to be common after numerous echo soundings became available. Their origin posed something of a problem for marine geologists for many years. In the absence of reliable data, it had long been assumed that the ocean bottom must be a nearly featureless plain. The discovery that large canyonlike features often occur along continental margins was one of several shocks that has gradually led to the realization that the ocean bottom is subjected to many processes quite capable of moving sediment and carving canyons.

These canyons are typically V-shaped, follow curved paths, and have tributary channels leading into them. Some are often as large as the Grand Canyon of the Colorado River. In a few instances, submarine canyons are obviously associated with large rivers on the continent—for instance, offshore from the Hudson and Congo Rivers. In those cases, the upper portions of the valley could have been cut during times of lowered sea level when a river flowed through the canyon, so their origin posed no problem. How, then, does one explain the portion of a canyon that swept out to the edge of the continental slope where it merged into much shallower channels with low levees on each side? These channels often run for tens or hundreds of kilometers across the fan of sediment that forms the base of the slope. Many canyons have no obvious connection to present-day rivers, although they may have been associated with rivers no longer in the region. Some canyons were cut through sedimentary rocks which are relatively soft. Others were found to be cut through harder granite. Some sediment commonly occurs on canyon floors.

These canyons, it is now generally agreed, are cut by *turbidity currents*, flows of sediment-laden water that move along the ocean bottom carrying material from the continental shelf to the deep ocean floor. On the continental slope, they appear to reach relatively high speeds—10–50 kilometers per hour—capable of eroding rock. When the flows reach the base of the slope, they move out onto the sediment fans that form the continental rise.

Turbidity currents can be triggered by earthquakes or by sudden large discharges of river-borne sediment during floods. The process of canyon cutting is still active in many areas, especially where the continental shelf is relatively narrow and sediment can be readily transported to the head of the canyon. Such canyons have been especially closely studied in Southern California, where their upper portions lie quite close to the beach.

Continental rises (see Fig. 2–16) are generally smooth-surfaced accumulations of sediment at the base of continental slopes. They cover the transition between the continent and the ocean basin. Low hills are present, perhaps formed by slumping of sediment from the continental slope. Channels a few meters deep, apparently routes of sediment movement, cross the rise. These channels are often connected to submarine canyons.

Continental rises are apparently formed by sediments carried by turbidity currents and deposited at the base of the slope. Deposition is caused by the reduction in current speed when it flows out onto the gently sloping ocean floor. Initially each canyon has a fan-shaped deposit of sediment at its mouth, similar to streams flowing out of the mountains onto a flat desert floor. Eventually individual fans build sideways and coalesce to form a continuous sediment wedge whose surface slopes as much as 1 meter in 40.

Typically, the continental rise contains sediment deposits a few kilometers thick, and tens of kilometers wide. One estimate of the total volume of sediment in the world ocean gave the following results:

Continental shelf	1.2 billion cubic kilometers
Continental slope and rise	1.6 billion cubic kilometers
Deep-ocean bottom	0.8 billion cubic kilometers

These data indicate that about 40 percent by volume of the sediment in the ocean lies in the continental rises at the base of continental slopes. Altogether, deposits on the continental margin account for about 78 percent of the total recognized sediment deposits in the world ocean.

MARGINAL-OCEAN BASINS

Marginal-ocean basins are large depressions in the ocean bottom lying near continents. They are separated from the open ocean by submarine ridges, islands, or parts of continents. Despite their proximity to continents, marginal basins are usually more than 2 kilometers deep, so that their bottom waters are partially isolated from ocean waters at comparable depths outside the basin. Marginal basins have a total area about $\frac{1}{20}$ that of the continents.

These small seas are, in a sense, transitional between continents and deep-ocean basins, as their location would suggest. Studies of their crustal structure has shown that they commonly have a typical oceanic crust overlain by thick sediment deposits. Being near the coast, they receive the discharge of many large rivers and the sediment brought by these rivers to the ocean is deposited in the basins, resulting in generally smooth bottoms and sediment deposits many kilometers thick. These deposits contain about $\frac{1}{6}$ of all known ocean sediments.

Nearly all marginal-ocean basins are located in areas where crustal deformation or mountain building is active. They may eventually be changed and incorporated into the adjacent con-

FIG. 2–17

The Red Sea with the Gulf of Aqaba and the Gulf of Suez. The dark band on the right is the Nile Valley and the river can be seen as the light meandering line. Note the straight coastlines on either side of the Red Sea. The absence of clouds is an indication of the region's aridity. The photograph was taken from Gemini XII at an altitude of 280 kilometers (175 miles)—portions of the spacecraft are visible on the right side. (Photograph courtesy NASA)

tinental block. Three kinds of associations are typical of marginal-ocean basins. The most common is that of a basin associated with island arcs and submarine volcanic ridges which partially isolate surface waters and may completely isolate the subsurface waters. The many small basins around Indonesia and along the Pacific margin of Asia are examples of this association.

A second type of marginal sea lies between continental blocks —for instance, the Mediterranean Sea between Europe and Africa, the Black Sea (now almost completely landlocked), and the Gulf of Mexico and Caribbean Sea between North and South America (see Fig. 2–6).

The third type is the long, narrow marginal sea formed where oceanic rises intersect continental blocks. The Red Sea (shown in Fig. 2–17) and the Gulf of California are examples of such basins.

Marginal seas are subjected to more changeable weather conditions than occur in the open ocean, owing to the proximity of large land masses. Marginal seas in high latitudes, such as the Sea of Okhotsk, become cold enough to freeze over during winter months. In mid-latitudes, where the world's deserts are located, marginal seas such as the Mediterranean Sea and Persian Gulf are strongly heated and evaporated by the sun so that their waters are much warmer and saltier than the world-ocean average. (Think of the effect a day's exposure to the summer sun has on a swimming pool as compared to a large, deep lake.)

Because of their marked response to seasonal weather changes. water circulating through marginal seas has a recognizable influence on adjacent ocean basins. As we have already mentioned, this is especially true for the Atlantic Ocean, which has many marginal seas and is a relatively small ocean basin.

In Europe and North America, shallow seas occupy depressed areas at the edge of continental blocks, such as Hudson Bay in Canada. The North Sea of northern Europe is another such flooded continental area, leaving only England and her associated islands above sea level at the present itme.

COASTAL OCEAN

The ocean overlying a continental margin—which we shall call *coastal ocean* (see Fig. 2–15)—is profoundly affected by adjacent land masses. Processes that occur in the open ocean are distinctly modified near continental blocks. In the first place, continents stand as massive barriers to currents moving east or west across deep ocean basins: they deflect these east–west water movements, causing north–south-directed boundary currents that parallel continental margins. Thus currents of cold water moving across the open ocean may be deflected toward warmer regions, where their waters are heated. Other boundary currents carry warm waters poleward. Heat transport in the ocean and throughout the earth is fundamentally influenced by this process.

The coastal ocean is much shallower than the open ocean. Thus any process that is sensitive to water depths has a different effect on coastal waters than it would have on water in the center of an ocean basin. For instance, cooling of waters in autumn and winter and warming in spring and summer are much more pronounced in the coastal ocean than in the deep ocean. The shallowness of the coastal ocean also favors more active mixing of waters, because currents flowing over irregularities in the continental shelf and slope cause mixing throughout the overlying waters. This shallowness, combined with strong tidal currents and strong, seasonably variable winds, results in much greater mixing of surface and bottom waters in coastal areas than is known to exist in the open ocean.

Furthermore, the histories of the two parts of the ocean are strikingly different. The open ocean has existed in one form or another for billions of years and has always been connected to other parts of the world ocean. The coastal ocean, on the other hand, has a much more complex history. Those parts of the coastal ocean lying above continental shelves achieved their present form only about 3000 years ago, when sea level rose to its present height. Many of the processes now active in coastal areas are the result of this short history. In many areas the coastal ocean has not yet fully adjusted to recent changes.

The world ocean as a whole changes little, if at all, with time. We call this a *steady state* condition. This apparent lack of change makes it possible to combine data taken years apart in a given area to study processes in the deep ocean. Partial isolation from the major ocean basins by virtue of the complex topography of continental margins means that the coastal ocean is significantly affected by seasonal changes in temperature and river discharge. Substantial changes may occur within a few hours owing to tidal currents. Therefore, it is not feasible to posit a steady state for the coastal ocean, as we do for the open ocean. Studies of coastal phenomena must include seasonal effects that can usually be ignored in open ocean studies.

From a human point of view, the coastal ocean is by all odds the most important. Man's contact with the ocean and utilization of marine resources takes place almost entirely in the coastal ocean. Virtually all mineral resources extracted from the ocean bottom come from continental margins—in general, superficial deposits remaining from the last rise in sea level or submerged equivalents of deposits well known on the continents.

Food from the sea is taken largely from the coastal ocean. These shallow waters support more marine life per unit of surface area than any other part of the ocean, because coastal circulation regimes continually restore essential nutrient salts to sunlit surface waters, where growth of marine plants takes place. This process also takes place in the deep ocean, but much more slowly.

Finally, present and potential problems of the ocean are predominantly found in coastal waters. These problems arise in part from the distinctive features of the continental margin previously discussed; for instance, restricted circulation in coastal regions retards dilution and dispersal of wastes discharged from the continents. Basically, then, coastal problems are intensified as a result of population concentration along the borders of continents.

We shall repeatedly find that the coastal ocean is a distinctive and important part of the ocean. Therefore, in the discussions that follow, note carefully the processes active in the coastal ocean and compare them to situations that prevail in the open ocean.

SUMMARY OUTLINE

Earth—the water planet
- Concentric spheres
    - Core—39.1% of earth mass; metallic, high-density interior
    - Mantle—68.1% of earth mass; intermediate density, rock-like shell
    - Crust (lithosphere)—0.4% of earth mass; rocky outer shell
    - Hydrosphere—0.027% of earth mass; waters of earth's exterior, 85% in ocean
- Hypsographic curve—shows two distinct levels on earth
    - Continents—upper surface between 0 and 1 km above sea level
    - Ocean basin—bottom between 3 and 5 km below sea level
- Difference between elevation of continents and depth of ocean basin is small compared to radius of earth

Distribution of land and ocean
- Continents and ocean basins unevenly distributed
- Continents and ocean basins tend to be antipodal—i.e., occur opposite each other
- Northern Hemisphere is the land hemisphere; Southern Hemisphere is the oceanic
- Continents are relatively isolated, generally triangular-shaped
- North–south continents form barrier to east–west ocean currents
- Land dominates between 50° and 70°N
- Ocean dominates between 50° and 65°S; 52% of ocean lies in equatorial regions
- Continental islands and microcontinents are fragments of continental blocks

Pacific Ocean
- Largest ocean basin; approximates water-covered earth; contains more than 50% of surface water on earth
- Bordered by mountain building, isolates land from ocean, many marginal seas
- Many volcanoes and volcanic islands; atolls common in western Pacific
- Pacific islands—continental, volcanic, low-carbonate, high-carbonate

Atlantic Ocean
- Long, narrow, parallel sides; it connects North and South Polar regions
- Well-developed mid-ocean ridge system divides basin
- Relatively shallow ocean
- Many marginal seas; receives about 68% of world's drainage

Arctic Ocean
- Forms nearly landlocked arm of the Atlantic Ocean
- Shallow, with wide continental shelves and three ridge systems
- Covered by ice most of the year

Indian Ocean
    Southern Hemisphere; smallest basin; flat ocean bottom
    Major rivers discharge into northern section
    Contains extensions of oceanic-ridge system
Sea level
    Shoreline marks usual boundary between ocean and land
    Sea level can change and thereby substantially change location of shoreline
    Glaciation has caused change of sea level from -130 m to $+50$ m
    Sea level reached present position about 3000 years ago
Ocean depths
    Soundings of ocean depths provide information about bottom features
    Three major ocean basin provinces: continental margin, oceanic rise, ocean basin
    Continental margin—2-km depth contour good indicator of continent edge
    Continental shelf—submerged top of continental edge
        Extends from shoreline to shelf break (130 m)
        Similar to coastal plain
        About 70% of shelf is covered by relict sediments
    Continental slope—narrow, steeply inclined margin of continents
        Formed by faulting and slumping of rocks in many places
        Cut by submarine canyons probably formed by turbidity currents
    Continental rise—sediment accumulations at base of continental slope
        Contain about 40% of identified sediment in ocean
Marginal-ocean basins
    Usually formed by partial isolation resulting from tectonic processes
    Usually have distinct oceanographic characteristics—highly variable
Coastal ocean
    Overlies continental margin, affected by presence of land
    Currents deflected to form boundary currents—generally north–south
    Shallower than open ocean, favors temperature changes, more active mixing
    Processes characteristically involve short time periods and distances
    Important for man—prime area of contact, major source of food

SELECTED REFERENCES

BLOOM, A. L. 1969. *The Surface of the Earth*. Prentice-Hall, Inc., Englewood Cliffs, N.J. 152 pp. An introduction to geomorphology; Chapters 1 and 6 on surface energetics and shorelines are particularly pertinent.

HOLMES, ARTHUR. 1965. *Principles of Physical Geology*, 2nd ed. Ronald Press, New York. 1288 pp. Classical treatment of the principles of physical geology; advanced.

STRAHLER, ARTHUR N. 1963. *The Earth Sciences*. Harper & Row, New York. 681 pp. Comprehensive, with emphasis on geomorphology.

WENK, EDWARD. 1969. *The Physical Resources of the Ocean. Scientific American*, Sept. 1969. 221(3):166–76.

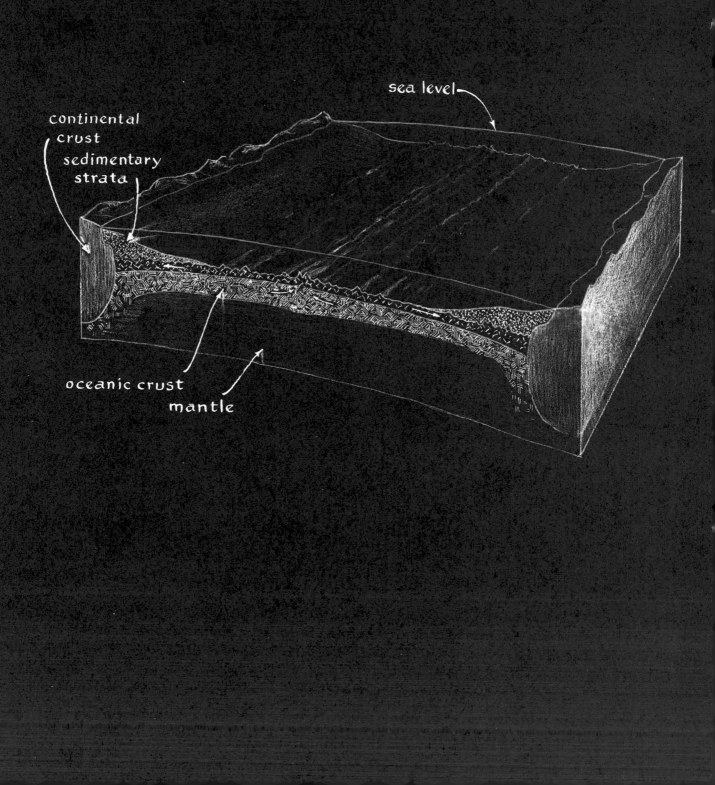

DEEP OCEAN FLOOR

Block diagram showing the earth's crust beneath a generalized ocean basin and the adjoining continental blocks.

three

Having discussed the coastal ocean that overlies submerged continental margins, we now turn to consider the deep ocean floor, charted in Fig. 3–1; this region constitutes 56.0 percent of the earth's surface. To simplify this discussion, we shall take the seaward margins of the continental rises to be the beginning of the ocean-basin province. This is usually near the 2-kilometer depth contour that we used to delimit the edge of the continental blocks in Chapter 2, as illustrated in Fig. 2–15.

The two major provinces of the deep ocean floor are the mid-oceanic rises (or ridges) and the deep-ocean basin. Although boundaries between these two provinces are difficult to define, for our purposes we shall place them where the outermost slopes of the mid-oceanic rise gradually merges with the flatter ocean bottom. Another criterion that could be used to distinguish the two provinces is the structure of the earth's crust. This works well with respect to the boundary between the thick crust of the continental blocks and the much thinner crust of the oceanic basins; unfortunately, however, our sketchy knowledge of these regions makes this an unreliable criterion at the present time.

OCEANIC RISES

Oceanic rises and *ridges* cover 23.1 percent of the earth's surface (Table 3–1), an area comparable to the land which covers 29.2

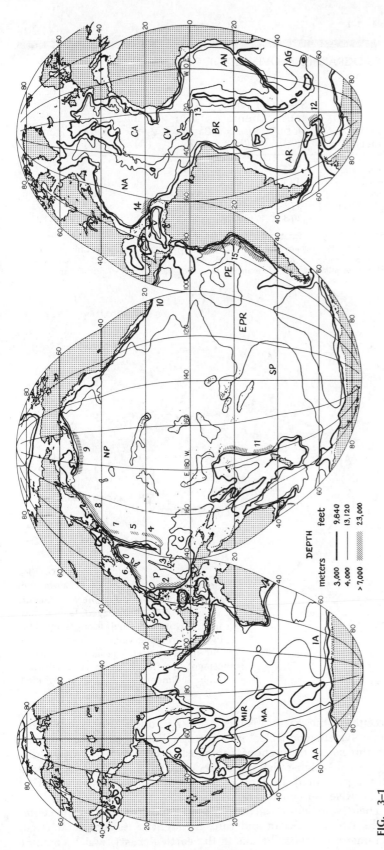

FIG. 3–1
Bathymetric chart of ocean-bottom topography.

Basins

A   Arabian
AA Atlantic–Antarctic
AG Agulhas
AN Angola
AR Argentine
BR Brazil
C   Caroline
CA Canaries
CV Cape Verde
IA Indian–Antarctic
MA Madagascar
NA North American
NP North Pacific
P   Phillipine

PE Peru
SC South China
SO Somali
SP South Pacific
V   Venezuela

Ridges

EPR East Pacific Rise
MAR Middle Atlantic Ridge
MIR Mid-Indiar Ridge

Trenches

1   Sunda (Java)
2   Philippine
3   Palau
4   Marianas
5   Bonin
6   Ryukyu
7   Japan
8   Kurile
9   Aleutian
10 Middle America
11 Tonga–Kermadec
12 South Sandwich
13 Romanche
14 Puerto Rico
15 Peru-Chile

DEPTH	
meters	feet
3,000	9,840
4,000	13,120
>7,000	23,000

Table 3–1

PHYSIOGRAPHIC PROVINCES OF THE OCEAN.

Ocean[†]	Shelf and Slope (%)	Conti- nental Rise (%)	Ocean Basin (%)	Volcanoes and Volcanic Ridges (%)	Rise and Ridge (%)	Trenches (%)
Pacific	13.1	2.7	43.0	2.5	35.9	2.9
Atlantic	19.4	8.5	38.0	2.1	31.2	0.7
Indian	9.1	5.7	49.2	5.4	30.2	0.3
World ocean	15.3	5.3	41.8	3.1	32.7	1.7
Earth's surface	10.8	3.7	29.5	2.2	23.1	1.2

*After H. W. Menard and S. M. Smith, "Hypsometry of ocean basin provinces," *Journal of Geophysical Research*, 71 (1966), 4305.
†Includes adjacent seas; for example, Arctic Sea included in Atlantic Ocean.

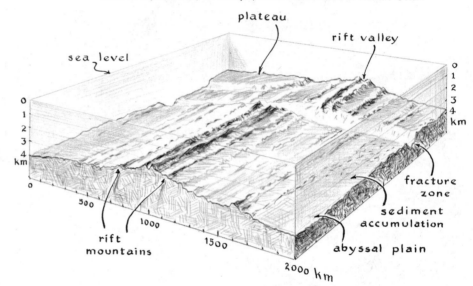

FIG. 3–2
Schematic representation of ocean-bottom topography of a portion of the Mid-Atlantic Ridge, typical of a rugged, faulted mid-ocean rise. Note the rough topography along the fracture zone where two segments of the rise are offset. The fault along which movement occurred is known as a transform fault. Sediment forming the abyssal plain is deposited on the margin of the rise.

percent of the earth (Table 2–1). Although they comprise a single system connected in all ocean basins, oceanic rises are not the same throughout the world. Two types can be distinguished—high-relief (e.g., the Mid-Atlantic Ridge, schematically represented in Fig. 3–2) and low-relief (e.g., the East Pacific Rise, shown in Fig. 3–3).

The *Mid-Atlantic Ridge* is located almost exactly in the center of the Atlantic Ocean and runs throughout its length. It is cut by numerous *faults* (breaks in the earth's crust), which combine with great numbers of volcanoes to create a rugged topography standing 1 to 3 kilometers above the nearby basin floor. The ridge is 1500 to 2000 kilometers across.

A steep-sided valley, which commonly occurs along the central portion of mid-oceanic rises, is especially well developed

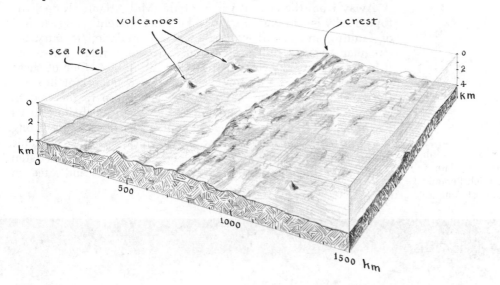

FIG. 3–3
Schematic representation of ocean-bottom topography along a portion of the East Pacific Rise, an example of a broad, low rise with relatively subdued relief.

on the Mid-Atlantic Ridge, which is where this formation was first recognized. Because its steep sides are apparently formed by faults, the feature is also called a *rift valley*. Rocks dredged from the valley floor have evidently solidified from lavas (molten rocks) fairly recently, within the past few million years. The valley is usually 25 to 50 kilometers wide, and 1 to 2 kilo-

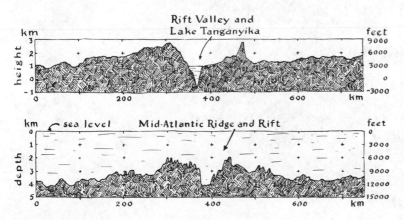

FIG. 3–4
Profiles of the rift valley of the Mid-Atlantic Ridge and the Tanganyika rift of East Africa. Note the similarity in form and size of the two features. (After Holmes, 1965)

meters deep. It is bordered by steep mountains, also formed by faulted blocks; these peaks come to within 2 kilometers of the ocean's surface. The closest terrestrial analog to conditions on the Mid-Atlantic Ridge is found in some fault-bounded valleys of East Africa shown in Fig. 3–4.

Away from the central valley of the Mid-Atlantic Ridge, the topography is still quite rugged. There is apparently a great deal of faulting and insufficient sediment accumulation to bury the irregular terrain. Fault systems cut across the ridge and offset it in a series of segments, as indicated in Fig. 3–5. One of these transverse faults is the Romanche Deep, an important link for the flow of deep-ocean water from the western to the eastern Atlantic basin. There are numerous and frequent shallow-focus earthquakes (occurring within 60 km of the earth's surface) along the Mid-Atlantic Ridge as well as volcanoes and volcanic islands, including the Azores, Iceland, Ascension, and Tristan da Cunha; the distribution of these features is indicated in Fig. 3–6.

FIG. 3–5

Artist's rendition of the bottom of the North Atlantic Ocean. Elevations and depths are noted in feet. (Painting by Heinrich Berann, courtesy Aluminum Corporation of America)

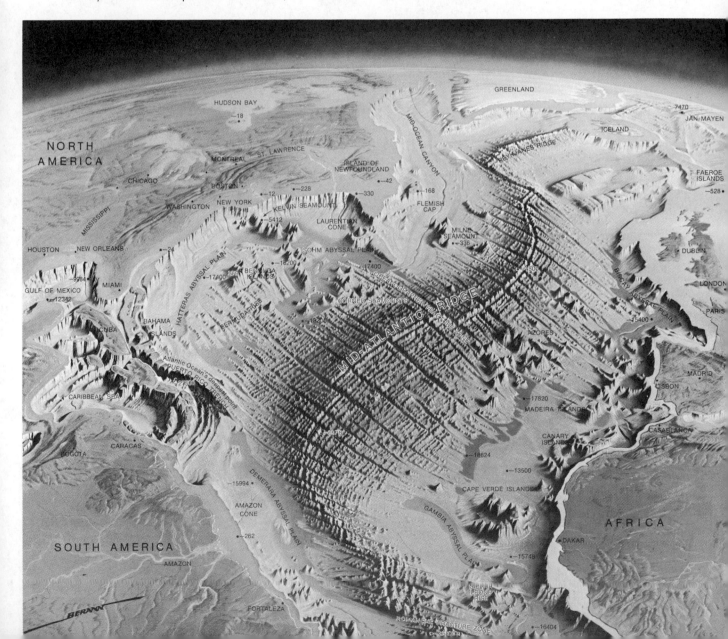

The Mid-Atlantic Ridge extends into the Arctic Ocean, intersecting the Eurasian continent in Siberia, near the mouth of the Lena River (see Fig. 2–10). Two other ridges that cross the Arctic Basin, parallel to the active ridge system, are apparently inactive at present.

The *East Pacific Rise* lies near the eastern margin of the Pacific Basin. It is not rugged like the Mid-Atlantic Ridge, but is a vast, low bulge on the ocean floor, approximately equal in size to North and South America combined. The rise stands about 2 to 4 kilometers above the adjacent ocean bottom and varies from 2000 to 4000 kilometers in width. Intersecting the North American continent in the Gulf of California, its continuation reappears off the Oregon coast and extends into the Gulf of Alaska. The two segments of the rise system are connected by the San Andreas fault in the state of California. Movement along this fault caused the San Francisco earthquake of 1906.

In the Indian Ocean, the ridge system—the *Mid-Indian Ridge*—resembles that in the Atlantic, being extensively faulted and having a rugged topography. On the northern end, the system intersects Africa–Asia in the Red Sea area, and part of the active system apparently extends under the East African rift valleys. On the south, the Mid-Indian Ridge system branches to join the system of active ridges and rises that circle Antarctica. Near Madagascar a low section of the ridges permits deep water to move between the deeper parts of the Indian Ocean and the Atlantic Ocean. The Mid-Indian Ridge separates

FIG. 3–6
Distribution of earthquakes (1961–67) and active volcanoes. Note the correspondence between earthquake epicenters and the boundaries of the large crustal blocks shown in Fig. 4–10. (Compiled from Bryan Isacks, J. Oliver, and L. R. Sykes, "Seismology and the New Global Tectonics," *Journal of Geophysical Research*, 73 (1968), 5855–5900, and from Holmes, A. 1965)

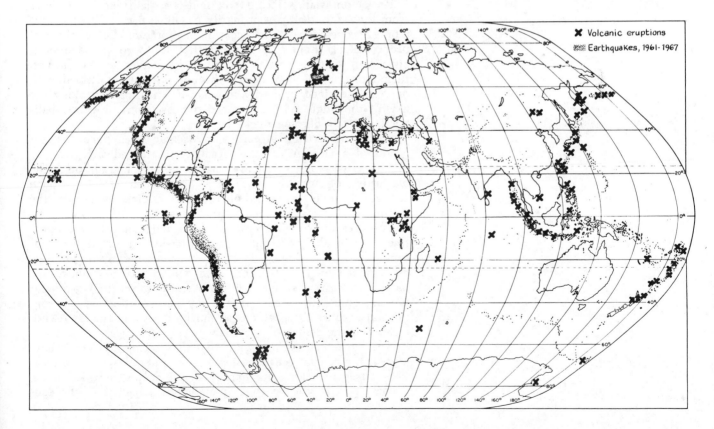

two distinctly different ocean areas. Microcontinents are abundant west of the ridge but absent to the east.

The mid-oceanic rise is offset at intervals throughout its length by intersecting *fracture zones*, long, narrow bands of steep mountains (see Fig. 3–5). This is especially noticeable in the Atlantic, but examples can be found in all ocean basins. These offset segments usually connect with bands of irregular topography that extend on both sides of the rise. Many of these zones also show evidence of volcanic activity.

Oceanic rises are areas of active earthquake activity (see Fig. 3–6), a narrow belt of frequent shallow-focus earthquakes occurring along their crests. In the North Atlantic, where the ridge is about 1500 kilometers wide, the belt of earthquake *epicenters* (point on the earth's surface above the earthquake focus) is about 150 kilometers wide. Detailed surveys show that these earthquakes coincide more or less with the axis of the rift valley. Earthquakes have been used to pinpoint active ridge systems in little known ocean areas. Such information is especially useful where other information is lacking or where relationships are unclear, as in the Arctic Ocean.

Mid-oceanic ridge systems are also sites of high heat flow from the earth's interior—about 1 to 3 microcalories per square centimeter per second (μcal cm^{-2} sec^{-1}). These "hot spots" do not correspond as closely to the location of rift valleys as do the sites of earthquakes. Instead, areas of unusually high heat flow (up to 8 μcal cm^{-2} sec^{-1}) tend to form narrow bands, suggesting narrow intrusions of molten rocks (*magma*).

Rocks on islands located on mid-oceanic ridges and those from the ridge itself are primarily basalt, typical of rocks commonly recovered from the ocean floor. Some of the geophysical properties of the ridges suggest that these rocks have been altered in such a way that earthquake waves travel through them more slowly than they do through normal basalt. Sediment deposits, confined primarily to small basins, are thin along the axis of the ridge. At increasing distances from the ridge, sediment deposits become thicker.

Studies of crustal structure in oceanic rises have not yet produced a clearcut picture. Available data indicate that the oceanic crust is not unusually thick under the rise although the properties of the various layers have been changed. Such changes include alteration of rocks extending even into the mantle, well below normal oceanic crust.

The presence of ridge systems in all major ocean areas effectively divides these areas into smaller basins. In the Atlantic, for example, this division of the ocean floor is especially noticeable, as indicated by Fig. 3–7. The Mid-Atlantic Ridge stands as a barrier to east–west flow of deep water or bottom sediment moving across the basin. In the Pacific the effect is less significant because the Pacific is so much larger and the rise is not located in the center of the basin.

Another effect of the rises is that they create a large shallow area in the central ocean basins. This, combined with the location of trenches along the margins of the basins, means that the deepest part of the ocean is immediately adjacent to the continental blocks and that some of the shallowest parts of the ocean occur along its center.

The shallowness of the mid-ocean areas affects sediment accumulation. Distinctive types of carbonate-rich sediment, not

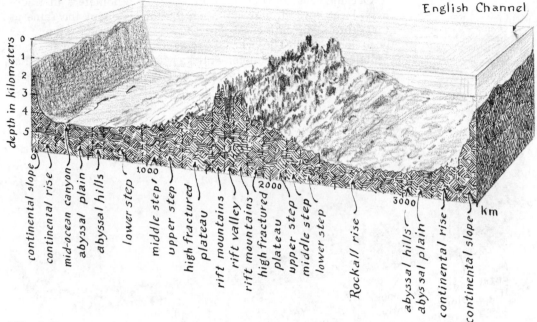

English Channel

depth in kilometers

continental slope
continental rise
mid-ocean canyon
abyssal plain
abyssal hills
lower step
middle step
upper step
high fractured plateau
rift mountains
rift valley
rift mountains
high fractured plateau
upper step
middle step
lower step
Rockall rise
abyssal hills
abyssal plain
continental rise
continental slope

FIG. 3–7
Diagrammatic view of the North Atlantic
ocean floor between Canada and Great Britain.
Note the great exaggeration in vertical relief.

found in deep-ocean basins, tend to accumulate in shallow areas, particularly in the small, fault-bounded basins characteristic of mid-oceanic rise systems. Furthermore, the presence of barriers to the movement of sediment along the bottom (e.g. turbidity flows) prevents continental sediment from reaching many areas of the ocean bottom. For instance, the segment of the East Pacific Rise off the coast of Washington and Oregon blocks movement of sediment along the bottom so that none of it reaches the deep North Pacific ocean floor. Thus the accumulation of sediment on the landward side of the ridge has caused it to be hundreds of meters shallower than the seaward flank of the rise.

OCEAN BASIN

The *ocean basin* proper covers about 29.5 percent (Table 3–1) of the earth's surface (roughly comparable to the 29.2 percent of surface that projects above the sea as land). Most of our information about this portion of the crust is sketchy, since it has been derived from indirect measurements and is subject to the limitations of the instruments used and the interpretation of data obtained. The simplicity of our present picture of the ocean basin (Fig. 3–8 is an example) is probably more a reflection of our ignorance, not because there is lack of complexity in such a large part of the earth's surface.

Oceanic crust is thin, typically about 7 kilometers in thickness. Heat flow through the ocean bottom is about the same as that through continental crust—on the average about 1.3 μcal cm^{-2} sec^{-1}. The ocean floor is covered by a thin layer of sediment, about 300 meters thick. Much of this sediment has accumulated particle by particle, so that deposits are draped

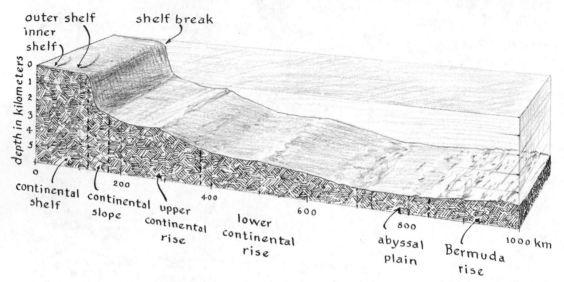

FIG. 3–8
Block diagram showing the Atlantic Ocean floor in an area known as the North American Basin.

over pre-existing topography in much the same way a gentle snowfall blankets and blurs the sharp features of a landscape. By contrast, large turbidity flows near continents can completely bury original ocean bottom features with a load of sediment like an avalanche.

There is no deep-ocean analog to the vigorous erosion by glaciers, wind, or rivers that takes place on land. To be sure, there are areas of strong currents, but they rarely seem to cut into rock surfaces. This is probably a consequence of the absence of chemical alteration of the rocks at the water–sediment interface. On land, such alteration is due to the combined effect of plant action and the chemical reactions of soil and rock with water and atmospheric gases. Ocean features, therefore, have a much longer existence than land features. Submarine volcanoes formed more than 25 million years ago are still very much in evidence—for instance, as foundation for the atolls at the northwestern end of the Hawaiian Island chain. Rocks have been recovered from submarine volcanoes that apparently formed about 80 million years ago, older than any known feature on land.

Ocean-bottom topography is formed by volcanic activity and by tectonic processes. For example, all ocean basins have *fracture zones*—long, narrow bands of volcanoes and mountainous terrain formed by fault-bounded blocks (see Fig. 3–5). These fracture zones are caused by faults in the ocean bottom caused by offsets in mid-ocean ridges, and are nearly perpendicular to them. In the Pacific Ocean, the fracture zones are remarkably straight, a few thousand kilometers long and about 100 to 200 kilometers wide. Each consists of individual ridges and troughs that are several hundred kilometers long and a few tens of kilometers wide. Some have cliffs or ridges up to a few kilometers above the nearby ocean floor. The greatest depths in the central Pacific Ocean (more than 6 kilometers) occur along these fracture zones.

Most earthquake activity associated with fracture zones is restricted to portions between segments of active mid-oceanic ridges. Elsewhere, fracture zones exhibit little earthquake activity; nor does there appear to be much volcanic activity associated with most of them.

In some areas, fracture zones offset continental margins. The sudden bend in the coast between Cape Cod and New Jersey is thought to be caused by a major fracture zone. This zone can be traced through the deeper ocean basin where it is marked by a group of *seamounts* known as the New England Seamounts.

Low *abyssal hills* (less than 1000 meters high) cover about 80 percent of the Pacific Ocean Basin floor, and about 50 percent of the Atlantic, including portions of the Mid-Atlantic Ridge. They are probably also abundant over much of the Indian Ocean Basin, although data there are more sketchy. These hills have an average relief of about 200 meters with diameters of about 6 kilometers, and are the most common topographic feature on the earth's surface. Many appear to be small volcanoes covered with a thin layer of sediment that is slightly thicker in the valleys than on the hills themselves. Some abyssal hills may be formed by intrusions of molten rock that push up the overlying sediment. Some are probably formed by faulting of oceanic crust associated with movements of the sea floor and the mantle below.

ABYSSAL PLAINS

Since the late 1940's, the availability of modern echo-sounding equipment has greatly accelerated exploration of the deep-ocean basins. Among the discoveries resulting from these explorations are immense areas of exceedingly flat ocean bottom lying near the continents that have been called *abyssal plains* (see Fig. 3–8). Somewhat similar features near volcanic-island groups, especially in the Pacific, are known as *archipelagic aprons*.

Abyssal plains are among the flattest portions of the earth's surface. They are defined as having slopes of less than 1 in 1000, equivalent to a slope of 1 meter per kilometer. Areas of comparable flatness are found in the High Plains area of the midcontinental United States, where the surface slopes about 1.5 in 1000 (8 feet per mile). Abyssal plains commonly occur at the seaward margin of the continental rise. In fact, there is usually a gradual merging of the deep-sea fans that make up the continental rise with the adjacent abyssal plain.

Deep-sea channels, a few meters deeper than their leveelike margins, extend across the plains from the end of the submarine canyons and deep-sea fans. For instance, Cascadia Channel in the Northeast Pacific, it was found, extends more than 500 kilometers from the base of the continental slope near the mouth of the Columbia River and crosses through a gap in the mid-oceanic rise, finally extending out onto the adjacent Tufts Abyssal Plain. Similar channels have been described in the Atlantic and Indian Ocean. Deformation of the ocean bottom may cut off a channel from its source of sediment, as apparently has happened in the vicinity of the Alaskan Trench. Even Cascadia Channel may not be actively transporting sediment because of deformation of part of its passage through the mountainous terrain near the mid-ocean rise. Most abyssal plains are heavily covered with sediment derived from the continents, to which we attribute their nearly flat topography.

Abyssal plains are especially common in marginal seas, such as the Gulf of Mexico, Caribbean, and the many marginal basins of the Pacific borderland. The Sigsbee Abyssal Plain has been explored extensively; it lies at the base of the Mississippi Cone built by sediment deposited from the discharge of the Mississippi River. Sediment that escapes deposition within the river delta or on the cone is deposited on the abyssal plain along with a minor amount coming from the Campeche Bank to the south. During the last glacial period (Wisconsin stage) sediment accumulated on the plain at the rate of about 60 centimeters per thousand years. In the past 10,000 years (since the retreat of the ice), sediment has accumulated more slowly, about 8 centimeters per thousand years.

Small hills first discovered by echo soundings have recently been studied by means of deep-drilling techniques. As originally postulated, these hills proved to be salt domes similar to those found on the Texas–Louisiana coastal plain. This fact suggests that thick lenses of salt lie at some depth beneath the abyssal plain. Some petroleum was also detected on the cores, indicating that petroleum accumulations are not limited exclusively to shallow-water areas.

SUBMARINE VOLCANOES

Volcanoes and *volcanic islands* are among the most conspicuous features of ocean basins. Many are marked above sea level by islands or atolls, but the majority are not visible at the surface. It is estimated that there are probably 10,000 volcanoes in the Pacific alone. Considering that the Pacific is about half the world ocean, it is likely that there are perhaps 20,000 volcanoes scattered across the ocean bottom. We have no data to indicate how many former volcanoes are now buried by later flows of volcanic material or covered by sediment.

Projecting usually 1 kilometer or more above the surrounding sea floor, volcanoes are detected by echo sounders on ships passing over the area. These are usually circular, or nearly so, with relatively steep sides (shown in Fig. 3–9) and relatively small summit areas. In other words, they resemble cones on the ocean bottom, a shape characteristic also of volcanoes on land. Their volcanic nature has been demonstrated by dredging of volcanic rock from their slopes. Some volcanoes have been at the sea surface long enough to have been worn down by waves, and their summits are usually covered by water-worn cobbles and sands. Some have been inhabited by shallow-water organisms, whose remains have been recovered by dredging.

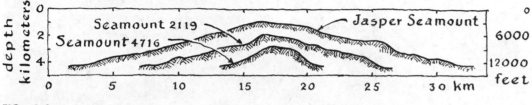

FIG. 3–9
Profiles of several dormant submarine volcanoes surveyed by echo sounders. Similarity in slopes suggest that little change in slope occurs during development of submarine volcanoes. (After Menard, 1964)

Despite the common occurrence of volcanoes on the ocean floor, there have been relatively few observations of submarine volcanic eruptions. Ships and planes have reported instances of discolored water in areas of known volcanic activity. Although few have been observed, one is shown in Figure 3–10. Most of our knowledge about behavior of lava in the ocean comes from observations of lavas from volcanic islands entering the ocean.

Volcanic lava is commonly at temperatures between 900° and 1200°C when it emerges from the volcano. As molten rock, it flows for a while, then solidifies first at the surface and finally throughout the flow. Further cooling turns the lava into volcanic rock.

There are two modes of volcanic eruption in the ocean. Some lavas form tranquil flows in which the surface cools first, forming an insulated cover for the still-molten material in the interior of the flow. When this type of lava flows into the ocean it forms "pillow lavas," rounded masses of volcanic rock. Despite the high temperatures of the lava, there is little explosive activity. Such tranquil flows appear to be quite common in the deep ocean. Underwater cameras have often recorded the presence of the characteristic "pillows" on the ocean bottom. High pressure at great depths seems to favor such tranquil flows.

The other type of marine volcanic eruption is explosive, producing large amounts of volcanic ash, small pieces of volcanic rock, and glass formed by quenching of molten rock. Certain lavas that form blocky, irregular flows rather than tranquil smooth-surfaced flows on land seem to be the type that causes explosive eruptions when they occur under the ocean. Such explosive eruptions seem to be rare in the deep ocean but ap-

FIG. 3–10
Eruption of Kovachi, a submarine volcano
in the Solomon Islands, South Pacific, in
October, 1969, caused shock waves and water
eruptions approximately 30 seconds apart (a).
Explosions ejected water and steam 60 to
90 meters into the air. The sea surface was
discolored for 130 kilometers (80 miles).
(Photograph courtesy Dr. R. B. M. Thompson
and Smithsonian Institution Center for Short-
Lived Phenomena) A new island (b) was
built by Kovachi in March, 1970. A similar
island formed by an eruption in 1961 has since
been eroded by waves and is now completely submerged. (Photograph courtesy
Smithsonian Institution Center for Short-Lived
Phenomena)

parently most common at depths less than 1.5 kilometers. Fragmentation of the rock favors chemical alteration, probably contributing dissolved iron and manganese to the ocean water.

Volcanoes on the ocean bottom occur over about half the Pacific Basin. They are commonly grouped into volcanic provinces, often covering up to 10 million square kilometers. The Hawaiian Islands form a long chain; other island groups are more circular. Location of these volcanic provinces seems to correspond to deep faults, major breaks in the oceanic crust through which molten rock can move up to the ocean bottom. Eventually the load of volcanic rock on the ocean bottom causes it to be depressed under the volcanoes, and there is often a compensating low rise of the ocean bottom beyond the depression.

Once formed, volcanoes persist for a long time. They are alternately active for extended periods during which a volcanic cone builds up. During periods of quiesence, volcanoes often subside slightly as the crust and mantle adjust to the added load of volcanic rock. Periods of activity last for millions to tens of millions of years. Certain areas of the southwestern Pacific Ocean appear to have undergone an extended period of volcanic activity during which immense volumes of lava poured out onto the ocean floor, beginning about 100 million years ago and lasting perhaps 30 million years. After cessation of activity, the region including its volcanoes gradually sank. This area, now marked by an abundance of low coral islands called the *Darwin Rise*, is thought to be an ancient oceanic rise, now inactive. Many extinct volcanoes are known to occur in other ocean basins—for instance, the Bermuda Islands, shown in Fig. 3–11.

Not all volcanic activity forms volcanoes. There are large areas on the ocean bottom where large smooth plains of volcanic rock occur. Apparently lava flowed out in large volumes, covering all previous topography over extensive areas of the ocean bottom and leaving a bare rock surface, now slightly

FIG. 3–11

The Bermuda Islands are solidified sand dunes built on a wave-eroded extinct volcano. The shallow top of the volcano (now forming a bank) was exposed during period of lowered sea level and carbonate sands were blown across the bank forming the dunes now changed to soft, crumbly limestone. Reefs built primarily of calcareous algae can be seen in the lower portion of the photograph. (Photograph courtesy Bermuda News Bureau)

covered by sediment. Probably these lavas were at first quite fluid. There may have been several centers of eruption, perhaps long breaks forming linear fissure eruptions. These *archipelagic plains* (or *aprons*), as they are called, surround volcanic groups or islands extending tens or hundreds of kilometers from the volcanoes themselves. The volumes of lava that form them are immense. About 80 percent of the volcanic rock in the Pacific is thought to occur in these vast archipelagic plains. Only about 15 percent occurs in volcanic ridges and large volcanic island groups.

Large ridges built of volcanic rock, nearly perpendicular to the mid-oceanic rise, occur in many parts of the ocean basin; together with volcanoes they account for 3.1 percent of the ocean floor. These ridges are especially important in the Atlantic where they divide the basin into a series of smaller basins (see Fig. 2–14). Two of these ridges in the Atlantic are large enough to significantly influence movement of deep and bottom waters in the basin. In the southeast Atlantic Ocean, the Walvis Ridge extends from the Tristan da Cunha–Gough Island area to southwest Africa. This sharp ridge comes to within 2 kilometers of the surface and prevents the cold, dense water moving along the bottom from the Antarctic region from flowing northward into the eastern portion of the deep Atlantic basin. Instead, these bottom waters must flow northward along the western side of the basin until they reach the Romanche Trench, which is deep enough to permit them to flow through the ridge and then southward into the eastern side of the South Atlantic basin.

At the northern end of the Atlantic, the massive volcanic ridge stretching from Greenland to Iceland and eastward to the Faroe and Shetland Islands inhibits the flow of cold Arctic waters into the North Atlantic basin. Some Arctic water does overflow but much less than would be expected in the absence of the ridge. Because of this ridge, the flow of deep water from the Arctic to the Atlantic basins is substantially diminished.

ISLAND ARCS AND TRENCHES

Island arcs, as indicated in Fig. 3–12, are large groups of volcanic islands at the boundaries between continental blocks and ocean basins. Deep (>6 kilometers) narrow trenches and associated low swells commonly occur on the seaward side of island arcs; the systems have regular arcuate shapes, convex toward the ocean side.

Island arcs and trenches are among the most active portions of the earth's crust. The following manifestations of dynamic processes associated with island arcs and trenches attest to their uniqueness in the ocean.

The greatest depths in the ocean basins occur in the trenches.

The largest gravity anomalies measured on the earth's surface are associated with trenches.

Numerous active volcanoes and many others recently active occur in island arcs.

Intense shallow seismicity occurs beneath island arcs.

Nearly all earthquakes deeper than 300 km occur beneath island arcs.

These earthquakes occur in zones dipping towards adjacent continental blocks and extend to depths of about 700 km.

The deep-ocean trenches exceed 6 kilometers in depth (see Table 3–2). Deepest is Challenger Deep, a part of the Marianas Trench (see Fig. 3–1), where a depth of 11,022 meters has been recorded. Obtaining accurate depth measurements in trenches is not easy, and depth figures are subject to revision as new techniques are developed or old ones improved. In general, the

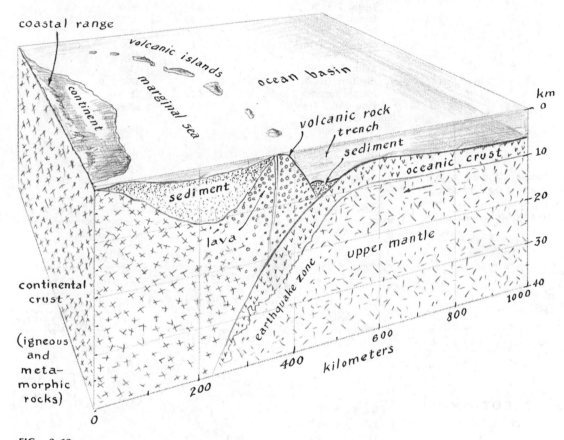

FIG. 3–12
Schematic diagram of island arcs and trench at boundary between continent and ocean basin.

trenches are hundreds of kilometers wide and about 3 to 4 kilometers deeper than the surrounding ocean bottom. Individual trenches have lengths of thousands of kilometers.

The sides of the trenches often consist of a series of steep-sided steps, suggesting that extensive faulting has taken place, causing blocks of the earth's crust to subside. The result is a narrow, deep, V-shaped profile which is almost invariably flat-bottomed due to sediment deposits. Large gravity anomalies are associated with major trenches. The disturbance of the earth's gravity has been attributed to large-scale downbuckling of the crust beneath the trench.

Deep troughs also occur on the ocean bottom. These are generally less deep than the trenches and not always associated with island arcs. Some deep areas are apparently associated with

Table 3–2

CHARACTERISTICS OF TRENCHES*

	Depth (km)	Length (km)	Average width (km)
Pacific Ocean			
Kurile–Kamchatka Trench	10.5	2200	120
Japan Trench	8.4	800	100
Bonin Trench	9.8	800	90
Marianas Trench	11.0	2550	70
Philippine Trench	10.5	1400	60
Tonga Trench	10.8	1400	55
Kermadec Trench	10.0	1500	40
Aleutian Trench	7.7	3700	50
Middle America Trench	6.7	2800	40
Peru–Chile Trench	8.1	5900	100
Indian Ocean			
Java Trench	7.5	4500	80
Atlantic Ocean			
Puerto Rico Trench	8.4	1550	120
South Sandwich Trench	8.4	1450	90
Romanche Trench	7.9	300	60

*After R. W. Fairbridge, "Trenches and Related Deep Sea Troughs," in R. W. Fairbridge, (ed.) *The Encyclopedia of Oceanography*, Reinhold Publishing Corporation, New York, pp. 929–38.

FIG. 3–13
Aleutian Islands, a simple island-arc system with trench on the convex, seaward side. These islands are built of volcanic rock from ancient eruptions. Some of the volcanoes are still active.

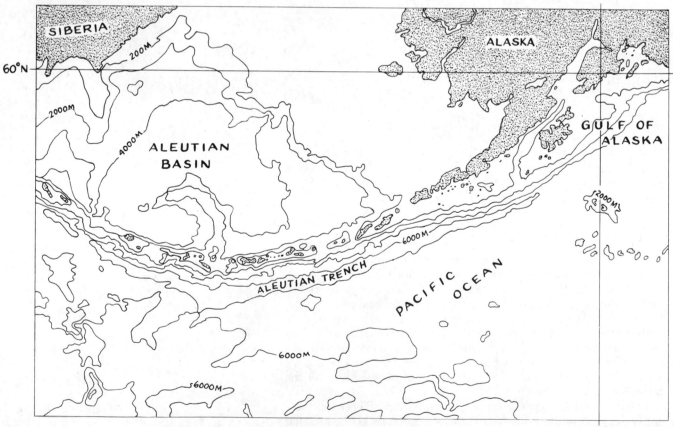

fracture zones that cut the mid-ocean rises and ridges, such as the Romanche Trench in the Atlantic.

Some island-arc systems seem to be relatively simple. For example, the Aleutian Islands, southwest of Alaska, are a single chain with an associated trench on its outer or seaward side, as shown in Fig. 3–13. West of the Aleutians, the Kamchatka Peninsula consists of a group of similar such volcanoes but there a platform of older, greatly altered rocks is exposed above sea level, as it is in the Japanese Islands to the south. Probably such older, deformed rock underlies many of the world's island arcs but is usually submerged and thus not directly observable.

The complexity of some island-arc systems is illustrated by the Indonesian area, as shown in Fig. 3–14, where several chains of active or recently active volcanoes form complexly twisted arcs. The associated Java (Indonesian) Trench has several associated island groups including the island of Timor, one of the few places on earth where deep-ocean sediments are now exposed on land. The West Indies of the western Atlantic is another example of a complex island-arc system. Deep-ocean sediments are exposed on the island of Barbados in the West Indian system.

CORAL REEFS

Not all ocean-basin structures are solely the result of geologic processes. Coral reefs, built by organisms, are spectacular features of shallow waters, scattered over about 190 million square kilometers of open tropical and subtropical oceans, as shown in Fig. 3–15.

Reefs are rigid, wave-resistant structures built by carbonate-secreting organisms of which coral is the most conspicuous but not always the most important contributor. Coralline algae, foraminifera, and mollusks are also significant contributors of carbonate to reefs. The structure formed by the intergrown skeletons must withstand wave attack, including storms, and also the continual boring, rasping, and nibbling of coral-eating animals. The framework is typically quite cavernous, providing protection as well as living space for many organisms.

Reefs require shallow, warm, sunlit waters of near-normal salinity. They grow best where average annual water temperatures are around 23 to 25°C (73° to 77°F). Some reefs occur where the average temperature of the coldest month is 18°C (64°F) or higher. In such marginal areas, many of the organisms typical of tropical areas are absent. Coral reefs are most abundant on the western side of ocean basins where surface waters are warmer than on the eastern side, especially in the tropical Indian and Pacific Oceans. (Atolls in the central South Pacific are shown in Fig. 3–16). Reefs rarely form near the mouths of large rivers, since low-salinity water and suspended sediment apparently inhibit development of reef-building animals.

The nature and history of the foundation often determine the shape and size of the reef. A stable foundation is an essential requirement, permitting reefs to grow in shallow water even

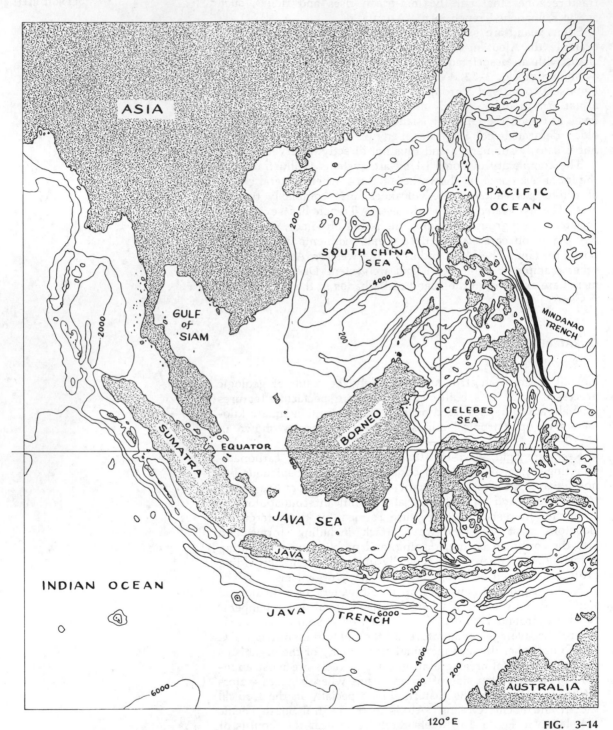

Labels within the figure:
ASIA

PACIFIC
OCEAN

SOUTH CHINA
SEA

200

4000

200

MINDANAO
TRENCH

GULF
of
SIAM

2000

BORNEO

CELEBES
SEA

SUMATRA

EQUATOR

JAVA SEA

JAVA

INDIAN OCEAN

JAVA TRENCH 6000

4000

2000

200

AUSTRALIA

6000

6000

120° E

FIG. 3–14

Indonesia, a complex island-arc system. Active
volcanoes occur on the larger islands. Some
smaller islands near the trench are built
of deformed sediment. The island of Timor in
the group is one of the few places where
deep-sea sediment is exposed on land.

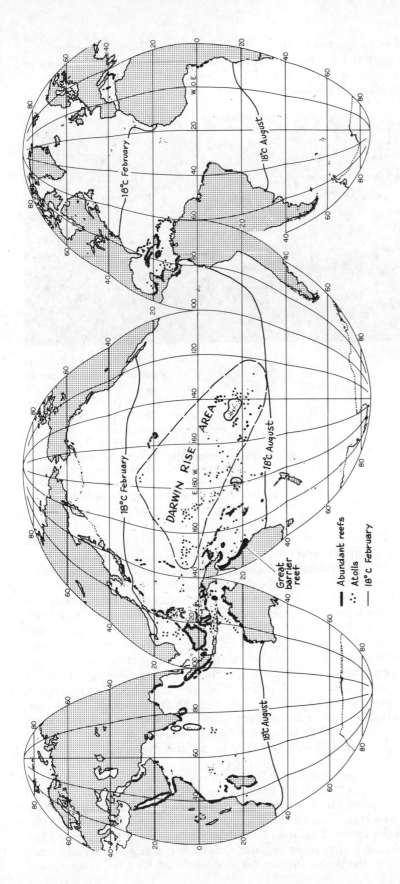

FIG. 3–15
Distribution of coral reefs and areas of abundant atolls. The contour lines show the limits of waters that never get colder than about 18°C.

though located in the middle of an ocean basin. Reef-building organisms seem to grow best within a few tens of meters of low-tide level where sunlight is available, permitting the growth of algae which add carbonate to the reef.

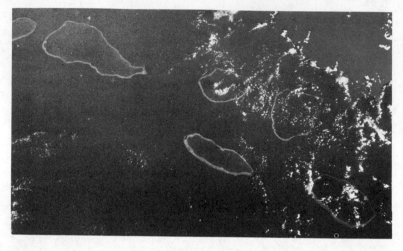

FIG.   3–16
Six atolls in the Tuamoto Archipelago, part of French Polynesia, in the Central South Pacific Ocean (16°S, 145°W). The white bands are reefs surrounding lagoons. Islands occur on the reefs, often at the sharp bends in the atoll, and can be recognized because they appear darker owing to the vegetation on them. Faint lines transverse to the reef mark the deep-water passes through which ships enter the lagoons. The large white masses are clouds. (Photograph courtesy NASA).

Three types of reef associations commonly occur, with much gradation between them as indicated by Fig. 3–17. *Fringing reefs* grow along rocky shores of islands and continents, with breaks at river mouths. Elongate *barrier reefs* parallel shorelines for great distances, commonly at some distance from shore. A lagoon locally too deep for coral growth separates the reef from the island or continental shore, as in the case of the Great Barrier Reef off Australia. A third type of reef is the *atoll*, a reef surrounding a lagoon. Atolls are commonly irregular in shape, often generally elliptical in outline. The lagoon is typically about 40 meters deep and connects to the adjacent ocean through *passes*, interruptions in the reef through which ships can navigate. Although there are often small sand-gravel islands on top of the reef or within the lagoon, there is no evidence of the rocky island that commonly forms the substrate for fringing or barrier reefs.

From his observations of coral reefs during the voyage of the *Beagle* (1831–1836), Charles Darwin formulated a theory of atoll formation. He theorized that a part of the Western Pacific Basin —sometimes called the Darwin Rise—as well as smaller areas in the Indian Ocean had subsided at some unspecified time. The islands which serve as foundations for many coral reefs sank below the water surface (or sea level has risen to cover them).

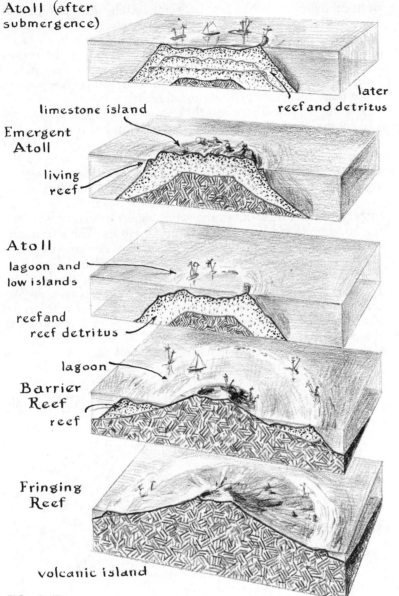

Atoll (after submergence)

later reef and detritus

limestone island

Emergent Atoll

living reef

Atoll

lagoon and low islands

reef and reef detritus

lagoon

Barrier Reef

reef

Fringing Reef

volcanic island

FIG. 3–17
Stages in the transition from fringing reef to barrier reef to atolls.

The fringing reef which ringed the island below sea level continues to grow upward as the island sinks. At first it forms a barrier reef offshore, and finally a ring-shaped structure with a lagoon in the center which marks the spot where the island sank. Sand or debris from the reef may form islands alongside the reef, within the lagoon.

While Darwin's theory was widely debated following its publication in 1842, the first attempt to verify it occurred in 1896, when a hole was drilled at Funafuti, an atoll in the Equatorial Pacific. The attempt was unsuccessful, but direct evidence that atolls rest on submerged volcanic islands came in 1952 from deep-drilling operations on Eniwetok Atoll in the Marshall Islands. There the drill penetrated 1.2 kilometers of limestone and carbonate sediment to recover volcanic rock on which the atoll was built. In 1965, the atoll at Midway Island in the northwestern Hawaiian Islands was drilled and samples of the original volcano recovered. Thus Darwin's brilliant theory was proven more than 100 years after its formulation.

Darwin was unaware of the sea-level changes caused by Pleistocene glaciations. If he had been, he might have modified his theory to account for the complications they have caused in the general picture of atoll formation. Although much of the Pacific Ocean bottom has been submerged over the past several tens of millions of years, the atolls have been exposed above sea level, forming high-carbonate islands when sea level was much lower. At these times the sands were exposed to weathering and some sand was changed into limestone. The island surfaces were often greatly modified by erosion due to waves and rivers. When sea level rose again, it submerged the island and coral grew on the modified platform. The complexities of these foundation structures accounts for some of the coral-reef patterns observed in lagoons at present. It also accounts for the altered carbonate sediment and limestone encountered beneath the atolls during drilling operations.

Figure 3–18 shows a typical island–reef–lagoon structure. Generally, the outer margin of the reef is quite steep. As the outer slope rises from the depths, the abundance and type of coral and algae changes in response to changes in the living conditions, especially light intensity. There appears to be a

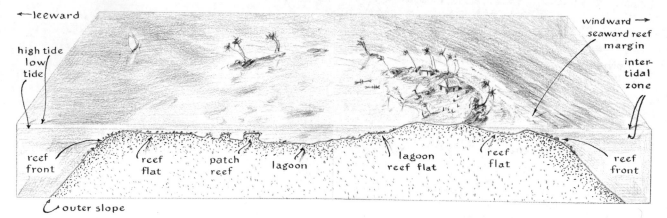

FIG. 3–18

Typical features of an atoll and associated reefs. Note the more extensive development of coral on the windward side. Various explanations have been offered for this phenomenon—for instance, that breaking waves and water movements on the windward side provide a better food supply. (After J. I. Tracey, Jr., P. E. Cloud, Jr., and K. O. Emery, *Conspicuous features of organic reefs*, Atoll Research Bulletin 46, Pacific Science Board, National Academy of Sciences, National Research Council, Washington, D. C., 1955)

marked decrease in the abundance of coral and coralline algae below about 20 meters.

The *reef front*—the upper seaward face of the reef—is often the zone of most intensive coral and algal growth, covering all the surface with organisms. Those that grow in the path of the breakers have heavy skeletons or other structures, which are highly resistant to waves both in architecture and materials. Often the reef front is cut by a series of grooves alternating with spurs of reef growing seaward. This *spur-and-groove structure* effectively absorbs wave energy, thereby protecting the reef from damage by breaking waves. It is commonly well developed on the windward side, but often more irregular on the protected leeward side.

Grooves often extend into the reef flat forming *surge channels*, in which incoming waves throw water high into the air as the wave advances. Carbonate sediment and skeletal debris of animals, swept by waves from the reef flat above, slides down channels in the steep reef front, to collect in areas of more gentle slope.

Deep, irregular relief on the reef face greatly increases the surface available for attachment of organisms. Corals, algae, and sponges grow among and around the spurs, often forming tunnels and caves in which fish, crayfish, and other organisms find shelter from the turbulent water, pronounced temperature changes and certain predatory animals commonly found on the exposed reef surface.

The reef flat is commonly just below low-tide level. Wave pounding is reduced here, and thus more delicate coral structures can survive. The abundance of life present gives rise to stiff competition for growing space. In a few places where waves constantly break on the reef and throw water higher, certain hardy, soft-bodied algae, barnacles, and coralline algae form rimmed pools that project above sea level perhaps 10 centimeters or more. Water splashed into these pools by breaking waves is held temporarily until it drains out slowly, keeping plants and animals within the pool from drying out. Where the coralline algae grow extensively they form a purplish-red ridge, known as the *algal ridge* (or *Lithothamnion ridge*) at the seaward edge of the reef flat.

On top of the flattish reef structure there are commonly low-carbonate sand and gravel islands, shaped by waves, and winds. These islands are usually a few meters high, although wind-blown deposits of carbonate sand can form dunes as high as 10 meters or more. During violent storms, sand, gravel, coral rubble, and blocks from the reef or lagoon can be thrown onto the island forming ramparts (small ridges). Beaches of sand and occasionally gravel surround these islands. These islands are commonly covered by dense growth of brush, (unless the brush is controlled) a few trees, and nests of many sea birds.

On the lagoon side, there is a similar situation to that observed on the sea side—a reef flat with sparse coral growth. Here, reduced wave action is less effective in scouring the reef surface, so it is often sandy and provides less opportunity for coral growth. At the edge of the reef flat, the bottom slopes down toward the lagoon.

The lagoon is a shallow basin, usually sheltered by the surrounding reef. Typically the lagoon floor is about 40 meters deep, nearly flat, and covered with poorly sorted sand and gravel sized fragments broken off the reef. Within the lagoon there are scattered small reefs, called *patch reefs*, (shown in Fig. 3–19), that have built up from the lagoon floor. The tops of these reefs are covered with coral and algal growth. Some patch reefs are quite extensive, making it difficult to take a small boat through the lagoon. Others are quite small.

FIG. 3–19
Edge of a patch reef in the lagoon at Midway Island, an atoll in the Central North Pacific. The irregular surface consists of coral and encrusting calcareous algae. The lagoon floor is about 5 meters deep and is covered with carbonate sand and larger pieces broken off the patch reef. (Photograph courtesy U. S. Geological Survey—J. I. Tracey, Jr. photographer)

SUMMARY OUTLINE

Deep ocean floor, beginning at seaward margin of continental rises, covers 56.0% of the earth

Oceanic rises—comprise a single system throughout all ocean basins

Mid-Atlantic Ridge—rugged, highly faulted, with rift valleys

East Pacific Rise—low bulge on the ocean floor, little faulting, no rift valley

Mid-Indian Ridge—resembles Mid-Atlantic Ridge

Characteristics of oceanic rises:

Comparable in size to continents

Many volcanoes and shallow-focus earthquakes

High heat flow through ocean floor, usually in narrow bands

Basaltic rock on islands

Ocean basin

Thin crust (about 7 km) covered by thin sediment layer, typically accumulated particle by particle and covering older topography

Fracture zones of rough topography cross ocean basins roughly perpendicular to oceanic rise; earthquakes occur between offset segments of oceanic rises

Abyssal hills are extremely abundant

Volcanoes and volcanic islands are typical of ocean basins

Abyssal plains—among flattest portions of earth's crust
  Occur at seaward margin of continental rise
Crossed by deep-sea channels, avenues of sediment dispersal
Especially common in marginal seas—e.g., Gulf of Mexico,
Caribbean

Volcanoes
Volcanoes usually circular, average 1 km above ocean bottom;
tops form islands when they penetrate ocean surface
Volcanic eruptions—rarely observed in deep ocean
  Tranquil flows—quiet eruptions, form pillow lavas
  Explosive eruptions—form volcanic glass, rare in deep ocean
Volcanoes cover about half of Pacific Basin, occur in groups,
persist for a long time
Archipelagic plains
  Formed by lava flows on bottom, around volcanic groups
  Generally smooth upper surface
  About 80% of volcanic rock in Pacific thought to occur in
  such plains
Volcanic ridges—usually transverse to mid-oceanic rise
  Divide deep ocean floor into small basins
  Obstruct flow of bottom waters
  Origin probably similar to fracture zone except for limited
  volcanic activity along fracture zone, substantial activity
  along volcanic ridge
Island arcs and trenches occur at ocean basin margins, especially
in Pacific
  Contain greatest depths in ocean, most volcanoes, most earth-
  quakes, most intensive mountain building
  Some island arcs consist primarily of volcanic islands, others
  have old, greatly altered rock exposed in some of the islands
  Ancient deep-ocean sediment found only in island-arc regions
Coral reefs—wave-resistant platforms built in shallow water by
carbonate-secreting organisms
  Corals and encrusting calcareous algae form the framework
  which traps other sediment
    Grow in shallow, warm, sunlit waters of near-normal salinity
    Reef form determined by history of platform
    Atolls result of submergence of platform, most common in
    Western Pacific
  Reef front—steep outer edge of reef, site of vigorous coral
  growth; spur-and-groove structure provides resistance to waves
  and increases surface available for attachment of organisms
  Reef flat—near low-tide level, sometimes has islands of sand
  and gravel
  Lagoon—shallow basin, sheltered by surrounding reef, with
  small reefs growing in it

SELECTED REFERENCES

HEEZEN, BRUCE C., MARIE THARP, AND MAURICE EWING. 1959. *The Floors of the Oceans*, Vol. I: *The North Atlantic*. The Geological Society of America, New York. 122 pp. Description of the North Atlantic Ocean bottom.

KEEN, M. J. 1968. *An Introduction to Marine Geology*. Pergamon Press, Elmsford, New York. 218 pp. Geology of the ocean bottom; intermediate in difficulty.

MENARD, H. W. 1964. *Marine Geology of the Pacific*. McGraw-Hill, New York. 271 pp. Thorough discussion of ocean-bottom geology in the Pacific basin; intermediate difficulty.

SHEPARD, FRANCIS P. 1963. *Submarine Geology*, 2d ed. Harper & Row, New York. 557 pp. General survey of marine geology; intermediate to advanced level.

TUREKIAN, KARL K. 1968. *Oceans*. Prentice-Hall, Englewood Cliffs, N.J. 120 pp. Elementary discussion of ocean-bottom topography.

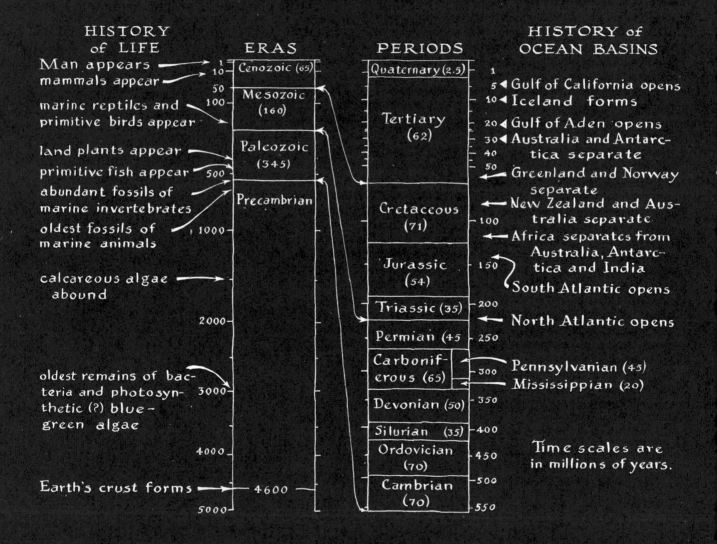

HISTORY of LIFE | ERAS | PERIODS | HISTORY of OCEAN BASINS

Man appears →
mammals appear

marine reptiles and primitive birds appear →

land plants appear →
primitive fish appear →
abundant fossils of marine invertebrates
oldest fossils of marine animals

calcareous algae abound →

oldest remains of bacteria and photosynthetic (?) blue-green algae

Earth's crust forms →

Cenozoic (65)
Mesozoic (160)
Paleozoic (345)
Precambrian

Quaternary (2.5)
Tertiary (62)
Cretaceous (71)
Jurassic (54)
Triassic (35)
Permian (45
Carboniferous (65)
Devonian (50)
Silurian (35)
Ordovician (70)
Cambrian (70)

Gulf of California opens
Iceland forms
Gulf of Aden opens
Australia and Antarctica separate
Greenland and Norway separate
New Zealand and Australia separate
Africa separates from Australia, Antarctica and India
South Atlantic opens
North Atlantic opens
Pennsylvanian (45)
Mississippian (20)

Time scales are in millions of years.

HISTORY OF OCEAN AND ATMOSPHERE

Diagramatic representation of the history of ocean basins and the history of life in the oceans.

FIG. 4–1
In seismic-reflection surveys, a ship drops an explosive charge at some preset point. Sound pulses reflected from different layers are picked up by sensitive microphones—*hydrophones*—towed by a ship. These weak signals are recorded and analyzed to determine depths to the reflecting horizons.

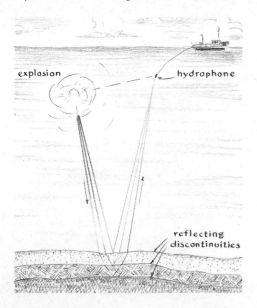

U p to this point, our discussion of ocean basins has been concerned primarily with *bathymetry* (relative elevations and depressions) of the ocean bottom. For many years this has been a prime source of information about the ocean bottom, providing most of the clues about its structure and origin. In recent decades, much has been learned about the subsurface structure of the ocean bottom, using direct and indirect means. Studies of reflection or refraction of waves passing through the ocean bottom from explosions or from earthquakes have provided much of this knowledge. This technique, known as *seismology*, provides information about thickness and distribution of different rock types through interpretation of observed behavior of sound waves passing through the oceanic crust.

Layers below the ocean bottom reflect sound waves. Measurements of elapsed time between transmission of the pulse and its echo indicate the thickness of sediment and rock layers of the ocean bottom. Modern equipment permits profiles of layer thickness to be made while the ship is moving, much as the bottom is profiled. The technique is known as *reflection seismology*, illustrated in Fig. 4–1.

Powerful sound sources, including explosives and electrical arcs, generate sound pulses strong enough to detect different layers and structures many kilometers below the ocean bottom. Surveys using this technique (*refraction seismology*) require either two ships, as shown in Fig. 4–2—one to provide a sound source and the other to receive the signal—or one ship and a set of instrumented buoys to transmit signals back to the ship. Information obtained in this way has been extremely helpful in gaining knowledge of the structure of the ocean basins and the submerged portions of continental blocks.

Two other physical properties of the earth have been studied extensively: its gravity and the flow of heat outward through its crust. Detailed study of gravity over a given area provides information useful for determining the distribution of materials of different densities. High-density rocks near the surface cause the local gravity to be slightly greater than normal. Conversely, low-density rocks in the same position cause the local gravity to be abnormally low. By itself this type of information can be used to deduce probable crustal structures. Gravity data are especially informative when combined with data indicating depths of boundaries between subsurface rock or sediment masses.

Heat flowing through the crust under continents and ocean basins is thought to come from radioactive decay of thorium, uranium, and potassium in rocks. Measuring temperature changes with depth below the ocean bottom permits calculation of the rate of heat transfer from the rocks below. Extremely high heat flows through the ocean bottom often indicate the presence of magma near the surface. Hence high flows measured near the crests of mid-ocean rises suggest that rocks at high temperatures occur near the surface (see Table 4–1). North of the Gulf of California and in Iceland, both located near the crest of an oceanic rise, there are large heat flows. Some of these areas may in the future be developed for geothermal power.

Direct studies of ocean-basin structures have been limited by the inability to use direct observation or retrieve samples buried beneath a few meters of sediment. Dredging of rocks on the ocean bottom has provided limited amounts of material for locations that were formerly poorly known. Deep-ocean drilling, however, has provided much richer sources of information, samples having been taken from considerable depths below the ocean bottom such as the core seen in Fig. 4–3. Both sample location and depth were known with certainty, providing information on ocean basins comparable to that long available to geologists working on land.

CRUSTAL STRUCTURE

The outer surface of the earth consists of rocks whose temperature is well below their melting point. Consequently, the earth's crust acts like a solid; it is fairly strong but brittle and easily broken. This outer surface has been called the *lithosphere* (from *litho*, meaning stone), and is thought to be between 60 and 100 kilometers thick. It includes both the crust and part of the upper mantle. The lithosphere comprises large plates that act as units on the surface of the earth. Below the lithosphere, rocks are so close to the melting point that they act more like tar.

Two types of crustal material are exposed at the earth's surface: *continental crust* and *oceanic crust*. Continents consist of

FIG. 4–2
Seismic-refraction studies commonly use two widely separated ships, one to release the explosive charge, the second to receive and record the signals. Sound waves from the source travel along the layers and their boundaries. They move up through the overlying layers to be received at a second ship.

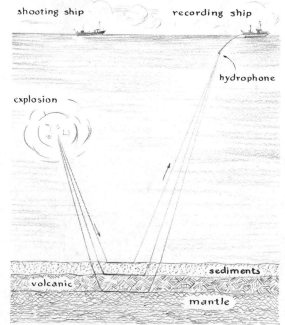

shooting ship   recording ship

hydrophone

explosion

sediments

volcanic

mantle

Table 4–1

HEAT FLOW THROUGH THE EARTH'S CRUST* (IN MICROCALORIES PER SQUARE CENTIMETER PER SECOND)

Continents		Oceans	
Precambrian areas (older than 600M yr)	0.9 ± 0.2	Trenches	1.0 ± 0.6
		Basins	1.3 ± 0.5
Paleozoic mountain areas (from 600 to 230M yr)	1.2 ± 0.4	Oceanic ridges	1.8 ± 1.6
Post Paleozoic mountain belts (younger than 230M yr)	1.9 ± 0.5		
Undeformed areas, post Precambrian (no deformation in past 600M yr)	1.5 ± 0.4		

*After W. H. K. Lee and S. Uyeda, "Review of Heat Flow Data", in *Terrestrial heat flow*, Geophysical Monograph American Geophysical Union Washington, D.C. no. 8, p. 87.

FIG. 4–3

A core of altered igneous rock (solidified from a melt) overlain by white marble (an altered sedimentary rock). The core was obtained by deep-sea drilling operations in water 4700 meters deep in the western North Atlantic Ocean. This rock sample was taken after drilling 450 meters of sediments almost 100 million years old. This provides a direct means of investigating ocean-bottom structure and ocean-basin history. (Photograph courtesy Deep-Sea Drilling Project, Scripps Institution of Oceanography, under contract to National Science Foundation)

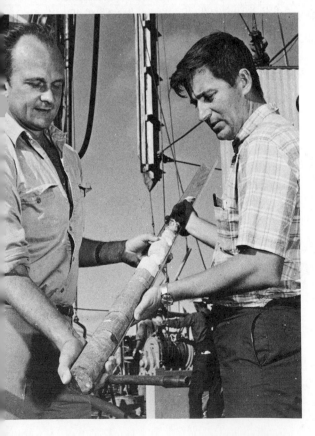

low-density rocks such as granite, often with a thin covering of sediments and sedimentary rocks. Their bulk density is about 2.7 grams per cubic centimeter. These rocks contain an abundance of silica and aluminum, and they are known as SIAL (SI = silica; AL = aluminum). The continents consist of an accumulation of continental crust, which is on the average about 35 kilometers thick. The continental blocks essentially float on the upper mantle.

The crust underlying ocean basins differs substantially from continental crust in composition and in thickness as indicated in Fig. 4–4. Oceanic crust consists primarily of dark-colored rocks containing much more iron and magnesium than do continental rocks, although silica is still the dominant constituent. This material is known as SIMA (SI = silica; MA = magnesium). Because of its higher proportion of iron and other heavy metals, the density is higher—about 2.8–3.0 grams per cubic centimeter. Some rocks have densities as high as 3.4 grams per cubic centimeter. A layer about 7 kilometers thick underlies the ocean basins. It, too, floats on the upper mantle.

Using the techniques described above, geophysicists have shown that several distinctive layers, listed in Table 4–2, are widespread under ocean basins. A layer of sediment (Layer 1), averaging about 0.5 kilometer thick, covers the ocean bottom. Below this is the *second layer*, rarely sampled, but assumed to consist of consolidated sediment or volcanic rock. Most of the oceanic crust is included in *Layer 3*, about 5 kilometers thick. This layer is thought to consist primarily of basaltic rock, perhaps slightly altered.

Relatively few studies have been made of oceanic rises and thus there is still substantial uncertainty about their structure. From the limited evidence available, it seems probable that the crustal structure may be different in various portions of the rise system. Near their crests, rises generally have little or no sediment cover. Sediments associated with the Mid-Atlantic Ridge usually occur in small fault-bounded depressions and appear to

Table 4–2

CRUSTAL STRUCTURE 83

SUBSURFACE LAYERS OF OCEAN BASINS—THEIR THICKNESS
AND PHYSICAL PROPERTIES

Layer	Probable Composition	Approximate Thickness (KM)	Probable Density (g/CM³)
Seawater		4–6	1.03
Layer 1	Sediment— unconsolidated	0.5	2.3
Layer 2 (basement)	Consolidated sediment or volcanic rock	1.7 ± 0.75	2.7
Layer 3 (oceanic layer)	Basalt, partially altered	5 ± 1.5	3.0
Layer 4 (mantle)	Ultrabasic rock		3.4

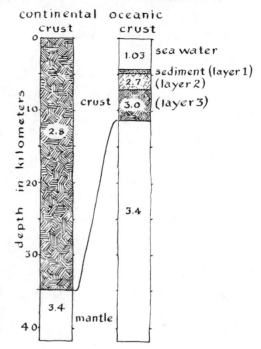

FIG. 4–4

Typical crustal structure under continents and ocean basins. The numbers give the density (in grams per cubic centimeter) for the individual layers. Data for buried layers of oceanic crust are obtained primarily from various geophysical techniques.

be locally derived. On the Mid-Atlantic Ridge there is evidence for a thick accumulation of volcanic rocks (shown in Fig. 4–5) but no evidence for such an accumulation on the East Pacific Rise (shown in Fig. 4–6). On one point the evidence seems reasonably certain: the structure of rises is definitely different from that of the ocean crust under most of the ocean basin.

Studies of trenches have shown exceedingly complex structures. Under the *Puerto Rico Trench* the oceanic crust can be seen to dip sharply, as Fig. 4–7 indicates. A thick mass of what is probably volcanic rock makes up the bulk of the island. Our picture of the *Peru–Chile Trench* (shown in Fig. 4–8) is also lacking in detail about the nature of this transitional zone. Both show, however, the abrupt transition from oceanic to continental or island-arc structures.

Separating crustal material from the mantle is the *Mohorovicic discontinutiy* (also known as the M-discontinuity, or Moho), which forms the upper boundary of the mantle. This was discovered when it was observed that compressional waves from distant earthquakes passing from the crust into the mantle suddenly change speed—from 6.1 to 6.7 kilometers per second in the crust to about 8.1 kilometers per second in the mantle below the Moho. The cause for this change has not yet been satisfactorily explained.

Since the lithosphere is about 60 to 100 kilometers thick, it is clear that it includes some of the upper mantle. Despite this major discontinuity in physical properties, the two parts of the earth's outer surface appear to be coupled together. Within the lithosphere, rocks are sufficiently brittle to break, causing earthquakes as they move past each other along faults. Most earthquakes occur within the lithosphere although deep-focus earthquakes have been detected as deep as 700 km.

Earthquakes occur most frequently along continental margins around the Pacific Ocean, along oceanic rises, and in certain "young" mountain ranges. They are generally rare in the middle of continents or ocean basins. Frequent earthquakes indicate

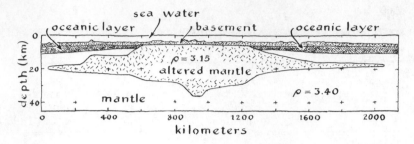

FIG. 4–5
Possible structure of the crust beneath the Mid-Atlantic Ridge, based on data from gravity and seismic investigations. The altered mantle material beneath the rise has compressional-wave velocities lower than normal for the mantle and a density, ρ, about 5 percent less than unaltered mantle material. (After M. Talwani, X. LePichon, and M. Ewing, "Crustal Structure of the Mid-ocean Ridges, 2," *Journal of Geophysical Research*, 70 (1965), 341–52.

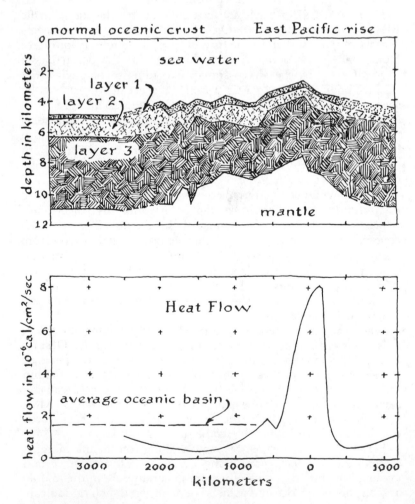

FIG.   4–6
Possible structure of oceanic crust under the East Pacific Rise as compared to the structure of normal oceanic crust. Note the thinning of Layer 3 over the crest. A heat flow profile for the rise shows a close correspondence between high heat flow and the rise crest. (Redrawn from Menard, 1964)

that the crust is actively moving and generally that rocks are being deformed. Earthquakes, along with volcanoes, are especially abundant in trench and island-arc areas.

Beneath the lithosphere is the *asthenosphere (aestheno =* weak). This zone extends from the base of the lithosphere down into the mantle to depths of about 250 kilometers. The asthenosphere occurs under both continents and ocean basins. It appears to be a part of the earth where silicate materials are very near the melting point, so that it acts more like a viscous liquid than a solid when subjected to stresses over millions of years. It has little strength over such long periods of time and responds to forces by flowing—that is, deforming plastically. Earthquakes can occur in the asthenosphere because the ma-

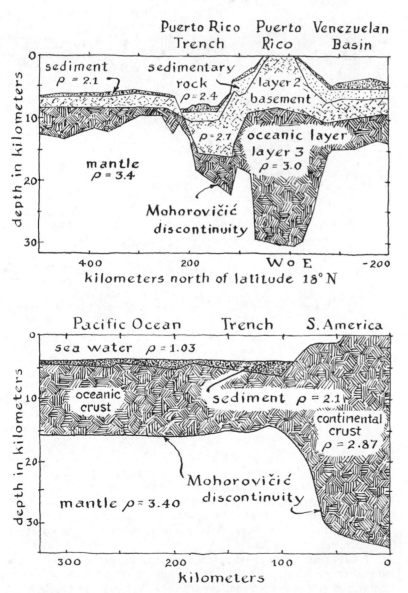

FIG.   4–7
Probable structure of the crust beneath the Puerto Rico Trench, based on seismic studies. Vertical exaggeration tenfold. Note that the oceanic crust dips below the trench and the thick accumulation of volcanic rock (part of Layer 2) forming Puerto Rico. [After M. Talwani, G. H. Sutton, and J. L. Worzel, "Crustal Section Across the Puerto Rico Trench," *Journal of Geophysical Research,* 64 (1959), 1545–55]

FIG.   4–8
Probable structure of the crust at the boundary between the Pacific Ocean and South America at approximately 53°S (data derived from gravity studies). Note the thick accumulation of sediment in this largely inactive segment of the Peru–Chile Trench. [After D. E. Hayes, "A geophysical investigation of the Peru–Chile Trench," *Marine Geology,* 4 (1966), 309–51]

terials can respond by breaking, much in the manner of cold tar. This layer was detected through observation of the low velocity of earthquake waves traveling through it.

Perhaps the best-studied portion of the transition from continental to oceanic crust is that of eastern North America, illustrated in Fig. 4–9. At this location there is a fairly abrupt transition from continent to ocean basin within a few hundred kilometers of the edge of the continental block. There a thick wedge of sediment has been deposited over the transitional area. Such a transition is thought to be typical of stable continental margins (those not affected by active mountain building for many millions of years).

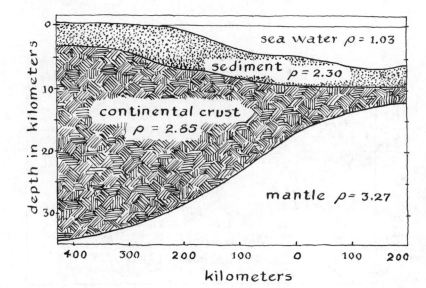

FIG. 4–9
Probable crustal structure at the transition between the North American continent and the Atlantic Ocean offshore from New Jersey. This structure is apparently typical for stable continental margins that have not experienced mountain building for millions of years. Note the sharp bend in the Mohorovicic discontinuity and the thick wedge of sediment at the transition. (After M. Ewing, and F. Press, "Geophysical Contrasts between Continents and Ocean Basins," in A. Poldervaart (ed.), *Crust of the Earth*, Geological Society of America, New York, 1955, pp. 1–6)

Except for the oceanic rises, the crustal structures discussed so far have been categorized as either oceanic or continental. The boundaries have been marked by either a trench or a thick wedge of sediment over a long, stable continental margin. The question then arises: Are there areas where one type of crust is changing to another?

In areas of active crustal deformation, marginal-ocean basins are often structurally part of the transitional zone. Many marginal basins have a normal oceanic crust (Layer 3) overlain by a thick sediment cover. In the Aleutian Basin, the sediment layer is about 6 kilometers thick; it is 10 kilometers thick in the Gulf of Mexico, 10 to 15 kilometers thick in the Black Sea, and 20 kilometers thick in the Caspian Sea.

SEA-FLOOR SPREADING

Many of the processes and features described so far can be explained by the hypothesis of *sea-floor spreading*, caused by formation of new crust at oceanic rises. Newly solidified crust is moved away from its site of origin as magma pushes up from below. Since the earth is not growing larger, oceanic crust must eventually be destroyed. According to contemporary hypotheses, this occurs in the trenches and areas of island arcs near ocean margins. Continents are moved more or less passively, like rafts being carried along on a slowly moving current.

Continental blocks are composed of material segregated from the upper mantle that has been accumulating at the earth's surface over billions of years. It seems probable that volcanoes have brought molten silicate rocks to the surface, as well as atmospheric gases and water vapor. This suggests that although large accumulations of water have existed at the earth's surface for billions of years, the oceans have gradually deepened as the mass of material in the continental blocks has increased.

It is estimated that the oceans have deepened at the rate of about 1 meter per million years, and continental materials apparently have accumulated at a similar rate. The oceans have never completely flooded the continents, nor have they ever completely disappeared from the earth's surface. Both ocean and atmosphere have been altered by the evolution of life, especially plants.

The present configuration of land, ocean basins, and mountain chains as illustrated in Fig. 4–10 results from the movements of about nine major *crustal blocks* and many smaller ones. Each moves with a motion that can be described as a rotation about a center that is in the North Atlantic near the southern tip of Greenland. Five units contain continental blocks. North and South America are in the same block, whereas the Indian subcontinent forms a different block from the Eurasian continent to which it is presently attached. The plates are bounded by oceanic rises, large faults, and either arc–trench complexes or belts of recently built mountains.

Deformation of crustal units tends to occur only at their edges. For example, nearly all the world's earthquake areas and volcanoes lie along edges of major blocks, or in some instances along fractures within these blocks (see Fig. 3–6). They are manifestations of movement at the boundaries of crustal units.

There is relatively little deformation in the interior of crustal blocks. Although their movements have been worked out in terms of simple patterns, the relationship of the blocks to the driving force is not obvious and likely not simple. Energy to move the blocks is thought to come from radioactive decay in the earth's interior. Heat released by such decay causes movements in the mantle and subsequently in the crust.

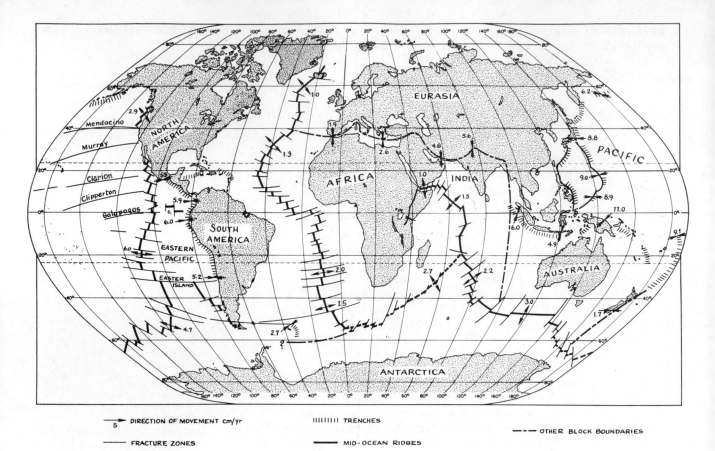

FIG. 4–10

Location of oceanic rises and trenches in ocean basins and the plates of the earth's crust. Rates and relative motions of the plates are also indicated. Compare with Fig. 3–6.

The simplest hypothesis of sea-floor spreading calls for heated—and consequently less dense—material to rise beneath oceanic rises, as shown in Fig. 4–11. As a result the overlying crust is pulled apart, permitting molten rock to move upward through cracks, forming either volcanoes or intrusions into overlying rocks. This lava (or magma), when solidified, is the material that forms the newly generated crust, primarily SIMA.

The volume of newly formed rocks must be compensated elsewhere by destruction of an equal amount of previously formed crust, apparently in the trench–island-arc systems, where the lithosphere is pulled down below adjacent blocks. Beneath trenches, the crust and its sediment cover are thought to be destroyed by melting and reassimilation into the mantle. (The details of this process are still unclear.) Some of the melted material moves through the crust, erupting at the surface in the volcanoes commonly found in island arcs. Shallow- and intermediate- and deep-focus earthquakes typical of the trench–island-arc areas are the result of plates cracking or interacting in some other way while moving past one another. Only trench–island-arc areas commonly have intermediate- (60-300 km deep) and deep-focus (300-700 km deep) earthquakes.

Movement of crustal units is slow in terms of human experience, but rapid when compared to earth history. Crustal movement away from the rise takes place at rates of 1.5 centimeters (⅔ inch) per year on the southern portion of the Mid-Atlantic Ridge and more than 5 centimeters (2 inches) per year on the rise encircling Antarctica. At these speeds it takes 30,000 to 100,000 years to move 1 mile. Remember that we are dealing with periods of about 200 million years since the present pattern of movement began. The most recent episode of spreading began about 10 million years ago, a relatively short time in earth's 4.5-billion-year history.

When continental blocks come together in the absence of trenches, and ocean basins disappear, crustal deformation takes place along the boundaries; some boundaries are shown in Fig. 4–11. Sediments originally deposited near continental margins are squeezed and deformed upward, forming mountain ranges. Additional mountain building results from volcanic action, through eruption of molten rock. Mountain ranges, including the Himalayan Range, are common in a band extending eastward from the Mediterranean Sea through northern India and into Southeast Asia. The Appalachian Mountains of the eastern United States most likely formed in a similar way about 350 million years ago.

Continental blocks also contain other ancient zones of weakness where they were at one time broken by faults. When subjected to strain by movements in the mantle below, these ancient weak zones (faults) may break and become active again, giving rise to occasional earthquakes that affect the interiors of the

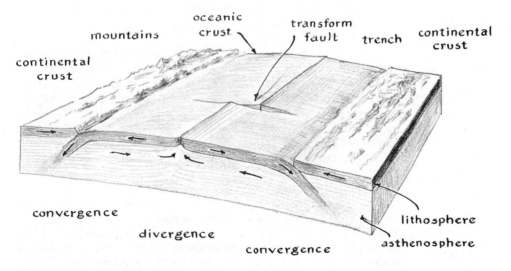

FIG. 4–11
Schematic representation of crustal blocks and their movements at oceanic rises and trenches. Note that the crustal plates move past each other along transform faults between offset segments of the rise. Elsewhere the blocks move in the same direction and with the same speed on different sides of the fault. The irregular topography along the inactive portions of the fault is called a fracture zone.

lithospheric plates. Also we now know that each of the continents consists of blocks that have been joined by earlier plate movements. These can be identified by geologists because the organisms contained in their sedimentary rocks indicate quite different depositional histories.

FRACTURE ZONES AND VOLCANIC RIDGES

As oceanic crust moves across the earth's surface, drawn away from an oceanic rise by movements in the mantle, two kinds of "tracks" are formed on the ocean bottom to mark the direction in which the crust moved: fracture zones and *volcanic ridges*.

A fracture zone, it will be remembered, is a band of rough topography and volcanic peaks on the ocean bottom, generally associated with offset segments of a mid-oceanic ridge. The fracture zone extends in a direction transverse to the ridge axis and parallels the direction of motion of the crustal blocks. On the portion of the fracture zone between offset segments of the ridge occur earthquakes, due to friction created by crust on either side of the fracture moving in opposite directions. Beyond the offset segments there are no earthquakes because crust on either side of the fracture zone is moving in the same direction.

Faulting associated with a fracture zone gives rise to mountainous topography. Reduced pressure near the fault causes rocks to melt at a lower temperature and also provides an avenue for lava movement to the surface. These scars created by faulting and volcanism persist long after the crust has moved away from the rise. They lengthen as the crust continues to move, sometimes extending hundreds to thousands of kilometers on either side of a ridge.

Volcanic ridges are associated with long-continued volcanic activity at a thermally active point on the earth's surface. A volcano may erupt at such a "hot spot" as the earth's crust moves over it, but it can only remain active while it is above the thermal area. As the crust moves away, it carries the now-extinct mountain of volcanic rock with it. A new volcano forms above the hot spot, and so on in sequence to create a volcanic ridge with mountains that may project above the sea surface as islands.

In the South Atlantic, volcanoes on the island of Tristan da Cunha have been active for millions of years. While the locus of volcanic activity has remained stationary, accumulations of volcanic rock have moved away with the sea floor on either side. The result is a pair of volcanic ridges—one, the Walvis Ridge, connecting Tristan da Cunha with southwestern Africa, and on the other side a less well-developed ridge extending westward to South America.

Another example is the Hawaiian Island chain. Here we see the tops of volcanoes formed progressively as the Pacific crust has moved northwestward through time. Mauna Loa and Kilauea are active volcanoes on the largest, most recently formed island of Hawaii, while there are indications of submarine volcanic activity at a location farther to the southeast. The oldest visible volcanic mountains in the chain are nearly eroded down to sea level; they are the small, steep pinnacles at the northwestern end. Beyond them, further to the northwest, atolls (Midway, Kure) mark the presence of former volcanic islands which are yet more ancient.

Submarine or insular ridges of this type are most common in the Pacific Ocean Basin. On the other hand, other oceans have submarine ridges and rises containing fragments of continental material which are also found near Pacific margins but never in the basin itself. Pacific volcanic ridges are generally associated with a long, parallel fracture zone, whereas the "microcontinental" type of ridge is not.

MAGNETIC STRIPES— THE EARTH'S GROWTH LINES

The earth has a magnetic field, as we all know from use of the magnetic compass to determine north. This magnetic field arises from circulation of core materials. The moon and planets, which apparently lack molten cores, exhibit little or no magnetic field.

Certain iron minerals in rocks respond to the earth's magnetism, and their internal structure retains a record of the direction of the magnetic field at the time of formation of the mineral. This is an especially valuable characteristic of iron-rich lavas, which solidify into rocks and record the magnetic field as minerals cool through a temperature range of about 600–700°C.

Studies of ancient volcanic rocks show that the earth's magnetic field has reversed itself many times in the past, as indicated in Fig. 4–12. During periods of so-called *normal magnetic orientation*, the north-seeking end of a compass needle would behave as it does now. During periods of *reversed magnetism*, the north-seeking end would point toward the south. Geophysicists can determine the permanent magnetism of a rock in the laboratory.

On a larger scale, instruments can be towed by aircraft or ships to map the earth's magnetic field over wide areas. Such surveys reveal the striped pattern illustrated in Fig. 4–13, due to the fact that the magnetic field over any ocean region is modified slightly by alternating bands of normal and reversed magnetic rocks. These magnetic stripes can be interpreted as growth lines of ocean basins. As newly formed, hot oceanic crust moves away from the rises, it cools below the 600–700°C range and records the magnetic field of the earth as it was at that time. Thus the ocean bottom acts like a gigantic tape recorder, the tape being the newly formed crust and the signal being the earth's magnetic field at the time of formation as shown in Fig. 4–14. Observed stripes in the magnetic field were not understood until the concept of sea-floor spreading provided an explanation.

The same pattern of striping provides evidence of the relative ages of various parts of the ocean basin. Data for interpretation of the pattern has been extracted from lavas of known ages that are on the continents and which indicate the magnetic field at the time of their formation. Careful dating of lavas and determination of their magnetic polarity has given us a dated

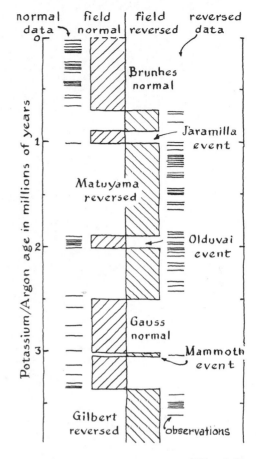

FIG. 4–12
Time scale for reversals of the earth's magnetic field. The polarity (direction of the magnetic field) is shown for the past 4 million years. [After A. Cox, R. R. Doell, and G. B. Dalrymple, "Reversals of the Earth's Magnetic Field," *Science*, 144 (1964), 1537–43]

record of reversals extending back several million years. Comparison of the pattern in the magnetic stripes with this dated pattern in terrestrial lavas permits ages of various parts of the ocean bottom to be estimated. On this evidence it is estimated that about 50 per cent of the deep-ocean bottom has formed during the past 70 million years.

Magnetic components in rocks also can be used to determine the probable position of the earth's magnetic pole relative to the rock's position when it formed. Thus it is possible to determine if a rock formed on the magnetic equator or at one of the poles, to take an extreme case. From such information we can reconstruct probable paths of continental blocks, assuming that the blocks and not the magnetic poles changed position. These studies provide independent evidence that pieces of continental blocks have moved substantial distances across the earth's surface since the various rock units formed.

Independent evidence such as this is always needed to check indirect determinations of earth history. If the interpretation of evidence is correct, it will usually be confirmed by the results of several other independent approaches. Furthermore, part of the value of a hypothesis such as sea-floor spreading is that it suggests new approaches to old problems. If the predictions of the hypothesis are borne out by other data, it provides an additional check on the validity of the hypothesis.

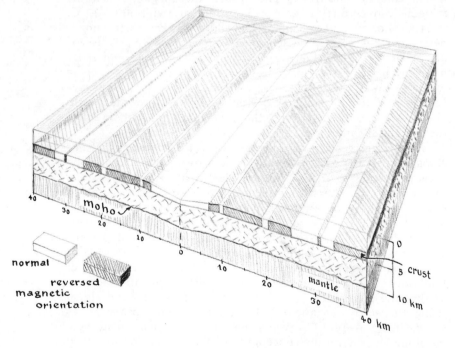

FIG. 4–13
Striped texture of the earth's magnetic field is shown by the results of a magnetic survey of the North Atlantic Ocean southwest of Iceland over a portion of the Mid-Atlantic Ridge. Areas of positive anomalies (unusually high intensity) are shown in white, areas of low intensity are shown in black.

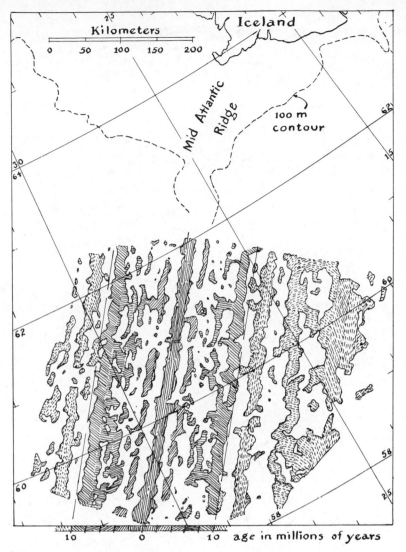

FIG. 4–14

Diagramatic representation of the magnetic field
at an oceanic rise showing the different magnetic orientations of the crust, resulting from
crustal formation when the earth's magnetic
field is normal and reversed. A constant
rate of spreading is assumed in estimated crustal
ages given at the bottom of the figure.

EVOLUTION OF OCEAN BASINS

The picture that has emerged from studies of crustal movements
suggests that about 225 million years ago, a single continent
formed all the land on earth. This ancient continent, sometimes
called *Pangaea*, stretched from present Antarctica to about the
present northern limit of Eurasia. About 200 million years ago
the present system of crustal movement began. At that time,
the beginnings of the Atlantic formed as parts of the ancient
continent split apart, much as the Gulf of California or Red
Sea are splitting apart today.

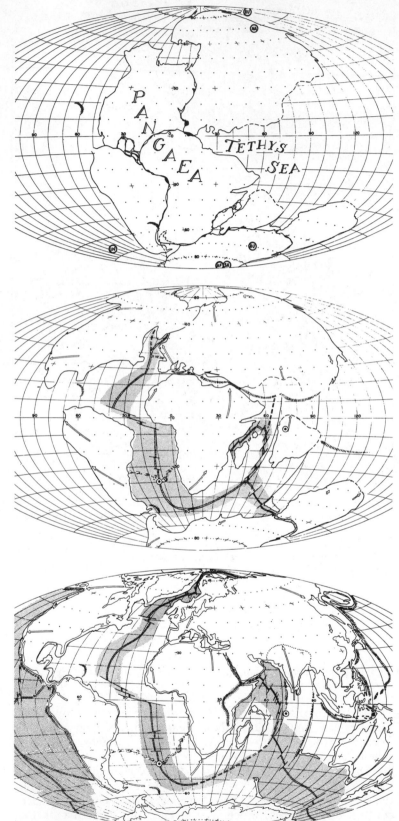

FIG. 4–15
Movement of continental blocks during the past 225 million years of earth history, following the breakup of the ancient continent, Pangaea (a). Note the closing of Tethys, an ancient marginal sea; the opening of the Atlantic Ocean (b); and the formation of the Indian Ocean. [R. S. Dietz and J. C. Holden, "Reconstruction of Pangaea; Breakup and Dispersion of Continents, Permian to Present," *Journal of Geophysical Research,* 75 (1970), 4939–56]

Tethys, an ancient seaway, now represented by thick accumulations of marine rocks, separated Africa from Eurasia along the equator. A trench system developed in this area and caused Africa and Eurasia to converge until today only the Caspian, Black, and Mediterranean Seas remain to mark the former location of Tethys. Between and around these remnants of seas are rugged, young mountain ranges, including the Himalayas, the highest mountains on earth. This area is unusually high because a portion of the Indian block was forced under part of the Eurasian block as a result of the collision. If the hypothesis is true, Asia has two thicknesses of crust beneath it in this area.

Before the present period of movement, it seems probable that there was only the ancestral Pacific Ocean with its marginal seas. During the past 200 million years the Pacific Basin has apparently shrunk and Tethys has essentially disappeared. The Indian Ocean was formed by the northward movement of Australia and the formation of the Indonesian island arc; these land masses block much of the water exchange between the Pacific and Indian Oceans. The Atlantic Ocean is the result of activity of the Mid-Atlantic Ridge. Finally, the major ocean areas around Antarctica have opened up to form the primary connection for the world ocean. In short, movement of continental blocks has remade the oceans. Although the evidence is hard to decipher, this may also have happened at various times during earth history.

The first step in *ocean-basin evolution*, outlined in Table 4–3,

Table   4–3

PROBABLE STAGES IN FORMATION OF AN OCEANIC BASIN

Stage	Process	Sediment Deposits	Typical Age (millions of yr)	Example
Embryonic	Uplift Rifting	None	10	East African Rift Valleys
Young	Rifting Slight lateral movement	Little	25	Gulf of Aden Red Sea Gulf of California
Mature	Extensive lateral movement	Moderate thicker along margins	100	Atlantic Ocean
Declining	Extensive trenches Overriding by crustal blocks	Extensive along margins	several hundred	Pacific Ocean
Closing (marginal basins)	Isolated basins cut off from communication with open ocean Coalescing of blocks Deformation of sediment and crust	Thick Deformed	several hundred	Black Sea Mediterranean Sea Caspian Sea

is the fracturing of a continental block which overlies some active oceanic rise. The continental block is first rifted, forming mountains as fault-bounded blocks move past one another vertically.

FIG. 4–16

The Red Sea (left) and the Gulf of Aden, which connects with the Indian Ocean, seen on the horizon. This seaway formed about 20 million years ago when Africa (the land mass on the bottom) and Eurasia (upper left) were rifted and subsequently split. Note the parallelism of the coasts of the Gulf of Aden. Clouds can be seen over the land especially in the lower portion of the photograph. The spacecraft can be seen in the lower left portion of the photograph. (Photograph courtesy NASA)

Subsequently, the continental block is pulled apart. This is apparently happening in the western United States along the Sierra Nevada, and also in eastern Africa where the East African Rift Valleys form a great mountain range running parallel to the coast and large lakes fill the valleys. Divergence in the movement of mantle material below the lithosphere causes opposite sides of a rift valley to move apart.

The Gulf of California in northwestern Mexico and the Gulf of Aden–Red Sea, shown in Fig. 4–16, are examples of areas where continental blocks are being pulled apart by processes acting in the mantle beneath continents. Such young ocean areas (probably less than 10 million years old) characteristically lack the thick sediment deposits of older marginal basins, and active volcanoes are common there.

Long after continental blocks have been pulled apart, the resulting ocean basin is symmetrically arranged on either side of the oceanic rise that caused the separation. This is easily seen in the Atlantic Ocean, where the continental margins and the geology on both sides show quite clearly that they were once joined. Sediments washed in from the continents over millions of years form thick deposits in marginal basins near the mouths of large rivers.

The subsequent history of an ocean basin is dependent on the location of its trench systems. A centrally located trench can cause the disappearance of the basin, as happened in the case of Tethys; movement of continental blocks finally closed the ocean basin, leaving only remnant seas and large inland lakes with no external drainage to the ocean.

The Pacific presents a still more complicated picture. Trenches along its margins may have totally resorbed portions of the original ocean-basin floor; this apparently has happened in the northeast Pacific to a block lying between the Aleutian Trench and a portion of the oceanic-rise system. Net motion of ocean-basin segments is a function of the relative amounts of motion on the rise system, where the crust forms, and the movement into trenches, where it is resorbed. Much remains to be learned about the history of the Pacific Ocean before we can develop a complete history of its basin.

ORIGIN OF OCEAN AND ATMOSPHERE

We know that the ocean is an ancient feature of the earth. But how the ocean and atmosphere formed and at what rate, and what controlled their composition are questions still actively debated by scientists; thus the answers offered here may well be challenged and even displaced by new hypotheses within even a few years. Satellite observations of the earth, powerful microscopes for studying ancient microfossils, and the recovery of rocks from the moon all have contributed to our present understanding.

Evidence for the age of the earth is indirect, coming from radiometric dating of the decay products of uranium and thorium isotopes which gradually accumulate after billions of years. Several lines of evidence suggest that the earth formed about 4.6 aeons ago. (An *aeon* is commonly used to denote 1 billion years).

There is no known record of the first aeon of earth history. All we can guess is that at some early time, the earth was apparently strongly heated, with coincident separation into three concentric layers. The heaviest material, primarily iron and nickel, sank to form the core. Heavy silicate rocks formed the mantle and the lighter materials rose to the surface, now segregated into the continents and oceanic crust.

Minerals from ancient rocks of Europe, North America, and Southern Hemisphere continents represent the first evidence of a solid crust on the earth. These rocks are 3.5 to 3.6 aeons old. Their formation, at the time of major changes in the earth's structure, may have been associated with such a cataclysmic event as the capture of the moon by the earth's gravitational field.

If the primordial earth had an atmosphere, it apparently was lost, perhaps as the result of the same heating that caused the segregation of core, mantle, and crust. At any rate, there is relatively much less rare noble gas (krypton, neon, xenon) in the present atmosphere than in the universe as a whole. This indicates that these gases were not retained in the abundance that one would expect from the retention of an ancient atmosphere. Thus, the present ocean and atmosphere must have been derived from later alteration of the earth rather than from cosmic sources. Both the water and gases must have been combined in rocks if they were not part of the early loss of gas from the earth's surface.

Based on our present knowledge of geologic processes in the ocean and on the continents, it seems probable that the ocean and atmosphere gradually formed by separation of water and gas from silicate rocks. Parts of the upper mantle or crust melted and the more volatile, water-rich portions of these molten rocks escaped as lava from volcanic eruptions or intruded overlying rocks, and in the process a part of their original water and gas content was lost to the atmosphere. Through time, the water condensed to fill depressions in the crust—the ocean basins—and the gases remain in the atmosphere. This process apparently continues today, although the amount of new (juvenile) water is too small to be detected with confidence. Fig. 4–17 illustrates several of the explanations that have been advanced to account for the accumulation of the ocean.

The ocean and atmosphere most likely originated as a result of the same processes that formed the continental and oceanic crust. Sea-floor spreading and continental drift are manifestations of this active crustal reworking. Erosion removes material from the continents, which is ultimately deposited in the ocean. There the sediment deposits are eventually reworked through sinking at trenches to be reassimilated and made into metamorphic or igneous rocks.

Assuming that this process can account for both ocean and atmosphere, what is the evidence of their early history? The earliest evidence for the existence of the ocean is the presence of recognizable water-laid sediments formed about 2.5 to 3 aeons ago (see Fig. 4–17). While this indicates the presence of abundant water on the earth's surface, it does not tell us about the chemical composition of this primordial ocean. Carbonate sediments about 2 aeons old suggest that ocean water has been salty for at least half of the earth's history. Very likely the major constituents of sea salt were not greatly different from those

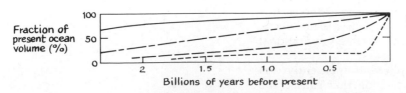

Fraction of present ocean volume (%)

Billions of years before present

FIG.   4–17
Several hypotheses of the rate of addition of water to the ocean during the past 2.5 billion years. The upper curve illustrates a rapid accumulation of the early ocean with slow accretion at present. The lowest curve illustrates a rapid accumulation in more recent geologic time. (Modified after Ph. H. Kuenen, *Marine Geology*, John Wiley & Sons, Inc., New York. 568 pp.)

now present in the ocean. Fossils about 0.6 aeon old suggest that since that time, the chemical composition of seawater has changed little.

Evidence suggesting the composition of the early atmosphere is extremely scarce and mostly indirect. Free oxygen almost certainly did not occur. This would account for the presence in some ancient sediments of minerals not stable in the presence of oxygen. Another significant bit of evidence is the existence of the iron-rich deposits which have been extensively mined as iron ore during recent centuries. These deposits could not have been formed on the earth's surface in the past 2 aeons. Iron apparently was present in ancient ocean waters in substantial quantities, whereas it is extremely rare in seawater now. In the presence of free oxygen, iron is highly insoluble in seawater and does not form large-scale marine sedimentary deposits. But in an ocean devoid of dissolved oxygen, iron would be more soluble and could form extensive deposits. Substantial amounts of free oxygen were apparently not introduced to the atmosphere until after the evolution of photosynthesizing plants which release oxygen in the process of making their food.

ORIGIN OF LIFE ON EARTH

For many years it was thought that living organisms could have existed on earth only during the last aeon or so. Detailed studies of sedimentary deposits have revealed traces of bacteria and colonial algae in rocks 2 aeons old. Subsequently, even older fossil fragments were discovered, suggesting the existence of one-celled green plants and bacteria for perhaps 3 aeons.

Animals have a shorter history. A complex, highly developed animal population bearing shells seems to have appeared suddenly about 0.6 aeon (600 million years) ago. Since that time, continued evolutionary changes have taken place, including the evolution of man during the last few million years.

All life, both plants and animal, is based on only six types of chemical compounds, as shown in Table 4–4. The six elements of which they are mainly composed are abundant in the ocean. Plant and animal protoplasm and body fluids are, in fact, strikingly similar in composition to seawater itself.

These facts suggest that life originated in the ocean. At first, however, there was no general agreement as to how this could have happened. Except for water, the major constituents of living systems are compounds (see Table 4–4) that today, are normally synthesized only by living organisms. They do not now form spontaneously by natural processes in the absence of life. To understand how the first organic materials were formed, we must consider the atmosphere of the primitive earth, in which nonbiological synthesis of organic matter could and did take place.

If we synthesize organic compounds in the laboratory today, we find that most of them quickly react with oxygen in the atmosphere—in other words, they "burn up" and release energy, forming water as a by-product. Thus evolution of life would have been difficult or impossible in an atmosphere such as we have now.

Table   4–4

PRINCIPAL CHEMICAL CONSTITUENTS OF ORGANISMS*

Compounds	Functions	Composition
Water	Universal solvent	Hydrogen, oxygen
Carbohydrates	Energy source, water-soluble	Hydrogen, oxygen, carbon
Fats	Energy storage, non-water-soluble	Hydrogen, oxygen, carbon
Adenosine phosphates	Energy transfer	Hydrogen, oxygen, carbon, nitrogen, phosphorus
Proteins	Structural; facilitation of chemical reactions (enzymes)	Hydrogen, oxygen, carbon, sulfur, (organized into 20 amino acid "building blocks")
Nucleic acids	Patterns for protein synthesis; genetic materials	Hydrogen, oxygen, carbon, nitrogen, phosphorus (organized into 5 nucleotide "building blocks")

*After McAlester, 1968.

The earth's *primitive atmosphere* probably accumulated slowly. Argon was derived from radioactive decay of potassium. Gases emitted from the earth's interior through volcanic activity included water vapor, carbon dioxide (CO_2), carbon monoxide (CO), nitrogen (N_2), sulfur dioxide (SO_2), hydrogen chloride (HCl), and a few others. This was a *reducing atmosphere* (one in which few hydrogen-accepting gases are present). Note that this atmosphere contained most of the elements essential to life (see Table 4–4).

At this time in earth history, large amounts of ultraviolet radiation from the sun reached the earth's surface. There was no ozone (a form of oxygen) in the upper atmosphere to absorb most of this type of radiation, as there is today. Electrical discharges from lightning storms almost certainly occurred from time to time. Radioactive materials, more plentiful then than now, gave off alpha, beta, and gamma radiation.

Scientists, trying to duplicate these conditions in the laboratory, have experimented with various mixtures of gases, subjecting them to the kinds of energy sources believed to have been present before the origin of life. In these experiments almost every important constituent of living systems has been synthesized, making it seem likely that the same thing happened in the primitive reducing atmosphere and ocean.

Picture, then, lakes and oceans in which were dissolved carbohydrates, amino acids, and hydrocarbons—"dilute organic soup," as it has been termed—at concentrations of a few parts per million. Many of these compounds (such as organic acids and amines) were surface-active—that is, partly oily or water-repellant (*hydrophobic*). Such molecules tend to accumulate at interfaces, such as the water surface or on bubbles in sea foam, and to adhere to mud particles. Lines of brownish scum from sea foam left on beaches by waves and tide today are examples of surface-active substances being concentrated at the shore. This process very likely permitted concentration of organic molecules by adsorption on shallow-water sediments.

Polymerization (joining together) of the concentrated organic compounds resulted in formation of more complex molecules. This process has been duplicated experimentally by exposing simple amino acids (adsorbed on clay) to ultraviolet light. Polymerized organic molecules tend to form discrete droplets in a water medium and to float free of the clay matrix where they formed. As insoluble *"coacervate"* (coagulated into a viscous liquid) drops, they are no longer subject to dilution by the water.

Many scientists believe that polymers thus formed were associated with certain phosphorus- or other metal-coordinated organic acids which had the property of transferring energy from one part of a system to another. In this way, energy absorbed or produced in one place can be utilized in another place, permitting balanced chemical reactions to proceed within the organic droplets. In modern living systems, coenzymes carry on energy transfer. Metal-coordinated organic acids mentioned above may thus be categorized as proto-coenzymes. It is possible that many of them could also absorb sunlight to provide energy for primitive metabolic systems, much as chlorophyll absorbs energy for plant photosynthesis.

The molecular structure of such coacervates can remain stable while undergoing chemical changes only if the reactions involved are self-perpetuating. Processes of synthesis and disintegration must be interconnected and coordinated so as to recur over and over, making possible self-reproduction of the system. Thus we may envision a colloidal coacervate which accumulates proteins and other compounds from the environment, grows in size, and then splits into two or more fragments, which then repeat the process.

Here *natural selection* becomes a factor in the development of organisms. Those systems which do not duplicate themselves successfully are eventually destroyed. Those that duplicate repeatedly survive, and undergo modifications.

In addition to favoring efficient reproduction, natural selection also favors improved metabolic processes, such as accumulation and release of energy. The earliest form of metabolism probably involved assimilation of ready-made organic compounds available in the surroundings. These served as building-blocks for the growing, developing systems, sunlight providing the necessary energy. This is somewhat analogous to animal (*heterotrophic*) nutrition today: animals ingest foods ready-made by plants or other animals. Today, however, energy for growth and development is provided by oxidation of these organic materials. In a reducing atmosphere, before the development of plant photosynthesis put oxygen in the air, the first metabolism could not have required free oxygen.

As the development of living systems proceeded, the supply of organic materials produced by nonbiological means may have been greatly reduced. *Autotrophic systems*, which could utilize solar energy to produce organic compounds from inorganic ones, as plants do today, would thus have had a significant advantage. Whether for this reason or some other, production of food and oxygen by plant photosynthesis came to dominate the metabolic process. This has been the situation on earth from that time to the present.

Increased production of free oxygen by plants caused important chemical changes in the ocean and atmosphere, and also in soils. Oxygen was undoubtedly toxic to early forms of life, so that the presence of oxygen-acceptors such as reduced (ferrous) iron compounds in the marine environment was essential to primitive plants.

Eventually, with the development of enzymes to mediate the toxic effect of oxygen on living cells, photosynthesizing organisms became more abundant. The ocean became saturated with oxygen, which then began to escape into the atmosphere. Ferric (oxidized iron) sediments on land date from that period, about 1.3 aeons ago. Iron in the ferric state is highly insoluble and does not readily enter the ocean, and ferric sediments did not form in marine environments after oxygen became abundant in the atmosphere.

Accumulation of free atmospheric oxygen had two important effects. First, the ultraviolet radiation from the sun broke up oxygen molecules (O_2) to form atomic oxygen (O) and ozone (O_3). Ozone in the atmosphere effectively filters out incoming ultraviolet light, thus shielding living organisms from this dangerous radiation and permitting life to exist on land and in shallow water.

The second effect was to provide oxygen now required for life processes of multicelled organisms. The fossil record seems to indicate that primitive plant and bacterial forms existed for 1 or more aeons before a heterotrophic mode of life became common on earth. Perhaps it took that long for an adequate supply of atmospheric oxygen to be built up by plant photosynthesis. It has been postulated that the seemingly explosive proliferation of shelled organisms about 0.6 aeon ago (Cambrian period) resulted from the availability of oxygen to permit development of such "physiological luxuries" as connective tissue and shells, which can be preserved as fossils.

Some paleontologists have suggested that small communities of animals may have developed much earlier, dependent on concentrations of algae (primitive plants) as a source of oxygen and carbohydrate food. Blue-green algae today are quite resistant to high levels of ultraviolet radiation, suggesting that their development predates an oxygen atmosphere. In any case, whereas plant life seems to be at least 2 aeons old, and abundant animal life about 0.6 aeon old, life on land seems only to have become plentiful during the past 0.4 aeon (400 million years).

GLOMAR CHALLENGER In the late 1960's and early 1970's a deep-sea drilling project was undertaken jointly by five major oceanographic institutions: Scripps Institute of Oceanography (University of California), Woods Hole Oceanographic Institution, Lamont–Doherty Geological Observatory of Columbia University the Institute of Marine Sciences of the University of Miami and the Department of Oceanography University of Washington. The objective was to study oceanic crust by taking long cores of sediment and rock from the deep ocean floor. A research vessel, the Glomar Challenger, was specially designed for the project to maintain position and to drill in very deep water.

Position is maintained by means of a sonar beacon; dropped to the ocean floor, its emitted signals are picked up by hydrophones below the ship's hull. Computers calculate the ship's position with respect to the beacon and automatically hold her on station while drilling is in progress. Samples of sediment and rock have been recovered from more than 1000 meters below the ocean bottom, and the vessel can operate in waters nearly 7 kilometers deep.

Analyses of cores from this deep-sea drilling project have provided important evidence supporting the theory of sea-floor spreading. Absence of sediment over mid-ocean ridges and progressive thickening of sediment toward ocean basin margins suggest that ridges are sites of new-crust formation. Increasing age of the crust in direct proportion to its distance from the ridge is indicated by its increasingly thick sediment cover. (Photograph courtesy Global Marine, Inc.)

103

Techniques for studying ocean structure
        Bathymetry
        Geophysical techniques: seismology, measurements of gravity and heat flow
        Deep-ocean drilling—obtaining samples of oceanic crust for direct studies

Crustal structure
        Lithosphere—outer portion of the earth composed of rocks well below melting point; includes crust and upper mantle (60–100 km thick)
            Continental crust—approximately 35 km thick, mostly granite (SIAL)
            Oceanic crust—approximately 7 km thick, mostly basalt (SIMA)
            Mohorovicic discontinuity—separates crust and mantle
        Asthenosphere—rocks near melting point, deform much like fluids
        Layered structure of oceanic crust
        Transition between oceanic and continental crust in trenches and at continental margins

Sea-floor spreading
        Formation of new crust at rises, destruction at trenches
        Nine major plates of lithospheric material move as rigid units
        Deformation and mountain building occur at edges
        Blocks bounded by rises, trenches, or transform faults
        Movements at rates of 1 to 5 centimeters per year
        Present continents formed from sections of crustal blocks that were separate at one time

Fracture zones and volcanic ridges
        "Tracks" of crustal block movement marked by fracture zones and volcanic ridges
        Fracture zones—rough topography associated with offset segments of an oceanic ridge
        Volcanic ridges form as crust moves over a thermally active zone
        Volcanic island chains also form as crust moves over a thermally active zone

Magnetic stripes—the earth's growth lines
        Magnetic components in rocks record orientation of ancient magnetic field
        Magnetic stripes record development and indicate relative ages of oceanic crust
        About one-half of deep-ocean bottom formed in past 70 million years

Evolution of ocean basins
        Ancient continent, Pangaea, began to break up about 200 million years ago
        Former seaway, Tethys, has disappeared
        Pacific has shrunk, Atlantic has grown, India and Australia have moved northward
        Continents broken up at divergence zones, such as rift valleys or oceanic rises
        History of ocean basin determined by relative location of rises and trenches

Origin of ocean and atmosphere
        No record of first aeon of earth history (probable crust formation and core separation); crust apparently formed about 3.6 aeons ago
        Atmosphere and ocean formed later; ocean existed about 3 aeons ago

Water and atmospheric gases derived from rocks
Gradual release of free oxygen by plants to the atmosphere
Origin of life on earth
Developed in reducing atmosphere, perhaps 3 aeons ago, probably in ocean
Nonbiological synthesis of organic compounds
Organized into reproducing structures, capable of storing and transmitting energy
Heterotrophs—utilize and require ready-made food
Autotrophs—manufacture food, from sunlight and inorganic materials; photosynthesis produces carbohydrates and releases oxygen
Multicelled organisms with connective tissue appeared suddenly about 0.6 aeons ago; development may be related to oxygen accumulation in atmosphere

SELECTED REFERENCES

ANDERSON, D. L. 1971. "The San Andreas Fault." *Scientific American* (Nov. 1971). 225(5): 52–68. Details of crustal movements along a major transform fault.

BARGHOORN, E. L. 1971. "The Oldest Fossils." *Scientific American* (May 1971). 224(5): 30–42. Evidence on earliest stages of evolution.

EICHER, D. L. 1968. *Geologic Time*, Prentice-Hall, Inc., Englewood Cliffs, N.J. 149 pp. Introduction to geologic time and the record preserved on land.

HOLMES, A. 1965. *Principles of Physical Geology*, Rev. ed. Ronald Press, New York. 1288 pp. Thorough discussion of geologic processes on an advanced level.

MCALESTER, A. L. 1968. *The History of Life*. Prentice-Hall, Inc. Englewood Cliffs, N.J. 151 pp. Origin of life discussion on intermediate level, primarily from fossil record.

MENARD, H. W. 1964. *Marine Geology of the Pacific*. McGraw-Hill, New York. 271 pp. History of Pacific Ocean basin; intermediate to advanced level.

MENARD, H. W. 1969. "The Deep-Ocean Floor." *Scientific American* (Sept. 1969). 221(3): 126–145. Summary of dynamics of ocean bottom.

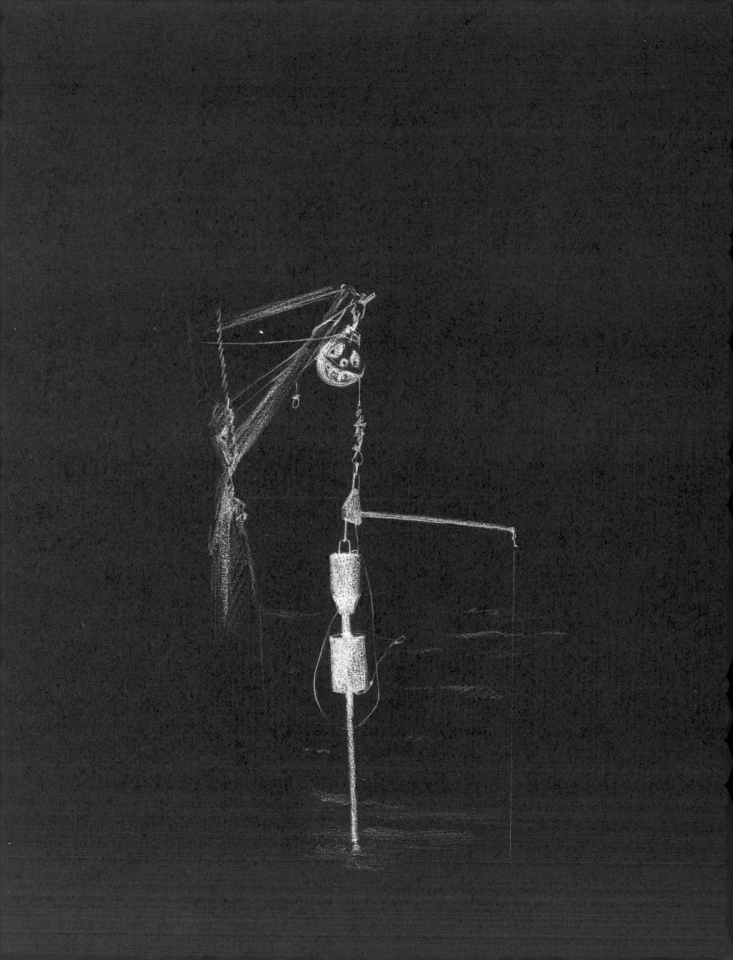

SEDIMENTS

Swedish Deep-Sea Expedition (1947–48)   No part of the earth seems so hostile to man's investigations as the deep ocean floor. An expedition which did much to unlock the secrets of that forbidding region was organized by Hans Pettersson, first director of the Oceanographic Institute at Göteborg, Sweden.

Starting in the early 1930's, as Professor of Oceanography at the University of Göteborg, Pettersson worked at promoting the idea of a great Swedish deep-sea expedition that would circle the world and make oceanographic history. Göteborg scientists spent years during World War II inventing the Kullenberg piston coring device, capable of taking a 21-meter core of undistorted sediment. Such a core contains a record of conditions on earth during the evolution of man. In addition, the technique called *seismic profiling* was developed for measuring sediment thickness anywhere on the ocean floor.

After the war, the 1500-ton training schooner *Albatross* was donated by a Swedish shipping company for such an expedition, and the Royal Society of Göteborg provided financing. The Swedish Deep-Sea Expedition of 1947–48 circumnavigated the earth in the tropic zone, taking deep-sediment cores that scientists from all over the world still study today. The ship's echo sounder recorded a continuous profile of the ocean floor; water samples taken at all levels revealed distributions of physical, chemical, and biological conditions; and a seismic profile of sediment thickness was made throughout the voyage. Pettersson's vision and perseverance had enlisted some of the best scientists of his time in a nongovernmental cooperative research project that stands as one of the high points in the history of oceanographic research.

five

Most of the ocean bottom is covered by sediment deposits—unconsolidated accumulations of particles brought to the ocean by rivers, glaciers, and winds, mixed with shells and skeletons of marine organisms. Only a few areas of ocean bottom are devoid of sediment cover. Some are swept clean by strong currents, other areas are newly formed and have not yet accumulated much sediment. Deposits vary greatly in thickness but are usually thickest near continents, especially in marginal ocean basins where barriers interfere with sediment transport into the main ocean basin.

Throughout its history, the ocean has been the final repository for debris produced on the earth's surface. The sediment deposits now recoverable provide an historical record for the ocean basins during the past 150 million years. Records of preceding ages have apparently been destroyed by crustal assimilation.

Furthermore, oceanic sediments tell us much about oceanic processes, such as nutrient supply to organisms and the effect of climatic changes on oceanic circulation. In short, the study of oceanic sediment provides us with a time dimension for ocean studies.

**ORIGIN AND CLASSIFICATION
OF MARINE SEDIMENT**

Much information about sediments comes from the study of individual particles. For purposes of study, sediment particles are classified in two different ways—by origin and by size. Classified by origin, sediment particles are divided into the following groups:

Lithogenous particles (derived from rocks): primarily silicate mineral grains (listed in Table 5–1) released by the breakdown of silicate rocks on the continents during weathering and soil formation; in the open ocean, volcanoes are locally important sources of lithogenous particles

Biogenous particles (derived from organisms): the insoluble remains of bones, teeth, or shells of marine organisms. Some typical particles are illustrated in Fig. 5–1, 5–2, and 5–3.

Hydrogenous particles (derived from water): formed by chemical reactions occurring in sea water or within the sediments; examples of the most common hydrogenous particles are manganese nodules illustrated in Fig. 5–4.

Sediment particles are also classified according to size, as in Table 5–2. Particle size is an important property of sediments

Table 5–1

COMMON MINERALS IN MARINE SEDIMENTS AND THEIR CHEMICAL COMPOSITION

Silicates—primarily lithogenous, derived from decomposition of silicate rocks:

Quartz	SiO_2
Opal	$SiO_2 \cdot H_2O$, glasslike substance formed by diatoms, radiolaria
Feldspars	—alumino-silicates with Na, K, and Ca in variable proportions
Micas	—primarily muscovite (a K, Al, hydrous silicate)
Clays	—similar to micas but with more variable composition and structure
Illite	—primarily altered, fine-grained muscovite
Montmorillonite	—a group of minerals formed commonly from volcanic glass
Chlorite	—Fe, Mg, Al hydrous silicate, commonly formed during weathering
Kaolinite	—Al silicate formed during tropical weathering
Gibbsite	—Al oxide, formed during tropical weathering
Zeolites	—hydrous alumino-silicates, similar to feldspars in structure.

Oxides, hydroxides—occur primarily in Fe–Mn nodules:

Hematite	—Fe_2O_3, probably red coloring on deep-sea clays
Goethite	—$FeO(OH)$, constituent of Fe–Mn nodules
Lepidocrocite	—$FeO(OH)$, another mineral form of Fe in nodules
Pyrolusite	—MnO_2, occurs in Fe–Mn nodules

Carbonates—primarily formed by marine calcareous organisms:

Calcite	—$CaCO_3$, forms foraminiferal shells, coccoliths
Aragonite	—more soluble form of $CaCO_3$, in pteropod shells

Phosphates—rare in sediments, common in bones and teeth:

Apatite	—$Ca_5(PO_4)_3F$

Sulfates:

Barite	—$BaSO_4$, found in certain primitive organisms, and sediments rich in organic matter
Celestite	—$SrSO_4$, found in shells of certain radiolaria, rarely in sediment

Biogenous minerals in boldface. Hydrogenous minerals are in italics. All others are lithogenous in origin.

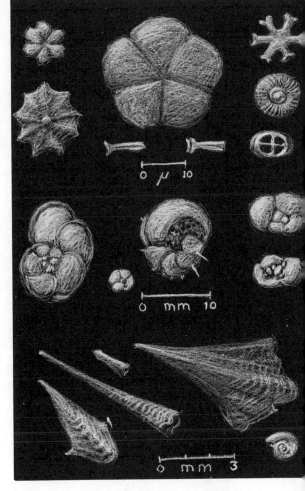

FIG. 5–1

Some common calcareous biogenous constituents in marine sediments: discoasters and coccoliths (top), foraminifera (middle) and pteropods (bottom).

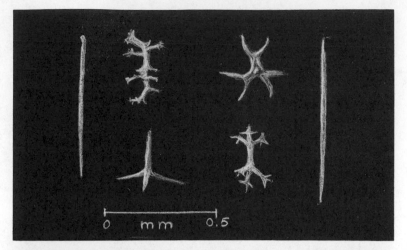

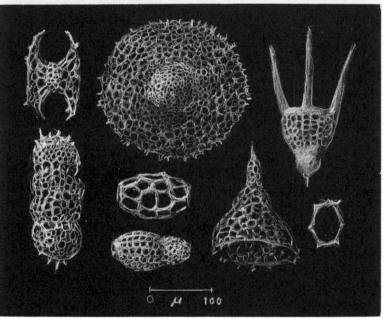

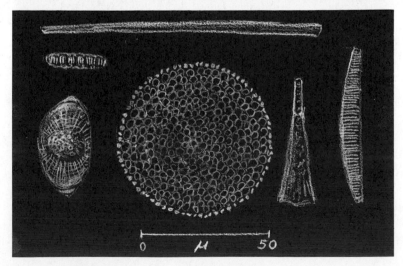

FIG. 5–2
Some common siliceous biogenous constituents
in marine sediments: sponge spicules (top)
and Radiolaria (middle, bottom).

because it determines the mode of transport and also determines how far particles travel before settling to the bottom. Typical sizes of particles from various sources are indicated in Fig. 5–5.

If a sediment consists of a single size of particle, we say that it is *well sorted*. A *well-sorted sediment* is one in which a limited size range of particles occurs, the other-sized particles have been removed, usually by mechanical means. Beach sand is a typical well-sorted sediment, in which particles have been segregated by size as a result of wave or current actions. A *poorly sorted sediment* contains particles of many sizes, usually indicating that little mechanical energy has been exerted to sort the different particle sizes. Glacial marine sediments, derived from mechanical wearing down of rocks by glaciers, are poorly sorted. Commonly they contain particles ranging in size from boulders to fine silt (see Table 5–2).

FIG. 5–3
Fish-bone fragments from deep-ocean sediments.

Table 5–2

CLASSIFICATION OF SEDIMENT PARTICLES BY SIZE (WENTWORTH SCALE)

Size Fraction	Particle Diameter (millimeters)
Boulders	>256
Cobbles	64–256
Pebbles	4–64
Granules	2–4
Sand	0.062–2 (62–2000 μ)
Silt	0.004–0.062 (4–62 μ)
Clay	<0.004 (<4 μ)

Sands usually occur along coastlines where waves and currents remove the finer particles, depositing them in deeper water. Effects of currents are shown in Fig. 5–6. Sands are rare in the deep ocean, although, as we shall see, they do occur there. *Mud* —a mixture of silt- and clay-sized particles—is the most common type of sediment in the deep ocean, as indicated in Table 5–3.

FIG. 5–4
Manganese nodule having whale's earbone at its center (a); typical North Pacific nodule (b); manganese encrusted slab of volcanic rock from the South Pacific (c). (Challenger Reports)

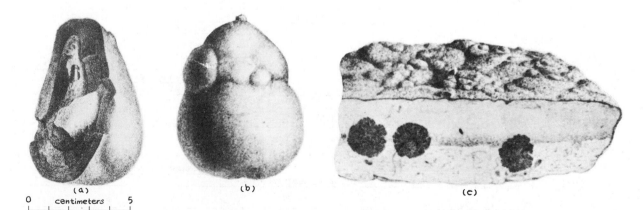

Table 5–3

ABUNDANCE AND COMPOSITION OF DEEP-OCEAN SEDIMENT*

Constituents	Calcareous Mud (%)	Siliceous Muds		Brown Mud (%)
		Diatomaceous (%)	Radiolarian (%)	
Biogenous:				
Calcareous	65	7	4	8
Siliceous	2	70	54	1
Lithogenous and hydrogenous	33	23	42	91
Abundance in deep-ocean sediments	48	14		38

*Recalculated after R. R. Revelle, *Marine Bottom Sediment Collected in the Pacific Ocean by the* Carnegie *on its Seventh Cruise*, Carnegie Institution Publication 556, part 1 (1944).

Lithogenous constituents of marine sediment are derived from breakdown of silicate rocks that are exposed to the atmosphere. Most rocks were originally formed at high temperatures and pressures in the absence of free oxygen and with little liquid water. When these rocks (both igneous and metamorphic) are altered by the complex of chemical and physical processes known as *weathering*, certain soluble constituents are released. These are carried in solution to the oceans. The remaining

FIG. 5–5
Grain-size distribution of common marine sediments and sediment sources.

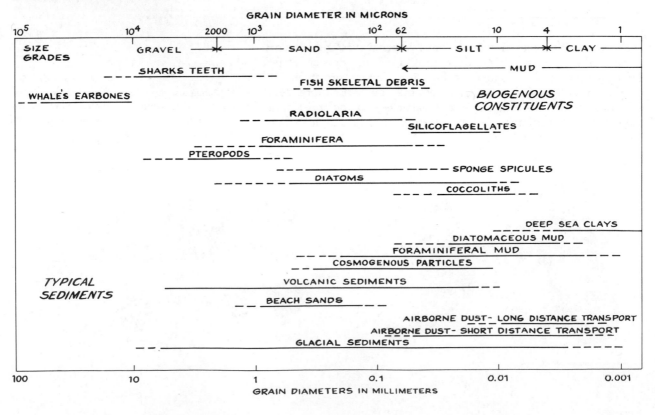

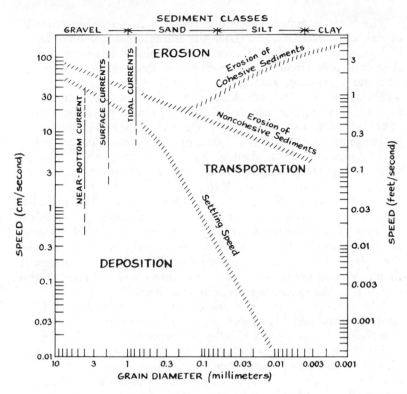

FIG. 5-6
Current speeds required for erosion of sediments and transport of grains as compared to settling speeds for sediment grains. (Note use of logarithmic scale.) [After B. C. Heezen and C. D. Hollister, "Deep-sea Current Evidences from Abyssal Sediments," *Marine Geology*, 1 (1964), 141]

altered rock material is often broken up by physical processes into sand- and clay-sized grains, which are also carried downstream by running water.

Some minerals—for example, quartz and some of the feldspars—are quite resistant to chemical alteration by air and water. These usually enter the ocean virtually unaltered and are common constituents of sands and silts. Some products of rock weathering and the slightly altered remains of micalike minerals form the clay minerals, common in fine-grained deep-sea clays. These particles are so small that they could not be studied by the optical and visual techniques employed in the *Challenger* expedition. For many years it was assumed that the particles were amorphous (having no internal structure). Modern physical techniques, including X rays, have shown that clay minerals have distinctive structures which are useful to identify sediments derived from various sources.

Most dissolved substances are removed from seawater by biological processes. Elements not used by organisms generally remain dissolved in the ocean for tens to hundreds of millions of years, whereas elements used by organisms remain a few tens or hundreds of years. Materials most commonly used by marine plants and animals are calcium carbonate, silica, and calcium phosphate, all of which are used to build shells, bones, or other skeletal structures.

Not all marine organisms form shells or skeletons, and many are so fragile that they are rarely, if ever, preserved as fossils. In general, only the sturdiest materials survive the posthumous descent to the bottom because of consumption and digestion by larger animals, decomposition by bacteria, and dissolution by seawater.

The most common *biogenous* remains in sediments consist of calcium carbonate, primarily the mineral *calcite*. The chemically similar mineral *aragonite* is more soluble in seawater and therefore less frequently preserved in deep-ocean sediments. The most common among the larger biogenous particles are shells of *Foraminifera*, which are one-celled animals, pictured in Fig. 5–1. Among the minute calcium carbonate particles, *coccoliths*—platelets formed by tiny, one-celled algae—are common in carbonate-rich sediments. Shells of *pteropods*—floating snails —form rare deposits at shallow depths, primarily in the Atlantic Ocean; at greater depths, their fragile aragonitic shells are broken up and usually dissolved.

Siliceous shells of *diatoms, Radiolaria,* and *silicoflagellates* are also common biogenous constituents. Of these, the shells of diatoms (one-celled algae) and Radiolaria (one-celled floating animals) are the most important. Siliceous *sponge spicules*— the spikey skeletal elements shown in Fig. 5–2—are also found in many deep-ocean sediments, but these never dominate the deposits to the extent that diatoms and Radiolaria do.

The earliest systematic work on sediment was done by Sir John Murray and his associates based primarily on sediment samples collected by the *Challenger* expedition. Much of their work was based on visual observation of the samples, and their classification scheme—still widely used—was based on the abundance of large, easily recognized biogenous constituents. Murray classified any sediment containing more than 30 percent biogenous constituents by volume as an *ooze*: a "diatom ooze" if diatoms were dominant, a "foraminiferal ooze" if shells of Foraminifera were the most abundant carbonate constituent. For our purposes, we shall call such sediments *diatomaceous muds* if diatom shells constitute more than 30 percent by volume, or *radiolarian muds*, or *foraminiferal muds*, using the same standard. If none of these biogenous constituents dominate the sediment, then we shall refer to the deposit as a *deep-sea mud*, sometimes also called a *red clay* because of its typical red or red-brown color, especially in the Atlantic.

Other biogenous remains are less common. Bones are rather fragile, open structures made up of minute calcium phosphate crystals embedded in an organic matrix. When bones are broken and eaten, bacterial activity decomposes the organic matrix, leaving only tiny fragments (such as shown in Fig. 5–3) of phosphate crystals which commonly dissolve. Only the most resistant bones, such as sharks' teeth or whales' earbones, remain for long periods on the ocean bottom; they are recovered by dredging and sometimes found in the centers of manganese nodules as those in Fig. 5–4.

Hydrogenous constituents are formed through chemical processes occurring in seawater itself. There is no obvious indication of biological activity involved in formation of manganese nodules (also known as iron-manganese nodules, because of their high iron content). These elements, originally brought to the ocean by rivers and possibly from volcanic activity on the ocean bottom, apparently precipitate in the presence of the abundant dissolved oxygen typically found in deep-ocean waters. Their precipitates form tiny particles that are swept along by currents until they come in contact with a surface. Since they tend to adhere to solid surfaces, the particles form a stain or coating around a solid nucleus—such as a fragment of volcanic rock or other debris. In the deep ocean, where sediment accumulates slowly, sharks' teeth and whales' earbones are common examples of such biological residues. Through time a Fe–Mn coating builds up slowly around the original nucleus, so that eventually a nodule is formed.

Nodules form by accretion of layers; commonly these are alternately iron-rich and iron-poor. These nodules also contain at least small amounts of most particles commonly found in seawater. Presumably these, too, have been incorporated in the nodules by striking the surface and adhering to it. An interesting aspect of such nodules is their high copper, cobalt, and nickel content. Such nodules may absorb metals directly from seawater, although the evidence is by no means clear. Regardless of the cause, the high concentration of these trace elements in manganese nodules makes them attractive as a possible source of copper and other scarce metals.

Manganese nodules form at a rate of only a few atomic layers per day, one of the slowest chemical reactions known. On the Blake Plateau of the continental rise off southeastern North America, the bottom is covered with such nodules, which seem to form at the rate of about 1 millimeter per million years. In other areas, the nodules seem to form 10 to 100 times faster. In isolated or nearly isolated marine basins, such as the Baltic Sea, manganese nodules form at rates of 20 to 1000 millimeters per 1000 years. The speed record for nodule growth is the manganese coating that formed on a naval artillery shell fragment dredged off Southern California; it formed at the rate of about 10 centimeters per 100 years.

Forming so slowly, manganese nodules cannot grow to any substantial size in areas which receive large amounts of other sediments, because in such cases they are buried too quickly. Consequently, such nodules are rare in the Atlantic, which is well supplied with lithogenous sediment. In the Pacific, where sediment accumulates slowly, nodules are quite common on the bottom. Manganese nodules are estimated to cover 10 to 20 percent of the deeper parts of the Pacific, based on analyses of deep-ocean photographs and sediment samples taken from these areas.

Drilling in deep-ocean basins has penetrated sediment layers rich in iron and manganese oxides. These are often several meters thick and apparently formed just above the rocks of the oceanic crust. The metals are thought to have migrated upward during cooling of the recently formed rocks beneath. Metal-bearing solutions that reach the water–sediment interface in this way may eventually be incorporated in manganese nodules or other metal-rich deposits.

Cosmogenous sediment is composed of particles and objects that fall to the earth from outer space. Some large objects survive the descent through the atmosphere and are recognizable as meteorites. A much larger percentage do not survive as recognizable particles but fall as bits of cosmic dust, much of it thought to have been derived from meteorites.

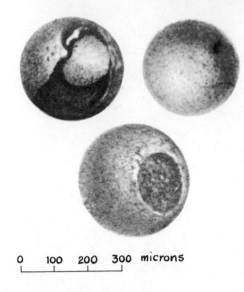

0   100   200   300   microns

FIG. 5–7
Cosmic spherules from deep-ocean deposits.
(Challenger Reports)

Cosmic spherules (shown in Fig. 5–7), first recognized in deep-ocean sediment by Sir John Murray, are magnetic; largely composed of iron or iron-rich minerals, they are about 200 to 300 microns in diameter. Such spherules also occur in glacier ice.

Since the Industrial Revolution, man himself has supplied substantial amounts of sediment entering the ocean. Usually the sediments are derived from various wastes, both municipal and industrial. Because of their high level of industrialization along coasts, the United States and Northern Europe are the largest contributors of such *man-made sediments*. Most of these deposits appear to remain on continental shelves near coastal urban regions.

SEDIMENT BUDGET

Rivers are by far the largest source of sediment to the ocean, responsible for about 20 billion tons each year; this figure is broken down by continent in Table 5–4. Beginning with the con-

Table 5–4

RIVER DISCHARGE OF SUSPENDED SEDIMENT FROM THE CONTINENTS*

Continent	Area Draining to Ocean (10^6 km^2)	Annual Suspended Sediment Discharge $\left(\dfrac{10^9 \text{ tons}}{\text{yr}}\right)$	$\left(\dfrac{\text{tons/km}^2}{\text{yr}}\right)$†
North America	20.4	2.0	96
South America	19.2	1.2	62
Africa	19.7	0.54	27
Australia	5.1	0.23	45
Europe	9.2	0.32	35
Asia	26.6	16.0	600
Total	100.4	20.2	

*From J. N. Holeman, "The Sediment Yield of Major Rivers of the World," *Water Resources Research*, 4 (1968), 737.

†Tons of sediment per square kilometer of drainage area per year.

tribution of lithogenous sediment, we find that only the eleven rivers listed in Table 5–5 contribute more than 100 million tons of sediment each year. Of these, the four largest sediment contributors are the Hwang Ho (Yellow), Ganges, Brahmaputra, and Yangtze, which together supply about 25 percent of the total sediment discharge from the continents. Two of these rivers discharge into the northern Indian Ocean and two into the western Pacific Ocean. Except for the Ganges–Brahmaputra system, and the Amazon and Congo rivers (which drain into the equatorial Atlantic), the world's major sediment-producing rivers drain into marginal seas rather than into open coastal-ocean areas. Consequently, most of the sediment discharged by rivers into the ocean can not reach the deep ocean floor.

Table 5–5   SEDIMENT BUDGET   117

MAJOR SEDIMENT-PRODUCING RIVERS*

| River | Drainage Area (10^3 km^2) | Average Annual Discharge | | |
		Water ($\frac{km^3}{yr}$)	Sediment ($\frac{10^6 \text{ tons}}{yr}$)	($\frac{tons}{km^2}$)†
Yellow (Hwang Ho)	666	47	2100	2900
Ganges	945	368	1600	1500
Brahmaputra	658	382	800	1400
Yangtze	1920	683	550	550
Indus	957	174	480	510
Amazon	5710	5680	400	70
Mississippi	3180	559	340	110
Irrawaddy	425	425	330	910
Mekong	786	346	190	480
Colorado	630	5	150	420

*After J. N. Holeman, "The Sediment Yield of Major Rivers of the World," *Water Resources Research*, 4 (1968), 737.

†Tons of sediment per square kilometer of drainage area per year.

Most lithogenous sediment entering the ocean is derived from semiarid regions, usually from mountainous areas, as indicated in Fig. 5–8. Rainfall in such areas is usually inadequate to support an erosion-resisting cover of vegetation, but sufficient to erode soil. In Asia and India, the long continued intensive agricultural development has further increased sediment yield. Tropical areas usually supply relatively little sediment per unit of drainage basin because of the heavy vegetation which inhibits the movement of the particles. Also, intensive weathering of rocks in tropical climates tends to break them down completely forming soluble components rather than to leave discrete mineral particles. In polar regions, low temperatures retard chemical decomposition of rocks because water, of major importance in weathering, is frozen for much of the year. Furthermore, the recent glaciation of these high-latitude regions has left many lakes which serve as sediment traps, preventing its transport to the ocean.

Some lithogenous sediment is transported by wind; unfortunately, we do not yet know how much sediment is involved, but the figure is perhaps 100 million tons per year. Winds can pick up and transport sediment where they are strong and the vegetation sparse, as in deserts and on high mountains. Wind transport of sediment to the deep ocean seems to be especially important in the middle latitudes where there are few rivers owing to the arid climate. Wind transport seems to be less important in the Southern Hemisphere, where there is less land.

The amount of biogenous sediment formed each year (indicated in Table 5–6) can also be estimated on the basis of river-water composition. In order to estimate the amount of biogenous sediment, we assume a steady state—in other words, that conditions in the ocean are not changing through time. If

Table 5–6

SEDIMENT SUPPLY FROM VARIOUS SOURCES

Source of Particles	Rate of Supply (g/yr)
Lithogenous—river-transported	2×10^{16}
Biogenous*	
Siliceous	0.5×10^{15}
Calcareous	1.5×10^{15}

*Assuming that all the SiO_2 and Ca brought to the ocean each year by rivers is removed from solution by organisms to be incorporated in sediment deposits.

the chemical composition of seawater is constant, then the amount of dissolved calcium, silica, and phosphate brought to the ocean each year must equal the amount removed during the year by organisms and incorporated into the sediment. Remember that this takes into consideration only the amount actually reaching the sediment, not the amount removed by organisms from seawater, most of which is returned to the water by dissolution of particles as they sink to the bottom. This approach indicates that about 1.5 billion tons of calcium carbonate is deposited in sediments each year; most of this is probably removed by foraminifera. About 0.5 billion tons of siliceous sediment forms each year. In total, this amounts to about 10 percent of the total sediment load brought to the ocean each year,

FIG. 5–8

Major source areas of lithogenous sediment. The rivers shown discharge more than 100 million tons of suspended sediment each year. [Data from J. N. Holeman, The sediment yield of major rivers of the world *Water Resources Research*, 4 (1968), 737]

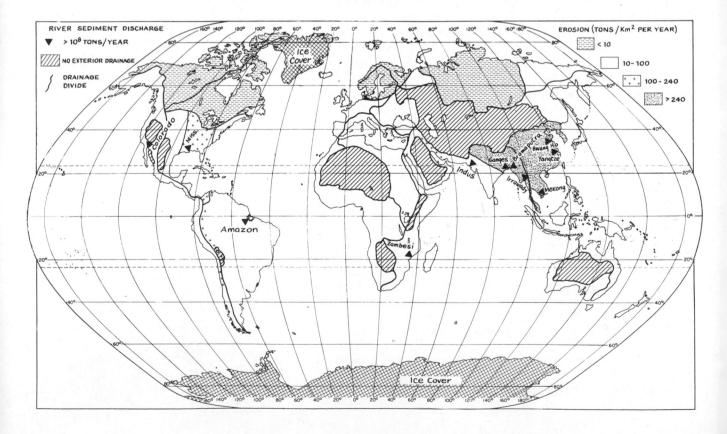

sediments.

SEDIMENT TRANSPORT

Transport of sediment particles is determined primarily by physical characteristics. Particle size (see Fig. 5–5) and current speed (see Fig. 5–6) are the two most important variables. Vertical transport (sinking of particles through a column of water to the ocean floor) is the easiest to visualize.

The settling speed of a particle through seawater is controlled primarily by particle diameters: large particles fall faster than small ones. For particles whose diameter is less than about 0.2 millimeter, the settling speed is predicted by *Stokes' law,* as follows:

$$V = d^2(\rho_s - \rho_f)g/18\eta$$

where V is the settling velocity in cm/sec; d is particle diameter in cm; ρ_f is the fluid density (usually 1.03 g/cm^3); ρ_s is the particle density (usually 2.65 g/cm^3); η is the fluid viscosity (0.014 poises at 10°C); and g is the acceleration of gravity (980 cm/sec^2). Particles with diameters larger than about 0.2 mm fall somewhat more slowly than predicted by Stokes' law, since there is additional resistance to their settling owing to turbulence generated around the particles.

To demonstrate what this means for settling times of various sized sediments, the following settling times were calculated for different sized grains:

	Settling velocity (cm/sec)	Time to settle 4 km
Sand (100 μ)	2.5	1.8 days
Silt (10 μ)	0.025	185 days
Clay (1μ)	0.0025	51 years

From this one would predict that sands could settle out near the point where they were discharged to the ocean, silts could be transported moderate distances, and clays could theoretically be transported throughout the world ocean before settling out of the 4 kilometers of sea water.

Now let us consider another aspect of sediment movement: lateral transport. Figure 5–6 summarizes available estimates about the current speeds necessary to suspend sediment particles and to transport them. Note that for noncohesive sediments, the speed necessary to move sediment particles steadily decreases as particle size decreases, so that even weak currents can erode and transport fine-grained sediment. If, however, particles are cohesive (stick together), the force required to erode the deposit first decreases as the grain size gets smaller and then increases, so that erosion of cohesive silts and clays may require currents just as strong as those necessary to move sands and gravels. Once eroded and set in motion, particles tend to settle out. This tendency is opposed by the upward component of random water movements existing during turbulent flow. Even at current-speed less than necessary to erode the sediment, particles may still be transported some distance before settling out of the water.

Both surface and tidal currents (see Fig. 5–6) are often strong enough to erode and transport most common sediment types. Even deep-ocean currents appear to be strong enough to erode and transport noncohesive, fine-grained bottom sediment. The effects of erosion are seen in the turbid water commonly near the bottom, both on the continental shelf and on the deep-ocean bottom. The importance of this process cannot be evaluated as yet. If common, it could lead to the erosion, transportation, and redeposition of sediment over large parts of the ocean bottom that have always been thought to be far from such sediment-transport processes.

Sediment is initially transported by rivers either suspended in the water (usually about 90 percent of the river's sediment load), or rolling along the bottom as part of the so-called "bed load." The bed load is commonly deposited near the head of the estuary, at a point where the downstream flow of river water encounters the upstream near-bottom currents of coastal-ocean water that are drawn into the river mouth by estuarine circulation. The bed load at that point is deposited because the currents are no longer strong enough to transport the relatively large grains that constitute the bed load.

The *suspended load* is carried farther down the estuary, but eventually much of it is deposited either as a result of lower current speed or of a process known as *flocculation*. Many of the minute particles commonly carried in the suspended load have electrical charges associated with them as a result of the structural alterations of silicate minerals that occur in weathering. This is especially common among clay minerals where metals normally in the crystal structure have been removed, causing electrical imbalance in the mineral lattice. While in the fresh water of the river, these electrical charges act to keep small particles apart. In the presence of dissolved salts, however, particles bearing dissimilar charges are attracted to one another, so that they form larger masses (through flocculation) which rapidly settle out of the water. Flocculation is thought to remove part of the fine-grained sediment load from rivers as salt concentrations increase toward the mouth of the estuary.

Organisms also cause sediment deposition. Many marine and estuarine organisms obtain food by filtering water. Mineral particles removed along with their food are compacted into fecal pellets and then excreted. Oysters, mussels, clams, and many other attached and floating animals effectively remove particles from suspension, bind them into fecal pellets which accumulate in sediment deposits. Plants growing along the banks and on tidal flats also trap sediment by providing quiet sheltered areas where these small particles can settle out. The plants then prevent erosion of sediment deposits when tidal currents or heavy waves scour the area.

As a result of these processes (discussed in Chapter 11), most sediments are trapped and deposited near river mouths. Estuaries unfilled by sediment are effective sediment traps. For example, rivers of the U.S. Atlantic coast transport an estimated 20 million tons of sediment each year to their estuaries; virtually none of this load reaches the continental shelf. Thus the continental shelf of eastern North America is covered primarily by sediment laid down about 3000 or more years ago.

Deposits of sand and silt that were formed under conditions no longer existing in an area are known as *relict sediments*. They are recognized by distinctive features—such as fossil remains of organisms that no longer inhabit the area. For example, oyster shells found far out on the continental shelf indicate deposition at a time of greatly lowered sea level when these areas were near the shore. Mastodon teeth in a marine sediment, or other remains of extinct land animals, indicate that the area was once dry land and that its sediments date from that period. Other characteristics of relict sediments include iron stains or coatings on grains that could not have been formed under present marine conditions.

Where rivers supply sediment to the coastal ocean, these relict deposits may be buried. Alternatively, they may be reworked by the action of waves during periods of rising sea level so that the characteristic features of a relict deposit are destroyed or masked. At the present time, about 70 percent of the world's continental shelves are covered by relict sediments.

When the sediment load of a river is large enough, it may fill its estuary; then the sediment load can be carried out into the ocean. The Mississippi River is a good example: its large sediment load has not only filled the estuary but also created the Mississippi Delta, a bed of trapped sediment that extends into the Gulf of Mexico. Estimates for different rivers suggest that anywhere from 10 to 95 percent of the sediment load may be deposited in a delta.

Winds transport significant amounts of sediment to the ocean. Fine silt particles, having diameters of less than 20 microns, may be carried great distances. This is especially important in the *jet streams* of mid-latitudes, where strong winds at high altitudes provide extremely rapid transport. In 1956, for instance, ash from volcanic eruptions on the Kamchatka Peninsula, northeast of Japan, was detected over England at heights of 18,000 meters (55,000 feet) five days after the eruption. Similar transport speeds have been reported for radioactive debris from atmospheric tests of nuclear weapons in the 1950's and early 1960's. Particles in the lower atmosphere are commonly removed by precipitation; for example, they serve as condensation nuclei for raindrops and snowflakes and are thus carried back to earth.

ACCUMULATION OF SEDIMENT
IN THE OCEAN

Much of the sand carried by a river is deposited near its mouth, often forming beaches. Sand is also moved along the coast by longshore currents. This is especially obvious on the Pacific coast of the United States, where the largest beaches and dunes in Washington and Oregon are associated with the Columbia River mouth. Sands are common in areas of strong wave action because finer-grained materials are removed.

Silts and clays are carried farther seaward and may be deposited at depths where wave action is not strong enough to stir up sediment or erode the bottom. On many continental shelves this occurs at depths of 50 to 150 meters.

Modern sediments are relatively uncommon on the outer edge of continental shelves. Several explanations have been advanced for this fact. Possibly *internal* (underwater) *waves* are too strong along the continental-shelf break to permit sediment accumulation. Also, the relatively sudden water movements associated with the passage of storms over the coastal ocean may cause relatively strong currents, especially at the continental-shelf break.

Sands and silts deposited on the continental shelf commonly accumulate at rates of more than 10 cm/1000 yr. This is relatively rapid sedimentation, in the sense that particles have insufficient time to react chemically with the seawater or with near-bottom waters as they lie on the ocean bottom before being buried. Coastal-ocean sediments therefore retain many of the characteristics that they acquired during weathering. The colors of these rapidly accumulating deposits tend to be rather dark, gray, greenish, or sometimes brownish, and they contain abundant organic matter (1 to 3 percent is not uncommon).

Clays and very small particles that escape the river mouth can be carried great distances. Even relatively large clay particles (1 micron) require about 50 years to settle through 4 kilometers of water. Smaller particles (<0.5 μ), which are more typical of the particles in the deep ocean, theoretically require up to 1000 years to sink to the bottom. It appears that most of these particles are removed from suspension within a few hundred years. It seems probable that part of this removal is due to filtration of seawater by planktonic organisms which ingest particles and incorporate them in fecal pellets. These relatively heavy pellets sink at rates of 10 meters or more per day.

Deep-ocean sediments commonly accumulate at rates of less than 1 cm/1000 yr. Thus the particles not only spend hundreds of years suspended in ocean water, but then remain exposed on the ocean bottom for many more hundreds of years before they are finally covered by enough sediment to seal them off from contact with near-bottom waters. Burrowing organisms further increase exposure of the particles by stirring up the sediment after deposition.

As a result, there is usually ample time for particles in deep-ocean sediments to react with seawater. Abundant dissolved oxygen in deep-ocean waters causes the grains to be highly oxidized. Iron is converted to the ferric state (iron rust) giving the sediments their typical reddish color. The presence of dissolved oxygen also favors utilization of organic matter by organisms or its destruction by inorganic processes. Consequently, deep-ocean sediments commonly contain less than 1 percent organic matter.

Over most of the deep ocean, the sediment cover is relatively thin—usually between 0.5 and 1 kilometer, averaging about 600 meters. Thin sediment deposits generally occur on mid-oceanic ridges or rises, where large areas may be completely swept clear of sediment, apparently by strong currents. Sediments do accumulate there in certain protected "pockets." Sediment thickness increases away from the ridges and is generally greatest near the continents along the ocean-basin margins. In general, these sediments are flat-lying and in many areas apparently consist of alternating layers of turbidites (deposits formed by turbidity currents) and particle-by-particle accumulations. Such stratified sediments are especially common in the Atlantic and Indian Oceans, and in parts of the Pacific. There is little evidence for disturbance of these sediments since their deposition many millions of years ago.

Sediment accumulation and its eventual fate in the trenches is not yet entirely clear. Judging from the thickness of horizontally bedded sediment, there is relatively little sediment in the trenches: less than 500 meters in the South Sandwich, Tonga, and Philippine Trenches; less than 1 kilometer in the Java and Puerto Rico Trenches. It is possible, however, that much thicker sediment accumulations exist in or near the trenches and that they have been so severely deformed that they cannot be identified by geophysical techniques.

For many years, oceanographers believed that the deep-ocean bottom was covered solely by very fine-grained deposits. In general, data from the *Challenger* expedition confirmed this belief, although there were a few instances of coarse-grained sediment samples which were thought to result from mislabeling of samples early in the cruise. Expanded exploration of ocean-bottom sediments showed conclusively that coarse-grained sediments including sands were locally quite common, especially near continents and in the Atlantic Ocean. The question thus arose: How do these sediments get into the deep ocean?

Several lines of evidence indicate that turbidity currents transport large amounts of sediment from continental margins to the deep-ocean bottom. Perhaps the most compelling evidence for turbidity currents came from a study of the November 18, 1929, Grand Banks earthquake which occurred on the continental slope south of Newfoundland. A puzzling phenomenon was the subsequent breaking of submarine telegraphic cables downslope from the earthquake epicenter; the cables were broken in a definite sequence from top to bottom. Apparently the disturbance caused by the earthquake was enough to stir up sediment and form a dense sediment-laden mass that moved downslope (like a gigantic submarine avalanche) at speeds up to 100 kilometers per hour (60 mph). The rapidly moving mass tore out each cable in turn as it passed down the slope. On the deep ocean floor, the turbid mass deposited a thin layer of sediment over a large area of ocean bottom.

Turbidity currents have not been directly observed in the deep ocean but have been known to occur in lakes since the mid-nineteenth century. In the United States, such currents have been studied extensively leading up to Hoover Dam, where they transport sediment from the point at which the sediment-laden Colorado River enters the lake to the area just behind the dam.

Deposits formed by turbidity currents have distinctive properties that occur widely, not only in deep-ocean sediments but also in rocks thought to have originally been deposited in ancient deep-ocean areas. Among these features is *graded bedding*, in which the base of a single unit contains large sediment grains—commonly sand overlain by progressively smaller-grained deposits until the top of the unit is silt- and clay-sized particles. These units are often repeated many times in a single sediment core.

When the current first passes, it is moving at high speeds and carrying sand-sized particles. Rear sections of the flow move progressively more slowly, transporting finer particles which are deposited on top of the heavy ones. The result is a rapidly formed sediment deposit showing distinct gradation in particle size, changing from coarsest grained at the bottom to the finest at the top.

A distinctive aspect of transport by turbidity currents is that they frequently carry shells of shallow-water organisms and land-plant fragments into the deep ocean. Plant fragments—some still green—have been dredged from the ocean bottom following cable breaks. Turbidite layers containing remains of shallow-water organisms often alternate with layers of sediment that accumulated slowly particle-by-particle and contain organisms or exhibit other features typical of deep-ocean sediments.

Turbidity currents are most active where continental shelves are narrow, least active on wide continental shelves. In the Atlantic, sediment deposited by turbidity currents has buried most of the older topography, so that these currents can now flow from the edge of the continents almost to the mid-ocean ridge. Likewise, sediment brought into the northern Indian Ocean can flow unimpeded over great distances of the deep-ocean bottom. In the Pacific, however, trenches and ridges along the margins of the ocean basin prevent movement of turbidity currents to the deep ocean floor.

Besides carrying sediment and burying ocean-bottom topography, turbidity currents are also thought to cut the submarine canyons that indent the margins of most continents. Whether turbidity currents are the sole agent responsible for these large features has been debated extensively, but there is good reason to suppose that they serve in many areas to remove the large volumes of sediment that would otherwise fill these canyons completely. Evidence of this comes from breakage of submarine cables off the mouths of such large, sediment-carrying rivers as the Congo.

Over large parts of the deep ocean, sediment accumulates particle-by-particle. Deposits formed this way differ substantially from those formed by turbidity currents. These slowly accumulating deposits, also known as *pelagic deposits*, tend to blanket the original ocean-bottom topography, preserving the original topography in blurred outline. A comparison has often been made with fallen snow on land. Sometimes, as with a heavy accumulation of snow, deposits on sides of hills slump down into nearby valleys. Where bottom currents are strong, pelagic deposits may fail to accumulate.

The effect of various transport processes is reflected in the distribution of marine sediments. Since the mid-nineteenth century, thousands of samples have been collected, analyzed, and studied. Figure 5–9 indicates the state of our present understanding of sediment distributions in the deep ocean. Some deposits were formed by processes still active in the ocean; others, formed at some earlier time, remain exposed on the ocean bottom but do not reflect present oceanic processes.

In those parts of the continental shelf where the temperature during the coldest month of the year is 18°C or warmer, reef-building corals can grow. Many continental shelves within this belt are covered by biogenous sediment usually consisting of shallow-water calcareous algae, shallow-water foraminifera, and coral. In the United States such sediments occur around the southern tip of Florida. They are also common on the shallow Bahama Banks, Bermuda (Fig. 3–11), and off the Yucatan Peninsula of Mexico, and are especially extensive around the northern coast of Australia, where the *Great Barrier Reef* (150 km wide) extends for about 2500 kilometers.

Ocean margins are usually covered by thick deposits of lithogenous sediment, much of it having accumulated fairly rapidly. Near volcanoes, the ocean bottom also receives substantial amounts of volcanic rock fragments, usually the relatively large particles (>50 μ in diameter) formed during eruptions. Sediments deposited in the Indonesian area, around the Aleutians off Alaska, and near the many island arcs of the western Pacific contain abundant volcanic debris. Volcanic sediments are not restricted to the Pacific Ocean but occur in all ocean basins.

Volcanic fragments smaller than about 10 microns may be carried great distances by winds. If a volcanic eruption is powerful enough to inject material into the stratosphere, fine ash may be transported around the world. Particles eroded by winds from soils in arid or mountainous regions generally remain in the lower atmosphere and are removed fairly quickly, forming a "plume" of such material downwind of the arid continent. This is readily observed in the North Atlantic, where Saharan dust is widely dispersed.

In polar ocean regions, glacial marine sediments are the most common type of sediment on the ocean bottom. These poorly sorted deposits contain all sizes of particles—from boulders to silt (see Fig. 5–5). They were deposited by melting ice, either icebergs from continental glaciers or sea ice. Sea ice commonly contains sediment only if it has gone aground and incorporated material from the shallow continental shelf, which happens in the Arctic Sea. When such sea ice melts, its sediment load is deposited on the bottom, where it mixes with the remains of marine organisms. Some glacial marine sediments now exposed on the ocean floor may have been deposited during the past period of glaciation. Some is doubtlessly still being deposited.

Sediments that accumulate fairly rapidly near continents (indicated in Table 5–7)—transported either by running water, turbidity currents, or ice—cover about 25 percent of the ocean

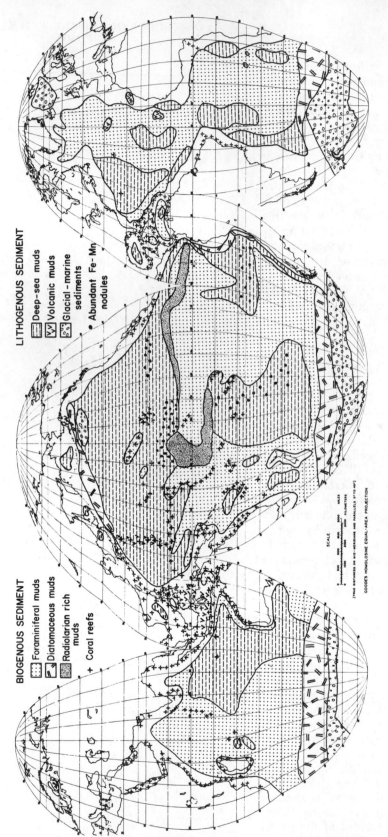

BIOGENOUS SEDIMENT

::: Foraminiferal muds

⌁ Diatomaceous muds

▦ Radiolarian rich
   muds

+ Coral reefs

LITHOGENOUS SEDIMENT

▦ Deep-sea muds

∨∨ Volcanic muds

◦◦ Glacial-marine
    sediments

• Abundant Fe-Mn
  nodules

SCALE

MILES
0   500 1000 1500 2000
0   1000 2000 3000 KILOMETERS

(TRUE DISTANCES ON MID-MERIDIAN AND PARALLELS 0° TO 40°)

GOODE'S HOMOLOSINE EQUAL-AREA PROJECTION

FIG. 5-9

Distribution of sediment in the deep ocean.

Table 5–7

DISTRIBUTION OF SEDIMENT 127
DEPOSITS

TYPICAL SEDIMENT ACCUMULATION RATES

Area	Average (Range) Accumulation Rate (cm/1000 yr)
Continental margin	
Continental shelf	30 (15–40)
Continental slope	20
Fjord (Saanich Inlet, Brit. Col.)	400
Fraser River delta (Brit. Col.)	700,000
Marginal ocean basins	
Black Sea	30
Gulf of California	100
Gulf of Mexico	10
Clyde Sea	500
Deep-ocean sediments	
Foraminiferal muds	1 (0.2–3)
Deep-sea muds	0.1 (0.03–0.8)

bottom. Accumulations are thickest in marginal-ocean basins which, although they account for only about 2 percent of the ocean area, contain about one-sixth of all identifiable oceanic sediment. The remainder of the ocean floor is covered by pelagic sediments that accumulate slowly—between 1 and 10 millimeters per thousand years.

More than half of the deep-ocean bottom is covered by biogenous sediments, either calcareous or siliceous (see Table 5–7). Formation of these deposits is governed by three fundamental processes: *productivity* of overlying surface waters, *destruction*, and *dilution*.

In those areas where nutrients (phosphate and nitrate) are resupplied to the surface layers and there is sufficient sunlight to support photosynthesis, marine life is usually abundant. Such conditions are found in the areas of deep-water upwelling that occur along western continental boundaries, at the equator, in high latitudes where a permanent density stratification is not present, and where currents diverge. Areas of upwelling and current divergence are often rather small, but biogenous deposits have been reported from the upwelling areas off South America and Southwest Africa. These areas are too small to be seen on Fig. 5–9. Diatomaceous sediments predominate in the high latitudes both in the North Pacific and surrounding Antarctica. Absence of predominantly diatomaceous sediment in the North Atlantic is presumably caused by mixing of diatom shells with abundant lithogenous sediment from neighboring continents. Furthermore, surface waters of the North Atlantic are not overly productive. In equatorial regions of the Pacific, sediment accumulations may be markedly thicker than is usual on the deep ocean floor—that is, in excess of 600 meters and in some areas (120° to 170°W) over 1 kilometer thick. Accumulation rates of about 15 millimeters per thousand years are considered to be typical in the most productive parts of the equatorial Pacific, compared to about 1 millimeter per thousand years elsewhere. Radiolaria often dominate these sediments.

A second factor controlling distribution of biogenous sediment is *destruction* of skeletal remains. Phosphatic and siliceous constituents apparently dissolve at all depths in the ocean as indicated in Fig. 5–10, so their survival depends in large measure on how sturdy or well constructed the fragments were initially. Only the most resistant bones and most heavily silicified shells of radiolaria and diatoms are common sedimentary constituents. Thin or slightly silicified shell fragments are easily broken down and readily dissolved; they do not survive the long settling time and the long period of exposure at the sediment surface.

The effect of particle destruction (see Fig. 5–10) can most readily be seen in the distribution of calcareous deep-ocean sediment. Because of its relatively high temperatures, surface ocean waters are usually saturated with calcium carbonate, so that calcareous fragments there are not dissolved. At mid-depths, the lower temperatures and higher CO_2 contents of seawater cause dissolution of calcareous fragments, but rather slowly. Below about 4500 meters, waters are rich in dissolved CO_2 and able to dissolve calcium carbonate much more readily. This depth relationship is markedly shown in the calcium content of deep-ocean sediments when plotted as a function of depth, as in Fig. 5–11. Carbonate-rich sediments are common at relatively shallow depths in the ocean but almost completely absent below about 6000 meters, as Fig. 5–12 indicates. For instance, aragonitic pteropod shells are found only on the tops of banks in the Atlantic. Distribution of carbonate-rich foraminiferal muds also reflects this depth zonation. Note that these sediments are most common on the shallower parts of the ocean bottom: on the Mid-Atlantic Ridge, the East Pacific Rise, and the Mid-Indian Ridge (see Fig. 5–9).

An additional factor controlling distribution of biogenous sediment is *dilution* by lithogenous sediment. Remember that we defined biogenous sediment as consisting of more than 30

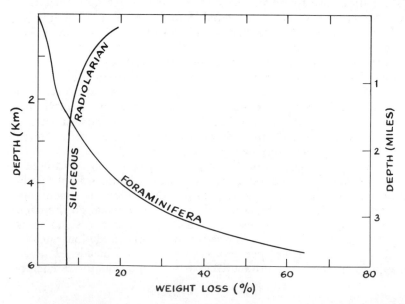

FIG. 5–10

General relationship between depth and weight loss in particles due to dissolution after 4 months of submergence at various depths in the Central Pacific Ocean. [After W. H. Berger, "Foraminiferal Ooze: Solution at Depths," *Science*, 156(3773) (1967), 383–85]

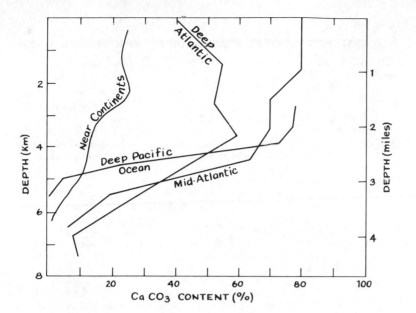

FIG. 5–11
Distribution of calcium carbonate in sediment deposited at various depths in the deep ocean and near the continents.

percent by volume of a biogenous constituent. We thus change the classification of such a deposit if large amounts of lithogenous sediment are added. Near continents where material derived from soil erosion is supplied at a relatively rapid rate, the probability of biogenous sediment formation is slight except on the tops of banks or ridges where turbidity currents cannot cover them.

Deep-sea muds (also known as *red clays* or *brown muds*) cover most deeper parts of the ocean basins. They contain less than 30 percent biogenous constituents and are derived primarily from the continents. This very fine-grained fraction of deep-ocean sediment is especially interesting because it contains

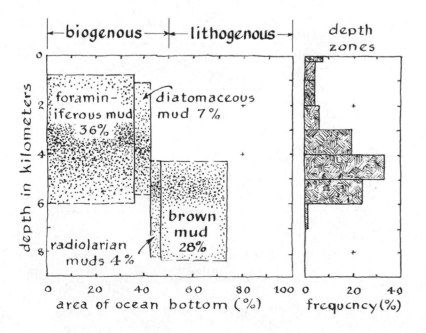

FIG. 5–12
Depth distribution of deep-ocean sediment. Minimum, maximum and average depths are shown for each major sediment type. Frequency of ocean bottom at various depth zones is also shown.

Table 5–8

COMMON CLAY MINERALS AND THEIR DISTRIBUTION IN MARINE SEDIMENT*

Mineral	Origin	Area of Abundance in Ocean
Chlorite	Abrasion of metamorphic and sedimentary rocks, especially by glaciers in polar regions	Polar and sub-polar regions
Montmorillonite group of minerals	Chemical alteration of volcanic ash; may form in place after deposition in sea water	South Pacific and Indian Oceans near mid-ocean ridges
Illite (hydro-mica group of minerals)	Chemical and mechanical alteration of mica minerals in rocks on continents	Near continents, especially common in North Atlantic
Kaolinite	Forms by decomposition of silicate minerals under intensive weathering in the formation of tropical soils	Equatorial Atlantic, and Indian Ocean near Australia

*Data from J. J. Griffin, H. Windom, and E. D. Goldberg, "The distribution of clay minerals in the World Ocean," *Deep-Sea Research* 15 (1968) 433–59.

a distinctive suite of layered silicates derived from different climatic zones (see Table 5–8). These mineral grains are too small to be studied visually and require examination by X ray techniques which differentiate them on the basis of their internal crystal structure. While present in the muds on continental shelves and rises, the clay mineral fraction is most important in the deep ocean, where it constitutes 60 percent or more of the nonbiogenous fraction of these sediments, as indicated in Fig. 5–13.

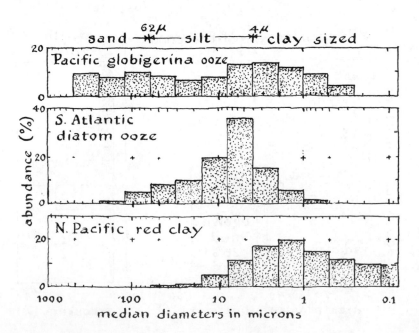

FIG. 5–13
Grain-size distribution in typical deep-ocean sediments. (After Sverdrup et al., 1942)

Clay minerals are also interesting because they are small enough (<2 μ) to be transported by all sediment-moving processes—winds, rivers, and glaciers. With their distinctive composition, the clay minerals permit us to distinguish between each of these transport mechanisms in their importance in various parts of the ocean.

Studies of the clay-sized fraction of Atlantic Ocean sediments have demonstrated that tropical rivers contribute distinctive clay minerals (*kaolinite, gibbsite*) that typically form in the highly weathered tropical soils. These clays are widely distributed, as Fig. 5–14 shows, across the tropical ocean, carried by the equatorial-current system. *Illite*, a group of minerals derived from slightly altered micas that occur commonly in many different kinds of rocks, is also a common constituent of deep-ocean sediment. Radiometric dating of illite grains show that they have ages of formation quite similar to nearby continental rocks. This is further evidence to indicate that they were derived from the land.

River-transported illite is abundant in the Atlantic Ocean basin, as Fig. 5–15 indicates. In the Central Pacific, where river-transported sediment does not occur, illite is also a common constituent. Presumably it has been carried there by winds.

Another clay mineral, *montmorillonite*, comes in part from certain soils but is also formed in the ocean from the in-place alteration of volcanic glass from eruptions. In the Central Pacific where submarine volcanism is common and other sediment sources far removed, this mineral occurs in great abundance, often to the exclusion of other clay minerals. The relatively high concentrations of montmorillonite near the Mid-Indian Ridge

FIG.   5–14

Distribution of kaolinite in the fine-grained portion (<2 μ) of deep-ocean sediment. [After J. J. Griffin, H. Windom, and E. D. Goldberg, "Distribution of Clay Minerals in the World Ocean," *Deep-Sea Research*, 15 (1968), 433–59]

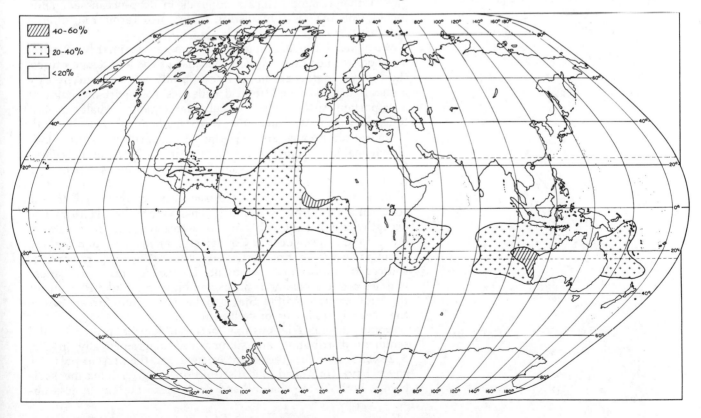

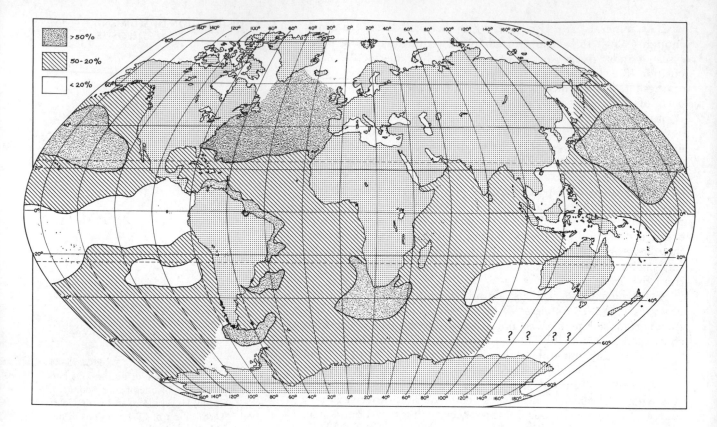

FIG. 5–15

Distribution of illite in the fine-grained portion ($<2\,\mu$) of deep-ocean sediment. (After Griffin et al., 1968)

may also be due to the high incidence of vulcanism in that region. Other common silicate minerals in deep-ocean sediments include the *zeolites (phillipsite)*, which also form from altered volcanic glass.

At higher latitudes, typical clay minerals (primarily *chlorite*) are those formed in arctic soils. Because of the low temperatures, these clay minerals are much less radically altered than are clay minerals derived from tropical soils. Ice transport of sediment is important in the polar and subpolar regions, usually in latitudes between 50° and 90° in both hemispheres. In mid-latitudes, jet streams are important transport routes for wind-blown material from the continents; in the equatorial regions (between 15°N and 15°S), trade winds transport sediment in a westerly direction.

In general, river-transported sediment is restricted to the continental margins, and most remains on the shelf or rise except where turbidity currents are active and can carry material out into the deep ocean. Conversely, manganese nodules are common in the central portions of the Pacific, far from major sources of lithogenous sediment. In the Atlantic, manganese nodules are less common; an exception is the Blake Plateau off the southeastern United States, where the Gulf Stream keeps the ocean bottom free of sediment.

Some general relationships between marine processes and sediment distribution are summarized diagramatically in Fig. 5–16. First note that the recent ice ages and the present polar ice cover have controlled the distribution of glacial marine sediments, both on the shallow and deep-ocean bottom. A less obvious effect of the present climatic regime is the formation of cold, dense bottom waters in the Antarctic that probably influence the dissolution of carbonate sediments in the deepest part of the ocean, causing dissolution in the deepest portions.

Effects of high productivity can be seen on the deep-ocean bottom and on the continental shelf. On the deep-ocean bottom, bands of diatomaceous muds in high latitudes and radiolarian muds in equatorial regions directly reflect the high biological productivity of surface waters in these regions. On the continental shelf, the abundance of recent carbonate sediment in tropical waters is again a consequence of the locally high carbonate productivity.

FIG. 5–16

Schematic representation of the general distribution of sediment in the deep and coastal ocean. [Modified in part after K. O. Emery, *Relict Sediments on Continental Shelves of World*, American Association of Petroleum Geologists Bulletin, 52(3) (1968), 445–64]

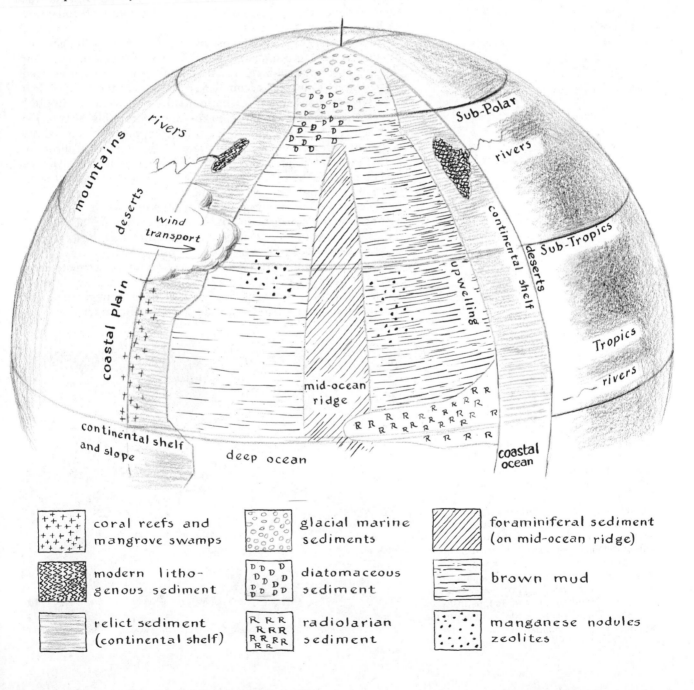

coral reefs and mangrove swamps

modern lithogenous sediment

relict sediment (continental shelf)

glacial marine sediments

diatomaceous sediment

radiolarian sediment

foraminiferal sediment (on mid-ocean ridge)

brown mud

manganese nodules zeolites

Transport of lithogenous sediment from the continents to the deep-ocean bottom is inhibited by the topography of most continental margins and the lack of modern sediment brought to the continental shelf by rivers. This is in part a consequence of the recent changes in sea level following the last retreat of the glaciers.

CORRELATIONS AND AGE DETERMINATIONS

So far we have considered sediment distributions over particular parts of the ocean bottom. This leads to questions about the ages of deposits and correlation of specific zones of equal age within sediment sequences.

In a few instances it is possible to determine sediment age directly when the sediments record yearly changes in sediment supply. These annual layers, called *varves*, are usually preserved only where sediment accumulation is rapid and where low dissolved-oxygen content of bottom waters prevents or inhibits the activity of burrowing organisms that thoroughly disturb the sediment as they feed on its organic component. Such varved sediments form in certain fjords and in the Gulf of California. A few areas on various continental slopes where oxygen-deficient waters contact the bottom are also known to have varved sediment. Where these occur, the time of sediment deposition can be determined by counting back from the present surface much as one would count growth rings on a tree stump. This type of deposit is rare in the open ocean, and other methods must be used to determine sediment age.

The most common techniques used to date or correlate sedimentary deposits employ biogenous remains such as those in Fig. 5–17. Fossils show characteristic changes through time that paleontologists use to assign ages to well-characterized as-

FIG. 5–17
Fossil remains of coccoliths and discoasters (calcareous nanoplankton) from sediment cores. In this electron photomicrograph, the fossils are magnified several thousand times; they range from 2 to 25 microns in diameter, equivalent to fine silt particles. The star-shaped fossils in the upper portion of the photograph are discoasters, the remains of extinct organisms thought to have been related to coccoliths. Disks and rings are coccoliths, an abundant constituent of nannoplankton now living in the ocean. (Photograph courtesy of Deep-Sea Drilling Project, Scripps Institution of Oceanography, under contract to the National Science Foundation)

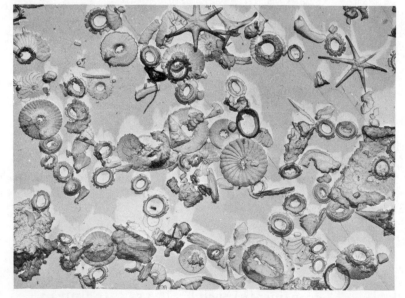

semblages of fossils. Time zones are commonly defined in terms of presence or absence of several different fossils. Microorganisms are preferred for work with marine sediments, since a large number can be recovered even from a small volume of sediment. Foraminifera are most often used although Radiolaria and coccoliths (shown in Fig. 5–18), and even pollen grains, have been used to correlate deposits.

From studies of fossil assemblages one can establish time zones and correlate deposits in widely separated areas. Eventually the history of an entire ocean basin over the past 200 million years may be worked out from clues buried in its sedi-

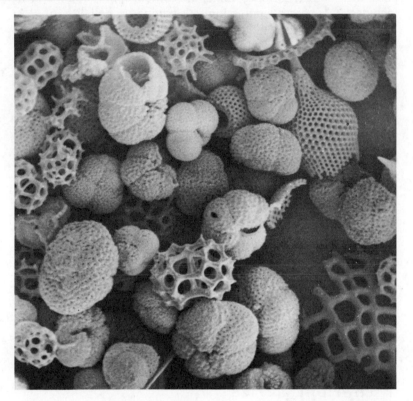

FIG. 5–18
Skeletons of foraminifera and Radiolaria taken from sediment in the western Pacific Ocean in waters 2660 meters deep. The core penetrated 131 meters into the ocean bottom. The fossils have been magnified about 160× by an electron scanning microscope. (Photograph courtesy Deep-Sea Drilling Project, Scripps Institution of Oceanography, under contract to the National Science Foundation)

ments. Where absolute ages in years are needed, radioactive substances are used as clocks. Radionuclides decay at known, unvarying rates; the time elapsed since deposition is reflected in the amount of decay product and the amount of the original radionuclide present in a deposit. Depending on the age of the sediment, different radionuclides, depicted in Fig. 5–19, are used.

For recently deposited sediment, carbon-14 is a useful radionuclide. Cosmic rays continually bombarding the earth form carbon-14 in the upper atmosphere, and all living organisms have carbon-14 atoms amidst the carbon of their tissues. When an organism dies, no further carbon is taken up by the tissues or exchanged with atmosphere or ocean water in the form of CO_2. Instead, the amount of carbon-14 in the organisms begins to decrease at a steady rate. One-half of it is gone in 5600 years (the "half-life" of this radionuclide), and after 11,200 years only one-quarter of the original amount remains. Thus by comparing the amount of carbon-14 per unit amount of carbon in a fossil with that in organisms presently living under similar conditions, it is possible to determine the time elapsed since that organism was alive. Assuming that the organism died and was incorporated in the layer when it formed, we can assign an age to the deposit. Although extremely useful for recently deposited sediment, carbon-14 has too short a half-life to permit dating of sediments older than about 30,000 years.

Dating older sediments requires the use of radionuclides with longer half-lives. Potassium-40 is extremely useful for this purpose. Most rock-forming minerals contain potassium, which includes 1.2 percent potassium-40 having a half-life of 1.3 billion years. When potassium-40 decays, 11 percent of the daughter product is argon-40, an element that is not present in the mineral when it forms. Thus, a comparison of the amount of argon-40 in a mineral with its potassium content provides a measure of the time elapsed since that mineral formed. The more argon-40 in a mineral, the greater its age. To use this radionuclide to date sediment deposits, it is necessary to date only minerals that formed when the deposit was laid down. Volcanic ash and glauconite (a micalike mineral) are often used in dating deep-ocean deposits. The age of a mineral must be further interpreted to calculate sediment age. Because of its long half-life, potassium-40 is useful only for sediments older than about 100,000 years. In general, older sediments may be dated more easily than younger ones.

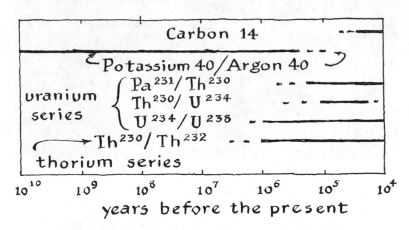

FIG. 5–19
Radionuclides used for determining sediment ages and approximate time periods for which each technique is useful.

Uranium and thorium, radionuclides with long half-lives, also occur in marine sediments and have been used to date sediment deposits. Each has many "daughter" products, formed by decay of the "parent" radionuclide. During sedimentary processes, the original relationship among these daughter products is disturbed, and equilibrium is reestablished only after a considerable period of time. The exact time depends on the nuclide involved. Careful studies of the relative abundance of these various radionuclides provide a useful measure of the time elapsed since the sediment formed. The various radionuclides involved in these decay schemes provide dating techniques for time spans longer than those measurable by carbon-14 but generally shorter than the useful range of potassium-40.

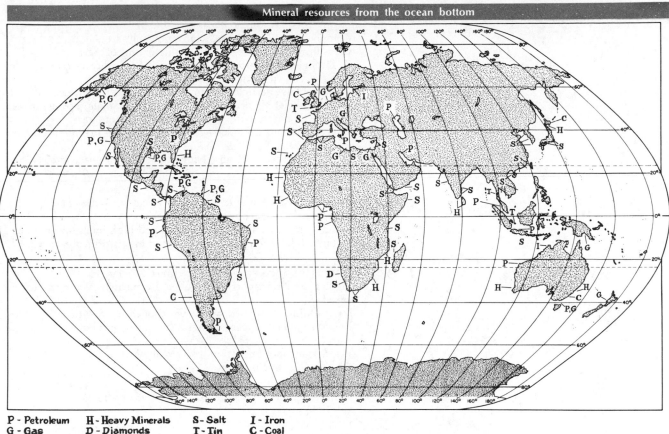

P - Petroleum   H - Heavy Minerals   S - Salt   I - Iron
G - Gas   D - Diamonds   T - Tin   C - Coal

There are three modes of mineral occurrence in sediment—surficial deposits recoverable by simple dredges; fluids or soluble minerals that can be recovered through drilling operations; and hard minerals in sediment deposits or rock. Development of these resources has been largely restricted to the submerged continental margins, where the ocean is relatively shallow and land-based processing plants are close at hand.

Throughout the 1960's, most of the world's recovery of mineral resources from submerged lands took place in the vicinity of the United States. There has been little mining of the ocean bottom, except for extensions of a few land deposits which have been exploited by tunneling out under the adjacent ocean. Techniques for drilling in deep-ocean waters, explored in the 1960's, have developed rapidly since that time.

Petroleum and natural gas account for most of the mineral products recovered from the ocean bottom. Together with sulfur (which is produced after first being melted in rocks) and water, most of the minerals have been produced as fluids. Improved technology has permitted petroleum production to expand into deeper waters. Petroleum has been detected in deep-ocean sediments and may be produced from great depths in the future.

Next in terms of value are sand, gravel, and shell produced from shallow continental-shelf deposits. These are especially useful where land deposits are commonly covered by the city itself. Carbonate-shell deposits are used for making cement in some areas. Diamonds, gold, tin, and other minerals have been produced from now-submerged river channels and beaches on the continental shelf.

Surficial deposits of the deep ocean have been seriously considered as mineral sources. Deposits of manganese nodules may be mined for their copper and nickel as well as their iron and manganese. Nodules containing phosphate may be produced from shallow, isolated banks where the nearby land is deficient in materials for fertilizers.

137

SUMMARY OUTLINE

Marine sediment—derived from continents, covers ocean bottom

Origin and classification of marine sediment

Composition—lithogenous, biogenous, hydrogenous

Classification by size—sand, mud, sorting of particles

Lithogenous particles—weathering of silicate rocks yields dissolved constituents, sands, and clays

Biogenous particles—elements removed from seawater to build skeletons of various plants and animals; sediment consists of oozes, with larger fragments more rare

Hydrogenous particles—formed by Fe–Mn precipitation from seawater

Minor sources—cosmogenous debris from space; man's contribution

Sediment budget

Rivers transport about 20 billion tons of sediment per year, plus dissolved fraction

Wind transport important in desert and high mountain regions

About 2 billion tons of biogenous sediment formed each year

Sediment transport

Settling velocity—controlled by particle size

Erosion and transport—controlled by particle size and current speed

Most sediment trapped near river mouth or in estuary

Relict sediment covers about 70% of continental shelves of world

Wind transport important for small particles in mid-latitudes; removal by precipitation

Accumulation of sediment in the ocean

Continental-shelf sediments—deposited at rates of about 10 cm/1000 years

Deep-ocean sediments—deposited at rates of 1 mm to 1 cm/1000 years, 600 m average thickness; thick deposits seldom found in trenches

Turbidity currents

Dense, sediment-laden mixtures of water moving along the bottom

Form characteristic deposits—graded bedding, displaced shallow-water plant and animal remains

Important means of sediment transport across continental shelves, especially narrow shelves

Bury ocean-bottom topography forming smooth plains

Distribution of sediment deposits

Biogenous sediment on tropical continental shelves

Volcanic sediment near areas of volcanic activity, transported by water and by stratospheric winds

Rapidly accumulating lithogenous sediment covers about 25% of ocean bottom

Slowly accumulating sediment (pelagic) covers about 70% of ocean bottom, both biogenous and lithogenous

Accumulation of biogenous sediment controlled by production, destruction, dilution, and removal of biogenous particles

Clay minerals—layered silicates from different climatic zones

Important in the deep ocean; moved by all transport processes

Contributed by rivers in the Atlantic, by winds and volcanoes in the Pacific

Manganese nodules in the central Pacific

Map shows worldwide distributions of sediment types

Correlation and age determination
Varve counting and paleontological techniques provide relative ages
Radiometric techniques provide absolute ages

SELECTED REFERENCES

ARRHENIUS, G. 1963. "Pelagic Sediments," in M. N. Hill (ed.), *The Sea*. Vol. 3. Interscience, New York, pp. 655–727. Thorough discussion of deep-ocean sediments and their origin; technical.

SHEPARD, F. P. 1963. *Submarine Geology*, 2nd ed. Harper & Row, New York. 557 pp. General treatment of sediments in ocean, emphasizing continental-shelf and adjacent environments; technical, descriptive.

TUREKIAN, K. K. 1968. *Oceans*. Prentice-Hall, Inc., Englewood Cliffs, N.J. 120 pp. Emphasizes chemical aspects of sedimentary processes; elementary.

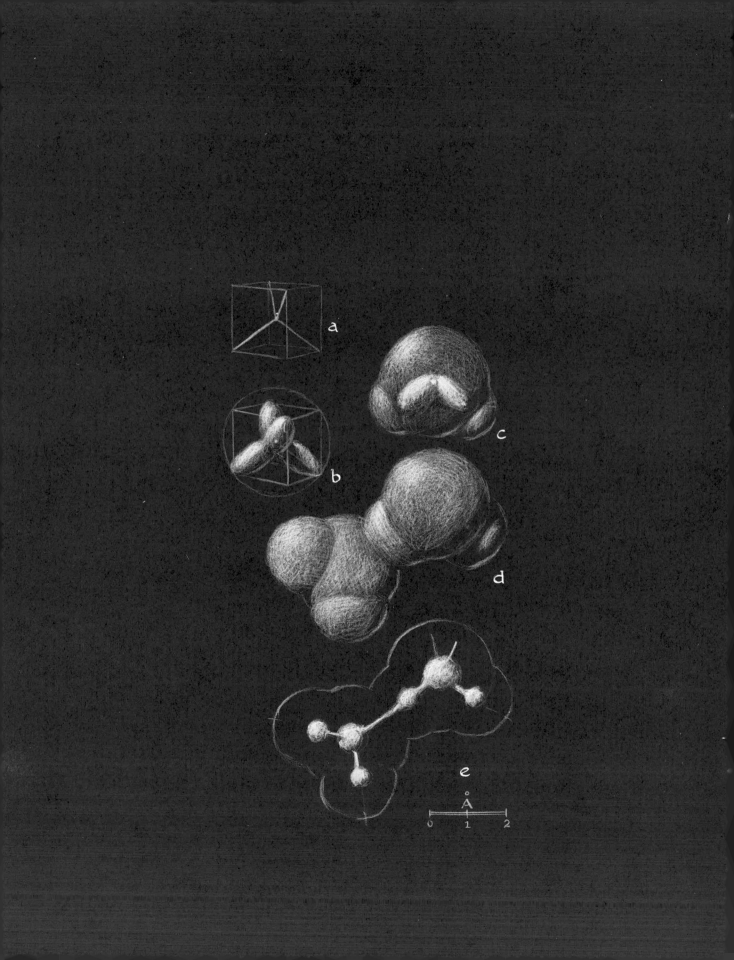

a

b

c

d

e

Å

0   1   2

SEAWATER

Different ways of representing water molecules. To demonstrate the structure of a single molecule, bond directions (a) and (b) and the electron cloud (c) are shown. Two molecules bonded together as they would occur in ice appear as electron clouds (d) and as "Tinker-Toys" (e).

In previous chapters we have examined the earth's surface and specifically discussed ocean basins—the great depressions between continents where seawater has collected over billions of years. The characteristic feature of the ocean, however, is not the basin that contains it but seawater itself. To understand the global importance of the ocean we must understand the physical and chemical properties of water and seawater.

In this chapter we shall first consider pure water and then the salts dissolved in ocean waters. Some properties of water are little affected by addition of the sea salts. An example is water's capacity to absorb and give up large quantities of energy as heat with little change in temperature. Other characteristics of seawater, such as its suitability for marine life, are strongly influenced by the dissolved materials. Sometimes even minute differences in chemical composition can alter its biological properties. Processes of living organisms in turn affect the abundance of dissolved gases and salts in seawater. Finally, some of the important properties of ocean water are controlled in part by the structure of the water and the behavior of dissolved salts—specifically transmission of sound necessary for echo sounding or using sound to locate fish or submarines.

Seawater is a complex substance. On the average, it is about 96.5 percent water containing 3.5 percent salt, a few parts per million of living things, and perhaps an equal amount of dust. In fact, we are describing a dirty "living soup."

MOLECULAR STRUCTURE OF WATER

Two hydrogen atoms and one oxygen atom combine to form one water molecule. Water's unusual properties result from the structure of this molecule, specifically the associated electron cloud. This cloud has a definite shape, forming a four-cornered molecule, pictured in Fig. 6–1 and perhaps best described as resembling the four-pronged "jacks" that children play with. The oxygen atom is relatively large and accounts for most of the molecule's volume. The two hydrogen atoms are much smaller and are partially buried in the electron cloud of the oxygen atom, forming two of the four "prongs" of the molecule.

The other two "prongs" are formed by unshared electrons of the oxygen atom. The hydrogen atoms are on one side of the molecule, the unshared electrons on the other side. Thus, viewed on an atomic scale, a water molecule is essentially negative on one side (the side with the unshared electrons) and positive on the other (the side with the hydrogen atoms). A molecule whose electronic charges are separated is called a *polar molecule*. It responds to electrical charges in its vicinity. The unshared electrons on one molecule tend to form rather weak, so-called *hydrogen bonds* with hydrogen atoms of an adjacent molecule. This tendency for hydrogen bonding between water molecules gives water many of its unusual properties, ranging from its large capacity for absorbing heat, its abnormally high boiling point and freezing point, and its outstanding capacity to dissolve crystals that are held together by *ionic bonds*; many of these properties are listed in Table 6–1.

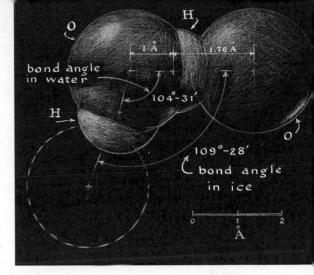

FIG. 6–1

Schematic representation of the water molecule. Note the separation between the negatively charged portions of the molecule and the positively charged portions.

Table 6–1

COMPARISON OF PHYSICAL PROPERTIES OF WATER AND A "NORMAL" LIQUID (n-heptane)*

	Water $(H_2O)_n$	Normal heptane (C_7H_{16})
Molecular weight	$(18)_n$	100
Density (g/cm³)	1.0	0.73
Boiling point	100°C	98.4°C
Melting point	0°C	−97°C
Specific heat (cal/g per °C)	1	0.5
Heat of evaporation (cal/g)	540	76
Melting heat (cal/g)	79	34
Surface tension, 20°C (dynes/cm)	73	25
Viscosity, 20°C (poise)	0.01	0.005

*Modified from Dietrich, 1963.

Crystals and molecules are attracted to one another and held together by various forces. The weakest of these results from fleeting electronic interactions between molecules. These so-called *van der Waals bonds* form, break up, and reform easily; although present in water, they are greatly overshadowed in importance by the stronger hydrogen bonds.

Ionic bonds form between adjacent molecules (or atoms) which have lost or gained electrons thus acquiring either a positive charge, where an electron has been lost, or a negative charge, where an electron has been gained. A familiar example of ionic bonding is the association between sodium (Na^+) and chlorine (Cl^-) to form table salt, sodium chloride. The bond is due to strong attraction between closely spaced molecules of opposite charge. To get some idea of the strength of these bonds, consider the amount of energy required to break each type of bond in one *mole* of a substance, e.g., water. One mole of water weighs 18 grams, or about ½ ounce—that is its formula weight in grams, oxygen having an atomic weight of 16 and each hydrogen atom having an atomic weight of 1:

van der Waals bonds:	0.6 kilocalorie per mole
hydrogen bonds:	4.5 kilocalories per mole
ionic bonds:	10's of kilocalories per mole

At temperatures and pressures characteristic of the earth's surface, water is one of the few substances that we commonly see in all three forms of matter—solid, liquid, and gas. In fact, the range in earth-surface temperatures is controlled in many open areas by large quantities of water changing from one state to another. Before considering that part of the story, let us first understand the differences between the states of matter, using water as our example.

Water vapor—the gas—is probably the easiest to understand. In water vapor, each molecule exists separately and is little affected by other molecules. Therefore, each molecule is free to move with little restriction, except for the walls of the container in which we enclose it. This freedom to move accounts for the characteristics of a gas: it has neither size nor shape but expands readily to fill any container in which it is placed.

Water molecules striking the sides of a container exert *pressure*. Pressure can be increased in two ways: we can put more material in the container or shrink it, thereby increasing the rate of molecules striking the sides of the container; alternatively, we can make the molecules move faster and strike the container walls more frequently. The first of these we do by placing more material (even another gas) in the container, the second by increasing the temperature, which makes molecules move more rapidly.

At the opposite extreme in internal order is the solid—in this case, *ice*. Solids have a definite size and shape. Most are also crystalline—that is, they have a definite internal structure. Usually solids break, or perhaps bend, when enough force is applied. This resistance to flow, or deformation, results from the orderly internal structure of the solid. Each atom, or molecule, has a position which it occupies for long periods of time. These atoms or molecules cannot move readily from position to position, nor rotate in a given position. But at room temperatures there is always some vibration of atoms in their fixed position. Some small amount of movement is possible if there is a hole nearby into which a molecule can move. Just as with water

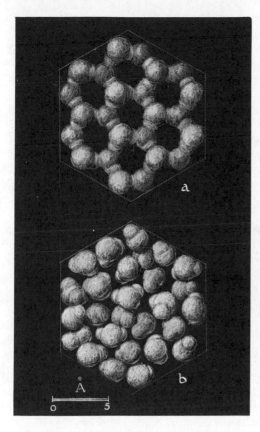
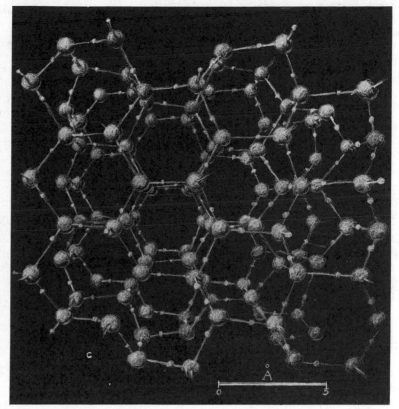

FIG. 6–2
Crystal structure of ordinary ice showing the fixed positions of water molecules (a). Note the six-sided rings formed by 24 water molecules. In the same volume of liquid water, 27 molecules would be present (b). The ice lattice is shown in (c).

vapor, the movements of the atoms or molecules increase with increasing temperature, so that movement between positions also increases with rising temperature.

In ice, the water molecules are held together by hydrogen bonds, as shown in Fig. 6–2. The large oxygen atoms have definite positions, and the hydrogen atoms in each molecule are also oriented in a regular manner. The oxygen atoms form six-sided puckered rings which are arranged in layers, each layer a mirror image of the adjacent layer. This forms a fairly open network of atoms and thereby gives ice a somewhat lower density (approximately 0.92 grams per cubic centimeter) than that of liquid water (approximately 1 g/cm³) because the molecules are not as tightly packed together.

Despite the openness of the internal structure of ice crystals, impurities such as sea salts are not readily incorporated into the holes. Consequently salt is excluded from ice formed from seawater. The excluded sea salts remain in pockets of unfrozen liquid (brine).

Liquid water has several anomalous physical properties (see Table 6–1) intermediate between those of solids and gases. It

FIG. 6–3

Schematic representation of liquid water at atmospheric pressure. Note the two types of water structures: areas of bonded molecules (ordered regions) and free water molecules between them.

FIG. 6–4

Melting and boiling points of water and chemically similar compounds. (After Horne, 1969)

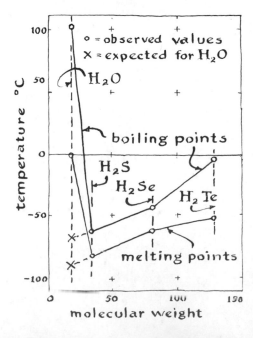

apparently consists of a mixture of two different forms of molecular aggregates, as shown in Fig. 6–3. The first component consists of clusters of hydrogen-bonded water molecules. These clusters form and reform very rapidly—10 to 100 times during one-millionth of a microsecond (10^{-10} to 10^{-11} second). Although the lifetime of any individual one is extremely short, clusters persist long enough to influence the physical behavior of water. These structured portions of water are less dense than the unstructured portion. In some respects, this lowered density of the structured portion arises from the same causes that make ice less dense than liquid water. It seems likely, however, that the ice structure is not exactly the same as that of the flickering clusters of molecules. Although the clusters form and reform rapidly, they persist in liquid water until broken up by external forces. The clusters apparently disappear when the pressure exceeds about 1000 atmospheres (a pressure reached only in the few deepest trenches of the ocean floor) or at temperatures exceeding 100°C (at atmospheric pressure), where liquid structures break up and molecules escape to form a gas.

The other constituent of liquid water is unstructured or "free" water molecules which surround the structured portions. These molecules move and rotate without restriction. Interactions with nearby molecules are weaker than in the structured portion. The "free" water portion of liquid water is denser than the structured portion, since the molecules fit more closely together. The relative proportion of the structured and unstructured portions of liquid water vary with changes in temperature, pressure, and salt content and composition.

Because of its unusual molecular structure, liquid water is strikingly different from hydrogen compounds of elements with chemical properties similar to oxygen, as indicated in Fig. 6–4. The melting point and boiling point of water occur at temperatures 90° and 170°C higher, respectively, than might be predicted from studying the chemical behavior of these other compounds. If water were a "normal" liquid, it would occur only as a gas at earth-surface temperatures and pressures.

TEMPERATURE EFFECTS ON WATER

Temperature affects the internal structure of water and its properties. Much of the heat energy absorbed by water is used up in changing the internal structure so that water temperature rises less than other substances, after absorbing a given amount of heat. The large amount of water on earth acts as a climatic buffer preventing the wide variations in surface temperature experienced by our neighbors in space. For example, surface temperatures on the moon go from about +135°C at noon during the lunar day to about −155°C during the lunar night. On earth, the highest temperature ever recorded was 57°C, at Death Valley, California (July 10, 1913), and the lowest was −68°C, at Verhoyansk, in Russia (February, 1892).

Earth's limited temperature range is controlled primarily by the abundance of water on the earth's surface, since much of the incoming solar radiation goes into evaporating water and melting ice. Let us see what happens to water during these changes and how this affects the earth's heat balance.

First, we must define a measure of heat, the *calorie*, as the amount of heat (a form of energy) required to raise the temperature of 1 gram of liquid water by 1 degree Celsius (1°C). This means that we can change the temperature of 1 gram of water by 50°C by supplying 50 calories of heat. Alternatively, we would change the temperature of 50 grams of water by 1°C with the same amount of heat.

Breaking bonds in ice and liquid water requires energy, usually heat. Conversely, heat is released upon their reformation. Consider what happens when 1 gram of ice just below the freezing point is heated slowly, keeping track of the amount of heat added.

As we add heat to ice, its temperature increases about 2°C for each calorie of heat we add (see Fig. 6–5). When the ice reaches its melting point, 0°C, the temperature remains constant and instead the ice begins to melt to liquid water. As long as ice and liquid water exist together in a container, the temperature of our system remains fixed at 0°C. After adding about 80 calories per gram, the last bit of ice melts, leaving only liquid water.

As we continue to apply heat, the temperature rises again, but at a slightly lower rate: 1°C per calorie of added heat. This rate of temperature change holds nearly constant between 0°C and 100°C. At 100°C, the boiling point, the temperature rise again ceases and gas (water vapor) forms. At the boiling point, the same situation occurs that we experienced at the melting point. The temperature remains fixed as long as liquid and gas are present. After adding about 540 calories per gram, the last of the water evaporates. If we have been careful to capture the vapor (steam) and continue heating it, we find that the temperature rises much more rapidly: about 2°C per calorie of heat added—approximately the same as we experienced when heating ice.

Heat added to the water is taken up in two forms. One is *sensible heat*, heat that we detect either through our sense of touch or with thermometers. This change in temperature results from the increased vibration of molecules or their motion in the gas state.

The other form is called *latent heat*, which represents the energy necessary to break bonds in the water structures. As we have seen, at both the melting point and the boiling point of water there was no change in temperature as long as two states of matter existed together. The added energy went into breaking the bonds in the disappearing phase. This is called latent heat because we get back exactly the same amount of heat when the process is reversed. Condensing water vapor to form liquid water releases 540 calories per gram at 100°C; freezing water at 0°C releases 80 calories per gram. The difference between the latent heats of melting and evaporation arises from the fact that only a small number of the hydrogen bonds are broken when ice melts. All bonds are broken when water evaporates.

Although ice freezes at 0°C and water boils at 100°C, it is possible for molecules to go from vapor to solid or liquid at other temperatures. The processes involved are similar to those

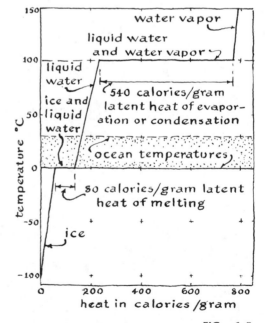

FIG. 6–5
Temperature changes when heat is added or removed from ice, liquid water, or water vapor. Note that the temperature does not change when mixtures of ice and liquid water or liquid water and water vapor are present. (Redrawn from Gross, 1971)

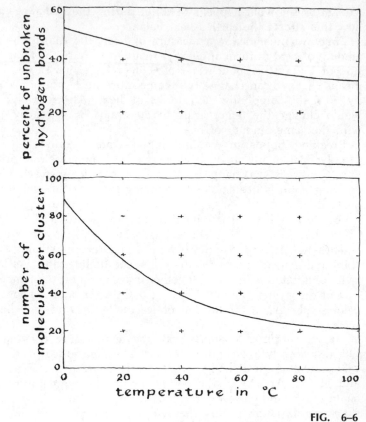

FIG. 6–6

Effect of changes in temperature on the relative abundance of unbroken hydrogen bonds, cluster size, and the number of molecules per cluster. [Data from G. Nemethy and H. A. Scheraga, "Structure of Water and Hydrophobic Bonding in Protein," *Journal of Chemical Physics*, 36 (1962), 3382–3400]

described before, but the amount of heat involved is different. For example, the latent heat of evaporation changes as follows:

Temperature (°C)	Latent heat of evaporation (cal/g)
0	595
20	585
100	539

It takes more energy to remove a water molecule from liquid water at 0°C or 20°C than it does at 100°C. Molecules vibrate less vigorously at lower temperatures and hence it takes more energy to break the bonds. This is an important process in the ocean since most water evaporates from the ocean surface at temperatures around 18 to 20°C. If water had to reach 100°C before evaporating, we would have no water vapor in the atmosphere.

Both cluster size and number of molecules per cluster decrease with increasing temperature, as Fig. 6–6 illustrates. It is interesting to note in this figure that a large number of hydrogen-bonded water molecules remain bound together even after ice melts at 0°C. There is some evidence that subtle changes in structure may occur at other intermediate temperatures.

DENSITY

Whether a substance sinks or floats in a liquid is determined primarily by its *density* (mass per unit volume, expressed in grams per cubic centimeter). A substance less dense than its surroundings tends to move upward; in other words, it floats. If its density is greater than its surroundings, it displaces less mass than its own mass, settles out of the liquid, and sinks. The sinking rate is determined by its relative density and the resistance it experiences while moving through the water. A small object generally experiences more drag per unit mass than a large object, and thus a small particle takes much longer to sink through the fluid. A large object experiences relatively less drag and thus sinks more rapidly.

A *density-stratified system of fluids* is one in which lighter fluids float on heavier ones. A mass of fluid may sink through materials less dense than itself until it reaches a zone where the fluid below is more dense and the fluid above less dense. At this point, there is no force acting on the fluid and it tends to remain at that level. Furthermore, vertical displacement of the fluid will be countered by forces tending to keep it at the same relative position. For this reason the density of fluids and solids determines in most instances whether a body of fluid or a solid remains in the upper layers or sinks to the bottom. Density differences drive currents in the ocean and atmosphere, as we shall see in later chapters.

Factors controlling seawater density include temperature, salinity, and pressure. Temperature influences density by increasing the vibrations and movements of atoms and molecules. As temperature increases, molecules vibrate or move more, and thus effectively occupy more volume. For a constant mass, the density decreases as the volume increases. Remember that density is a ratio; it decreases either because of a decrease in the numerator or an increase in the denominator.

Ice, water vapor, and normal seawater behave like most materials, becoming less dense with increasing temperature. Therefore, in any stable structure of these substances the least dense material will generally be the warmest—if salinity is constant—and will occur at the top.

Pure liquid water exhibits a well-known anomalous density maximum at 4°C that is worth mentioning here. It is an important factor in the fresh-water lakes of cold regions but not in the ocean. The anomalous density maximum does not occur in seawater of salinity greater than $24.7^0/_{00}$.

Ice at 0°C has a density of about 0.92 gram per cubic centimeter. Like most substances, ice becomes less dense as its temperature increases, illustrated in Fig. 6–7. Since ice is about 8 percent less dense than liquid water, it floats on water. When liquid water at 0°C is warmed, its density increases slightly until it reaches a density maximum at 3.98°C. After that temperature

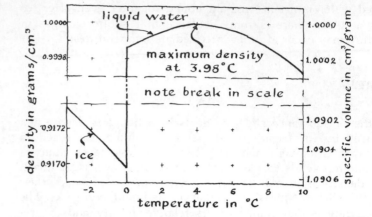

FIG. 6–7
Effect of temperature on density, expressed in g/cm³, and specific volume, expressed in cm³/g, for both ice and pure liquid water.

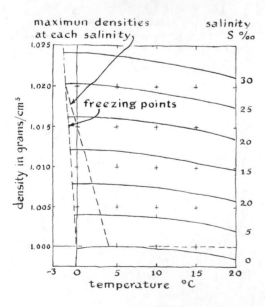

FIG. 6–8
Effect of salinity on temperature of maximum density and initial freezing temperature of seawater.

is reached, the density decreases as temperature increases further. In a fresh-water lake, water cooled to 4°C sinks to the bottom and there is protected against further cooling. This keeps the bottom of the lake ice-free and supplies large amounts of oxygen, dissolved in the water when it was in contact with the atmosphere, to bottom-dwelling animals. After the basin is filled with 4°C water, further cooling forms a less dense water layer on the lake surface, which eventually freezes if the weather remains cold long enough. If water behaved like most fluids and its density steadily increased with decreasing temperature, the coldest water would be found in the bottom of the lake and ice would also sink. This mode of freezing, with ice sinking to the bottom, would be far more efficient than freezing from the top. Most lakes at high altitudes and in cold climates would likely remain frozen solid during the summer.

As we have previously seen, temperature has a profound effect on the relative proportions of the structured and "free" portions of liquid water. Since these two constituents have different densities, the temperature of water affects its density. One explanation for the anomalous density maximum in pure liquid water is that there is a relatively rapid increase in the amount of structured portion of water below 4°C. This makes water less dense as the denser "free" portion of the mix is diminished.

Either adding salt to water (as in Fig. 6–8) or increasing the pressure causes a downward shift in the temperature of maximum density. The amount of salt in seawater is sufficient to eliminate this anomalous feature. Furthermore, adding salt to water increases its density, and the salinity of seawater is commonly the dominant factor controlling its density.

COMPOSITION OF SEA SALT

Virtually anyone with access to seawater can make his own crude salinity determination. The simplest way, though not the most precise, is to place a known amount of seawater in a pan and let it evaporate. The result is a gritty, bitter-tasting salt that never completely dries. The weight of the salt in a known volume of seawater provides a crude measure of salinity.

Determination of the composition of this salt residue as shown schematically in Fig. 6–9 has occupied many excellent chemists since the first analyses were made in 1819. By now, at least traces of nearly all the naturally occurring elements have been detected in sea salts as shown in Fig. 6–10, and the job is still not finished. Most of the analyses have been made on surface waters collected near the coast. Such waters are most likely to exhibit variability in the chemical composition of their dissolved salts. Furthermore, the development of each new analytical instrument opens new horizons for additional, more detailed studies of ocean chemistry.

Six constituents (Cl^-, Na^+, SO_4^{2-}, Mg^{2+}, Ca^{2+}, K^+) comprise 99 percent of sea salts. The other elements present in sea salt add up to only 1 percent. As Fig. 6–9 shows Na^+ and Cl^- alone make up 86 percent of sea salt.

To express the amount of dissolved salt present in seawater, oceanographers commonly use two concepts—chlorinity and salinity, both based on chemical determinations.

Since the oceans are well mixed, sea salts have a nearly constant composition. Therefore, we can use the most abundant constituent, Cl^-, as the index of the amount of salt present in a

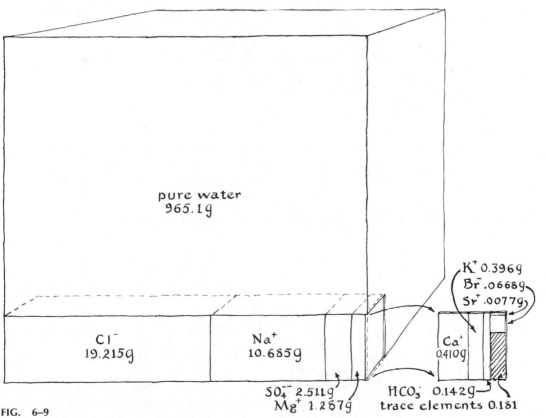

FIG. 6–9

Major and minor constituents of seawater. The numbers indicate the amount of each constituent in grams, contained in a kilogram of seawater (S = 34.7%).

volume of seawater. *Chlorinity* (*Cl*) is defined as the amount of chlorine, in grams, in 1 kilogram of seawater (bromine and iodine are replaced by chlorine). Chlorinity is expressed as parts per thousand, or parts per mille, written $^0/_{00}$.

Chlorinity indicates the amount of chlorine. Oceanographers commonly convert this to salinity, a measure of the total amount of dissolved salt. *Salinity* (*S*) is defined as the *total amount of*

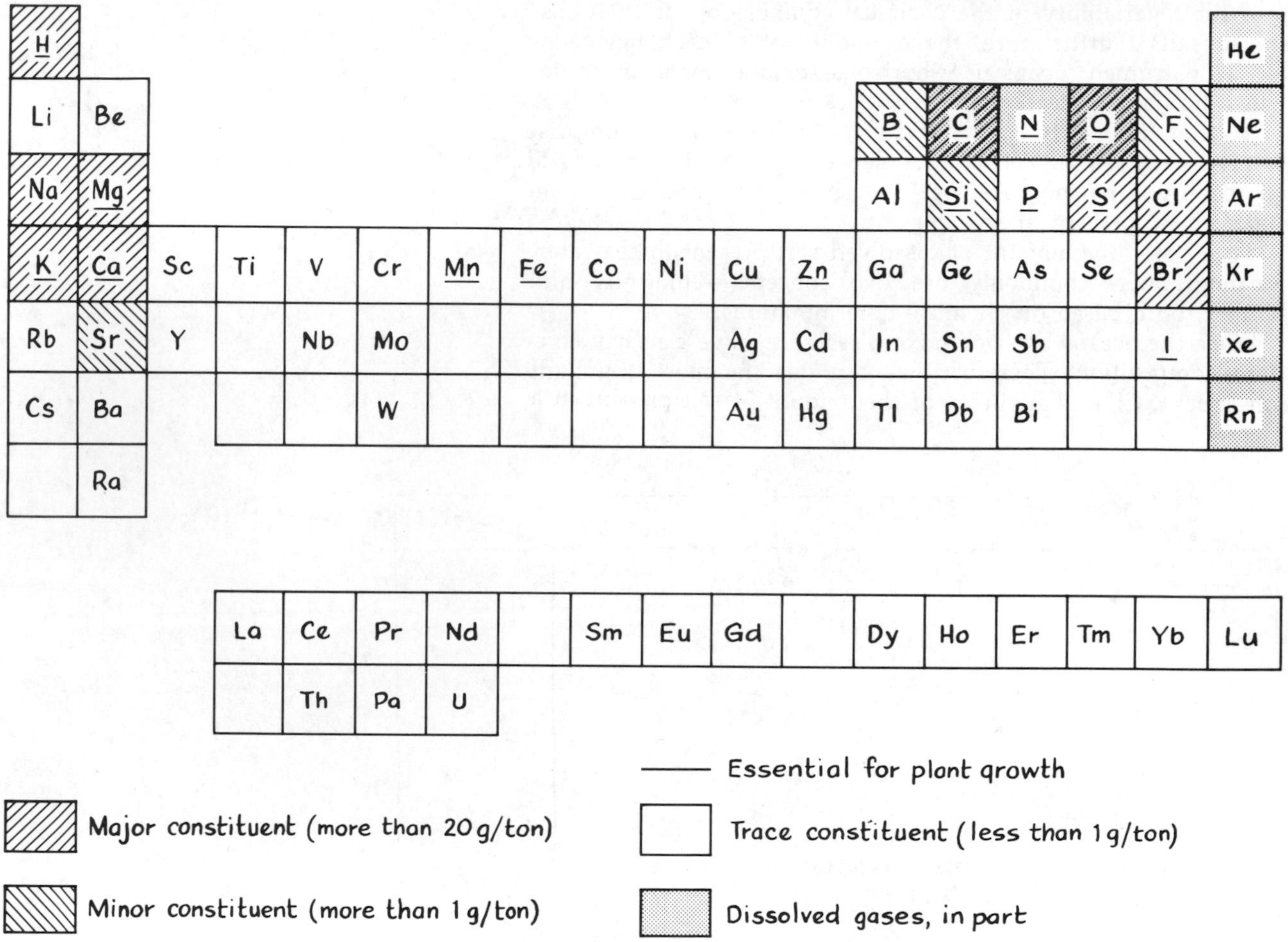

H																	He
Li	Be											B	C	N	O	F	Ne
Na	Mg											Al	Si	P	S	Cl	Ar
K	Ca	Sc	Ti	V	Cr	Mn	Fe	Co	Ni	Cu	Zn	Ga	Ge	As	Se	Br	Kr
Rb	Sr	Y		Nb	Mo					Ag	Cd	In	Sn	Sb		I	Xe
Cs	Ba			W						Au	Hg	Tl	Pb	Bi			Rn
	Ra																

La	Ce	Pr	Nd		Sm	Eu	Gd		Dy	Ho	Er	Tm	Yb	Lu
	Th	Pa	U											

—— Essential for plant growth

▨ Major constituent (more than 20 g/ton)

▧ Minor constituent (more than 1 g/ton)

☐ Trace constituent (less than 1 g/ton)

▦ Dissolved gases, in part

FIG. 6–10
Elements detected in seawater. Most non-conservative elements are involved in biological processes (underlined). (Redrawn from Gross, 1971)

solid material, in grams, contained in 1 kilogram of seawater (iodine and bromine are replaced by chlorine and all organic matter destroyed). Salinity is also commonly expressed as parts per thousand.

Salinity (*S*) and chlorinity (*Cl*) are both measures of the saltiness of seawater. This relationship can be expressed mathematically by

$$S(^0/_{00}) = 1.80655 \times \text{chlorinity}$$

For example, if a seawater sample had a measured chlorinity of

$20^0/_{00}$, its salinity would be calculated as follows:

$$S = (20 \times 1.80655) = 36.13^0/_{00}$$

Using chemical techniques, oceanographers can determine chlorinity within $\pm 0.01^0/_{00}$. Laboratory instruments (salinometers) for measuring conductivity by physical means can determine chlorinity to within $\pm 0.005^0/_{00}$ or better. For this reason, among others, salinometers are commonly used for salinity determinations on oceanographic ships and at shore-based laboratories.

While the relative proportions of major elements in sea salts change little in the ocean, physical processes can change the amount of water in seawater. Therefore, salinity and concentrations of the so-called *conservative properties*, which are not involved in biological processes, are changed by:

1. evaporation and precipitation (rainfall, snow);
2. formation of insoluble precipitates whereby formerly dissolved elements or compounds settle out of seawater;
3. mixing of water masses having different salinities;
4. diffusion of dissolved materials from one water mass to another; and
5. movement of water masses within the ocean.

Salinity of seawater is most variable near the air–sea interface, at boundaries of ocean currents, and in coastal areas. Even though the composition of seawater changes little, those slight deviations that can be detected are extremely useful to trace movements of water masses from their source to areas where they mix or otherwise lose their identity.

Many elements are present in variable concentration in seawater; these are the *nonconservative properties*. The variability of their concentrations is caused primarily by biological processes. Some nonconservative elements are incorporated in nonliving particles which settle out of the ocean within a few months to a few years, to be incorporated in sediment or eventually returned to the water following decomposition of the particles.

Nonconservative constituents associated with inorganic particles are commonly removed fairly near their point of entry into the ocean. For example, Al is usually associated with rock or soil particles most of which settle out near the mouth of the river that brought them to the ocean. If the particles are wind-transported and enter the ocean through its surface, the elements they contain are likely to be more widely dispersed than if brought in by rivers. Wind-blown particles from continents are most abundant within a few hundred kilometers of the coast downwind from the source.

Elements used by organisms commonly exhibit a characteristic distribution pattern. Such elements are usually depleted near the surface, where they are taken up by organisms growing in the *photic zone*, the sunlit surface zone where photosynthesis takes place. After an organism dies, it sinks out of the photic zone to decompose in deeper water or on the bottom. When the organism decomposes, the element is released. This gives rise to the typical variation in abundance of such nutrient elements as P, N, and Si, which are depleted at the surface and enriched at intermediate depths (see Chapters 14 and 17).

This characteristic depletion of nutrients in surface waters limits biological activity in many parts of the world ocean. When the nutrient elements are brought back to the ocean surface, in areas with plenty of sunlight, intensive biological production usually occurs.

SALT COMPOSITION AND RESIDENCE TIMES

Each year, rivers bring about 4 billion tons (4×10^{15} grams) of salts to the ocean. Nearly all the NaCl in river water is re-cycled sea salt that fell on land in rain, coming from sea-salt particles derived originally from the sea surface. The remaining salts dissolved in river water come from the chemical breakdown (weathering) of rocks on land. Although this amount of salt is only about one-thousandth of the total amount of salt in the ocean, it would seem reasonable to find that seawater is getting saltier as the earth grows older.

All our data, however, indicate that the salinity of seawater has changed little. Direct measurements of salinity go back only a few hundred years and indicate no appreciable change. But this short period of observation is inadequate to decipher the ocean's billions of years of history.

Marine fossils preserved in rocks provide the best long-term information available to us. Fossils in these rocks—once marine muds—are usually found to be similar or closely related to organisms that now live in open-ocean waters. This bit of evidence, combined with some chemical data, suggest that the salinity of the open ocean has changed little, if at all, during the past half-billion years of ocean history. Salinity of seawater before the time when animals developed preservable skeletons remains essentially unknown.

Since rivers are delivering salt to the ocean while seawater salinity remains unchanged, salt must be removed at about the same rate at which it is supplied, an example of the *steady-state condition* of the world ocean which is changing little—if at all—through time. Some of these salt-removing processes apparently involve complex chemical reactions with sediments or particles suspended in the ocean. In other cases, seawater is evaporated in isolated basins to form salt deposits in arid regions.

A useful concept for analyzing substances in seawater is *residence time*, the time required to replace completely the amount of a given substance in the ocean. We can develop this concept in either of two ways, using either the rate of salt addition by rivers or the rate of removal of elements incorporated in sediments depositing on the ocean bottom. Using the second option, we can define residence time (T) as:

$$T = \frac{M}{R}$$

where M is the total amount (in grams) of the substance in the ocean and R is the rate of removal by sediment (in grams per year). For example, we can calculate the residence time for Na^+ in seawater as follows:

$$T_{Na} = \frac{M_{Na}}{R_{Na}} = \frac{1.5 \times 10^{22}\ g}{5.7 \times 10^{13}\ g/yr} = 2.6 \times 10^8\ yr$$

Sodium's calculated residence time of 260 million years is one of the longest residence times for an element in the ocean. Other elements, such as Al, have residence times of about 100 years.

Residence time of an element in the sea is apparently related to its chemical behavior. Such elements as Na^+ which are little affected by sedimentary or biological processes generally have residence times of many millions of years. Elements used by organisms or readily incorporated in sediments tend to have much shorter residence times, ranging from a few hundred to a few thousand years.

We can even define a residence time for water. There is a net removal, due to evaporation, of a layer of water about 10 cm thick from the ocean surface each year. This water falls on the land and returns to the ocean through river runoff. Recall that the ocean has an average depth of about 4000 meters. From this we obtain an approximate residence of 40,000 years for water.

SALT IN WATER

Natural waters are rarely pure, and usually contain some salt. River waters contain on the average about 0.01 percent dissolved salts, and average seawater contains about 3.5 percent ($35^0/_{00}$) of various salts. Even rainwater usually contains small amounts of salts and gases that it has dissolved while falling through the atmosphere. In large measure these impurities are a consequence of the remarkable solvent powers of water.

Sodium chloride (common table salt) is the most abundant constituent of sea salts. (In fact, many experimentalists use NaCl alone to make a simplified seawater for experimental work, thereby avoiding some of the problems of making precise measurements in our dirty "living soup" of seawater.) Sodium chloride crystals consist of sodium ions (Na^+) and chloride ions (Cl^-) tightly bound together by ionic bonds. These bonds are the result of strong attraction of unlike charges on nearby ions. The charges arise because the ions have lost or gained an electron. Sodium readily loses an electron, forming a sodium ion. Chlorine, on the other hand, readily accepts an extra electron, forming a chloride ion, Cl^-. Water molecules disrupt these bonds by orienting themselves around ions and effectively shielding adjacent ions from the influence of each other. As a result, we see that the crystal has disappeared and we say that it has *dissolved*.

Dissolving salt in water affects its behavior in several ways. We have already discussed the effect on temperature of maximum density. Among the other properties affected by salts are:

1. *temperature of initial freezing*—decreased by increased salinity;
2. *vapor pressure*—decreased by increased salinity; and
3. *osmotic pressure* (water molecules move through a semipermeable membrane from a less saline solution to a more saline solution; the opposing pressure necessary to top this movement is called the osmotic pressure).

Consider the temperature of initial freezing. Pure water freezes at 0°C and the temperature of the water-ice mixture remains fixed until there is only ice. In seawater, the temperature of initial freezing is lowered by increased salinity. Furthermore, salts are excluded from the ice and remain in the liquid, causing the brine to become still more salty. Hence the temperature of freezing for the remaining liquid is still lower and the temperature of the system must drop before additional ice forms. Consequently, seawater has a lower temperature of initial freezing and no fixed freezing point as observed in pure water .

Salts also affect the internal structure of water. Some dissolved ions such as Na^+ and K^+ favor the unstructured portions while other ions such as Mg^{2+} favor the structured portions. The effect of changing the relative proportions of these constituents can be seen in changes of such properties as *viscosity*, the internal resistance of a liquid to flow, as indicated in Fig. 6–11.

After a crystal has dissolved, water molecules remain associated with ions, forming an envelope or cloud that actually moves with the ion. It is postulated that about 4 water molecules are associated with each sodium ion, about 2 molecules with each chloride ion. Ions with higher charges, such as Mg^{2+} or SO_4^{2-}, also have these so-called *hydration atmospheres* (or *sheaths*) usually containing more water molecules. Like other structured parts of liquid water, the hydration atmospheres are affected by temperature and pressure. They are apparently destroyed by pressures exceeding 2000 atmospheres, but are not completely destroyed by temperatures above 100°C.

Hydration sheaths affect the water around them. Some, such as NaCl and especially $MgSO_4$, favor the structured portion of water and actually increase viscosity, a property of water that responds sensitively to the relative abundance of the structured portion. Salts such as KCl seem to favor the unstructured portion of the water and might be called *"structure breakers."* They cause an initial reduction in the relative viscosity of water. Dissolved gases and insoluble or partially soluble substances also affect the relative abundance of structured and "free" water nearby.

Materials that dissolve completely forming separate ions in water are called *strong electrolytes.* Sodium chloride is an example. After a crystal dissolves, the ions form hydration sheaths that move more or less independently, so that ions formed by strong electrolytes behave like separate ions. In general, strong electrolytes have little effect on water structure as shown by their slight influence on water viscosity (see Fig. 6–11).

Not all materials separate completely into individual ions when they dissolve, and we say that they are *weak electrolytes.* Magnesium sulfate ($MgSO_4$) is an example of a weak electrolyte. Only a fraction of the material dissociates into individual

FIG. 6–11

Effect of various salts on viscosity of water at 25°C. (After Horne, 1969)

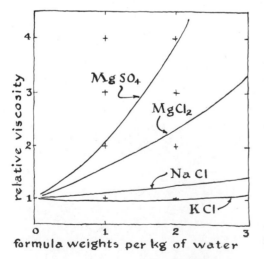

Table 6–2

SALT IN WATER 157

CHEMICAL SPECIES OF SOME MAJOR CONSTITUENTS IN SURFACE SEAWATER*

Cations	Free Ion (%)	Combined with:		
		Sulfate (%)	Bicarbonate (%)	Carbonate (%)
Na$^+$	99	1.2	0.01	—
Mg^{2+}	87	11	1	0.3
Ca^{2+}	91	8	1	0.2
K$^+$	99	1	—	—

Anions	Free Ion (%)	Combined with:			
		Na (%)	Mg (%)	Ca (%)	K (%)
SO$_4^{2-}$	54	21	21.5	3	0.5
HCO$_3^-$	69	8	19	4	—
CO$_3^{2-}$	9	17	67	7	—

*From R. M. Garrels and M. E. Thomson, A chemical model for sea water at 25°C and one atmosphere pressure. *American Journal of Science*, 260 (1962), 57–66.

ions, as shown in Table 6–2. About 11 percent of the Mg ions remain associated with sulfate ions. Sulfate ions are partially associated with Mg^{2+}, partially with Na$^+$.

Still another mode of association is possible in seawater—the formation of *ion pairs*, associations of ions that remain together much of the time even though they retain their individual hydration sheaths. In human terms we could think of it as a liaison, rather than a marriage. Ion pairs can have charges where the valence (amount of positive or negative charge) of individual members of the pair do not completely cancel.

Complex ions are still more strongly attracted to each other, so much so that their hydration sheaths merge to form a single sheath, as represented in Fig. 6–12. The ion is a definite, identifiable chemical species that may give the solution some characteristic features. The separation between complex ions and ion pairs is of concern to marine chemists but need not concern us.

Using the knowledge we have gained of the behavior of ions in water and seawater, it is possible to predict the probable chemical forms of the various elements dissolved in seawater. Obviously, we have many gaps in our understanding, but many of the essential features can be indicated. We have already seen the probable chemical species of the major elements. Table 6–3 lists all the elements detected in seawater and their probable chemical form.

Since we have considered primarily those materials dissolved in seawater, it is worth recalling that certain substances do not remain dissolved but form solids and precipitate out, the particles usually settling through the water to the bottom. If Ca^{2+} and CO$_3^{2-}$ are added in quantity to surface seawater, they form a precipitate, CaCO$_3$, which settles to the bottom. Over most of

FIG. 6–12

Schematic representation of hydration sheaths of ions formed by strong electrolytes (NaCl), ion pairs (Fe^{+++}NO$_3^-$). (After Horne, 1969)

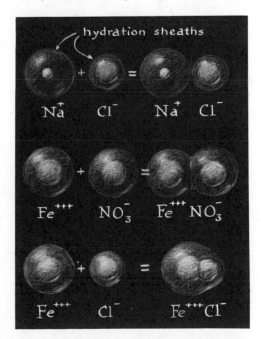

Table 6–3

ELEMENTS DETECTED IN SEAWATER*

Element	Chemical Form	ppm	Residence Time (Years × 1000)
Ag	$AgCl_2{}^-$	0.0003	2100
Al		0.01	0.1
Ar	Ar	0.6	
As	AsO_4H^{2-}	0.003	
Au	$AuCl_4{}^-$	0.000011	560
B	$B(OH)_3$	4.6	
Ba	Ba^{++}	0.03	84
Be		0.0000006	0.15
Bi		0.000017	45
Br	Br^-	65	
C	CO_3H^-, organic C	28	
Ca	Ca^{2+}	400	8000
Cd	Cd^{2+}	0.00011	500
Ce		0.0004	6.1
Cl	Cl^-	19000	
Co	Co^{2+}	0.00027	18
Cr		0.00005	0.35
Cs	Cs^+	0.0005	40
Cu	Cu^{2+}	0.003	50
F	F	1.3	
Fe	$Fe(OH)_3$	0.01	0.14
Ga		0.00003	1.4
Ge	$Ge(OH)_4$	0.00007	7
H	H_2O	108000	
He	He	0.0000069	20000
Hf		<0.000008	
Hg	$HgCl_4{}^{2-}$	0.00003	42
I	I^-, $IO_3{}^-$?	0.06	
In		<0.02	
K	K^+	380	11000
Kr	Kr	0.0025	
La		0.000012	0.44
Li	Li^+	0.18	20000
Mg	Mg^{2+}	1350	45000
Mn	Mn^{2+}	0.002	1.4
Mo	$MoO_4{}^{2-}$	0.01	500
N	Organic N, $NO_3{}^-$, $NH_4{}^+$	0.5	2.5
Na	Na^+	10500	260000
Nb		0.00001	0.3
Ne	Ne	0.00014	
Ni	Ni^{2+}	0.0054	18
O	OH_2, O_2, $SO_4{}^{2-}$	857000	
P	PO_4H^{2-}	0.07	
Pa		2×10^{-9}	
Pb	Pb^{2+}	0.00003	2
Ra		6×10^{-11}	
Rb	Rb^+	0.12	270
Rn	Rn	6×10^{-16}	
S	$SO_4{}^{2-}$	885	
Sb		0.00033	350
Sc		<0.000004	5.6
Se		0.00009	
Si	$Si(OH)_4$	3	8
Sn		0.003	100
Sr	Sr^{9+}	8.1	19000
Ta		<0.0000025	
Th		0.00005	0.35

Table 6–3 (continued)
DISSOLVED GASES   159

Ti		0.001	0.16
Tl	Tl^+	<0.00001	
U	$UO_2(CO_3)_3{}^{4-}$	0.003	500
V	$VO_5H_3{}^{2-}$	0.002	10
W	$WO_4{}^{2-}$	0.0001	1
Xe	Xe	0.000052	
Y		0.0003	7.5
Zn	Zn^{2+}	0.01	180
Zr		0.000022	

*After Horne, 1969.

the ocean, removal of these materials is by biological processes ($CaCO_3$ is used to form skeletons by many marine organisms). Inorganic precipitation seems to be a relatively rare occurrence at this time, although it may have been much more common at earlier stages of the ocean history. At the present it appears that the Bahama Banks, east of Florida, represent one of the few areas of the world where this occurs, although some $CaCO_3$ precipitation may occur in isolated seas and lagoons.

Seawater is warmed as it flows slowly over the Bahama Banks, causing two effects as the water warms: the solubility of $CaCO_3$ in water is diminished, and the solubility of carbon dioxide decreases so that some escapes to the atmosphere. While in solution, each Ca^+ ion is balanced by the presence of two $HCO_3{}^-$ ions. If one of these is lost by dissociation and escape of CO_2, then $CaCO_3$ (lime) precipitates as a solid coating on rounded grains (oölites) or as tiny needles of the mineral aragonite.

An increasingly common method of determining the salinity of seawater is to measure its conductivity. Conductivity of water is controlled by ion movement through the water. The more abundant the ions, the greater the transmission of electricity and the higher the conductivity. In seawater, the relative abundance of the major ions is nearly invariant, and the conductivity provides a precise means of determining salinity. The major ions conducting the electrical charge are:

Cl^-	64%
Na^+	29
Mg^{2+}	3
$SO_4{}^{2-}$	2

In fresh waters the composition of dissolved salts is more variable, and conductivity does not provide a precise measure of salt content. In fresh water the use of this method requires careful study to insure that the changes in conductivity are not caused by changes in chemical composition but are caused by changes in their concentration.

DISSOLVED GASES

Atmospheric gases are soluble in water, passing into a dissolved

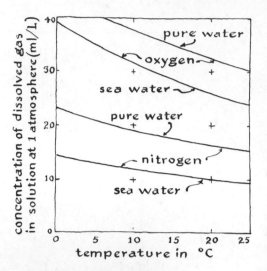

FIG. 6–13

Solubility of oxygen and nitrogen in pure water and in seawater. Note that both gases are more soluble in pure water.

FIG. 6–14

Relative abundance (by volume) of gases in dry atmosphere.

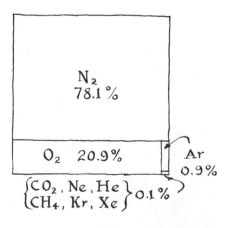

state at the air–water interface. Conversely, molecules of gas also pass through the interface in the opposite direction—back into the atmosphere. Water can only dissolve a certain amount of any substance under given conditions of temperature and pressure, and when that limiting value is reached the amount of gas going into solution is the same as that going out. At this point the water is *saturated* with the gas, which is then said to be present in *equilibrium concentration*.

Temperature, salinity, and pressure all affect the saturation concentration for a gas. In the normal range of oceanic salinity, temperature is the dominant factor, as indicated in Fig. 6–13. In general those gases such as nitrogen, oxygen, or the rare gases that do not react chemically with water become less soluble in seawater as temperature or salinity increase. Seawater at all depths is saturated with atmospheric gases. Exceptions are dissolved oxygen and carbon dioxide, which are involved in life processes.

Nitrogen, the most abundant gas in the atmosphere, as Fig. 6–14 indicates, is also a common gas dissolved in seawater, as shown in Fig. 6–15. It is not involved in biological processes (with a few minor exceptions), and thus remains very near saturation throughout the ocean. Sometimes, in deep water, more nitrogen is present than can be accounted for by the temperature, salinity, and pressure of the water. We then say that the water is *supersaturated* with nitrogen, a condition which may be explained as follows.

Since the solubility of a gas in seawater is determined primarily by temperature, pressure, and salinity, and since exchange of dissolved gas occurs at the sea surface, it is the conditions at the surface that control the amount of gas dissolved in water. As a water mass moves away from the surface where it last exchanged atmospheric gases, its temperature and salinity can change as a result of mixing with other water masses. Also, depth changes are accompanied by pressure changes. The observed slight supersaturations of nitrogen and rare gases in the deep ocean can thus be accounted for by considering these changes in physical conditions of the water since it was last at the surface.

There is, to be sure, some slight production of nitrogen by bacteria living in oxygen-deficient basin waters. These bacteria break down nitrate ions (NO_3^{2-}) from seawater and release nitrogen gas (N_2). In the absence of dissolved oxygen, most biological processes proceed slowly, and the volume of ocean area deficient in oxygen is small, so this process is a minor factor in the dissolved-nitrogen picture. There is also some formation of nitrate from dissolved nitrogen by bacteria, but again it appears to play a minor role.

A small amount of helium, a rare gas, is produced by radioactive decay in rocks and sediment at the sea floor. Otherwise, the rare gases—Argon (Ar), Krypton (Kr), and Xenon (Xe)—behave identically to nitrogen, allowing for their slightly different solubilities.

Oxygen is perhaps the most variable of the dissolved gases in seawater. As all the other gases, it enters the ocean primarily through the surface. Oxygen is also produced by photosynthesis of plants in the sunlit layer of the ocean, usually restricted to a zone a few tens of meters deep just below the ocean surface. In the deep ocean, dissolved oxygen is supplied by the sinking of cold, oxygen-rich waters in the Antarctic and Arctic regions. Oxygen is used by both plants and animals at all depths in respiration.

In those parts of the ocean where the rate of oxygen consumption by marine life exceeds the rate of resupply of dissolved oxygen, certain bacteria are able to derive their necessary oxygen by breaking down compounds dissolved in seawater. First the nitrate ions are broken down, releasing nitrogen. There being little nitrate in most ocean waters, this does not last long. Next, different organisms break down SO_4^{2-} to obtain oxygen and release hydrogen sulfide, H_2S, as a metabolic by-product, causing the familiar rotten-egg smell sometimes present in coastal marshes at low tide.

The carbon dioxide cycle in seawater is the most complicated of all the dissolved gases. Carbon dioxide occurs as a dissolved gas, as a weakly dissociated acid, and as the dissolved ions of carbonate (CO_3^{2-}) and bicarbonate (HCO_3^-). Hydrogen ion (H^+) in seawater, the concentration of which determines acidity of the water, is freely accepted and given up by both carbonate and bicarbonate ions. A system containing these ions is thus said to be *buffered* against sudden, sharp changes in acidity, at least on a short-term basis.

Carbon dioxide is highly soluble in seawater and comes to the ocean through the sea surface; it is also produced at nearly all depths by the respiration of plants and animals. A small amount of carbon dioxide is taken up by certain marine organisms to form their calcareous shells. Shells formed in surface waters dissolve in intermediate and deep waters after the organism dies and sinks below the surface. On the ocean bottom less than about 4 kilometers deep (the depth varies in different ocean basins) the rate of supply of carbonate shells is greater than the rate of dissolution. However, the rate of resupply is much too slow to prevent the dissolution of carbonate in the cold, CO_2-rich waters deeper than the 4-kilometer level. There the sediments are nearly devoid of carbonate because of its rapid dissolution; this is the red mud area of the deep-ocean basins.

PHYSICAL PROPERTIES OF SEAWATER

Pressure in the ocean increases with depth. An increase in depth of 10 meters raises the pressure by 1 atmosphere. Pressures in the deepest trenches are around 1100 atmospheres.

Viscosity—resistance of a fluid to flow—in "normal" liquids increases as pressure increases. In a simple way, we can imagine that this results from the increased resistance of atoms or molecules sliding past each other as they are pressed more closely together.

Liquid water responds in exactly the opposite way from a "normal" liquid; its viscosity decreases with increasing pressure. It reaches a minimum viscosity at pressures of about 500 to

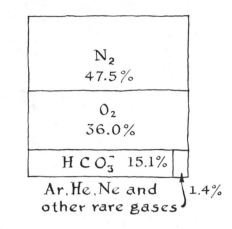

FIG. 6–15
Relative abundance (by weight) of gases dissolved in surface seawater, $S = 36^0/_{00}$, $T = 20°C$. Note the abundance of HCO_3^- and rare gases compared with their abundance in the atmosphere (see Fig. 3–14).

1000 atmospheres, corresponding to depths of 5000 to 10,000 meters.

Ease of molecular movement in water (and therefore its viscosity) is related to the degree of binding of molecules and the amount of structuring of the water. The simplest explanation of the pressure effect on water is that increased pressure in the depth range to 5000 meters or more favors the destruction of the structured portions, thereby increasing the ease of movement of molecules. This decreases its viscosity. At pressures exceeding 1000 atmospheres, the structured portion of the water has been completely destroyed, and water then acts like a "normal" liquid. Its viscosity increases with further pressure increase.

Air trapped in seawater near the surface forms bubbles. These range in size from a few millimeters down to a few microns in diameter and are formed by breaking of waves and impact of raindrops. Bubbles provide an important route for gases in seawater to pass in and out of solution and for the production of sea-salt particles that serve as condensation nuclei for raindrops and snow particles in the atmosphere.

Development and bursting of bubbles has been studied in detail by high-speed photography, as shown in Fig. 6–16. First a gas bubble forms (a) and rises toward the surface at rates of about 10 centimeters per second. Reaching the surface (b) the bubble has a cap or thin film of water that bursts (c), forming a thin spray of droplets (d) whose diameters are about 1 to 20 microns. Shortly thereafter a large droplet (e), about 100 microns in diameter, is formed by a jet rising rapidly from the bottom of the bubble and is ejected into the atmosphere at speeds of 10 meters per second. This rises 10 centimeters or more above

FIG. 6–16
Bubble bursting at the sea surface forming water droplets in the atmosphere. Many of these droplets evaporate, forming salt particles that serve as condensation nuclei for raindrops or snowflakes.

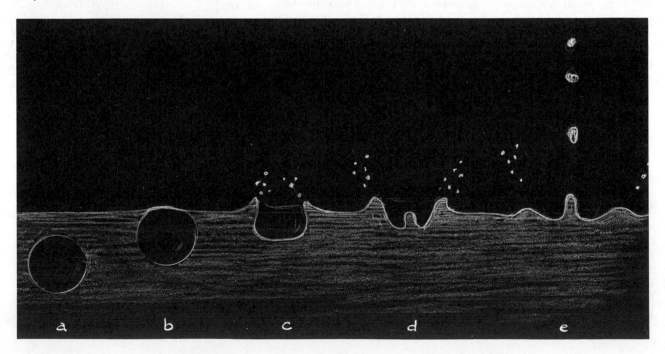

the sea surface. The smaller droplets are caught by wind and transported long distances in the atmosphere. The larger droplets rapidly fall back to the sea surface.

Air-borne droplets evaporate, leaving tiny salt particles that are carried by winds. The residence time of most such particles in the atmosphere is probably short. Some, however, are carried high into the atmosphere, where they eventually act as nuclei around which raindrops or snowflakes form. In this manner, escaped salt is carried back to the earth's surface. The total amount of salt moved by this mechanism is large, about 1 billion tons per year. This is about equal to the total sediment load of the world's rivers.

Composition of the salt particles closely resembles that of normal sea salt, although there is some enrichment in the more volatile materials that evaporate from the sea surface, such as boron and iodine.

Surface tension is the force necessary to break the filmlike surface layer of a liquid. Water has an anomalously high surface tension, such that a clean needle can float on a water surface. Surface tension causes an undeformed water drop to form a sphere. This may be seen by shaking a bottle of oil and vinegar —the vinegar forms droplets in the oil during the temporary period of suspension.

In water, high surface tension seems to result from strengthened local water structure near any interface, including the air–water boundary. Near an interface there is more hydrogen bonding of the water molecules, which tend to orient themselves with their oxygens pointing out of the liquid, as shown in Fig. 6–17. This structured zone appears to extend about 10 molecules deep into the liquid. Because of the highly structured nature of interface water, certain ions tend to be excluded so that the chemical composition of this zone may differ appreciably from that of

FIG. 6–17
Changes in water structure near the air–water interface.

bulk seawater. This selective exclusion likely affects the chemical composition of sea salts that pass into the atmosphere from bubble breaking and other processes. In fact, several elements are more abundant in these airborne sea salts than they are in normal seawater.

Surface films are sensitive to the presence of impurities. Many substances that are insoluble in water tend to collect at the interface. Many are molecules whose various components have different chemical behavior. For example, many surface-active agents—substances that concentrate at or change the properties of an interface—have one end that readily dissolves in water while the other end of the molecule (often a long, complicated structure) does not dissolve in water. The water-soluble end is called the *hydrophilic* ("water-loving") end, the insoluble end the *hydrophobic* ("water-hating") end; this relationship is illustrated in Fig. 6–18, if you look closely.

Surface-active agents change the surface tension of seawater, commonly lowering it. Surface "slicks" or films, where surface-active materials have been swept together by wind or current action, are common in coastal waters, especially where currents converge. These areas have relatively few small waves at the water surface, and their relative smoothness compared to adjacent waters has given rise to the descriptive term "slick." They commonly contain a wide variety of organic materials of an oily or otherwise hydrophobic nature and are of considerable biological significance in the ocean. The subject will be discussed in Chapter 14.

Detergents and soaps are familiar surface-active agents. Their long molecules surround dirt or grease, substances easily attached to the hydrophobic end of the surface-active molecule. The hydrophilic end is directed toward the surrounding water, and the whole aggregate moves as if it were soluble, so that the otherwise insoluble dirt or grease can "wash away" in the water.

FIG. 6–18
Changes in water structure in the vicinity of a nonpolar, hydrophobic material. Note that the polar water molecules exclude the nonpolar propane molecule from the water structure.

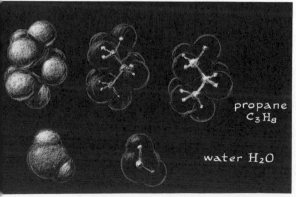

In the late 1960's, some 30 percent of the world's salt production came from the ocean. About 5 percent of the U.S. production at that time came from evaporating seawater in southern San Francisco Bay. Brine from large evaporating ponds is transferred to "pickle ponds" and finally to a crystallization pond, where the salt forms. The photo shows a harvesting operation after the last of the liquid, known as "bittern," has been removed and the layer of crystallized salt has reached a depth of about 10 centimeters.

Magnesium metal and bromine extracted from seawater provide a large fraction of these raw materials for industry: 70 percent of the world's bromine and 61 percent of its magnesium in 1968 came from chemical processing of seawater. One of the problems of seawater as a raw material is securing adequate supplies of undiluted, uncontaminated seawater. A more serious problem is that an extremely large volume of seawater must be pumped and processed to obtain enough material to make an operation economically feasible. Only the most abundant constituents of sea salts can be economically extracted because of these large power requirements.

The largest amount of material produced from the ocean is water itself. For the world as a whole, about 59 percent of its "manufactured" water comes from the ocean. Two different approaches are possible: removing the water from the salt (by distillation, for example) or removing the salts from the water (using various chemical processes such as ion exchange or reverse osmosis). Despite the large amount of energy required to extract the water, desalinization is still a common practice in coastal urban areas in arid climates, although there is little likelihood that water from the ocean will be widely used for irrigation of crops in the near future. (Photograph courtesy Leslie Salt Company)

SUMMARY OUTLINE   *Seawater*—a living soup, 96.5% water, 3.5% salts

Molecular structure of water

Water—a polar molecule

Many unusual properties result from formation of hydrogen bonds

Ionic bonds readily broken in water; weak van der Waals bonds also present

In water vapor, molecules move freely, exert vapor pressure

Ice, held together by hydrogen bonds, has properties of a solid, regular crystal structure; salt is excluded from structure

Liquid water an anomalous substance, containing structured and unstructured portions in a dynamic equilibrium

Temperature effects on water

Heat energy absorbed in changes of state with addition of heat

Latent heat utilized in breaking bonds; heat is released in condensation of water vapor and freezing of water

Cluster size and number of molecules per cluster decrease with increasing temperature

Density

Sinking (or floating) is determined by density relationships

Seawater density dependent on temperature, salinity, and pressure

Density decreases with decreasing temperature, but fresh water has an anomalous density maximum at 3.98°C, so that ice is less dense than very cold water

Salinity commonly the dominant factor controlling density of seawater

Composition of sea salt

Six elements comprise 99% of sea salts

Chlorinity—g/kg, ($^0/_{00}$) = amount of chlorine in seawater, a measure of salinity

Salinity—total amount of dissolved salts ($^0/_{00}$) in seawater

Conservative elements are those whose relative proportions in seawater are usually constant; their concentrations may change with addition or removal of water

Nonconservative elements—variable, generally low concentrations which depend on biological activity

Salt composition and residence times

Salts dissolved from continental materials enter seawater; these are eventually removed by incorporation in sediments

Residence time of an element in seawater is related to chemical behavior

Salt in water

Water shields ions of unlike charge from one another—therefore breaks ionic bonds

Temperature of initial freezing is affected by salinity changes

Different ions affect structuring properties of liquid water, cause formation of hydration atmospheres around the ion

Strong electrolytes dissolve completely; weak electrolytes dissociate in part only

Other modes of association—complex ions and ion pairs

Chemical form of elements in seawater can be predicted

Inorganic precipitation (e.g., of $CaCO_3$) may occur under specialized conditions

Electrical conductivity a measure of salinity of water

Dissolved gases

Equilibrium concentration depends on solubility of a gas

Saturation concentration affected by salinity, temperature, and pressure conditions at the surface

Nitrogen—near saturation throughout the ocean

Oxygen—produced by photosynthesis; utilized by plants and animals at all depths

CO_2—occurs in many forms, in a state of equilibrium; produced by respiration of plants and animals at all depths

Physical properties of seawater

Pressure and viscosity

Increase in direct proportion in "normal" liquids

At less than 1000 atmospheres pressure, increased pressure reduces water viscosity because it favors the unstructured mode

Bubbles and sea salt

Bubbles provide a route for solution–evaporation transfer of gases

Bursting bubbles release salts to the atmosphere

Surface tension

High surface tension results from strengthened local water structure near any interface

Surface-active agents change surface tension of water, cause hydrophobic substances to "dissolve" in water

SELECTED REFERENCES

GROSS, M. G. 1971. *Oceanography.* Second Edition. Charles E. Merrill Publishing Company, Columbus, Ohio. 150 pp. Elementary.

HORNE, R. A. 1969. *Marine Chemistry: the Structure of Water and the Chemistry of the Hydrosphere.* Wiley–Interscience, New York. 568 pp. Advanced treatment of the subject, intended to serve also as a reference handbook.

KUENEN, P. H. 1963. *Realms of Water: Some Aspects of its Cycle in Nature.* Science Editions (John Wiley), New York. 327 pp. Elementary to intermediate in difficulty; stresses hydrologic cycle.

MACINTYRE, FERRAN. 1970. "Why the sea is salt." *Scientific American,* 223 (Nov., 1970), 104–115.

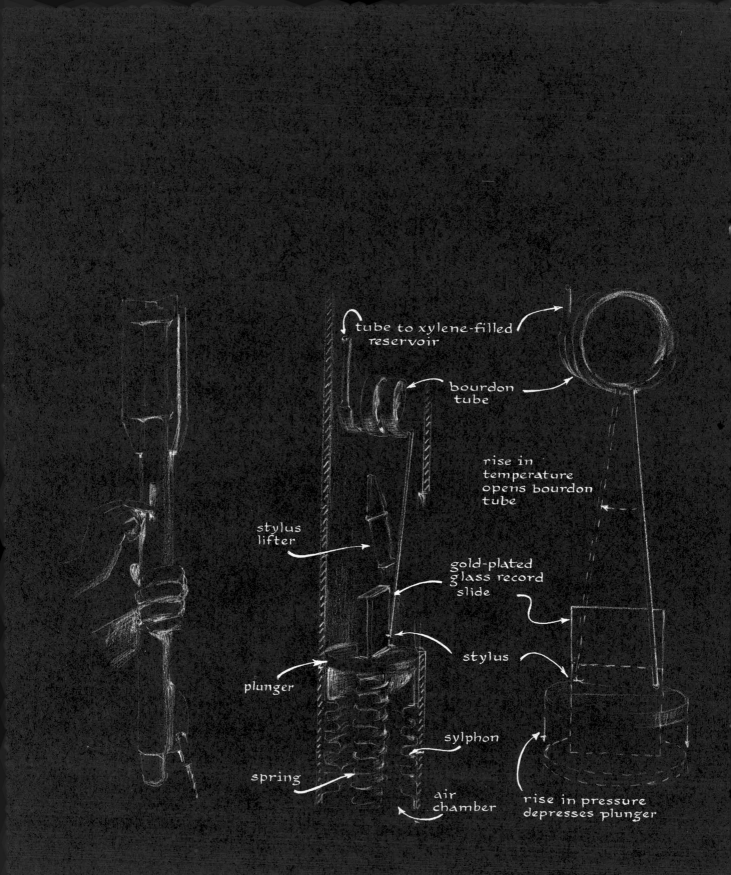

tube to xylene-filled
reservoir

bourdon
tube

rise in
temperature
opens bourdon
tube

stylus
lifter

gold-plated
glass record
slide

plunger

stylus

sylphon

spring

air
chamber

rise in pressure
depresses plunger

TEMPERATURE, SALINITY, AND DENSITY

The *bathythermograph* is a torpedo-shaped device used to measure temperature changes with depth below the surface while a ship is underway. A temperature-sensitive element causes a stylus to move. Increasing pressure compresses a bellows moving a metal-coated slide which is scratched by the stylus. A small graph is thus drawn of temperature at various depths.

seven

Having considered the physical and chemical properties of seawater, including the detailed structure of water and sea salts, we now turn to consider the behavior of ocean waters on a global scale. In this chapter we consider how the ocean is closely coupled to the overlying atmosphere. We also discuss incoming solar radiation, which is the major energy source for ocean processes.

Two different approaches are used. The first is the concept of the *budget*. We consider heat and water budgets for the earth, showing the major sources of each and what happens to both in the ocean. Regarding the water budget, it is important to remember that the amount of water on the earth's surface is nearly fixed, so that we need not consider new sources, only the redistribution of the water on the earth during the year (or perhaps some longer time period).

In the case of the heat budget, the underlying assumption is that the earth is getting neither hotter nor colder, but loses to space an amount of heat equal to that it receives during a single year, as Fig. 7–1 indicates.

To determine the routes followed by ocean water as a result of the major processes, we shall use another tool commonly employed by oceanographers—the *distribution of properties*. By charting distributions of water salinity and temperature, we can determine where various distinctive types of water form and how they move and mix in the ocean.

Finally, we shall consider sea-ice formation as an example of a process that, while it is only common in certain parts of the ocean, nevertheless exerts oceanwide influence. Sea-ice formation also conveniently illustrates some aspects of seawater chemistry previously discussed.

The ocean and atmosphere are closely coupled systems that are affected by the shape, distribution and size of large land masses. A complete description of either ocean or atmosphere would require detailed pictures of both oceanic and atmospheric conditions and current patterns throughout the entire earth over the course of an entire year. Despite the enormous advances in oceanography in the past century, such a complete descriptive series is still not available.

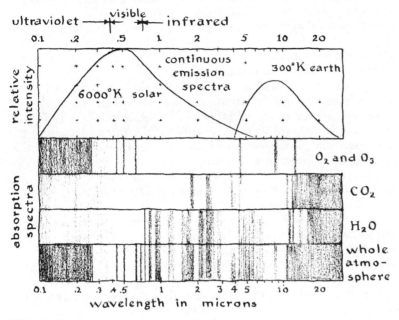

FIG. 7–1

Emission spectrum of the sun and earth. The earth radiates to space as much energy as it receives from the sun. The major difference is the shift from the relatively short-wave radiation of the sun, which we experience as visible light, to the longer-wave infrared radiation which we perceive as heat. The amount of heat received from the sun is balanced by the amount of heat radiated back to space from the earth. (After Weyl, 1970)

At present, we use averages to describe the ocean. This can be done in several ways. In this chapter we first examine the surface ocean and then consider a vertical cross-section. In the next chapter we shall graduate to a three-dimensional ocean extending over the entire earth, but shall continue to ignore changes in time. Ideally, we would like to include a time scale in our picture of ocean processes, and in fact this can be done for small ocean areas. Using several ships, aircraft, or satellites, it is sometimes possible to accumulate enough information to construct a picture of ocean conditions as they existed for a short time. Such an undertaking is called *synoptic oceanography*, and it represents the goal for many oceanographic investigations. Increasingly successful use of satellites and unmanned buoys offers substantial hope that this may be achieved over many ocean areas in the future, providing the nearly instantaneous pictures of the ocean like those we have long had for the atmosphere.

LIGHT IN THE OCEAN

Radiation from the sun striking the earth's surface is the source of virtually all the energy that heats the ocean surface and warms the lower portion of the atmosphere. Part of this incoming solar radiation is within the visible part of the spectrum and provides the energy needed by plants for photosynthesis, on land and in the ocean. After passing into the surface of the ocean, most of this energy is converted into heat, either raising water temperatures or causing evaporation.

The spectrum of radiant energy from the sun is filtered once as it passes through the atmosphere, as shown in Fig. 7–1, and is further filtered in the surface ocean. Within the first 10 centimeters of even pure water, virtually all the infrared portion of the spectrum is absorbed and changed into heat. Within the first meter of seawater about 60 percent of the entering radiation is absorbed, and about 80 percent is absorbed in the first 10 meters. Only about 1 percent remains at 140 meters in the clearest subtropical ocean waters.

In coastal waters, the abundance of floating marine organisms, suspended sediment particles, and various dissolved organic substances cause the absorption of light to take place at even shallower depths. Near Cape Cod, Massachusetts, for instance, only 1 percent of the surface light commonly penetrates to a depth of 16 meters. In such waters, the maximum transparency shifts from the bluish region typical of clear oceanic waters to longer wavelengths, as shown in Fig. 7–2. In very turbid coastal waters, the peak transparency occurs around 0.6 micron, in the yellow range. The Thames River water near New London, Connecticut, shows maximum transparency in the red, at wavelengths between 0.65 and 0.7 micron. In highly polluted waters, absorption of all light takes place within the upper few centimeters of the surface.

This *attenuation* (reduction of light intensity) is caused by absorption and scattering form particles and the molecules of the water itself. Absorption is caused primarily by abundant small particles suspended in water. Some of these are inorganic sediment particles; others are minute organisms and their wastes or decomposition products. Near coasts, the absorption of light increases owing to increased abundance of substances dissolved in the water. In many areas these dissolved substances consist of a stable, humuslike organic material, brought to the ocean by northern rivers, which imparts a yellowish color to the water. Elsewhere they are produced locally by decomposition of marine organisms, probably *phytoplankton* (minute floating plants).

Far from the coast, the ocean often has a deep luminous blue color quite unlike the greenish or brownish colors common to coastal waters. The deep blue color indicates an absence of colored particles or dissolved substances. In these areas, usually

FIG. 7–2

Spectrum of extinction for visible light in 1 meter of seawater in areas removed from sources of fresh-water discharge. (a) is the curve for extremely pure water in the open ocean; (b) is for relatively turbid tropical–subtropical ocean water; (c) is for ocean water in mid-latitudes; (d) is for clearest coastal water; and (e), (f), and (g) are for coastal waters of increasing turbidity. All wavelengths in the violet-to-yellow range are transmitted freely in open-ocean water. Near coasts, the presence of dissolved pigments ("yellow substance" of organic origin) strongly inhibits transmission in the violet and blue ranges. (After Dietrich, 1963)

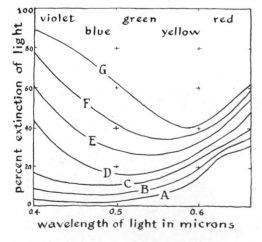

restricted to the central portions of the deep-ocean basins, the color of the water is thought to result from *scattering* of light rays within the water. Water molecules themselves are probably too small and too closely spaced to account for the scattering. Instead, it has been proposed that the blue color results when irregular molecular movements cause inhomogeneities in the water, which are themselves the scattering elements. A similar type of scattering is responsible for the blue color of the clean atmosphere.

One of the simplest ways to measure the attenuation of light in seawater uses the *Secchi disk*, a flat disk, 20 to 30 centimeters in diameter, with four alternating black and white painted segments. The disk is lowered over the side of a ship nearest the sun, on a line with marked distances, until it is no longer visible. The visible depth of the Secchi disk is a measure of water clarity. In open-ocean waters, Secchi disks may be seen at depths of tens of meters. In isolated heavily polluted parts of New York Harbor, the Secchi disk disappears from view at depths of about 25 centimeters.

Since sunlight sufficient for the human eye is often lacking below 30 meters, it is commonly necessary to employ light sources for underwater photography and viewing. Television cameras are extremely useful tools for ocean-bottom exploration, commonly having their own high-intensity light sources. Photographs may be taken in shallow water using standard flashbulbs; work at greater depths commonly requires special flashbulbs.

Incoming solar radiation striking the ocean surface is partly absorbed, as we have seen, and partly reflected. The relative amount of each component depends on the state of the sea surface and the angle at which the sun's rays strike the water. When the sun is directly overhead, only about 2 percent of the incoming radiation is reflected; the remainder enters the water. When the sun is near the horizon, nearly all the incoming radiation is reflected. Waves on the sea surface increase the amount of light reflected by as much as 50 percent. Where the sun is very near the horizon, waves serve to decrease the amount of reflection.

HEAT BUDGET AND
ATMOSPHERIC CIRCULATION

One way to study average conditions on the earth as a whole is to construct a *budget*—of energy, or of a substance such as water. Budgets indicate where materials or energy come from (*sources*) and where they go (*sinks*), but they often cannot give detailed information about the exact rate at which these transfers occur or the specific routes they follow. Thus our generalized picture may differ substantially from real conditions at any time or location.

First we consider the earth's *heat budget*. The earth is essentially a sphere which is exposed to the sun's radiation from a distance of 149 million kilometers (about 90 million miles).

At this distance the top of the earth's atmosphere receives 2 calories per square centimeter per minute. This amount of radiant energy is distributed over the earth as it rotates about its axis every 24 hours. Thus on the average, about 0.5 cal/cm²/min strike any given point at the top of the atmosphere, as Fig. 7–3a indicates, since the earth's surface is dark for part of the 24-hour day.

The earth's axis of rotation is inclined 23.5° to the plane of the earth's movement around the sun. This also affects the amount of energy received at the top of the atmosphere. Because of this inclination, the amount of radiant energy at the top of the atmosphere varies throughout the year for any point on earth, thus causing the change in seasons. Those points where the sun is directly overhead at noon receive the maximum amount of incoming solar radiation, abbreviated as *insolation*. Points north and south receive less, and no incoming radiation at all is received where the sun's rays just graze the earth's surface (see Fig. 7–3b).

On a featureless earth having no atmosphere, it would be

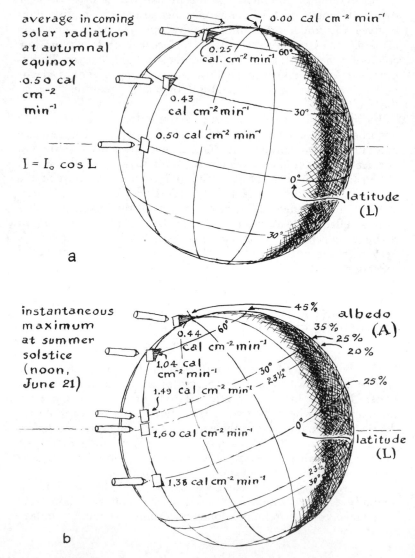

FIG. 7–3
Incoming solar radiation at the top of the earth's atmosphere. *A* shows the average solar radiation received over 24 hours at different latitudes in the Northern Hemisphere at the autumnal equinox, September 23; *B* shows the instantaneous maximum of incoming solar radiation at the earth's surface during the summer solstice when the sun is directly overhead at 23.5°N. Note the change in albedo between the equator and the pole.

possible to calculate the amount of energy received at any point
and from that figure to determine the probable temperature for
that point, based on its distance north or south of the equator
and the season of the year. On such a planet, the only other
factor to be considered would be the *albedo,* or local reflectivity,
because not all incoming radiation is absorbed: some is reflected
back to space.

Let us now consider the effect of adding a completely transparent atmosphere to our featureless earth, and receiving the
same amount of solar radiation. Incoming radiation would pass
undiminished through the atmosphere to be absorbed at the
earth's surface. The warmed earth would in turn warm the atmosphere above it, causing a simple planetary circulation system. The warmed atmosphere would locally be less dense and
therefore rise to the top of the atmosphere, where it would lose
heat by radiation and other processes and then sink back to the
surface to be warmed again.

Because of the curvature of the earth's surface, the amount of
incoming radiation is greatest just beneath the sun and decreases
as one travels north or south of that point. (During the 24-
hour-long polar day, the polar region receives more insolation
than the point immediately below the sun because of the longer
polar day.) Averaged over a year, the maximum amount of incoming solar energy is received in the low latitudes, so that
insolation is unevenly distributed over the earth's surface. The
atmosphere, on the other hand, tends to radiate energy back to
space at a nearly constant rate, as shown in Fig. 7–4. As a result,
the earth receives most of its heat between about 40°N and
40°S, whereas there is a net heat loss to space in the high lati-

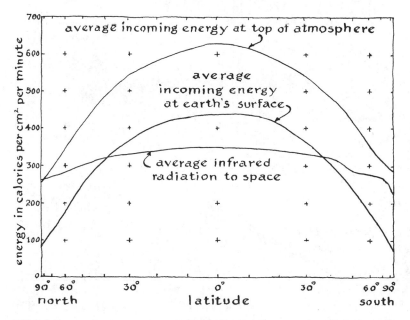

FIG. 7–4
Average solar energy reaching the top of the
atmosphere (upper curve); average solar
energy reaching the land or ocean (lower
curve); and the amount of energy lost to
space by infrared radiation.

tudes between 40° and 90° in both hemispheres. The necessary heat transfer from low to high latitudes takes place in the atmosphere and ocean. In this somewhat oversimplified picture, atmosphere and ocean are seen to function as a simple heat engine, transforming energy into motion. We experience the atmospheric motions as winds, the oceanic movements as currents.

Earth's atmosphere contains constant amounts of nitrogen and oxygen, plus small and variable amounts of dust, water vapor, carbon dioxide, and ozone. Although present in minute quantities, dust, carbon dioxide, and water absorb some incoming solar radiation; thus only about 47 percent of the solar radiation striking the top of the atmosphere actually reaches the earth's surface, as indicated in Fig. 7–5. The remainder is absorbed by the atmosphere or scattered and reflected back to space. Estimates of the earth's albedo range from 0.30 to 0.35; Fig. 7–5 is based on an albedo of 0.35. Clouds, covering on the average 50 percent of the earth's surface, account for about two-thirds of the back-scattered radiation.

Despite the large amount of heat received by the earth from

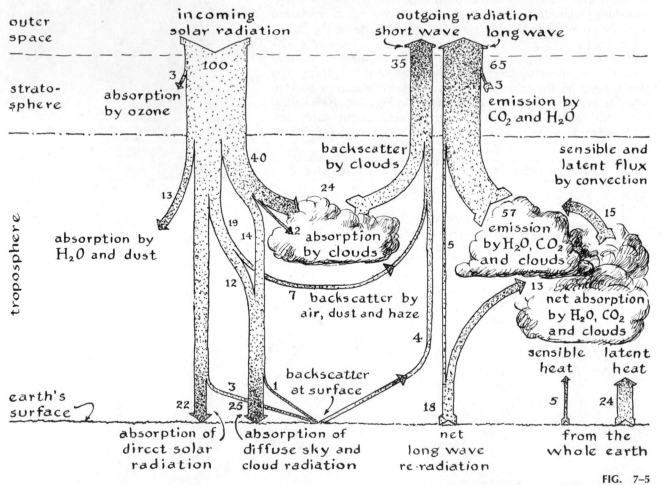

FIG. 7–5
Average annual heat budget of the earth showing the major interactions with incoming solar radiation and loss of heat from earth.

the sun each year, historical records indicate that the earth's surface-temperature changes over the past few thousand years have been so small that it is extremely difficult to document their occurrence. Therefore, we conclude that the earth is radiating back to space as much energy as it receives from the sun. The sun's surface, being extremely hot, radiates energy including the visible portion of the electromagnetic spectrum. Because the earth's surface is so much cooler, its radiation occurs in the *infrared* portion of the electromagnetic spectrum (see Fig. 7–1).

Of the energy lost to space, only about 5 percent is radiated back directly from the earth's surface (this is commonly called *backradiation*). About 60 percent is radiated back to space primarily from clouds (see Fig. 7–5). The atmosphere is nearly opaque to the earth's radiations, so that it acts much like a blanket in keeping the surface warmer than would be the case under a transparent atmosphere. A so-called "*greenhouse effect*" is caused by strong absorption of infrared radiation by carbon dioxide and water-vapor molecules. Without these trace constituents, earth-surface temperatures might drop as low as $-20°C$, which is the temperature at the top of the cloud layer. Instead, the average surface temperature overall is about $13°C$.

Relatively little heat is transferred to the atmosphere by direct heating (*sensible heat*). More than 80 percent of the atmospheric heating takes place through release of latent heat during condensation of water vapor, and virtually all of this water vapor comes from the ocean.

The ocean also transfers heat on the earth's surface. Although ocean currents move far more slowly than winds, water's large heat capacity (compared to air) enables it to carry far more heat by volume than air does. In the Northern Hemisphere, ocean currents are thought to carry about one-third the amount of heat that is carried by the entire atmosphere. In the Southern Hemisphere, oceanic-heat transport appears to be less important.

Atmospheric circulation is controlled primarily by minute amounts of water vapor, ozone, and carbon dioxide. Although nitrogen constitutes about 78 percent of the atmosphere by volume, it has little direct effect on circulation: carbon dioxide and water vapor together cause the "greenhouse effect," previously mentioned. Absorption of solar radiation (see Fig. 7–4) by ozone (O_3) causes warming of the lower stratosphere, forming a pronounced change in density (*tropopause*) that separates the *stratospheric* (high-altitude) and *tropospheric* (near-earth) circulation systems. In the stratosphere, gases are warmed by absorption of *ultraviolet* radiation; thus temperature of ionized gases increases with increased altitude. In the troposphere, gases are warmed by the earth; consequently temperature decreases with increased altitude. Our weather systems, with vigorous vertical convection and formation of clouds, all occur in the troposphere.

On a featureless, water-covered earth, the simplest atmospheric circulation would be a single cell. Warmed moist air would rise in the equatorial regions and move toward the poles in both hemispheres, carrying heat as both latent heat (water vapor) and as sensible heat (warm air). When the air was cooled and water vapor precipitated, it would sink and flow back toward the equator in a surface wind. As we shall see in the next chapter, such a simple cell does occur near the equator: the so-called *Hadley cell*, named for the British meteorologist who first postulated its existence. This cell includes the extremely persistent Trade Winds of low latitudes.

The earth, however, is of course not featureless, and the amount of heat that the atmosphere must transport from equator to poles is too great for a simple atmospheric pattern to maintain stability. Instead, the atmospheric circulation breaks down into several cells and exhibits a well-developed, wavelike pattern, especially in the mid-latitudes where the westerly winds prevail, as illustrated in Fig. 7–6. These large waves give rise to the series of *fronts* (sharp boundaries between two air masses) that dominate weather patterns through much of the United States and Europe. These waves (or fronts) commonly separate cold, dry polar air masses from warmer, more humid air masses of mid-latitudes.

So far our discussion of heat budgets has involved the entire earth; it is often necessary, however, to consider smaller regions. The procedure remains the same: we must account for all the sources and sinks. The principal difference is that we

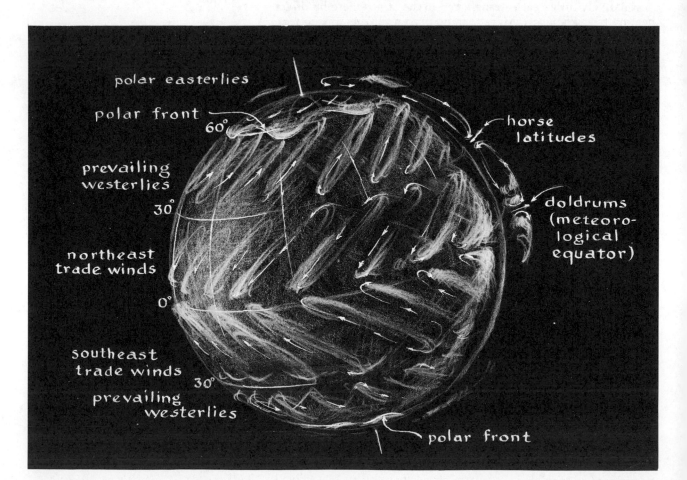

FIG. 7–6
Schematic representation of the planetary
circulation of the earth's atmosphere.

must also consider the movement of waters and the heat they bring into (or carry away from) the region.

A *regional heat budget* for a part of the ocean can be expressed mathematically as follows:

Heat gained = Heat lost
$$Q_s + Q_c = Q_e + Q_r + Q_h$$

where

Q_s = the energy gained from solar radiation

Q_c = the energy gained (or lost) by ocean currents

Q_r = the energy lost by radiation

Q_e = the energy lost through evaporation (this may be a positive quantity if more heat is gained from precipitation than is lost through evaporation)

Q_h = the energy lost by direct heating of the atmosphere (sensible heat)

The large input of solar energy into tropical regions causes high ocean-surface temperatures there as seen in Figs. 7–7 and 7–8. Maximum open-ocean temperatures, between 26 and 29°C, occur just north of the equator, at about 5°N (excluding isolated basins). Temperature distribution in the ocean is strongly modified by ocean currents and by the unequal distribution of land masses in the two hemispheres.

Note that the temperature difference between the warmest and coldest month of the year is small—less than 3°C—in the low latitudes (10°N to 10°S) and near the North and South Poles. In the mid-latitudes, around 30° north and south, the annual variation is largest: around 6°C. The average ocean-surface temperature is about 17.5°C, as Fig. 7–9 indicates, whereas average land temperature is about 14.4°C.

Despite the complications arising from unequal distribution of continents and their deflection of ocean currents, distribution of equal-temperature zones tends to parallel the equator. In general, *isotherms* (lines connecting points with equal surface temperature) are aligned nearly east–west, especially in the Pacific Ocean. In the narrow Atlantic and near the coasts, isotherms follow more complicated paths.

Below the surface layer of ocean waters, which are thoroughly mixed by winds and other processes, temperatures change little during the year. It is important to realize that the temperature changes shown in Fig. 7–7 and 7–8 are restricted to waters within a few hundred meters of the surface or less. Only at high latitudes do water temperatures remain nearly constant to great depth. Elsewhere water temperatures drop sharply as one proceeds below the surface zone.

WATER BUDGET

The amount of free liquid water at the earth's surface has remained essentially constant for the past 5000 years, since the retreat of continental glaciers from North America and Europe. At present about 97 percent of the surface water is in the ocean,

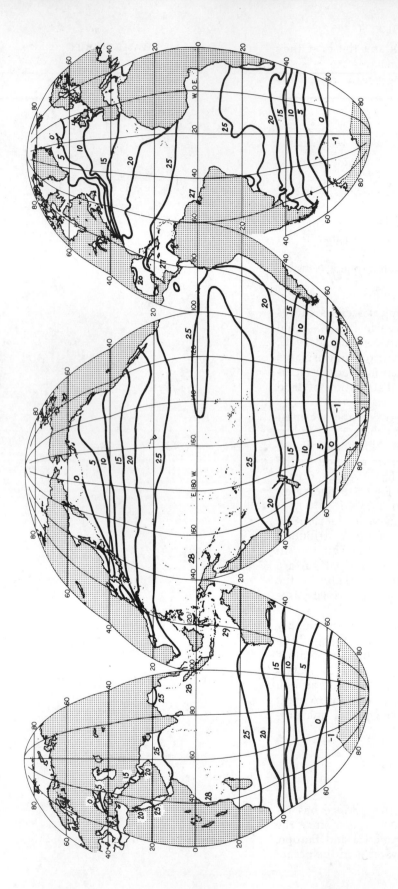

FIG. 7-7
Ocean-surface temperature in February. (After
Sverdrup et al., 1942)

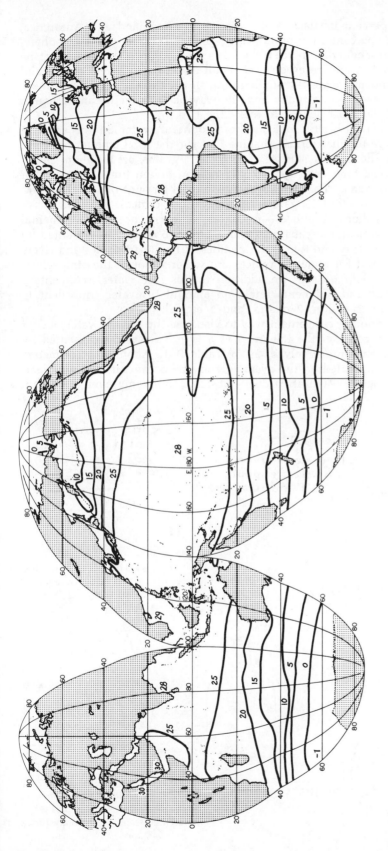

FIG. 7-8
Ocean-surface temperature in August. (After
Sverdrup et al., 1942)

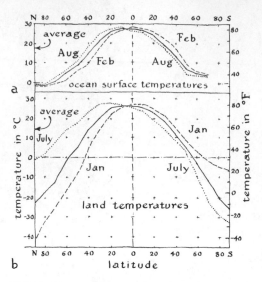

FIG. 7–9

Average and extreme temperatures of the earth's surface: (a). Average and extreme temperatures for the ocean surface. Note that the maximum variation occurs in mid-latitudes, around 40°N and 40°S. [After G. Wüst, W. Brogmus, and E. Noodt, "Die Zonale Verteilung von Salzgehalt, Niederschlag, Verdunstung, Temperatur und Dichte an der Oberfläche der Ozeane," *Kieler Meeresforschungen* 10 (1954), 137–61] (b). Average and extreme temperatures for land areas. Note that the extreme temperature variations occur in polar regions, in contrast to the ocean. (Data from *Smithsonian Physical Tables*, 1964)

2 percent in glaciers and icecaps in Greenland and Antarctica, 0.6 percent in ground water, 0.3 percent in the atmosphere, and about 0.1 percent in lakes and rivers. The ocean is the major water reservoir; the small amounts in lakes, rivers, and atmosphere are essentially in transit back to their source. Ground water also eventually finds its way back to the ocean, although taking much longer to make the trip.

Let us briefly look at the amount of water on the earth in more familiar terms. Imagine that for the total amount of water on the earth's surface, 1.4×10^{24} grams or 1 billion billion metric tons, each 10^{15} tons (1 million billion tons) is equivalent to 1 ounce. That would give us a total amount of water equivalent to about 11 gallons, 1 pint, and 9 ounces. Most of this— 11 gallons, 2 ounces—is contained in the oceans. Continental glaciers and icecaps hold about 1 pint, 6 ounces, represented by the icetray and two extra icecubes in Fig. 7–10. Lakes and rivers account for about ½ ounce, and water vapor in the atmosphere for about 6 drops. In addition to this free water, sedimentary rocks and sediments retain an amount of water equivalent to about 1 gallon, 2 quarts, and 1 cupful.

Another convenient way to visualize the earth's water budget is to consider the thickness of a layer of water represented by the amounts involved, as in Table 7–1. About 97 centimeters is evaporated on the average from the ocean surface each year. Actually, evaporation does not occur uniformly over the ocean,

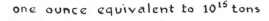

FIG. 7–10

Relative quantities of liquid water, ice, and water vapor at the earth's surface, plus water in sediments and sedimentary rocks.

Table 7–1

WATER BUDGET 183

WATER BUDGET OF THE EARTH*

	Precipitation		Evaporation	
	10^3 km³/yr	cm/yr	10^3 km³/yr	cm/yr
Ocean	320	88	350	97
Continent	100	67	70	47
Entire earth	420	82	420	82

*After Albert Defant, *Physical Oceanography*, Pergamon Press, Elmsford, N.Y., 1961, Vol. I; p. 235.

so our example is slightly misleading. Of this amount, 88 centimeters falls as rain on the ocean and is thus almost immediately returned. The remaining 9 centimeters falls on land and eventually makes its way into rivers to return to the ocean.

Yet another way of describing the water budget or the hydrological cycle is to express amounts in thousands of cubic kilometers, as is done in Fig. 7–11.

Evaporation from the ocean surface is most active in sub-

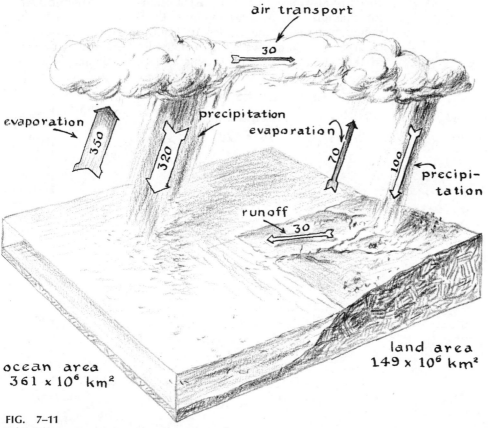

FIG. 7–11
Schematic representation of total amounts of liquid water involved in the hydrological cycle each year, in 10^3 km³.

tropical areas, where it is favored by clear skies and dry winds. These are the Trade Wind zones. The western sides of ocean basins are especially favorable for evaporation because dry winds blow from the continents.

Neither is precipitation uniform over the entire ocean, as Fig. 7–12 shows. Equatorial and subpolar or polar regions experience heavy precipitation; in the tropics particularly, large amounts of rainfall result from an abundance of warm, moist air rising in the equatorial atmospheric circulation. As rising air cools, it loses its ability to hold water vapor, and precipitation results. Satellite photographs, such as in Fig. 7–13, commonly show a line of clouds in tropical regions because of this circulation; this is known as the intertropical convergence. In the high latitudes, strong atmospheric cooling also favors precipitation.

Because of evaporation and precipitation distributions, salinity varies over the ocean surface. Highest salinities occur in open-ocean subtropical areas, where evaporation is pronounced and there are no significant sources of fresh water. Near coasts, rivers discharge fresh water back into the ocean, and salinities are lowered. The surface waters of high-latitude oceans are commonly low in salinity owing to relatively heavy precipitation and little evaporation.

Continents (especially those with mountain ranges) and ocean currents also affect salinity distributions. The Atlantic Ocean, for example, has a higher salinity than the Pacific. Mountain ranges, especially the Andes of South America, cause winds from the Pacific to give up their moisture before blowing across to the Atlantic. High mountains deflect winds to higher alti-

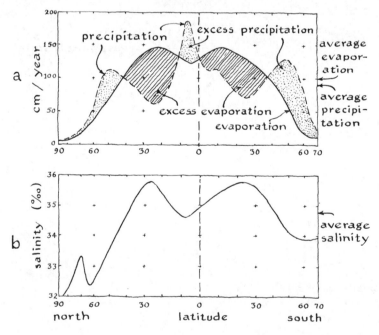

FIG. 7–12
Relationship between evaporation and precipitation at various latitudes (a) and the salinity of surface waters (b). (After Wüst et al., 1954)

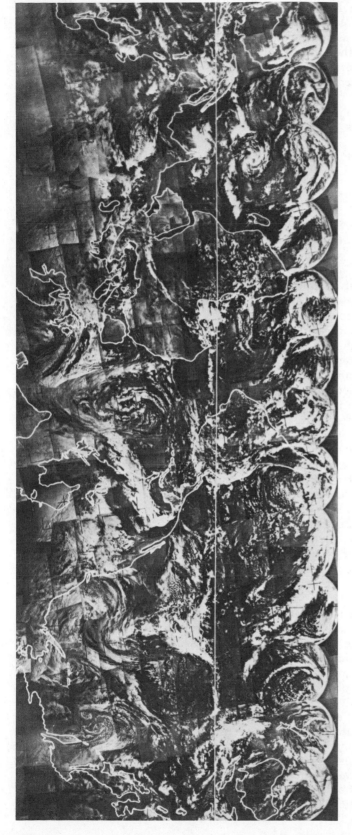

FIG. 7–13
Composite photograph (photomosaic) of the earth surface prepared from 450 individual pictures taken by NASA satellite Tirox IX during the 24 hours of February 13, 1965, the first complete picture of the world's weather. Continents are outlined by white lines. Large areas of white are clouds or the permanent ice covers on Antarctica (bottom margin). Note the large spiral weather systems, especially conspicuous in the Central Pacific. Also note the thin line of clouds along the equator, especially noticeable in the Atlantic. (Photograph courtesy NASA)

tudes, where they are cooled and their water vapor condenses to fall as rain on the mountains. On the other hand, the Atlantic's water vapor is carried to the Pacific by winds blowing across the Gulf of Panama.

A water budget can be constructed for a single region and expressed mathematically as:

Evaporation = Precipitation + Runoff + Contribution from currents

$$E = P + R + C$$

where

E = the amount of water lost through evaporation
P = the amount of water gained through precipitation
R = the amount of water brought to the region by rivers
C = the relative amount of fresh water moved by currents.

The last quantity (C) can be positive if water of lower salinity flows into the region or negative if more saline water flows into the region. Unless there is a pronounced salinity change, the amount of water gained by precipitation and river runoff will equal the amount lost through evaporation and carried away by currents.

To summarize, surface-water salinity in the open ocean is controlled primarily by the local balance between evaporation and precipitation. Where evaporation exceeds precipitation, surface waters are more saline than average ocean water ($S =$ 34.7$^0/_{00}$). This occurs principally in the center of the open ocean at mid-latitudes (see Fig. 7–14). Because these areas are far from the continents, there is no river runoff to compensate for the water lost through evaporation.

Where precipitation and river runoff from continents exceed evaporation, surface-water salinities are lower than for the average ocean. Low-salinity waters are typical of three regions: the tropics, owing to the large rainfall there; high latitudes, because of relatively high precipitation and limited evaporation; and along continental margins, owing to local rainfall and river runoff. Consequently, high surface-water salinities in the open ocean tend to occur near the centers of the ocean basins.

In coastal ocean areas, highest surface-water salinities occur in partially isolated basins of tropical areas, such as the Red Sea, the Arabian Sea, and the Mediterranean Sea. Here water losses due to evaporation are not replaced by river runoff or precipitation. The restricted communication with adjacent ocean waters prevents ready renewal of the surface waters. The result is salinity of surface waters of 40$^0/_{00}$ or more in basins such as the Persian Gulf and the northern Red Sea. These occur at mid-latitudes, around 30° north, where several of the world's great land deserts occur—such as the Sahara, and Mojave. The arid Australian continent also testifies to the low rainfall at comparable latitudes in the Southern Hemisphere.

Heat and water budgets are intimately related. Evaporation of water removes heat from the ocean surface, whereas precipita-

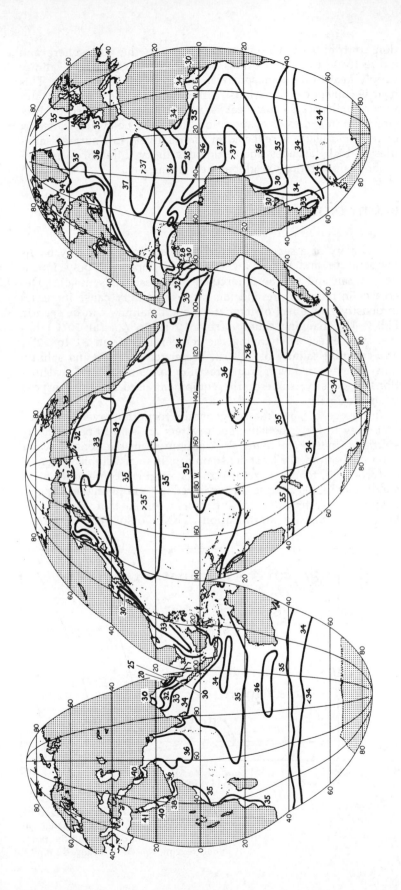

FIG. 7-14

Salinity of ocean-surface water in northern summer. (After Sverdrup et al., 1942)

tion contributes large amounts of heat to the atmosphere and indirectly to the ocean. Areas of relatively high surface salinity supply heat to the atmosphere in the form of water vapor. This heat is transported by winds to the high latitudes. There, water vapor condenses so that heat is given off, accompanied by precipitation in excess of evaporation. Thus the poleward transport of heat and water is reflected in the relatively low surface salinities at high latitudes and regional salinity differences are another manifestation of global heat transport.

DENSITY OF SEAWATER

The density of a seawater parcel is determined primarily by its temperature and salinity. (Pressure is important only in the deep ocean and will be ignored in our present discussion.) Decreases in temperature and increases in salinity cause increases in density. The effect of temperature and salinity can be seen in Fig. 7–15: changing salinity from 19 to $26^0/_{00}$ (at 30°C) has the same effect as changing the temperature from 31 to 12°C at a constant salinity of $20^0/_{00}$. Water temperature and salinity must be measured with great precision to insure accurate calculations of density; some of the equipment used for this purpose is shown in Figs. 7–16 and 7–17.

In the ocean, water masses generally form thin layers arranged according to water density; such waters are said to be *vertically stratified*. In the *stable-density* configuration, the densest water is on the bottom, the least dense on top. If such a stable system is disturbed, it tends to return to its original configuration. A stable situation can also be achieved by supplying energy to a stratified system and mixing the layers so that there is no longer any density stratification. This would be a *neutrally*

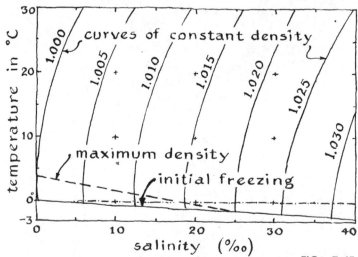

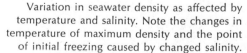

FIG. 7–15
Variation in seawater density as affected by temperature and salinity. Note the changes in temperature of maximum density and the point of initial freezing caused by changed salinity.

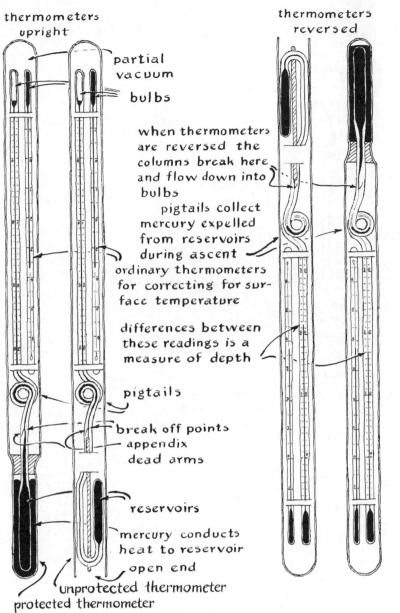

thermometers upright

partial vacuum

bulbs

when thermometers are reversed the columns break here and flow down into bulbs

pigtails collect mercury expelled from reservoirs during ascent

ordinary thermometers for correcting for surface temperature

differences between these readings is a measure of depth

pigtails

break off points
appendix
dead arms

reservoirs

mercury conducts heat to reservoir

open end

unprotected thermometer
protected thermometer

thermometers reversed

FIG. 7–16
Reversing thermometers used to obtain precise temperature measurements in the ocean.

stable system. If any one of the layers is more dense than the one beneath it, the system is unstable and the denser layer tends to sink to its appropriate density level, thus creating a stable structure.

Since ocean water normally has stable stratification, the most dense water might be expected to occupy the deepest part of a given ocean basin. Where such is not the case, this usually indicates that some physical barrier—say, a shallow sill—is preventing lateral spreading of dense water. For example, dense waters formed in the Arctic Ocean cannot readily flow into the North Atlantic because of submarine ridges stretching from Greenland to Scotland.

In the absence of such barriers, a water mass tends to seek

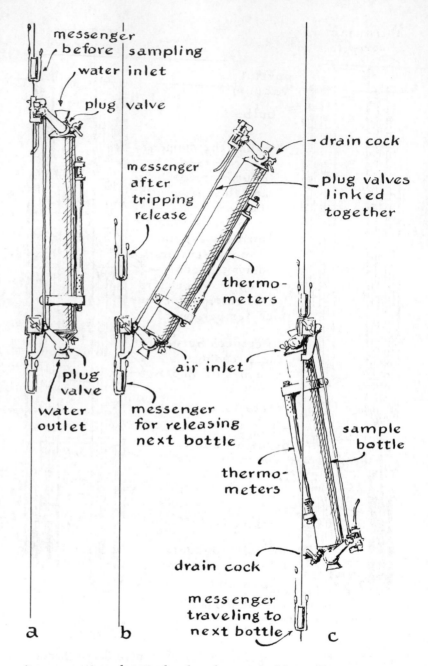

messenger before sampling

water inlet

plug valve

drain cock

plug valves linked together

messenger after tripping release

thermo-meters

plug valve

air inlet

water outlet

messenger for releasing next bottle

sample bottle

thermo-meters

drain cock

messenger traveling to next bottle

a   b   c

FIG. 7–17
Nansen bottles used to obtain water samples and to carry reversing thermometers.

its appropriate density level and to spread laterally. Because of this, lateral changes in water temperature and salinity are much smaller than vertical changes. In general, subsurface waters tend to move along surfaces of constant density rather than across pronounced density gradients.

Energy must be supplied if a density-stratified system is to be destroyed. Below the surface of the open ocean, such vigorous mixing is rare. In coastal waters, on the other hand, mixing is common because of tidal currents and movement of waters across rough bottom topography. Near the ocean surface, energy sufficient to cause mixing is supplied by the wind, passage of waves, and surface heating and cooling.

Well-developed density stratification of the open ocean inhibits strong vertical currents. Only in high-latitude areas is the density structure weak enough to permit widespread but probably weak vertical circulation through the ocean depths. Over most of the open ocean, vertical circulation occurs only within the surface layer.

This pattern is quite different from that of the atmosphere, where strong vertical circulation is a characteristic feature. The ocean is heated and cooled at its upper surface. Heating of the surface inhibits vertical circulation and powers a very inefficient heat engine. In contrast the atmosphere is cooled at the top by radiation to space and heated at the bottom through contact with land and ocean. Consequently, vertical circulation results and is quite well developed through the lower parts of the atmosphere.

Three general depth zones may be identified in the ocean: the surface, pycnocline, and deep zones, schematically represented in Fig. 7–18. The *surface zone* is in contact with the atmosphere and undergoes substantial seasonal changes because of seasonal variability in local water budgets (evaporation, precipitation) and heat budgets (incoming solar radiation, evaporation). The surface zone contains the least dense water masses in the ocean and is the ultimate source of all marine-food production because only there is the light available sufficient for photosynthesis.

The surface zone averages about 100 meters in thickness and comprises about 2 percent of the ocean's volume. Surface waters are generally quite warm—except at high latitudes, where extensive mixing with deeper layers prevents the development of a warm surface zone, even during the summer. Near the surface, waters have nearly neutral stability because of strong mixing, and vertical water movements are quite common. Winds, waves, and strong cooling of surface waters cause mixing in the open ocean.

In the *pycnocline zone*, water density changes appreciably with increasing depth and thus forms a layer of much greater stability than is provided by the overlying surface waters. The formation of a pycnocline zone may be the result of marked changes in either salinity or temperature with depth. Marked

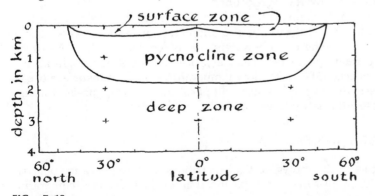

FIG. 7–18
Schematic representation of the horizontal layering of ocean waters.

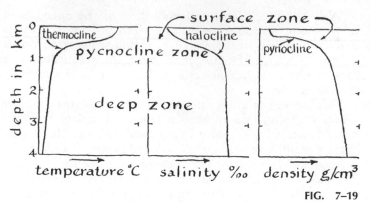

FIG. 7–19

Variations with depth in temperature cause the thermocline. Similar variations in salinity cause the halocline. The pycnocline can be caused by either independently or by both occurring together.

temperature changes with depth are common in the open ocean, as Fig. 7–19 indicates, and there the pycnocline coincides with the *thermocline*, or zone of sharp temperature change. Thermoclines are especially prominent in subtropical open-ocean areas, where the surface layers are strongly heated during much of the year.

At higher latitudes, there is less surface heating by incoming solar radiation. There, the lowered surface salinity caused by abundant precipitation causes a *halocline*, a marked change in salinity with depth. The halocline is well developed over much of the North Pacific Ocean, for example, where it is an important factor in the stability of the pycnocline zone.

Disturbances to the pycnocline zone are opposed by the extremely stable density structure. The pycnocline thus acts as the floor to the surface processes, because low-density surface waters cannot readily move downward through it. Deep-ocean waters move slowly upward through the pycnocline as part of the worldwide deep-ocean circulation. This is a significant factor in oceanic circulation at mid-latitudes and low latitudes. Seasonal changes in temperature and salinity are quite small in waters below the pycnocline.

The *deep zone* contains about 80 percent of the ocean's waters and lies at depths of about 2 kilometers or more—except at high latitudes, where deep waters are in contact with the surface. The relationship between deep-ocean waters and high-latitude areas accounts for the low temperatures of the subsurface ocean. Dissolved oxygen is also supplied to the deep ocean through these high latitude exposures of cold water.

WATER MASSES

Figure 7–20 shows vertical distribution of temperature and salinity in all three ocean basins. The profiles display temperature and salinity distributions along the western sides of the Atlantic and Indian Oceans and the central portion of the Pacific Ocean. Note both the similarities and the differences between this and

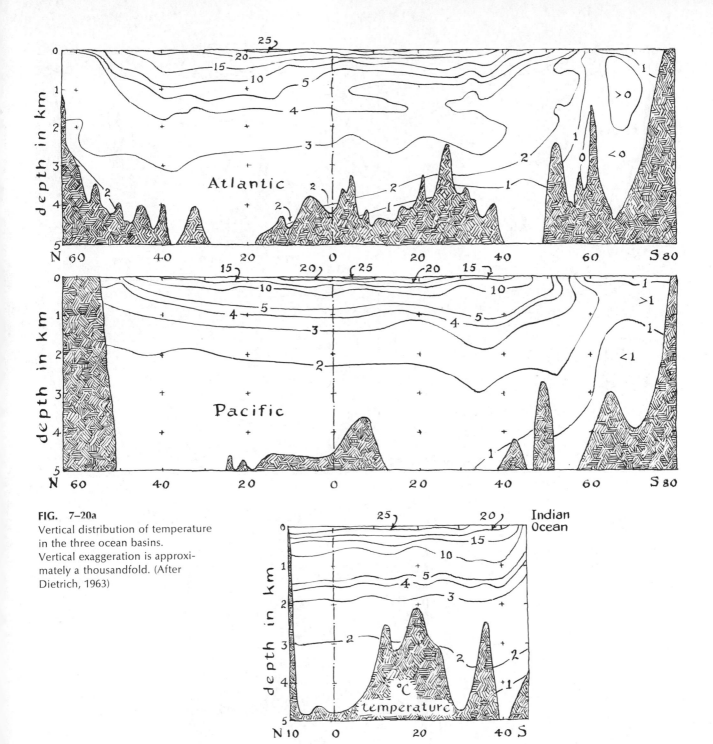

FIG. 7–20a

Vertical distribution of temperature in the three ocean basins. Vertical exaggeration is approximately a thousandfold. (After Dietrich, 1963)

the generalized diagram (Fig. 7–18) we previously considered. Note also the complexities of temperature and salinity distribution evident on even such a simplified diagram.

The base of the surface zone corresponds to the 10°C-isotherm contour on the vertical section. The base of the pycnocline can be taken as the 4°C isotherm. Note that there are

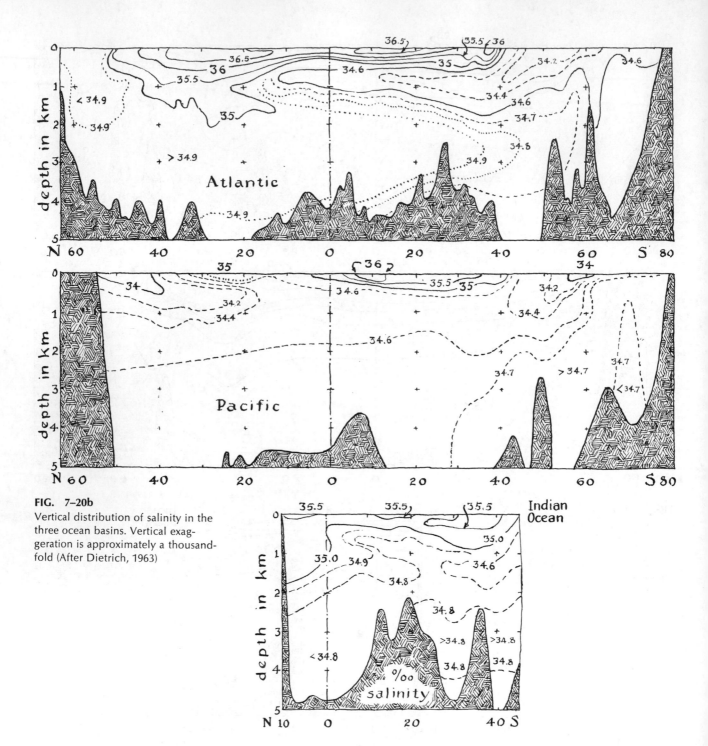

FIG. 7–20b
Vertical distribution of salinity in the three ocean basins. Vertical exaggeration is approximately a thousandfold (After Dietrich, 1963)

strong similarities between the distribution of pycnocline and surface zone in the three ocean basins, but that they are by no means identical. Also note that there are differences between hemispheres even within the same ocean basin.

The vertical temperature and salinity sections show something that we have not previously discussed: distinctly different strata or water masses exist at intermediate depths. For example, there is a relatively low-salinity water mass at about 1 kilometer extending from the Antarctic region into the South Atlantic Ocean. Through analysis of sections such as these, oceanographers are able to determine where water masses form and how they move through the subsurface ocean. This technique has been one of the most rewarding means of mapping subsurface currents. Such studies have demonstrated that the dominant movement of near-bottom currents is generally north–south, in contrast to the primarily east–west movements of currents at the surface and intermediate depths.

Another aspect of the ocean illustrated by the vertical section is that most of the ocean has much lower temperatures, shown in Fig. 7–21, than we might have predicted from examining the ocean surface alone. The average temperature of the entire ocean is 3.5°C, in contrast with the average ocean-surface temperature averages about 17.5°C. Thus it is the surface ocean that serves to buffer our climate over the short term. Most of the deep ocean is too isolated to have much effect on short-term climatic conditions.

Water masses with characteristic temperatures and salinities are formed in particular ocean areas, as listed in Table 7–2. If a

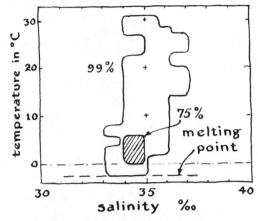

FIG. 7–21
Temperature and salinity of 99 percent of ocean water are represented by points within the large area enclosed by the 99 percent contour. The shaded area represents the range of temperature and salinity of 75 percent of the water in the world ocean. [After R. B. Montgomery, "Water Characteristics of the Atlantic Ocean and of the World Ocean," *Deep-Sea Research*, 5 (1958), 134–48)

Table 7–2

SOME MAJOR WATER MASSES OF THE OCEAN*

	Temperature (°C)	Salinity (⁰/₀₀)
Antarctic Bottom Water	−0.4	34.66
North Atlantic Deep Water	3–4	34.9–35.0
North Atlantic Central Water	4–17	35.1–36.2
Mediterranean Water	6–10	35.3–36.4
North Polar Water	−1–+10	34.9
South Atlantic Central Water	5–16	34.3–35.6
Subantarctic Water	3–9	33.8–34.6
Antarctic Intermediate Water	3–5	34.1–34.6
Indian Equatorial Water	4–16	34.8–35.2
Indian Central Water	6–15	34.5–35.4
Red Sea Water	9	35.5
Pacific Subarctic Water	2–10	33.5–34.4
Western N. Pacific Water	7–16	34.1–34.6
Pacific Equatorial Water	6–16	34.5–35.2
Eastern S. Pacific Water	9–16	34.3–35.1
Antarctic Circumpolar Water	0.5–2.5	34.7–34.8

*After Albert Defant, *Physical Oceanography*, Pergamon Press, Elmsford, N.Y., 1961, Vol. I, p. 217.

mass of surface water acquires a temperature and salinity that make it more dense than the water below, it will sink below the surface. From that point, until it returns again to the surface, the temperature and salinity of the waters change primarily by mixing with other water masses. In the deep open ocean, where there is little energy to cause mixing, this process can be very slow. Temperature and salinity can therefore be used in the open ocean as tags to identify individual water masses as they spread laterally between other masses, or move along the ocean floor.

Observed in combination, temperature and salinity are more informative than density alone. For this purpose, a *temperature–salinity (T–S) diagram* is constructed by plotting points corresponding to the observed temperature and salinity measurements for individual water samples. If conditions are constant and the water mass is homogeneous, the entire mass is represented by a single point on a *T–S* diagram. More commonly, there is some variation within the mass, as well as mixing with adjacent masses, resulting in a *T–S curve*, indicating a continuum of properties within the water column at each station, shown in Fig. 7–22. This *T–S* curve remains the same (or within

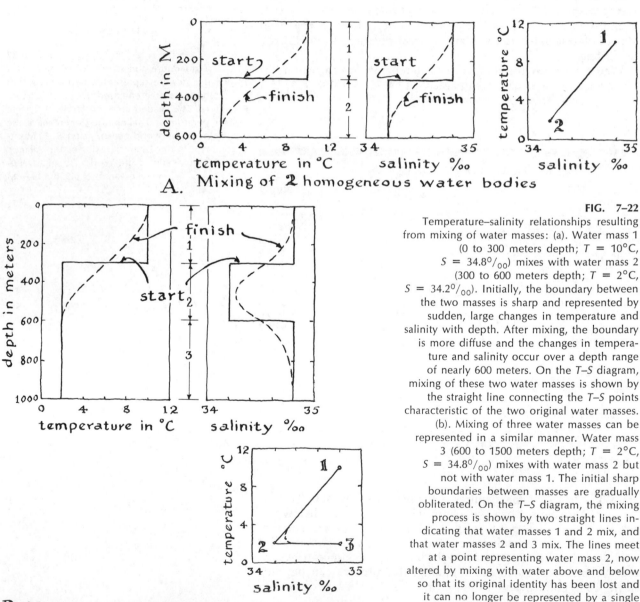

A. Mixing of 2 homogeneous water bodies

B. Mixing of 3 homogeneous water bodies

FIG. 7–22
Temperature–salinity relationships resulting from mixing of water masses: (a). Water mass 1 (0 to 300 meters depth; $T = 10°C$, $S = 34.8°/_{00}$) mixes with water mass 2 (300 to 600 meters depth; $T = 2°C$, $S = 34.2°/_{00}$). Initially, the boundary between the two masses is sharp and represented by sudden, large changes in temperature and salinity with depth. After mixing, the boundary is more diffuse and the changes in temperature and salinity occur over a depth range of nearly 600 meters. On the *T–S* diagram, mixing of these two water masses is shown by the straight line connecting the *T–S* points characteristic of the two original water masses.

(b). Mixing of three water masses can be represented in a similar manner. Water mass 3 (600 to 1500 meters depth; $T = 2°C$, $S = 34.8°/_{00}$) mixes with water mass 2 but not with water mass 1. The initial sharp boundaries between masses are gradually obliterated. On the *T–S* diagram, the mixing process is shown by two straight lines indicating that water masses 1 and 2 mix, and that water masses 2 and 3 mix. The lines meet at a point representing water mass 2, now altered by mixing with water above and below so that its original identity has been lost and it can no longer be represented by a single point on the *T–S* diagram.

a small range of difference) for each location at any point in time, because the water masses present have formed at a constant-source area and have the same properties when they reach any given station. Similar curves are often characteristic of large ocean areas.

Consider the two different water masses in Fig. 7–22(a), each with its characteristic temperature and salinity. As they mix, the newly formed waters will have temperatures and salinities that plot on the line connecting the two original points. Furthermore, a particular position on this line indicates the relative proportions of the waters that have been mixed. For example, a mixture of equal amounts of two different waters will fall midway between them on the line connecting the two points on the T–S diagram.

Adding a third water mass results in a more complicated diagram, but the same general relationship holds. From analyses of T–S curves it is possible to determine the sources of the water masses present at that station. Figure 7–23 shows diagrammatically how mixing of subsurface water layers might appear on a T–S diagram. There is a characteristic reversal point associated with the core (central depth) of water mass B. This can be distinguished until the mass has completely lost its

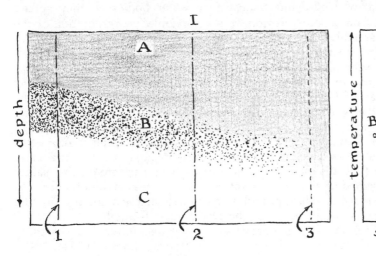

FIG. 7–23
Use of a T–S diagram to study subsurface water movements. Water mass B (intermediate density) flows beneath surface layer A and above the denser water mass C. A series of oceanographic stations taken at three locations show different but related T–S curves. At station 1, closest to the source of B, its characteristic temperature and salinity plot at point B on the T–S diagram. Farther from the source, the water mass mixes with those above and below, so it no longer has its original value although the general shape of the T–S diagram is still quite discernable. At station 3, farthest from the source, water mass B is no longer recognizable and masses A and C mix together, resulting in a continuum of T–S values intermediate between the two.

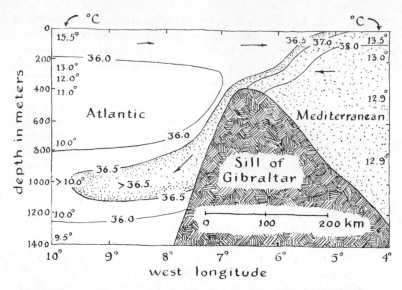

FIG. 7–24
Vertical distribution of salinity and temperature in a cross-section (east–west) through the Strait of Gibralter at the entrance to the Mediterranean. Arrows show the main direction of water movement (After G. Schott, *Geographie des Atlantisches Ozeans*, Boysen, Hamburg, 1942)

identity, although it becomes less and less obvious as mixing progresses.

The so-called *"core" technique* has been used to study movement of well-defined subsurface water bodies as they spread out from a source region. It is a horizontal representation of the most likely paths of water particles. It allows flow across density surfaces since the density of the core water mass usually varies. An example is the warm, saline water (6–10°C, 35.3–36.4⁰/₀₀) formed in the Mediterranean Sea which flows out through the Strait of Gibralter at depth and spreads laterally at mid-depths in the North Atlantic Ocean, pictured in Fig. 7–24. Its distinctive temperature and salinity permit it to be clearly distinguished from water masses above and below. As the water moves away from its source, it mixes with waters adjacent and gradually loses its distinctive characteristics. Comparison of the values observed at any given point where that mass is identifiable with the original values provides an estimate of the amount of mixing that has occurred since the water mass formed.

Each major ocean area has a set of distinctive *T–S* curves. A set of curves is shown for the North Atlantic in Fig. 7–25.

The *T–S* diagram serves other functions besides the tracing of water masses. It permits ready identification of errors in a set of observations. A plot of temperature and salinity observations for a given area should show water density increasing with depth, in a regular progression or sequential order. Points that fall far out of line on the diagram indicate a possible inaccuracy in equipment function or data-reduction technique. Internal-consistency criteria among curves permit a simple graphic check on the accuracy of oceanographic observations over a wide area.

SOUND IN THE SEA

The behavior of sound in the ocean is controlled by temperature, pressure, and salinity. Sound is a form of mechanical energy,

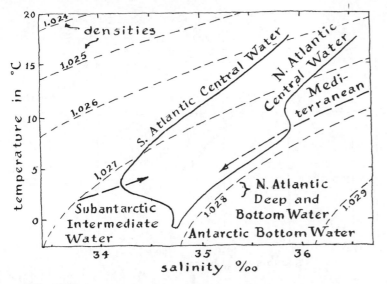

FIG. 7–25
Temperature–salinity curves for characteristic
water masses in the Atlantic. (Data from
Sverdrup et al., 1942)

consisting of the regular alternation of pressure or stress in an elastic substance. Of all the various forms of radiated energy, sound penetrates ocean water best, so that sounds made by organisms or physical processes can be detected at relatively great distances. Sound is also used to determine depth to the bottom, to detect other vessels or objects (including fish), and for emergency signaling using the technique known as SONAR, an acronym for "sound navigation and ranging," an underwater equivalent of radar.

In the ocean, sound travels at a speed of about 1480 meters per second. The speed of sound in water is increased by increasing hydrostatic pressure (which increases with depth), temperature, and salinity. In most of the ocean, the effect of salinity is least important, the effect of temperature the most important. Sound in water can be slowed by bubbles or excessive concentrations of dissolved gases in some bays and harbors.

Because of the pronounced effects of temperature on sound speed, there is substantial variation in the speed of sound with depth in the ocean, indicated in Fig. 26(a). Near the ocean surface the temperature effects dominate. At greater depths, the pressure effect becomes appreciable.

The intensity of a sound wave traveling through water is diminished as energy is absorbed. The absorption of sound in seawater is far greater than in pure water at frequencies less than 100 kilocycles per second (Kilohertz), but is essentially identical to pure water at higher frequencies, as Fig. 26(b) demonstrates. The presence of $MgSO_4$ is the cause of this anomaly. As previously pointed out, $MgSO_4$ in seawater is typically surrounded by a large, bulky hydrated aggregate. This complex aggregate is capable of absorbing acoustical energy over a large frequency range, which accounts for its unusual sound absorption in seawater.

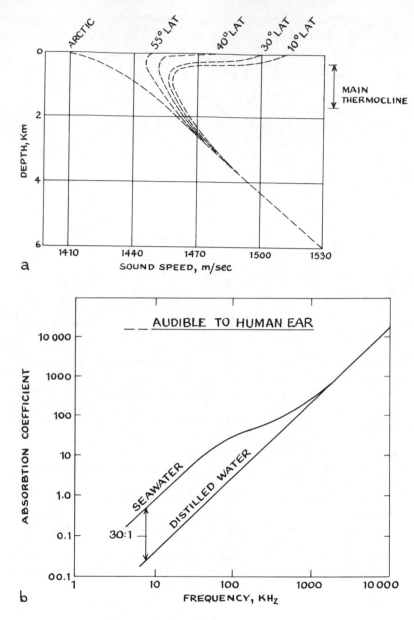

FIG. 7–26
Acoustical properties of seawater: (a) Typical profiles showing variation in speed of sound with depth in various latitudes. (b) Absorption coefficients in seawater and in distilled water. (After R. J. Ureck, *Principles of Underwater Sound for Engineers*, McGraw-Hill, New York, 1969)

Sound passing through water can be considered as traveling in paths, called *rays*. Some of these rays are direct—in other words, they pass directly from source to receiver. Others are indirect and may involve reflection from some sharp boundary such as the air–water interface or the water–sediment interface. This reflection of sound is the basis for the echo-sounding technique previously discussed.

As sound rays pass through water of varying sound speed (caused by variations in temperature, salinity and pressure) the rays are bent toward regions of lower sound speed. Thus sounds originating in the surface ocean tend to be bent or refracted downward into the depth zone of lowered sound speed. Sound originating below such a layer is likewise bent upward. Consequently the sound ray tends to be trapped as if it were in a channel. This *sound channel* is a typical feature of the open ocean at depths of around 1000 meters at mid-latitudes to near the surface in polar regions, and it greatly extends the transmission range of underwater sound. This sound channel has been used for long-range underwater signaling, called SOFAR ("sound fixing and ranging"). Emergency location of downed aircraft crews is just one example of its utility. Other sound channels occur near the surface and in shallow waters less than 200 meters deep, owing to temperature changes or physical boundaries.

The thermocline plays a critical role in sound transmission and reception from the surface. The strong decrease of temperature associated with the thermocline causes sound rays to be bent downward, which results in a significant loss of range for sound-ranging devices. Daily—even hourly—changes in the thermocline may reduce sound-detection ranges.

Marine organisms interfere with the operation of sound-ranging devices in several ways. Systems that passively listen to detect characteristic noises of ships passing or of other activities are especially troubled by the noises made by fishes—especially drumfish, grunts, snappers, and croakers, whose names suggest some of the sounds they make. These sounds have been confused with ships' engines, propellers, and other equipment. The utility of these noises to the fish is not always apparent. Mammals—including whales, dolphins, and seals—also produce noises apparently used for communication and their own form of sound ranging. Even the octopus dragging shells along the bottom has interfered with sound ranging.

More active sound-ranging systems are also affected by the presence of organisms. Many marine organisms have minute cavities that are highly efficient reflectors of sound energy. Consequently, they register as much larger targets on a detection device, and can completely disrupt the operation of the equipment.

SEA-ICE FORMATION

Because of low air and water temperatures in polar regions and in certain coastal areas, sea-ice formation is a conspicuous and important ocean phenomenon (shown in Fig. 7–27), illustrating temperature effects at high latitudes. Sea ice exists year-round in the central Arctic and in some Antarctic bays, extending in winter across the entire Arctic and far out to sea around the Antarctic continent. During most winters it can be seen to form as far south as Virginia on the Atlantic coast of the United States. During glacial times, it must have been even more widespread than at present as shown in Fig. 7–28.

In addition to its local importance, sea-ice formation influences the ocean far beyond the ice-covered area. For instance, Antarctic Bottom Water is thought to form through mixing of extremely cold surface waters and brines released by the freezing and melting of sea ice.

When seawater reaches its temperature of initial freezing,

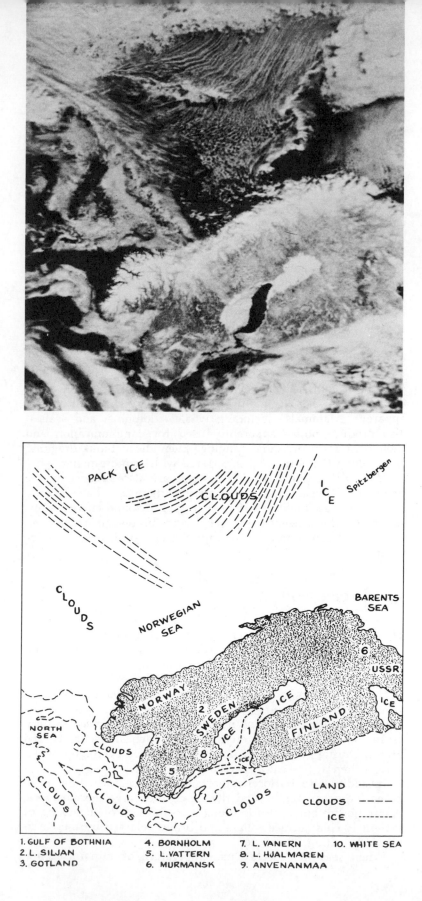

FIG. 7–27
Ice pack in the Arctic Ocean north of Scandinavia, which itself is covered with snow, as seen by the Nimbus-4 weather satellite on April 13, 1970. The pack ice is the irregular white area near the top of the photograph. The Gulf of Bothnia in the Baltic Sea (lower center) is ice-covered. The filmy streaks in the center of the photograph are clouds. (Photograph and explanatory diagram courtesy NASA)

1. GULF OF BOTHNIA   4. BORNHOLM   7. L. VANERN   10. WHITE SEA
2. L. SILJAN   5. L.VATTERN   8. L. HJALMAREN
3. GOTLAND   6. MURMANSK   9. ANVENANMAA

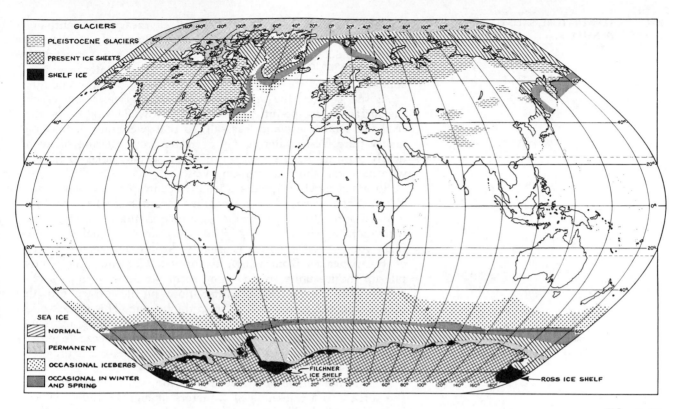

GLACIERS
- PLEISTOCENE GLACIERS
- PRESENT ICE SHEETS
- SHELF ICE

SEA ICE
- NORMAL
- PERMANENT
- OCCASIONAL ICEBERGS
- OCCASIONAL IN WINTER AND SPRING

FILCHNER ICE SHELF

ROSS ICE SHELF

FIG. 7–28

Distribution of sea ice and probable maximum extent of land ice during glacial Pleistocene times.

clouds of tiny ice particles form, making the water slightly turbid. The ocean surface becomes dull and no longer reflects the sky. Initially disklike, the ice particles grow and form hexagonal *spicules* 1 to 2 centimeters long. Sea salt is excluded, as it does not fit into the ice-crystal structure.

The water remaining becomes slightly saltier, so that its freezing point is lowered. As freezing continues, convection cells are set up at the underside of the ice layer. Water comes up from below to replace the increasingly saline brines, which tend to form finger-shaped parcels and stream downward.

Ice crystals form a layer of *slush*, which covers the ocean surface like a blanket. Some crystals grow downward, and cross-connections form between crystals. The result is a thin, plastic layer of ice in which small cells of seawater are trapped. One kilogram of newly formed sea ice contains about 800 grams of ice (salinity $0^0/_{00}$) and 200 grams of seawater (salinity $35^0/_{00}$). Therefore, the average salinity of newly formed ice is about $7^0/_{00}$, depending on the temperature at which it was formed. At temperatures near the freezing point, ice forms slowly so that brines can escape. Relatively little seawater is enclosed in the cells, and the salinity of the ice is low. At lower temperatures, ice forms more rapidly and much more seawater is trapped. In this case, ice salinity is higher, although always less than the seawater from which it formed.

As the weather grows colder and heat is continually removed from the sea surface, new ice is formed. Water in the cells freezes, and their interior walls become thicker. Concentrated brines remain in the ever-smaller cell cavities, and eventually certain salts begin to crystallize out. Sodium sulfate crystals form at $-8.2°C$, whereas sodium chloride does not crystallize above $-23°C$. Salt crystals and ice crystals may intergrow in a completely solid ice structure if temperatures are low enough.

As sea ice ages, the brine cells tend to migrate. Many reach the surface to escape into the seawater below. Through time the seawater retains less and less salt, so that eventually it may be nearly pure fresh water. Pools of water from melting of such nearly pure ice have saved the lives of several polar explorers who have ran short of drinking water.

Sheets of newly formed sea ice break into pieces under stress of wind and waves. These are called *pancake ice* because that term so aptly describes their appearance in the early stages of formation, as shown in Fig. 7–29. As pancakes coalesce they form young sea ice, which is usually flexible and retains traces of the pancakes from which it formed. Currents and winds pile up the ice units so that they form hummocks or pressure ridges. With the freezing of winter snows on the surface, growth occurs from the top as well as from the bottom. This thick, solid mass may break into large *floes* in response to winds and surface currents; these constantly move and shift, freeze together and break loose, buckle up and flatten out as the pack is dragged by currents flowing beneath.

During the course of a freezing season, sea ice may form at the surface to a thickness of 2 meters at high latitudes. When more than 1 year old, it can form fields of coherent ice. Where ice never completely melts away but continues to grow during subsequent seasons, as in the central Arctic Ocean, adjacent units are piled on top and an average thickness of 3 to 4 meters is achieved. During the summer months the central Arctic ice pack melts to a thickness of about 2 meters.

The behavior of oceanic pack ice should not be confused with

FIG. 7–29
Wave action on newly formed sea ice forms pancake ice. (Official U.S. Navy photograph)

the huge masses of land ice that cover Greenland and the Antarctic continent throughout the year. The latter sometimes extends out over bays and sheltered waters as a permanent floating *ice shelf*. Pieces of glacial ice sometimes break off in the ocean, particularly when glaciers flow out of valleys, as in western Greenland. Irregular pieces are known as *icebergs*, as shown in Figs. 7–30 and 7–31; larger, more regular, and usually flat-topped pieces are called *tabular* icebergs, as in Fig. 7–32; these may be tens to hundreds of kilometers long. Icebergs are rarely

FIG. 7–30
An iceberg in pack ice off the coast of Labrador in March, 1962. The blocks of ice and snow on top of the berg are as big as houses. (Official U.S. Navy photograph)

FIG. 7–31
An iceberg along the Grand Banks of Newfoundland is tracked by a U.S. Coast Guard airplane for the International Ice Patrol. (Official U.S. Coast Guard photograph)

FIG. 7–32
A tabular iceberg in Nelville Bay, Greenland. The icebreaker approaching the iceberg is approximately 90 meters long. The iceberg extends 24 m above the water, is about 1.1 km long and about 0.75 km wide. The iceberg extends about 150 m below the surface. (Official U.S. Coast Guard photograph)

more than 35 meters above water; 5 times as much is usually below water.

Icebergs are carried by currents out of the polar regions at speeds of about 15 to 20 kilometers per day. In the North Atlantic they are carried southward by the Labrador Current until they reach the vicinity of the New York–Europe shipping lanes. Icebergs from western Greenland travel up to 2500 kilometers, reaching 45°N, which is the vicinity of the Grand Banks.

Currents around Antarctica keep the icebergs relatively close to the continent. They are known to move up past the tip of South America to about 40°S in the Atlantic but rarely reach beyond 50°S in the Pacific Ocean. In 1894, a piece of ice was spotted at 26°30'S, only about 350 kilometers from the Tropic of Capricorn, which is commonly taken as the margin of the subtropical ocean. Icebergs commonly have about a 4-year life span in the ocean.

CLIMATIC REGIONS

FIG. 7–33
Oceanic climatic zones and their relationships to terrestrial climatic zones: (1). tundra, (2). boreal forest, (3). temperate, (4). desert, (5). savannah, (6). steppe, (7). tropical rain forest, (8). icecap. [After D. V. Bogdanov, "Map of the Natural Zones of the Ocean," *Deep-Sea Research*, 10 (1963), 520–23]

The open ocean can be divided into climatic regions. These regions, indicated in Fig. 7–33, generally exhibit characteristic ranges in surface-water characteristics. These regions also have relatively stable boundaries which change little through time. Note that the climate regions extend generally east–west. Both temperature and salinity of surface waters in these regions are related primarily to incoming solar radiation and the relative amounts of evaporation and precipitation.

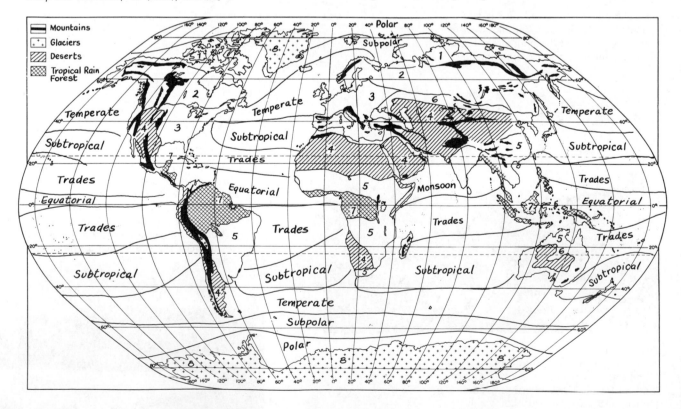

The Arctic Ocean and the band around Antarctica constitute the major *polar-ocean* areas. Ice occurs at the surface most of the year and therefore keeps surface temperatures at or near the freezing point despite the intense insolation during local summer. In winter there is no sunlight. The relatively low surface salinities, combined with the ice cover, act to inhibit vertical mixing, so that there is little mixing between surface and deeper waters.

Subpolar regions are affected by seasonal migration of sea ice. In winter, pack ice covers the area but it disappears during summer, except for isolated blocks and icebergs. There is enough heating to melt the ice and to raise surface temperatures to about 5°C. Abundant light and the stable layer formed by surface warming combine to provide good conditions for short but intensive phytoplankton (marine algae) blooms. This rich plant crop supports large populations of *zooplankters* (minute marine animals), birds, and whales. Diatom shells (frustules) sink to the bottom forming diatomaceous sediments.

The *temperate* region corresponds to the band of westerlies where there are strong storms, especially during winter, with gales and heavy precipitation. These strong winds cause the West Wind Drift, one of ocean's major currents. Extensive mixing resupplies nutrient elements from deep water to the surface layers, supporting a marine food chain that includes large populations of fish. Most of the world's major fisheries are found in this region.

Subtropical regions coincide with the nearly stationary high-pressure cells of the mid-latitudes. Winds are weak and therefore surface currents tend to be weak also. Clear skies, dry air, and abundant sunshine lead to extensive evaporation, so that surface salinities are generally high. The pycnocline is well developed and prevents ready exchange with water beneath. Plants growing in surface waters deplete the supply of nutrient elements, and since there is little circulation of deep water to surface layers, plant productivity is limited in this region. Surface waters and floating materials from a vast area tend to converge at the center of ocean basins in subtropical latitudes. For instance, the Sargasso Sea lies in the Atlantic's subtropical current system and derives its name from the floating seaweed that accumulates there.

The *tropical* (trade wind) regions are characterized by persistent winds blowing from the northeast in the Northern Hemisphere and from the southeast in the Southern Hemisphere. These winds cause well-developed equatorial currents. The trade winds also cause moderate seas, in contrast with the relative calm of subtropical latitudes. Waters in tropical oceans originate in subtropical regions and are therefore more saline than average seawater. Moving toward the equator, precipitation increases, causing a decrease in surface water salinity.

In *equatorial* regions, surface waters are warm throughout the year. Daily temperature changes equal or exceed annual temperature variations. Warm, moist air generally rises near the equator, causing heavy precipitation and relatively low surface salinities. In the Atlantic and much of the Pacific winds tend to be weak. Sailors named parts of this region the *Doldrums* because their ships were often becalmed there. The equatorial countercurrent is located in this belt of weak winds between the strong and persistent trade winds of each hemisphere. In the Indian Ocean, the monsoons complicate the current patterns. Substantial vertical water movement occurs along the equator; this supplies nutrients to the surface layers and causes the area to be quite productive of phytoplankton and other organisms that feed on them.

SUMMARY OUTLINE

Ocean and atmosphere—closely coupled systems

Approaches used: Budgets and distribution of properties. .

Light in the ocean
    Solar radiation the source of all energy for heating ocean and atmosphere
    Infrared (heat) absorbed at surface
        In coastal waters, turbidity of water accelerates absorption and scattering of solar energy; water is most transparent to blue-green light
        Open ocean has blue color resulting from scattering by the water

Heat budget and atmospheric circulation
    Inclination of axis causes seasons; rotation spreads insolation over surface
    Most heat received between 40°N and 40°S, lost to space in high latitudes
        Heat radiated back to space as infrared
        Latent heat of water dominates heat transport in atmosphere
    Atmospheric circulation forms cells
    Zones of equal temperature generally parallel equator

Water budget—intimately involved in heat transport
    97% of free water is in the oceans
    Most evaporation in subtropical areas; precipitation heaviest in equatorial regions and in high latitudes
    Salinity controlled by balance between evaporation and precipitation
        High salinity most common in center of ocean basins
        Low salinity in coastal regions due to river runoff and in high latitudes because of high precipitation
    Poleward heat transport reflected in low salinities

Density of seawater
    Controlled primarily by temperature and salinity; small pressure effect except at great depths in ocean
    Usually stable density structure in the ocean
        Vertical stratification inhibits vertical movements of water
        Surface, pycnocline, and deep zones can be distinguished
    Thermocline—sharp temperature gradient
    Halocline—sharp salinity gradient
    Pycnocline—sharp density gradient

Water masses
    Temperature and salinity useful to distinguish subsurface water masses (T-S diagram)

Sound in the sea

  Sound is a form of mechanical energy; travels about 1480 m/sec
  in the ocean; used for long-range detection of objects (SONAR)
  Speed increased by increasing pressure, temperature, salinity
  Sound channel at around 1000 meters in open ocean used for
  long-range underwater signaling (SOFAR)
  Many fishes and mammals make sounds in the ocean; reflection
  of sound energy by organisms interferes with signaling devices

Sea-ice formation

  Ice spicules form slush, then plastic layer with enclosed brine
  cells
  Salts crystallize from brines; cells migrate to water below
  Pancake ice forms and develops into floes as season advances
  Icebergs derived from ice on land

Climatic regions

  Recognizable in open ocean, generally east-west trending
  Surface waters respond to local climate, especially winds

PETTERSSEN, SVERRE. 1968. *Introduction to Meteorology,* 3rd ed. McGraw-Hill,
New York. 333 pp. General introductory text, elementary to intermediate in
difficulty.

PICKARD, G. L. 1964. *Descriptive Physical Oceanography: an Introduction.* Pergamon Press, Elmsford, New York. 199 pp. Elementary.

SELECTED REFERENCES

OCEAN CIRCULATION

Eighteenth-century navigational instruments: Hanging compass (upper); Log line and glass (lower) The position of a ship can be determined by *celestial navigation*—that is, by observing the sun, moon, or stars. When these are not visible, position can be determined by *dead reckoning:* estimation of a ship's progress by its presumed direction of travel and speed. The *magnetic compass*, indicating north, can be used to determine the ship's heading. Rate of travel is estimated by measuring the apparent speed and making allowances for known currents and the effects of wind and weather.

Measurement of speed was traditionally made using the *log line* and *glass*. The log, or log chip, is a thin wooden triangle weighted with lead on the curved edge so it will float with the point up, providing a point fixed in the water. This is attached to a line, often marked by knots at 15.7-meter intervals for measuring a nautical mile of approximately 1.85 kilometers (6076 feet). A glass designed like a small hourglass would then be used to indicate a 30-second time interval.

To measure the ship's speed, an officer of the watch would heave the log over the stern of the ship on the lee side. The line would run through his hand, and when a marker placed about 20 meters from the log chip appeared, he called to the glass-holder, "Turn." The glass-holder inverted the glass and called, "Done." When the sand had run out, the glass-holder called, "Stop." The officer then observed the point on the line which indicated, by the number of knots which had passed his hand, the velocity of the ship in "knots"—i.e., nautical miles per hour.

Modern taffrail logs, consisting of a spinner towed in the water and a recording instrument attached to the taffrail, continuously display the ship's speed and the distance it has traveled.

211

FIG. 8–1

Various types of drifting devices can be used to provide information about movements of waters. These devices are specially weighted so that they float near the bottom and thus provide information about near-bottom currents. Similar devices without weights are used to provide information about surface currents. (Photograph courtesy Woods Hole Oceanographic Institution)

Currents—large-scale water movements—occur everywhere in the ocean. The forces causing major ocean currents come primarily from winds and from unequal heating and cooling of ocean waters. Both are the result of unequal heating of the earth's surface. Ocean currents contribute to the heat transport from tropics to poles, thereby partially equalizing earth-surface temperatures. In this chapter we shall study these currents, including their locations and causal forces.

Ocean currents have been charted for hundreds of years. Modern techniques for current observation involve sophisticated buoys with meters that record the direction and speed of water movements and then relay the data to a ship that services the buoy. Other buoy systems transmit data back to shore stations. Some current-measuring systems use satellites as a communication link.

The bulk of current measurements upon which our current maps are based were made in past centuries by navigators who observed how a ship's course is deflected by movements of water through which it is passing. A navigator, having predicted that his ship would be in one location, after steaming for a time found that his final position differed from the prediction. Assuming no errors in navigation and in the absence of winds, the displacement (called the *set*) is caused by currents. Measuring and recording such deflections of ships' courses over many years permitted Matthew Fontaine Maury to compile the observations of navigators in the record files of the U.S. Navy and to draw early charts of ocean currents. Today, we still rely on similar information sources.

Many observations when averaged over many years provide a picture of the average currents. When the effects of variable wind and short-term tidal currents cancel out, only the average current can be detected. Minor variations and even some seasonal variations are difficult or impossible to detect with such data. Detailed knowledge of currents and their variability is obtained by employing more modern methods over large ocean areas.

The movement of floating objects also provides information about surface currents. Wreckage of ships or glass floats used in Japanese fishing nets may move across the Pacific, providing evidence of a current flowing eastward. The wreckage of the ship *Jeanette*, crushed in Arctic ice in 1881 north of the Siberian coast, was discovered near southwestern Greenland three years later. This provided compelling evidence of a current from Siberia toward Greenland and gave an indication of its average speed.

Drift bottles or floating plastic envelopes are often prepared with messages and numbered cards, as shown in Fig. 8–1, to provide similar data, but on a more regular basis. These bottles, released by ships, planes, or offshore installations, carried by surface currents, eventually float ashore to be picked up on the beach or by boats. The finder returns the numbered card and furnishes the data and location of its finding to the organization which released it. This information combined with the release data and location also provides information about surface currents. The data do not, however, indicate the exact path traveled by the bottle, but only its origin, end point, and the time elapsed between.

Direct current observations are made by current meters such as in Fig. 8–2, which work like wind vanes to indicate current direction. Propellers or rotors on the meter measure current speed. There are many ingenious methods for recording these data over long periods so that both short-term fluctuations and long-term average currents can be calculated.

Using meters, current speeds and directions can be determined at the surface and at all depths in the ocean. They can be suspended from anchored ships, but the ship's own motion in the water makes it difficult to interpret the results. Mooring of meters on buoys or fixed platforms is more satisfactory, although expensive in deep-ocean areas.

Two other ingenious methods are used to study currents below the surface. Near-surface currents can be studied with *parachute drogues*—parachutes in the water rigged to be moved by currents, as illustrated in Fig. 8–3. The drogue is attached to a surface float that can be tracked by ship. Continual tracking of the float provides a direct indication of current speed and direction at the depth at which the parachute was placed.

The *Swallow float* (invented by the English oceanographer John Swallow) is a cylinder whose density is carefully adjusted so that it floats at a preselected depth, where it moves with the water. In the float is a powerful sound source called a "pinger." A ship at the surface receives the "pings" and by recording their location traces the movements of the float as it moves in the

FIG. 8–2
Ekman current meter used to record current speed and direction. The meter is suspended below the surface on a wire hung from the ship. (Photograph courtesy Woods Hole Oceanographic Institution)

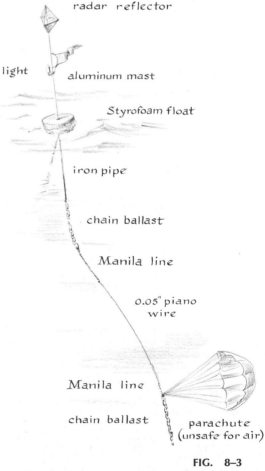

radar reflector

light

aluminum mast

Styrofoam float

iron pipe

chain ballast

Manila line

0.05" piano wire

Manila line

chain ballast

parachute (unsafe for air)

FIG. 8–3
Parachute drogue rigged to permit tracking movement of subsurface water masses. (Redrawn after Von Arx, 1962)

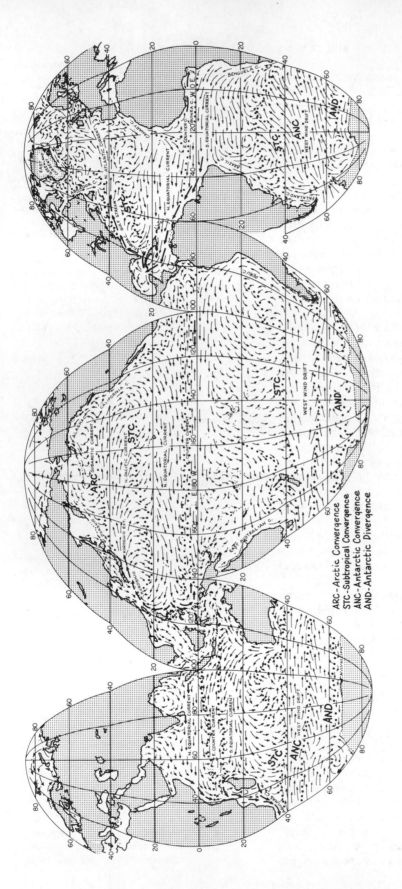

FIG. 8–4

Ocean-surface currents in February–March. (After Sverdrup et al., 1942) See back inside cover for a larger copy of this figure.

deep current. Just as in the case of the current meter, net currents can be calculated from long-term average movements of these floats.

All of these deep-ocean current-measurement techniques are expensive in terms of both equipment and ship time, and therefore we have limited data on subsurface currents. Lacking a source of subsurface-current information similar to that compiled for hundreds of years by surface ships, we find that most of the data on deep-ocean currents are deduced from changes in ocean-water properties, such as the temperature of deep waters or variations in dissolved-oxygen concentrations.

In a later section we shall see that it is also possible to determine current speed and direction indirectly. This so-called *geostrophic approach* uses precise temperature and salinity determinations to map density variations that can cause currents.

GENERAL SURFACE CIRCULATION IN THE OPEN OCEAN

The general surface circulation of the ocean is shown in Fig. 8–4. Circulation patterns are more or less similar in all three major ocean basins, despite their geographic differences. Much of the ocean surface in equatorial regions is dominated by the westerly flow of water in both the North and South Equatorial Currents, set up primarily by the trade winds. These are separate equatorial currents in the Northern and Southern Hemispheres, separated by a narrow current flowing eastward, the Equatorial Countercurrent, in the region of the light and variable Doldrums. Associated with each of these equatorial currents is a current *gyre*—a nearly closed current system. In each case, the current gyre is elongated east–west and lies primarily in the subtropical regions, centered around the 30°N and 30°S latitudes.

In addition to an equatorial current, each of these current gyres includes another east–west current flowing in a direction opposite to the equatorial currents. This is the West Wind Drift, the largest and most important current in the Southern Hemisphere. In the North Pacific, the West Wind Drift is also called the North Pacific Current. The North Atlantic Current is the continuation of the Gulf Stream System in the Atlantic and occupies a position similar to the West Wind Drift in the other ocean basins. The Indian Ocean has an equatorial current in the Southern Hemisphere. Near Asia seasonally variable winds caused by the heating and cooling of the land cause a seasonably variable current system—the monsoon currents, indicated in Fig. 8–5. In winter, a North Equatorial Current is present.

In the subpolar and polar regions of all the ocean basins, there are smaller current gyres. These gyres of the high latitudes circulate "opposite" to the subtropical gyres. Because of the position of the continents, subpolar gyres are well developed in the Northern Hemisphere, where their movement is counterclockwise (see Fig. 8–4). (Note the clockwise direction of the North Atlantic Current–Gulf Stream System–Northern Equatorial Current systems which form the North Atlantic subtropical gyre.)

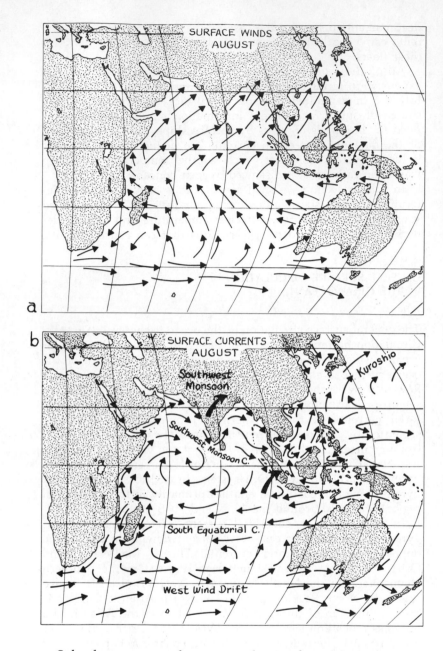

SURFACE WINDS
AUGUST

FIG. 8–5

Surface winds (a) and monsoon currents (b) in summer during the Southwest Monsoon.

Subpolar gyres can be seen in the Southern Hemisphere as well, but the basic circulation of that region does not favor their development. The Antarctic continent occupies a central position in the Southern Hemisphere circulation, so that the West Wind Drift essentially flows around it rather than being broken up and deflected toward the South Pole by continental land barriers. A few small, clockwise gyres can, however, be seen in the vicinity of Antarctica. (Note the counterclockwise direction of the West Wind Drift–Peru Current–Southern Equatorial Current system.)

Surface currents are changeable and far more complicated than can be shown in Fig. 8–4. Irregularities of the continental coastline cause local eddies in coastal-ocean currents, and sea-

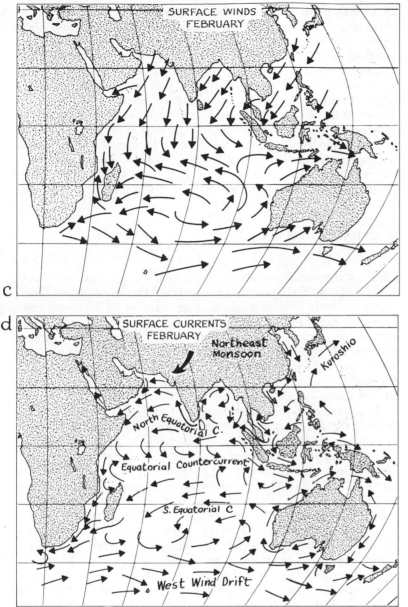

FIG. 8–5
(contd)

Surface winds (c) and monsoon currents (d)
in winter during the Northeast Monsoon.

sonal wind shifts cause major changes in currents, especially
common in the coastal ocean. The northward-setting Davidson
Current of the Washington–Oregon coast, for instance, is well
developed in winter, when strong winds from the southwest
hold fresh water along the coast and drive the surface waters
northward. During summer, when north winds tend to drive
water offshore, the Davidson Current disappears until the winds
shift back to the southwest. Along the Atlantic coast of the U.S.,
the extension of the Labrador Current brings cold northern
waters as far south as Virginia in winter. During summer, cold
waters from the Labrador Current extends no farther south than
Cape Cod, Massachusetts.

The most striking of seasonal current changes is the monsoon circulation of the Northern Indian Ocean. This seasonally variable current system is intimately associated with the monsoon winds. During summer (May to September), Asia is greatly warmed relative to the adjacent ocean. As the warmed continental air rises, it draws air from the Indian Ocean toward the land, as shown in Fig. 8–5(a). When this occurs, the Southwest Monsoon Current replaces the Northern Equatorial Current, as shown in Fig. 8–5(b), in the Indian Ocean. In winter (November to March), the reverse takes place: air over Asia is much colder than air over the ocean, so that the flow is from the land toward the ocean, as shown in Fig. 8–5(c); the Northern Equatorial Current reappears, as shown in Fig. 8–5(d), and the Indian Ocean circulation resembles that of the other ocean basins.

BOUNDARY CURRENTS

Most of the ocean surface is dominated by the east–west currents, such as the equatorial currents of the West Wind Drift. These currents involve east–west movements of large volumes of water generally remaining in the same climatic zone. The waters have ample opportunity to adjust to the temperature of the climatic regime in which the current is located. These currents are deflected when they encounter the continents, and the water flows are split into *boundary currents* that flow more or less north–south.

Boundary currents are important to us as land-dwellers. They include the strongest currents in the ocean (the *western boundary currents*); they also play a major role in transporting heat from tropics to polar regions. Furthermore, these currents form boundaries to the circulation of coastal waters. Where boundary currents are strong, the break between coastal- and open-ocean circulation is sharp, which is the case on the western side of ocean basins, especially in the Northern Hemisphere. Where boundary currents are relatively weak—as on the eastern side of ocean basins—the separation between coastal- and open-ocean circulation tends to be more diffuse.

The characteristics of boundary currents on different sides of the ocean differ substantially, as Table 8–1 indicates. Due to the rotation of the earth on its axis, there is a tendency for current gyres to be displaced toward the west. This results in deep, narrow, fast-moving currents on the western boundaries of ocean basins (Gulf Stream, Kuroshio) in contrast to relatively shallow, broad, and slow currents along the eastern boundaries (Canaries Current, California Current). This tendency is well developed and clearly discernable in the Northern Hemisphere but not so obvious in the Southern Hemisphere.

Surface water displaced to the west tends to pile up there, deepening the pycnocline. Phytoplankton grow in this accumulation of surface water and deplete the nutrient supply, which is not replaced by nutrient-rich waters from below the pycnocline. This tends to limit populations of marine organisms on the western side of ocean basins, except where vertical mixing is especially strong. In regions dominated by eastern boundary

Table 8–1

COMPARISON OF BOUNDARY CURRENT SYSTEMS IN THE NORTHERN HEMISPHERE

Type of Current (Example)	General Features	Speed (km/day)	Transport (Millions of Cubic Meters/sec)	Special Features
Eastern boundary currents California Current Canaries Current	Broad, $\approx$ 1000 km Shallow $\leqq$ 500 m	Slow, tens of km/day	Small typically 10–15	Diffuse boundaries separating from coastal currents Coastal upwelling common Waters derived from West Wind Drift
Western boundary currents Gulf Stream Kuroshio	Narrow $\leqq$ 100 km Deep-substantial transport to depths of 2 km	Swift, hundreds of km/day	Large, usually 50 or greater	Sharp boundary with coastal circulation system Little or no coastal upwelling; waters tend to be depleted in nutrients, unproductive Waters derived from Trade Wind belts

currents prevailing winds blow surface waters away from ocean-basin margins. Subsurface waters are drawn upward to replace them, in a process known as *upwelling* which we discuss in Chapters 11 and 12. For this reason, parts of the coastal ocean along eastern basin margins are extremely productive, as shown in Fig. 17–2.

The Gulf Stream System is a prime example of a western boundary current. Actually a complex of filamentary flows, eddies, and meanders (see Fig. 8–6), it includes several currents: the Florida Current (see Fig. 8–4), extending from the tip of Florida to Cape Hatteras, North Carolina; the Gulf Stream, extending from Cape Hatteras to the tip of the Grand Banks, Newfoundland; and its eastern extension forming the North Atlantic Current with its several branches and eddies.

Water moving in the Gulf Stream System comes from the equatorial currents, especially the Northern Equatorial Current. In the Atlantic substantial amounts of surface waters cross the equator. Part of the Southern Equatorial Current is deflected into the Northern Hemisphere by the eastern projection of South America. These currents feed the Gulf Stream System; details of this process remain to be worked out.

The Gulf Stream system forms a partial barrier between adjacent surface-water masses. Over the continental shelf and slope, the coastal-water masses typically exhibit seasonably variable salinity and temperature, while in contrast, waters in the midst of the Gulf Stream are fairly constant, with temperatures typically about 20°C or higher and salinities of around $36^0/_{00}$. There is some movement of surface waters toward the center of the North Atlantic subtropical gyre. This area, known as the *Sargasso Sea*, has indistinct boundaries formed by current systems. Average surface temperatures are 20°C or more and

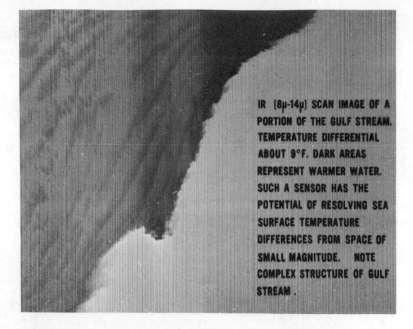

FIG. 8–6
Photograph using infrared temperature sensor shows that the edge of the Gulf Stream is marked by a sharp temperature change between colder waters from the coastal ocean (the lighter shade) and the warmer waters of the Gulf Stream itself. Note the complex structure of the small water parcels within the Gulf Stream (left side). An electronic sensor carried by low-flying aircraft detects infrared radiation coming from the surface layer, and an image is produced by electronic equipment. (Photograph courtesy NASA and U.S. Naval Oceanographic Office).

salinities are around 36.6⁰/₀₀. The relatively high salinity is due to evaporation in excess of precipitation.

In the core of the Florida Current speeds of 100 to 300 centimeters per second have been measured (corresponding to 2 to 6 knots). Current speeds are highest at the surface, in a band about 50 to 75 kilometers wide. Below the surface, at depths of 1500 to 2000 meters, current speeds are measured at 1 to 10 centimeters per second. This is truly a "deep-draft" current, and therefore does not encroach upon the continental shelf at this point. It flows along the edge of the North American continent until reaching Cape Hatteras, North Carolina where it moves seaward.

After passing Cape Hatteras, the Gulf Stream System tends to form a series of large curves or *meanders*, which form, detach, and reform in a complicated manner. The cause of these meanders is still not known, but may be related to the topography of the ocean bottom over which the current moves; if so, this would be one of the few cases in which ocean-bottom topography appears to influence ocean-surface currents. (In general, surface currents involve only water above the pycnocline—perhaps 100 meters thick—and are therefore unaffected by ocean-bottom topography.) After passing the Grand Banks, off Newfoundland, the flow forms the rather diffuse North Atlantic Current, a relatively shallow surface current.

Separating the Gulf Stream from adjacent, slower-moving water masses are a series of *oceanic fronts*. These are sudden changes in water color, temperature, and salinity. They are often spaced a few kilometers apart, but sometimes nearly overlapping. Even more obvious to an observer aboard ship is the change in color from greenish-gray or bluish coastal waters to the intense cobalt blue of the Gulf Stream waters. Often the front is also marked by choppy waters. Such fronts mark *convergences*—areas toward which surface waters flow from different or opposing directions, tending to sink at the region where they meet.

Eastern boundary currents are much less spectacular. The waters move much more slowly and the boundaries between the boundary current and coastal currents are more diffuse. Eastern boundary currents can flow over the continental margin. Like all boundary currents, they participate in the global heat transport by moving cooler waters toward the equator as in the case of the California Current.

LANGMUIR CIRCULATION

Momentum imparted to surface-ocean waters by winds blowing across the ocean also causes vertical movements in near surface waters. Heat, dissolved gases, and other substances are quickly mixed through surface waters to depths of a few meters to a few tens of meters. Part of this mixing results from wind-generated turbulence—random motion of water parcels of many different sizes.

Winds also cause cellular circulation patterns to be set up; these cells are known as *Langmuir cells* after their discoverer, Irving Langmuir. In Langmuir cells water moves with screwlike motions, in helical vortices alternately right- and left-handed (see Fig. 14–15). The long axes of these cells generally parallel the wind direction. The cells tend to be regularly spaced and often arranged in staggered parallel rows. Because of the counterrotation of the cells, alternate convergences and divergences are formed at the surface. Floating debris collects in the convergences, such as the Caribbean's floating sargassum weed whose distinctive pattern on the water surface can be seen by observers from high-flying aircraft. Between the lines of convergence are lines of divergences where water moves upward and along the surface of each cell toward the next convergence.

Langmuir cells form when wind speeds exceed a few kilometers per hour. The higher the wind speed, the more vigorous the circulation. When evaporation and cooling of surface waters tends to increase water density and favor convective movements, Langmuir cells can form at relatively low wind speeds. Increased stability resulting from surface warming or lowered salinity tends to inhibit cell formation, so that stronger winds are required before it can be set up.

Jet streams of water move downward under the convergences. In large lakes, where this type of circulation has been extensively studied, downwelling streams have been observed to extend 7 meters below the surface. Their speeds have been measured at about 4 centimeters per second. In the adjacent divergences, upwelling waters moved at speeds of about 1.5 centimeters per second.

The vertical dimension of these cells is dependent on wind speed and the vertical density structure of the surface layers. If the mixed layer is deep, there is relatively little hindrance to the vertical extent of the cells. If the mixed layer is shallow, the cells may not be able to penetrate the pycnocline. Such vertical water circulation may partially control the depth of the pycnocline. This is also an important mechanism for transporting heat, momentum and various substances from the surface to subsurface layers.

FORCES CAUSING CURRENTS

The circulation of the surface ocean is primarily wind-driven. The major prevailing wind systems supply much of the energy that keeps surface waters circulating, as is made clear by comparing the generalized surface winds in February, illustrated in Fig. 8–7(a), with the ocean-surface currents in February and March shown in Fig. 8–4. We have already discussed how seasonal wind changes cause surface-current shifts near Asia in the monsoon areas and along other coastal areas. (In the next section, we shall discuss how the wind causes water movements and how these are altered by the earth's rotation.)

In addition to the *drift currents* caused by the wind, other currents—called *geostrophic currents*—result from the distribution of water density, which is in turn controlled by water temperature and salinity. For our purposes it is convenient to discuss these two driving forces separately. In the ocean their effects combine to produce the currents we have previously described. Generally the currents below the depth to which wind effects penetrate are geostrophic currents.

Both drift and geostrophic currents move water horizontally. Other density-controlled currents are responsible for vertical

FIG. 8–7a

Generalized surface winds over the ocean in February.

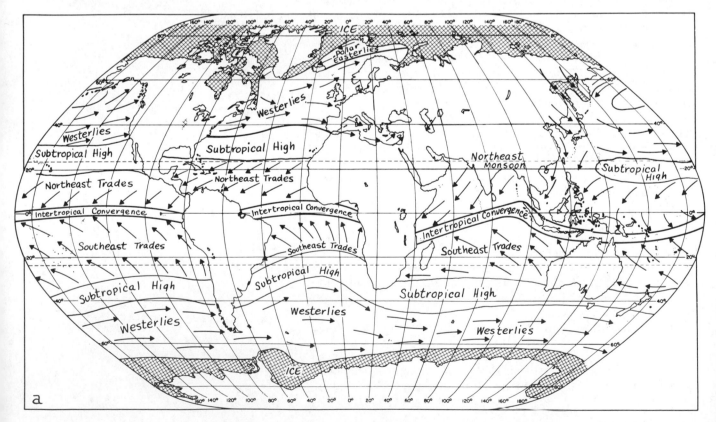

water movements that supply deep- and bottom-water masses to all ocean basins. Because both heat and salinity are involved, this circulation is referred to as *thermohaline circulation*. This circulation controls the vertical distribution of temperature and salinity in the ocean. The distinctions among the various types of currents are arbitrary and they are not easily separated.

The earth's rotation about its axis, which causes an apparent deflection in the trajectory of artillery shells and rockets, also acts as a modifying influence on winds and ocean currents. Although we know that the earth constantly rotates about its axis, it is difficult for us to appreciate all the implications of that fact. Except near the equator winds and ocean currents follow curved paths instead of the straight ones that a simple theory would predict for their movement if the earth were not rotating. To account for this discrepancy, we include the *Coriolis effect* (sometimes called the Coriolis force) in predictions of long-range wind or water movements. This effect does not set winds or waters in motion; it deflects them after they are in motion.

To understand the Coriolis effect, consider what happens to a rocket fired toward the North Pole from the equator. As it sits poised on the launching pad, the rocket is already moving eastward at a speed of approximately 1700 kilometers per hour by virtue of the earth's rotation (see Fig. 8–8). After launching, the rocket still moves eastward at this speed in addition to the speed imparted to it during the launch. If the rocket were aimed *along* the equator, it would move in a straight line with no deflection— that is, it would not demonstrate the Coriolis effect—just as it would have on a stationary earth.

FIG. 8-7b

Generalized surface winds over the ocean in August.

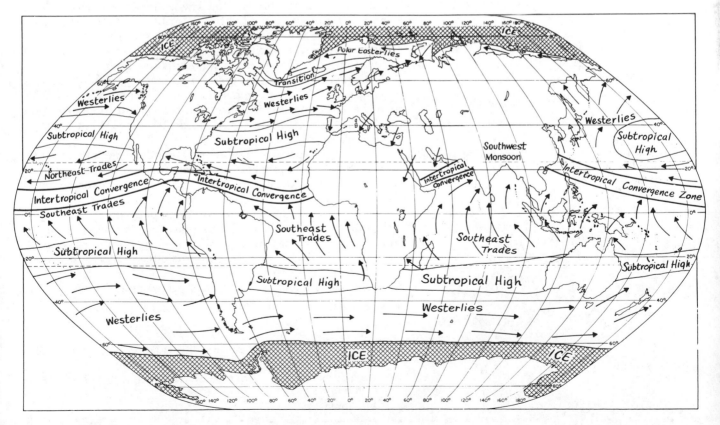

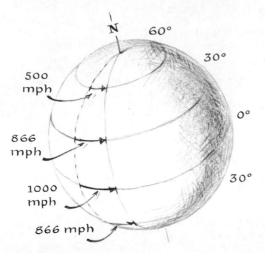

FIG. 8–8

Note the change in speed of the earth's surface moving in an eastward direction going from the equator to either pole. A rocket moving from the equator to the north pole would apparently be deflected to the right—the Coriolis effect.

FIG. 8–9

Some possible paths of particles moving freely over the earth's surface showing the influence of the Coriolis effect.

The rocket moving northward from the equator moves over portions of the earth where the surface is moving eastward at progressively slower speeds than that of the equator. At the Pole, the surface has no lateral eastward movement (just as there is no lateral movement at the center of a phonograph record, whereas the outer edge is moving rapidly) relative to the center. At New Orleans, for example, (approximate latitude 30°N), the earth's surface moves eastward with a speed of about 1500 kilometers per hour. But the rocket when it passes overhead still has a net eastward speed of 1700 kilometers per hour, and is therefore moving eastward faster than the earth beneath it. Consequently the rocket seems to veer to the right when an earthbound observer looks along the flight path. An observer on the moon, however, would see that the rocket was traveling a straight line, while the earth turned beneath it. As the rocket goes farther north, the apparent deflection increases and is a maximum at the poles. The Coriolis effect is also proportional to the speed of the rocket: the greater the rocket's speed, the greater the deflection in a given period because it has travelled farther.

A similar argument can be developed for the Southern Hemisphere (see Fig. 8–8). The deflection is still eastward but appears, to an observer on earth looking along the path of the rocket, like a deflection to the left.

To this point the discussion of the Coriolis force has explained its effect on particles or water masses moving north or south. But particles moving east or west are also deflected unless moving along the equator. To understand this we must consider a water covered rotating earth. The water is attracted to the surface by gravity but slightly deformed by the centrifugal force caused by the earth's rotation. Thus the water envelope is slightly flattened at the poles and widened at the equator in equilibrium with the speed of the earth's rotation.

The earth rotates toward the east so that a particle that starts moving eastward say at mid latitudes is moving slightly faster than the earth and this increases the centrifugal force acting on it. This in turn causes it to be deflected toward the equator or toward the right.

Conversely a particle moving westward is moving opposite to the earth's rotation and the centrifugal force is therefore slightly less. It is deflected away from the equator or again to the right. In short, the Coriolis effect always acts to deflect particles to the right of their original path in the Northern Hemisphere (to the left in the Southern Hemisphere). It arises from the earth's rotation and the movement of the particle. A particle at rest does not experience this effect nor does a particle moving along the equator.

The Coriolis effect (f) can be calculated as follows:

$$f = 2(\omega \sin \phi)\,\mu$$

where

ω = angular velocity of the earth, 7.3×10^{-5} sec^{-1}, corresponding to the complete rotation of the earth in 1 day

ϕ = geographic latitude; $\sin \phi = 0$ at the equator, $\sin \phi = 1$ at the poles

μ = horizontal velocity of the current

The Coriolis effect acts as a force operating in a direction perpendicular to the current. It causes the current deflections as previously mentioned. It could theoretically cause water parcels to follow spiral paths as they move in the general direction of the current as shown in Fig. 8–9, although these are not recognized in the ocean because other forces, including friction, affect the moving water.

Wind blowing across water drags the surface along so that a thin layer is set in motion. This layer in turn drags on the one beneath, setting it in motion. The process continues downward, involving successively deeper layers. Transfer of momentum between layers is inefficient and energy is lost. As a result, current speed decreases with depth below the surface. In an infinite ocean on a nonrotating earth, the water would always move in the same direction as the wind that set it in motion. A similar effect is seen in "storm surges"—large amounts of water piled up on a coastline by storm winds blowing the water ahead of them. Because the distances and times are relatively short, complications resulting from the earth's rotation are minimal.

Since the earth does rotate about its axis, however, movements of surface waters are deflected to the right of the wind in the Northern Hemisphere, as shown in Fig. 8–10. This tendency was observed by, among others, Fridtjof Nansen, who while studying the drift of polar ice when his ship *Fram* was

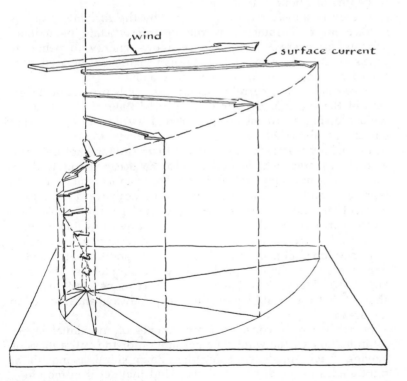

FIG. 8–10
Schematic representation of the Ekman spiral formed by a wind-driven current in deep water. Note the change in direction and decrease in speed with increased depth below the surface.

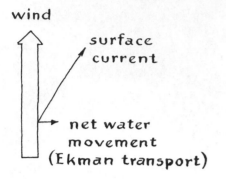

FIG. 8–11
Schematic representation of relationship of wind, surface current and net transport (also called Ekman Transport) in a drift current in the Northern Hemisphere.

frozen in polar ice during his 1893–96 expedition, found that ice moved 20 to 40° to the right of the wind.

Using Nansen's observations, the young physicist Walfrid Ekman showed mathematically that such effects can be explained using a simple uniform ocean with no boundaries as a model. Each layer, Ekman demonstrated, sets in motion the layer beneath, with the result that the latter moves somewhat more slowly than the former and that its motion is deflected to the right of the layer above. If this effect is represented by arrows (*vectors*) whose direction indicates current direction and whose length indicates speed, the change in current direction and speed with increasing depth forms a spiral, when viewed from above (see Fig. 8–10), now called the *Ekman spiral*. Figure 8–10 shows the Ekman spiral for the Northern Hemisphere. A similar spiral drawn for the Southern Hemisphere would exhibit the opposite sense of deflection, but current speeds would decrease at the same rate.

Wind effects penetrate to some considerable depth below the ocean surface, determined in part by the stability of the water column. The usual limit for wind effects (such as drift currents) is taken to be the depth at which the current arrow is exactly opposite to the surface-current arrow. At that depth, the current speed is about 4 percent of the surface current. Presence of a pycnocline may limit the depth of the drift current. Under a strong wind, wind-drift currents have been observed to extend to depths of about 100 meters.

Energy is transferred to this depth by the *turbulence* of the surface layer. Turbulence is the disorderly state of motion. Transfer of energy by turbulence allows energy to penetrate at least 100 times deeper into the ocean than could be accounted for by movements of water molecules alone.

Speed of surface currents set up by winds are about 2 percent of the speed of the wind that caused them. For instance, a wind blowing at 10 meters per second would cause a surface current of about 20 centimeters per second.

In shallow waters, wind-generated currents are not deflected as much as predicted theoretically for an ocean infinitely deep. The ocean is not completely uniform. For instance, the pycnocline represents an important barrier inhibiting downward transfer of momentum and materials from the surface into subsurface layers. Furthermore, the wind does not always blow long enough —probably a few days—from a single direction to establish a fully developed Ekman spiral. Under these conditions, the deflection is less than the 45° predicted by the simple case shown in Fig. 8–10. Nansen's original ice observations, for instance, showed that the ideal Ekman spiral is not always encountered in the ocean.

So far we have considered movements of individual layers set in motion by the wind. The net motion of the entire mass of moving water—predicted, using the Ekman spiral, as moving at right angles to the direction of the wind that set it in motion— is called the *Ekman transport*, represented in Fig. 8–11.

In coastal regions, the Ekman transport of surface waters can cause upwelling. Where the wind blows somewhat parallel to a coast, it can cause the surface waters to be blown offshore. These surface waters are replaced by waters moving upwards, typically from depths of 100 to 200 meters as seen in Fig. 8–12. These subsurface waters are characteristically colder and usually contain less dissolved oxygen than the surface waters and can thus be readily recognized. In addition these upwelled waters are usually rich in nutrients (phosphates and nitrates) necessary to support extensive growths of phytoplankton which feed other marine organisms. Consequently upwelling areas are highly productive of fish and other marine organisms.

Ekman transport can also cause surface waters to move toward a coastline. This leads to sinking or *downwelling,* the opposite of upwelling. The surface waters moving toward the coast tend to accumulate there depressing the pycnocline and to hold river discharge near the coast. Winter conditions along the Washington-Oregon coast leading to the Davidson current is one example of downwelling.

GEOSTROPHIC CURRENTS

Slight differences in water density set up forces strong enough to cause water movements. Most common are the *geostrophic*

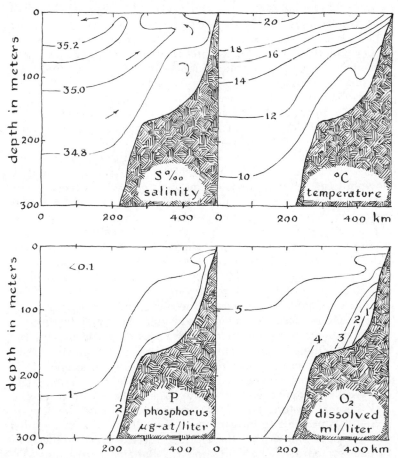

FIG. 8–12
Schematic representation of water properties in an area of wind-induced upwelling. Note that the upwelled water comes from a depth of about 200 meters.

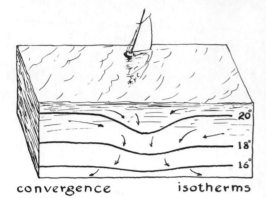

convergence   isotherms

FIG. 8–13
Schematic representation of a convergence in the open ocean. A slight hill of water forms and the surface layer thickens because of water accumulation.

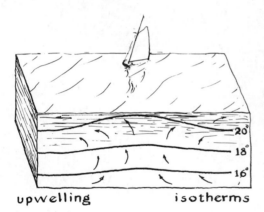

upwelling   isotherms

FIG. 8–14
Schematic representation of a divergence in the open ocean. Note that the surface layer is thinned.

currents, in which the flow of water in response to density-related forces is modified by the Coriolis effect.

If you examine the charts of prevailing winds (see Fig. 8–7) and remember that the net transport of surface waters is to the right of the wind in the Northern Hemisphere and to the left in the Southern, you can see that there is a distinct tendency for winds to move surface waters toward the subtropical regions. These regions, as we have already seen, are areas of light winds, and water tends to accumulate there. As Fig. 8–4 shows, these mid-latitude areas are zones of convergence for surface waters. Two things happen in an area of convergence, illustrated in Fig. 8–13: first, the water tends to pile up, forming a hill; and second, there is a depression of the local pycnocline through sinking.

In the reverse case—where surface water is blown away from an area—there occurs a *divergence*, illustrated in Fig. 8–14; a prominent divergence appears around Antarctica (see Fig. 8–4). In a divergence, subsurface waters move upward to replace the water moving away from the region.

As a result of the above water movements, the sea surface has a subtle topography. The net difference between the hills of water formed in the mid-latitude convergences on the western side of the ocean basins and the divergence surrounding Antarctica is about 2 meters, as Fig. 8–15 indicates. The distances involved are about half the earth's circumference, or at least 20,000 kilometers.

Despite this extremely low relief, water responds to these oceanic hills just as it would on land—by tending to run downhill. Let us consider a simple case of such a hill in the Northern Hemisphere and follow a water parcel along it as illustrated in Fig. 8–16, in order to see how the earth's rotation changes that path. Initially the water parcel moves downslope, just as it does on land. However, the Coriolis effect soon begins to change the path of the water parcel by deflecting it to the right. This continues until the water follows a path that allows it to flow downhill just enough to keep moving but where most of its motion is parallel to the side of the hill. If our hill were contoured to show lines of equal sea-surface elevation, the path of the water parcel would nearly parallel these contour lines. On a frictionless ocean, water movement would exactly parallel the side of the hill. This balance—between the gravitational force that pulls the water downhill and the Coriolis effect that deflects it—gives rise to geostrophic currents.

If it were possible to survey the elevation of the ocean surface precisely, oceanographers could map the major features of ocean currents using the topography, making allowance for the slight deviations arising from the frictional effects in the ocean. The currents would appear to be strongest on the steepest hills (where the lines of sea-surface elevation are closest together) and weakest where the slope of the sea surface was most gentle.

Where a western boundary current passes near land, it is sometimes possible to measure such sea-surface slopes. Detailed surveys have shown that the Florida Current (part of the Gulf Stream system between Cuba and the Bahama Banks) has sea-surface slopes of about 19 centimeters over a distance of approximately 200 kilometers. At this location current speeds have

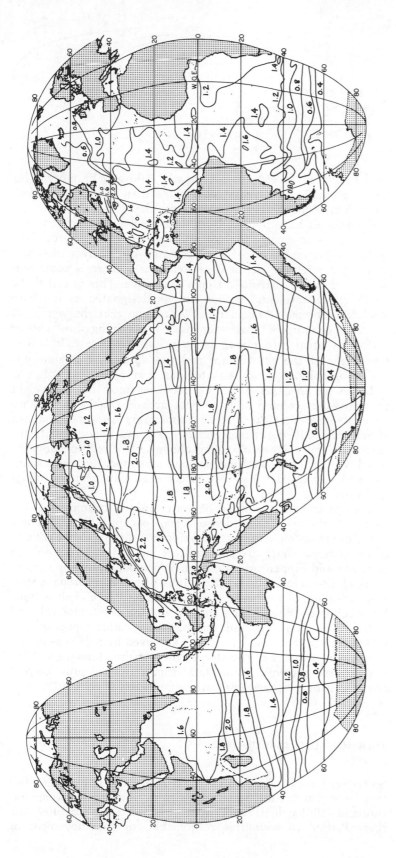

FIG. 8–15

Sea-surface elevations, in meters, above an arbitrary level surface. (After H. Stommel, "Summary charts of the Mean Dynamic Topography and Current Field at the Surface of the Ocean and Related Functions of the Mean Wind–Stress Studies on Oceanography" Studies on Oceanography ed. Kozo Yoshida, University of Washington Press, Seattle, 1964)

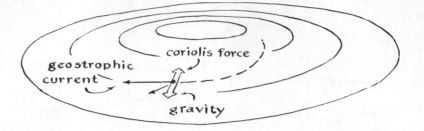

FIG. 8–16

Schematic representation of water flow and balance of forces in a geostrophic current in the Northern Hemisphere.

been measured at 150 cm/sec (nearly 3 knots) or more. This rapid current has a much steeper slope than currents in which waters move only a few kilometers per day.

Techniques have not yet been developed to the point where it is feasible to survey the entire ocean surface; therefore, the surface topography must be determined by indirect methods. Usually, a reference depth is chosen where there is some evidence to indicate that no currents are active. This is called the *depth of no horizontal motion* and is designated as the base level for subsequent calculations. Assuming that the weight of a water column of fixed dimensions overlying this chosen surface is equal throughout its area, it is possible to calculate the height of the sea surface and the *dynamic topography*— the topography to which a water parcel responds.

Let us see how this works. We have assumed that the weight of each water column of fixed base area is equal—that is, we have assumed that there is an equal mass of water and salt above each unit of our fixed surface. We can now calculate the height of the column of water above any given location for which we have precise measurements of temperature and salinity; knowing the density, we can then calculate relative topography (see Fig. 8–15).

Remember that the mass of water in the column is fixed. Thus a column of less dense water occupies more volume and thus stands higher above the reference surface than a column of more dense water. By this indirect method, it is possible to calculate and prepare maps of dynamic topography for various parts of the ocean. Furthermore, it is possible to prepare such maps for parts of the ocean well below the surface. This is done by selection of the reference surface and the depth interval used in the calculation. The resulting map of dynamic topography is interpreted in the same way that a surveyed map of sea-surface topography would be. The steeper the dynamic topography, the stronger the currents. Using data from various depths, current charts can be prepared for various subsurface circulation systems.

THERMOHALINE CIRCULATION

So far we have discussed only horizontal water movements, but there are also vertical water movements that control temperature and salinity distribution in most of the ocean's depths. The *thermohaline circulation* is responsible for the movement of water from polar regions—specifically, from the North Atlantic and Antarctic—and its circulation throughout all the ocean basins.

Thermohaline circulation is driven by density differences. When the density of waters at the surface equals or exceeds water density at depth, the water column becomes unstable and the more dense water mass sinks, displacing less dense waters beneath. As we know from our previous discussion of density and stability, sinking continues until the water mass reaches its appropriate density level. In a stable water column, the waters beneath will be slightly denser and the waters above will be slightly less dense; at this level, the newly implaced mass of water tends to spread laterally, forming a thin layer. Such processes give rise to the intricately layered structures discovered by precise measurements of temperature and salinity in many ocean areas.

In certain semi-isolated ocean areas, lowered temperature and increased salinity cause the formation of dense water masses. Being denser than all the other waters, they sink to the bottom of our stratified ocean. Low water temperatures are the result of intensive cooling in polar regions, where the surface waters reach the freezing point of seawater. Such a combination of conditions is found in the Atlantic Ocean near Greenland and near Antarctica in the Weddell Sea, as indicated in Fig. 8–17.

We have already learned that the Atlantic is the saltiest of the major ocean waters (see Chapter 2, p. 33). This salty water is carried into high latitudes by the Gulf Stream. Near Greenland it is intensively cooled to around $-1.4°C$. When surface waters reach a critical density, they may sink suddenly, as a mass (as they do in laboratory experiments), and flow along the bottom of the North Atlantic Basin, especially along the western side.

Submarine ridges between Greenland and Scotland, the region forming the entrance to the Arctic Basin, prevent any bottom waters formed in the Arctic Sea from entering the main part of the Atlantic. The Bering Sill is even more effective in isolating the Arctic from the Pacific Ocean.

Large quantities of Antarctic Bottom Water are formed in the Weddell Sea, a partially isolated embayment of the Antarctic continent. There, surface waters are cooled to temperatures of $-1.9°C$. At this low temperature and a salinity of $34.62^0/_{00}$, this water sinks to the bottom of the adjacent deep-ocean basin where it forms the densest water mass in the ocean. In the process of sinking it mixes with other waters and is warmed to $-0.9°$. After circulating around Antarctica, and mixing with other water masses, cold dense bottom waters move northward into the deeper parts of all three major ocean basins.

Using temperature as a tracer for this bottom water, we can follow Antarctic Bottom Waters (characteristics given in Table 7–2) as far north as the edges of the Grand Banks (45°N) in the North Atlantic. In the Pacific mixtures of North Atlantic Deep Water reach the Aleutian Islands (50°N). Deep water movement through ocean basins is controlled by ocean-bottom topography. For example, the Romanche Trench provides a path for Antarctic waters to flow into the deep basins of the eastern

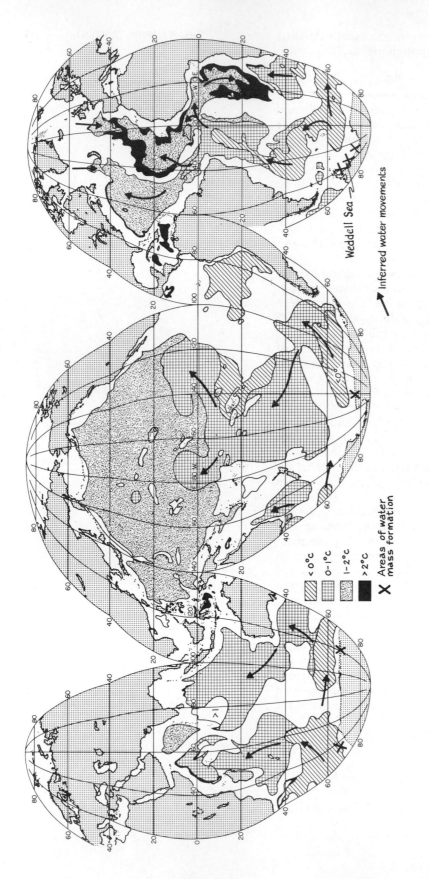

FIG. 8-17

Variation in temperature of bottom waters at depths greater than 4 kilometers and the inferred water movements. [After G. Wüst, "Die Stratosphäre. Deutsche Atlantische Exped. *Meteor,* 1925–27." *Wiss. Erg.* 6(1) (1935) 2. 288 pp.]

Weddell Sea

↗ Inferred water movements

< 0°C
0–1°C
1–2°C
> 2°C
X   Areas of water mass formation

portion of the South Atlantic; direct entry of this water mass into the eastern side of the South Atlantic is blocked, however, by the Walvis Ridge.

Movement of near-bottom currents tends to be much slower than that of surface currents. Speeds of 1 to 2 centimeters per second are typical—except along the western basin margins, where speeds of 10 centimeters per second have been calculated. This is another manifestation of the strong boundary currents along the western side of ocean basins.

Bottom waters are formed in large quantity, probably each year. To compensate for the production of bottom water there must be a gradual upward movement of waters toward the surface in all ocean basins. This upward movement opposes the downward movement of heat and dissolved gases from the surface. It also accounts for many features of the pycnocline.

Various techniques have been employed for estimating deep-water *residence times* (a measure of the time necessary to replace bottom waters by newly formed water masses). Radioactive carbon-14 has been used to determine the time elapsed since the dissolved-carbon compounds in the water were last at the surface. The resultant data suggest residence times ranging from 500 to 1000 years. If we take 1000 years to be a reasonable residence time for deep-ocean waters, this amounts to an upward movement of about 4 meters per year.

Other relatively dense water masses form in high latitudes of the North Atlantic and around Antarctica. There appears to be no dense-water formation in the North Pacific. Most of these high-latitude water masses have relatively low salinity, due to relatively high precipitation and little evaporation in the areas in which they form. Consequently, such surface waters do not achieve the high densities of the water masses formed from Atlantic surface waters.

ATLANTIC OCEAN CIRCULATION—
A THREE-DIMENSIONAL VIEW

The distribution and movements of water masses have been worked out using the distribution of temperature and salinity as previously explained. Plotting the distribution of these values as in Fig. 8–18 graphically demonstrates where different water masses form in the ocean. Using the geostrophic approximation discussed above, we can estimate their probable movement rates. Because the Atlantic Ocean is especially well known, we shall use it to illustrate the complexities of circulation throughout an entire ocean basin.

Over most of the ocean, the circulation of the surface waters is almost completely unconnected with the movement of subsurface waters. Figure 7–20b, in which salinity is plotted, shows that the relatively-high-salinity surface water (greater than $35^0/_{00}$ in the Atlantic) extends down to a few hundred meters at most. There is an exception in the North Atlantic (about 30°N), where the warm saline water from the Mediterranean Sea occurs at a depth of about 1 to 2 kilometers. If water temperature were plotted instead of salinity, we would have a similar picture, as seen in Fig. 7–20a.

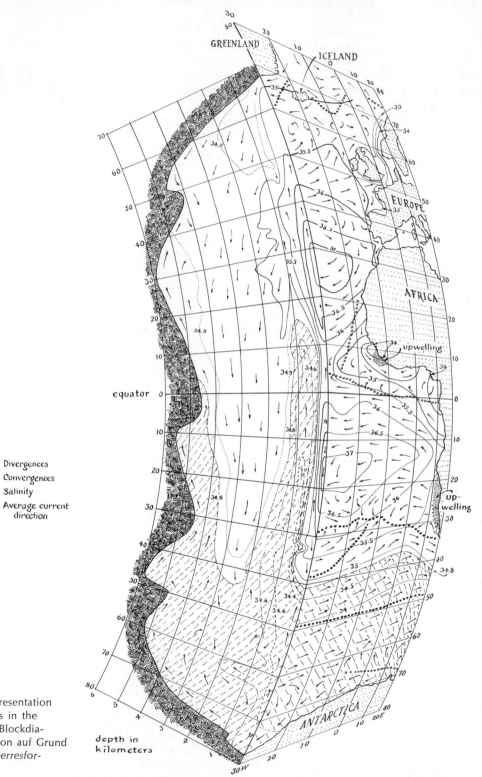

FIG. 8–18
Schematic, three-dimensional representation of surface and subsurface currents in the Atlantic Ocean. [After G. Wüst, "Blockdiagramme der Atlantischen Zirkulation auf Grund der *Meteor*-Ergebnisse," *Kieler Merresforschungen*, 7(1) (1950) 24–34]

In the West Wind Drift around Antarctica, there is little separation between surface and subsurface circulation. The small temperature or salinity change with depth does not hinder vertical circulation. This largest of all currents is the prime communication link for surface and subsurface waters of all oceans. Note that the waters in the West Wind Drift are among the lowest-salinity waters found in the Atlantic. (Waters less than $34.8^0/_{00}$ are shown by diagonal patterns in Fig. 8–18.)

Recall that the surface circulation is dominantly east–west, with north–south movements confined to the boundary currents along the continents. The subsurface circulation, on the other hand, is primarily a north–south circulation. As in the surface currents, the strongest near-bottom currents are associated with the western boundaries of the ocean basins. In these boundary currents, speeds of about 10 centimeters per second are common. Throughout most of the deep-water masses, the waters move about 1 centimeter per second.

We have previously discussed the water masses that flow along the ocean bottom because of their relatively high density. As can be seen in Fig. 8–17, cold bottom water from the Antarctic (called Antarctic Bottom Water) is a conspicuous feature of the deep circulation up to about 45°N. Deep waters from the North Atlantic (called North Atlantic Deep Water) are also conspicuous along the ocean bottom down to the point where they flow out over the Antarctic Bottom Water. North Atlantic Deep Water flows southward at depths between 2 and 4 kilometers, and mixes extensively with the waters circulating in the West Wind Drift before flowing into the deep Pacific basin.

A somewhat smaller intermediate water mass forms in the region of the Antarctic Convergence around 50°S. Because of the low salinities in that region, the water density is not high enough for it to sink to the bottom. Instead, the Antarctic Intermediate Water sinks about 1 kilometer and spreads northward, crosses the equator, and is recognizable to about 20°N.

Two points should be noted about the deep circulation. The first is that cold water flowing toward the equator (which eventually rises to the surface) is a form of heat transport. We have previously mentioned warm-water transport to high latitudes by surface currents. The return of cold water to low latitudes is the counterflow and it takes place in both—surface and sub-surface waters. It corresponds to the cold-air or cold-water return in a household furnace system.

The second point is that currents of the deep ocean transport water across the equator. The circulation of surface waters of the Northern and Southern Hemispheres is almost completely separate. Except for some South Atlantic surface water transported into the Northern Hemisphere where the equatorial current is deflected by the South American continent and some areas in the western Pacific, there is almost no movement of surface waters across the equator. Subsurface currents transport large amounts of water across the equator and balance any surface transport of water, including that by winds.

Before the American Revolution, Benjamin Franklin, noting that trans-Atlantic mail ships made the voyage to England more speedily than the voyage home, correctly deduced that a current flowing northward along the U.S. coast and eastward into the North Atlantic was responsible for the difference in sailing time. He published a map in 1770 showing his observations. About 200 years later, a team of five scientists under Dr. Jacques Piccard conducted a study of that powerful stream at first hand; in the 130-ton "submersible" Ben Franklin they drifted silently for 1500 miles in the depths of the Gulf Stream.

The mission was designed to study marine, biological, and acoustical properties of the current, and complete silence and immobility of the ship in relation to the water were essential. She drifted at depths of 200 to 700 meters, using her small motors only once—to avoid hitting coral formations. Midway in her voyage, the submersible somehow drifted out of the current and had to be towed back to its center by her escort ship. Variable-ballast tanks permitted her to avoid hitting bottom or rising to the surface.

The Ben Franklin remained sealed for 30 days, having descended off West Palm Beach, Florida, on July 14, 1969, and traveled to a point 300 miles off Halifax, Nova Scotia. Her crew made five trips to the bottom to photograph the terrain, and spent hundreds of hours observing the living world of the jetlike current. Over 5 million measurements were made of temperature, salinity, speed of sound, and depth, as well as studies of seawater composition and earth's gravity.

After the voyage, all aboard agreed that the experiment had raised more new questions than it had answered about the Gulf Stream, thus pointing the way toward the future studies of currents and ocean depths.

SUMMARY OUTLINE

Currents—large-scale water movements

Current observations
> Mapping of average current set from displaced ship's courses
> Floating objects—debris, drift bottles, parachute drogues, Swallow floats
> Direct current observations—current meters, buoys

General surface circulation in the open ocean
> Gyre—nearly closed set of currents, elongated east–west
> East–west currents, such as equatorial currents, West Wind Drift
> Boundary currents—eastern and western
> Seasonally variable currents—coastal, monsoon—controlled primarily by winds

Boundary currents
> Western boundary currents—Gulf Stream system Strong, narrow, deep
> Eastern boundary currents—California Current, Alaska Current; Broad, shallow, sluggish
> Oceanic fronts—sharp discontinuities in water properties, mark convergences

Langmuir circulation—organized set of helical vortices in surface waters
> Transport heat and momentum downward
> Mix surface waters, often to depths below wave affected zone

Forces causing currents

    Prevailing wind systems—primary driving force for surface currents

        Drift currents—caused directly by wind

        Geostrophic currents—caused indirectly by wind and its effect on density distribution

    Coriolis effect—deflecting force arising from earth's rotation

    Deflects currents to right in Northern Hemisphere, to the left in the Southern

Ekman spiral—systematic decrease in current speed and change in direction with increasing distance below surface

    Surface current flows 45° to right of wind in infinite, homogeneous Northern Hemisphere ocean, to the left in the Southern Hemisphere

    Surface current about 2 percent of speed of the wind causing it

    Net movement of surface layer is perpendicular to wind—known as Ekman transport

    Movement of surface waters offshore causing upwelling, vertical movements of subsurface waters to replace water moved seaward

    Sinking or downwelling is caused by waters blown toward coast

    Drift currents—primarily in surface layers, few tens or hundreds of meters

Geostrophic currents

    Caused by small difference in density, expressed as dynamic topography

    Involves balance between Coriolis effect and gravitational attraction

    Strongest currents where topography is steepest, weakest where slopes are gentle

    Dynamic topography computed from density distribution (controlled by temperature and salinity)

    Useful for surface and subsurface currents

Thermohaline circulation

    Vertical water circulation driven by density differences

    Bottom waters formed in North Atlantic and in Weddell Sea (Antarctica)

    Primarily a north–south circulation

    Gradually rises to surface through pycnocline over entire ocean

Atlantic circulation

    Intermediate waters form at high latitudes and flow toward equator

    Antarctic Bottom Waters flow generally northward

SELECTED REFERENCES

STOMMEL, HENRY. 1965. *The Gulf Stream: a Physical and Dynamical Description*, 2nd ed. University of California Press, Berkeley. 248 pp. Describes present understanding of Gulf Stream System; intermediate to technical level.

VON ARX, W. S. 1962. *An Introduction to Physical Oceanography*. Addison-Wesley, Reading, Mass. 422 pp. Physical and geophysical aspects of ocean systems are emphasized; little descriptive material; intermediate level.

WAVES

Storm waves at sea, showing chaotic ocean surface. The boundary between air and water becomes progressively more obscured as winds become stronger.

nine

Waves—disturbances of the water surface—can be seen at any ocean beach or along the shore of any large body of water. Seafarers have observed waves for thousands of years, with seasickness often the result. Despite an abundance of observations, understanding of sea waves has come slowly. The ancients knew that the waves were somehow generated by wind, but it was not until the nineteenth century that the first mathematical descriptions of waves were developed.

Waves are important to us in many ways. They make and remake ocean beaches each year and in the process provide entertainment for millions of surfers and swimmers. Effects of waves must be considered in the design and construction of docks, breakwaters, and jetties along the coast, and waves are all too often the cause of the failure or even destruction of these structures. Nor can waves be neglected when designing or operating ships.

IDEAL WAVES

A few minutes' observation of the ocean surface reveals a complex and continually changing pattern—a pattern that never exactly repeats itself regardless of how long we watch it. Ocean waves come in many sizes and shapes, ranging from the tiny ripples formed by a light breeze, through enormous storm waves, tens of meters high, to the tides (which are also waves, as we shall see in Chapter 10).

Because of their complexity, ocean waves usually do not lend themselves to accurate description or complete explanation in simple terms. Nevertheless, we commonly work with simplified explanations and descriptions that help us understand wave phenomena; moreover, most of the advances in the study of waves have come through use of appropriate simplifications. Initially our discussions will involve such simplified or *ideal waves*.

To start, let us consider simple *progressive waves* and their parts as they pass a fixed point—say, a piling. We can make such waves in a laboratory wave tank, or by steadily bobbing the end of a pencil in a basin of water or a still pond surface. Then we see a series of waves, each wave consisting of a *crest* —the highest point of the wave—and a *trough*—the lowest part of the wave. The vertical distance between any crest and the succeeding trough is the *wave height H*, and the horizontal distance between successive crests or successive troughs is the *wavelength L*; these wave constituents are illustrated in Fig. 9–1. The time (usually measured in seconds) that it takes for successive crests or troughs to pass our fixed point is the *wave period T*, which is rather easily measured with a stopwatch. We can express the same information by counting the number of waves that pass our fixed point in a given length of time. This gives the *frequency* $(1/T)$, which is expressed in cycles per second. For individual progressive waves, the speed (C, in meters per second) can be calculated from the simple relationship

$$C = \frac{L}{T}$$

where L is the wave length in meters, and T is the wave period in seconds.

Where the wave height is low, crests and troughs tend to be rounded and may be approximated mathematically by a *sine wave* (a mathematical expression for a smooth, regular oscillation). As wave height increases, sea waves normally have crests more sharply pointed than simple sine waves, and can be approximated by more complicated mathematical curves, such as the sharp-pointed, rounded-trough mathematical curve known as the *trochoid* (see Fig. 9–1).

Wave steepness (H/L)—the ratio of wave height to wavelength—is a measure of wave stability. It is also the factor that

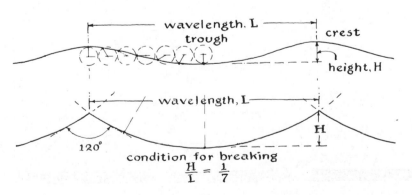

FIG. 9–1
Two idealized cases for simple waves, indicating their various parts. Relatively small waves can be described as simple sine waves. Larger waves tend to be more sharp-pointed than a simple sine curve. There are limits on the size to which a wave can grow. Waves commonly break when the angle at the crest is less than 120° or the ratio of wave height to wave length is $H/L = 1/7$.

determines whether a small boat will glide smoothly over low waves or pitch through steep, choppy waves. When wave steepness exceeds $\frac{1}{7}$, the wave becomes unstable and begins to break (see Fig. 9–1) by raveling of the oversteepened crest,

FIG.   9–2
Movements of water particles caused by the passage of waves in deep water (a) and in shallow water (b). Note that the particles tend to move at the surface in circular orbits which become smaller with depth. Near the bottom, the orbits are flattened. Little movement occurs at depths greater than $L/2$.

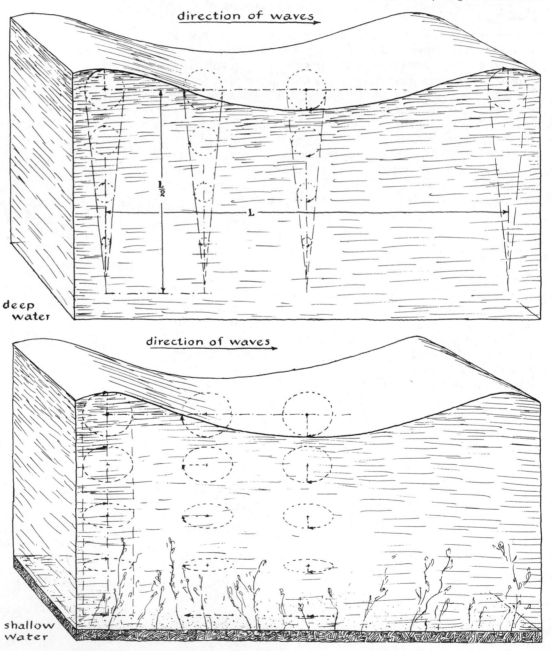

forming spilling breakers. The angle at the crest must be 120°
or greater for the wave to remain stable; we shall examine this
phenomenon in detail when we discuss breakers and surf.

So far, we have considered only the movement of the water
surface—the crests and troughs moving together that make up
a wave train. But what happens to the water itself as the wave
form passes? How is the motion of the water related to the
motion of the wave form? These questions have been studied
using wave tanks with bits of material floating on the surface,
or dyed bits of water or oil droplets below the surface.

When small waves move through deep water, individual bits
of water move in circular orbits which are vertical and nearly
stationary. The water moves forward as the crest passes, then
vertically, and finally backward as the trough passes; this series
of movements is diagrammed in Fig. 9–2. This orbit is retraced
as each subsequent wave passes, and after each wave has passed,
the water parcel is found nearly in its original position. Actually,
there is some slight net movement of the water because the
water moves forward slightly faster as the crest passes than it
moves backward under the trough. The result is a slight forward displacement of the water in the direction of wave motion
and perpendicular to the wave crests, as shown in Fig. 9–3.

It should be noted that if the water moved with the wave,
ships would never have been invented, for no ship ever built
could withstand the forces exerted by water movements on such
a scale. Many individuals have experienced this themselves.
When one floats in the ocean near the beach but beyond the
breakers, one experiences only a gentle rocking motion as the
waves pass, because there is little movement of the water. If,
however, one tries to stand where the waves are breaking, the
immediate pounding by the breakers demonstrates large-scale
rapid water movements, because in the breakers the water does
move with the wave form. Even large ships can be damaged
when hit by tons of water from a large, breaking wave.

In deep water (where the water depth is greater than $L/2$),
the water parcels move in nearly stationary circular orbits; such
waves, unaffected by the bottom are known as *deep-water
waves*. The diameter of these orbits at the surface is approximately equal to the wave height. It decreases to one-half
the wave height at a depth of $L/9$, and is nearly zero at a depth
of $L/2$ (see Fig. 9–2). At depths greater than $L/2$, the water is
moved little by wave passage; thus a submarine is essentially
undisturbed by waves when it is submerged to depths greater
than $L/2$.

In deep water, the *wave speed C*, in meters per second, may
be calculated by the following equations:

$$C = \frac{gT}{2\pi} = 1.56T \qquad C = \sqrt{\frac{gL}{2\pi}} = 1.25\sqrt{L}$$

where T is the wave period in seconds; L is the wavelength
in meters; and g is the acceleration of gravity, 9.8 m/sec². Of
the two equations, the first is the more useful, because wave
period is easily measured, whereas wave length is difficult to
determine at sea or on the beach.

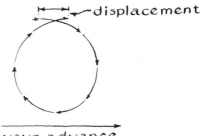
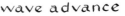

FIG. 9–3
Orbital motion and displacement of a water
particle during the passage of a wave.

Where water depths are less than $L/20$, the motion of the water parcels is strongly affected by the presence of the bottom (see Fig. 9–2); these waves are called *shallow-water waves*. The orbits of water parcels at the surface may be only slightly deformed, usually forming an ellipse (a flattened circle with its long axis parallel to the bottom). Near the bottom, wave action may be felt as the water particles move back and forth; vertical water movements are prevented by the proximity of the bottom. Sometimes we observe movements of water parcels in shallow waters as waves pass over them—for example, where bottom-attached plants are moved with the water as seen in Fig. 9–3. The speed of shallow-water waves, C, in meters per second, can be calculated by

$$C = \sqrt{gd} = 3.1\sqrt{d}$$

where g is the acceleration of gravity, 9.8 m/sec^2; and d is the water depth in meters.

So far, we have discussed the behavior of a simple (or ideal) wave, identical with others in a wave train. Although we rarely see only groups of simple uniform waves at sea, they are often useful in analyzing the more complicated waves that do occur. We assume that even the most complicated waves may, by appropriate mathematical techniques, be separated into several groups of simple waves which, when added together, reproduce the complexities of the original ocean waves. An observed profile of a sea is shown in Fig. 9–4(a). This group of waves has been analyzed to determine which wave frequencies occur in the wave spectrum; the various components (each a simple sine wave) are shown in Fig. 9–4(b). Combining these waves will result in a wave essentially identical to the one observed.

Now that we have considered individual ocean waves and calculated their speeds we may place them in context: in the ocean, waves usually occur as wave trains or as a system of waves of many wavelengths, each wave moving at a speed corresponding to its own wavelength. Suppose that we produce a wave train in a laboratory wave tank and then carefully observe the result. If we follow a single wave in the resulting wave train, we find that it advances through the group because the individual waves move at a speed double that of the group. The wave approaches the front, gradually losing height as it goes. Finally, it disappears at the front of the wave train, to be followed by other, later-formed waves that have also moved forward from the rear. New waves are continually forming at the back of the wave train, while others are disappearing at the front.

From this observation we can say that in deep-water waves, the wave energy travels at one-half the speed of the individual waves. In shallow water, individual waves are slowed down until the individual wave speed equals the group speed. In many ways, it is the group speed that is most significant, for individual waves rarely persist for long times in the ocean.
speed that is most significant, for individual waves rarely persist for long times in the ocean.

The formation and behavior of waves involve two types of forces: those that initially disturb the water and those that act

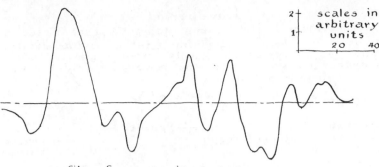

profile of waves in a sea

(a)

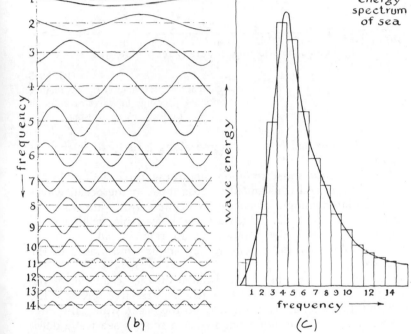

(b)

energy spectrum of sea

(c)

FIG. 9–4
An observed profile of waves in a sea (a). Such a
complicated wave pattern can be described as consisting
of many different sets of sine waves (b), all of
them superposed. The lower-frequency waves contain
more energy than the higher-frequency ones (c). The
energy in a wave is proportional to its height.

to restore the equilibrium or still-water condition. The disturbing
forces are most familiar to us. All of us have made small waves
by tossing a pebble into water. If the water surface was initially still, we observe a group of waves changing continuously
as they move away from the disturbance. Sudden impulses such
as explosions or earthquakes also cause some of the longest
waves in the ocean. If the disturbance, such as an explosion,
affects only a small area, the waves will move away from that
point, much as the waves moved away from our pebble. But if
the disturbance affects a large area, as a great earthquake can,
the resulting *seismic sea waves* (also called tsunamis) may be-

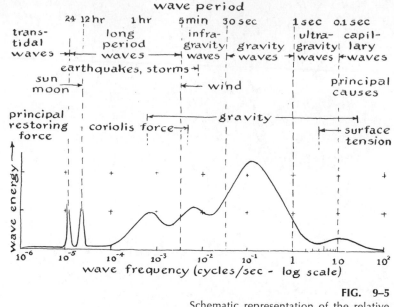

FIG. 9–5

Schematic representation of the relative amounts of energy in waves of different periods. (After Kinsman, 1965)

have as if they had been generated along a line rather than at a point.

Winds are the most common disturbing force acting on the ocean surface, and consequently cause most of the ocean waves (note the large amount of energy associated with wind waves as illustrated in Fig. 9–5). Winds are highly variable, so that wind waves vary greatly in different ocean regions, and with the seasons. (We shall have more to say about wind waves in the next section.)

The attraction of the sun and moon on the ocean causes the longest waves of all—the *tides*. Since the attractive forces of the sun and the moon act continuously on the ocean water, the tides are not free to move independently as a seismic sea wave does; such waves, where the disturbing force is continuously applied, are known as *forced waves*—in contrast to the *free waves*, which move independently of the disturbance that caused them. An explosion-generated wave is an example of a free wave. Wind waves have some characteristics of both free and forced waves.

Once a disturbance has formed a wave, restoring forces act to damp out the wave and to restore the initial or equilibrium state. Depending on the size of the wave, various forces may be involved. For the smallest waves (wave length <1.7 centimeters, period <0.1 second), the dominant restoring force is *surface tension*, in which the water surface tends to act like a drum head, smoothing out the waves. Such waves, called *capillary waves* are round crested with V-shaped troughs. For waves with periods between 1 second and about 5 minutes, gravity is the dominant restoring force; this range includes most of the waves we see. Because of the influence of gravity, these waves are known as *gravity waves*.

The largest waves—tides and seismic sea waves—involve substantial water movements. In addition to gravity effects, the Coriolis effect is important for waves whose periods exceed 5 minutes.

Waves transmit energy gained from the disturbance that formed them. This energy is in two forms. One-half is potential energy depending on the position of the water above or below the still-water level; the potential energy advances with the group speed of the individual waves. The rest of the wave energy —known as kinetic energy—is possessed by the water moving as the wave passes. There is a continual transformation of potential energy to kinetic energy and back to potential energy.

The total energy in a wave is proportional to the square of the wave height; in other words, doubling the wave height increases the energy by a factor of 4. An enormous amount of energy is contained in a wave. For example, a swell with a wave height of 2 meters has energy equivalent to 1200 calories per square meter of ocean surface; a 4-meter swell has 4800 cal/m^2. Nearly all the wave energy is dissipated as heat when the wave strikes a coastline.

FORMATION OF SEA AND SWELL

As we have learned from the preceding, waves are disturbances of the ocean surface which pass rapidly across the water, but with little accompanying water movement. Put in another way, waves are the manifestation of energy moving across the ocean surface. Now we shall see how the energy of the wind is supplied to the ocean surface and how wind waves are formed.

Wave formation by the wind is an easily observed phenomenon; in fact wind speed at sea can be estimated from wave conditions as indicated in Table 9–1. Even a gentle breeze results in the immediate formation of ripples or capillary waves, which form more or less regular arcs of long radius, often on top of earlier-formed waves. Ripples are thought by some oceanographers to play an important role in wind-wave formation by providing the surface roughness necessary for the wind to pull or push the water: in short, they provide the "grip" for the wind.

Ripples are short-lived. If the wind dies, they disappear almost immediately; but if the wind continues to blow, ripples grow and are gradually transformed into larger waves, usually short and choppy ones. These latter waves continue to grow so long as they continue to receive more energy than is lost through processes such as wave breaking. Energy is gained through the pushing-and-dragging effect of the wind. The amount of energy gained by the waves depends on factors such as sea roughness, the specific wave form, and the relative speed of the wind and waves. Choppy, newly formed *seas* provide a much better grip for the wind than smooth-crested *swell*.

The largest wind waves are formed by storms—often a series of storms—at sea. The size of the waves formed depends on the amount of energy supplied by the wind. The relevant factors operating here are wind speed; the length of time that the wind

Table 9–1

APPEARANCE OF THE SEA AT VARIOUS WIND SPEEDS

Wind speed

Beaufort number†	km per hour	Seaman's term	Effects observed at sea
0	under 1	Calm	Sea like mirror
1	1–5	Light air	Ripples with appearance of scales; no foam crests
2	6–11	Light breeze	Small wavelets; crests of glassy appearance, not breaking
3	12–19	Gentle breeze	Large wavelets; crests begin to break; scattered whitecaps
4	20–28	Moderate breeze	Small waves, becoming longer; numerous whitecaps
5	29–38	Fresh breeze	Moderate waves, taking longer form; many whitecaps; some spray
6	39–49	Strong breeze	Larger waves forming; whitecaps everywhere; more spray
7	50–61	Moderate gale	Sea heaps up; white foam from breaking waves begins to be blown in streaks
8	62–74	Fresh gale	Moderately high waves of greater length; edges of crests begin to break into spindrift; foam is blown in well-marked streaks
9	75–88	Strong gale	High waves; sea begins to roll; dense streaks of foam; spray may reduce visibility
10	89–102	Whole gale	Very high waves with overhanging crests; sea takes white appearance as foam is blown in very dense streaks; rolling is heavy and visibility reduced
11	103–117	Storm	Exceptionally high waves; sea covered with white-foam patches; visibility still more reduced
12	118–133		
13	134–149		Air filled with foam; sea completely
14	150–166	Hurricane	white with driving spray; visibility
15	167–183		greatly reduced
16	184–201		
17	202–220		

†Beaufort numbers, still used to indicate approximate wind speed, were devised in 1806 by the English Admiral Sir Francis Beaufort based on the amount of sail a fully rigged warship of his day could carry in a wind of a given strength. Modified from U.S. Naval Oceanographic Office, *American Practical Navigator* (Bowditch). Rev. Ed. H.O. Publ. No. 9, Washington, D.C. 1958, p. 1069.

blows in a constant direction; and the *fetch*—the maximum distance the wind flows in a constant direction; the process is illustrated in Fig. 9–6. Usually there will be waves present on the ocean at the time the new waves begin to form. Either the older waves will be destroyed by the storm, or newly formed waves will be generated on top of the old ones. There is continuous interaction between waves. Wave crests coincide, forming momentarily new and higher waves. Seconds later, the wave crests

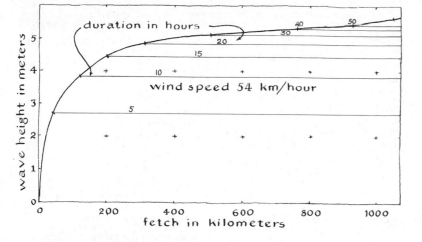

FIG. 9–6
Growth of wave height under a constant wind of increasing duration acting over different length and fetch. Note the rapid change in wave height during the first 10 hours as compared with the change between 40 and 50 hours. (After Sverdrup et al., 1942)

may no longer coincide, but instead cancel each other, the wave crests disappearing.

As the winds continue to blow, waves grow in size, as illustrated in Fig. 9–6, until they reach a maximum size—defined as the point at which the energy supplied by the wind is equaled by the energy lost by breaking waves, called *whitecaps*. When this condition is reached, we call the area in question a *fully developed sea*.

Given the wind speed, duration, and fetch, it is possible to predict the size of the waves generated by a given storm. Waves of many different sizes and periods are present in a fully developed sea, but waves with a relatively limited range of periods will predominate for a steady wind with a fixed speed (see Fig. 9–7). Such predictions are complicated by the fact that the wind never blows at a constant speed; it is just as likely to be gusty at sea as on land, and just as likely to change direction as not.

Initially, the waves in a sea are steep, chaotic, and sharp-crested, often reaching the theoretical limit of stability ($H/L = \frac{1}{7}$), when the waves either break or have their crests blown off by the wind; the factors affecting wave steepness are illustrated

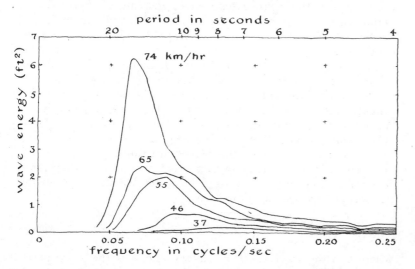

FIG. 9–7
In a fully developed sea, most of the wave energy occurs in a relatively restricted range of wave periods (or wave frequencies). Note that changes in wind speeds cause marked changes in wave energy and wave period. [After G. Neumann and W. J. Pierson, *Principles of Physical Oceanography*, Prentice-Hall, Inc., Englewood Cliffs, N.J., 1966]

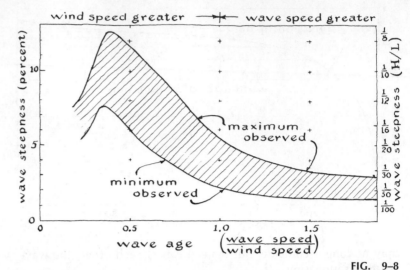

FIG. 9–8

Variation in wave steepness with wave age (wave speed/wind speed). The upper and lower curves are probable maximum and minimum values of wave steepness for a given wave age. (After H. U. Sverdrup and W. H. Munk, *Wind, Sea and Swell: Theory of Relations for Forecasting*, U.S. Naval Oceanographic Office H.O. Publ. 601, Washington, D. C., 1947)

in Fig. 9–8. As the waves continue to develop, their speed approaches, then equals, and finally exceeds the wind speed; as this happens, wave steepness decreases. As waves travel out of the generating area, or if the wind dies, the sharp-crested, mountainous, and unpredictable sea is gradually transformed into smoother, long-crested, longer-period waves—*swell*. These waves can travel far because they lose little energy due to viscous forces.

Let us examine some of the processes that cause waves to change from sea to swell, as they move through calmer ocean areas. One of these processes is the spreading of waves due to variations in the direction of the winds that formed them. Unless destroyed or influenced in some way by ocean boundaries, waves continue to travel for long distances in the direction that the wind was blowing when they formed. Since winds are rarely constant for long in terms of either speed or direction, waves formed in a storm will move away from the area and "fan out" (angular dispersion), as illustrated in Fig. 9–9, as they move. As this happens, the wave energy will be spread over a larger area causing a reduction in wave height.

At the same time that the waves are fanning out, they are also separating by wavelength, a process known as *dispersion*. Remember that the speed of deep-water waves C, in meters per second, can be calculated by

$$C = \frac{L}{T} = \sqrt{\frac{gL}{2\pi}} = 1.25\sqrt{L}$$

where L is the wavelength in meters, and T is the wave period in seconds. Thus longer waves tend to travel faster than shorter

steady wind   sea generated by a storm

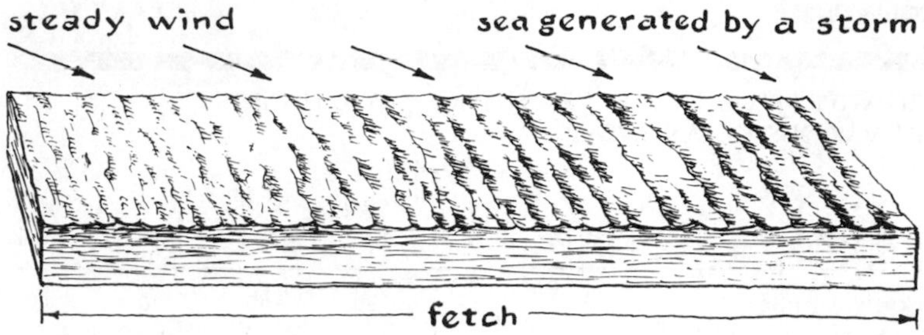

fetch

FIG. 9–9
Waves moving away from the generating area
fan out because of the variable direction of
the generating winds. At the same time, the
waves disperse because the longer waves move
at greater speeds than the shorter waves.

waves. As a result, complex waves of varying wavelengths
formed in the generating area are sorted as they move away
from the storm area, the long waves preceding the shorter
waves. Consequently, the first waves to reach a particular coast
from a distant large storm at sea will be the waves having the
longest periods. Island-dwellers, sensitive to the normal wave
period on their coasts, may be warned of approaching hurricanes by the arrival of such abnormally long waves, which
travel faster than the storm.

Swells can travel great distances, crossing entire oceans before
they encounter a coastline. For example, storms in the North
Atlantic form waves which end up as surf on the coast of
Morocco, about 3000 kilometers from their point of origin. On
extremely calm summer days, very-long-period swell strikes
the southern coast of England after traveling about 10,000 kilometers (roughly one-quarter of the way around the earth) from
storm areas in the South Atlantic Ocean. Similarly, in the
Pacific, waves from storms near Antarctica have been detected
on the Alaskan coast, more than 10,000 kilometers away.

WAVE HEIGHT AND WAVE ENERGY

Despite an abundance of data, observations of wave height often
leave much to be desired. An observer on a moving ship with
no fixed reference points for use in making estimates does not
always provide the most reliable information. Still, more than
40,000 observations made from sailing ships, as classified in
Table 9–2, indicate that about one-half of the waves in the
ocean are 2 meters or less in height. Only about 10 to 15 percent of the ocean waves exceed 6 meters in height, even in such
notoriously stormy areas as the North Atlantic or in the Roaring
Forties of the southern oceans.

How *does* one report wave height? When one looks out over
a stretch of water, there are waves present covering quite a
range of heights. It turns out, however, that the scene is not
totally random and can be described statistically. Detailed
studies of ocean waves show a nearly constant relationship between waves of various heights. One useful index is the one

Table 9–2

RELATIVE FREQUENCY, IN PERCENT, OF WAVE HEIGHTS
IN VARIOUS OCEAN AREAS*

| Ocean | Wave Height, Meters | | | | | |
	<1	1–1.5	1.5–2	2–4	4–6.1	>6.1
North Atlantic (Newfoundland to England)	20	20	20	15	10	15
North Pacific (Latitude of Oregon and south of Alaska Peninsula)	25	20	20	15	10	10
South Pacific (West Wind belt latitude of Southern Chile)	5	20	20	20	15	15
Southern Indian Ocean (Madagascar and Northern Australia)	35	25	20	15	5	5
Whole ocean	20	25	20	15	10	10

*After H. B. Bigelow, and W. T. Edmondson, *Wind Waves at Sea, Breakers and Surf* (U.S. Naval Oceanographic Office H.O. Publ. No. 602), Washington, D.C., 1947, 177 pp.

Table 9–3

WAVE-HEIGHT CHARACTERISTICS*

	Relative Height
Most frequent waves	0.50
Average waves	0.61
Significant (highest one-third)	1.00
Highest 10 percent	1.29

*U.S. Naval Oceanographic Office, 1958. p. 730.

based on the height of the *significant waves*—the average of the highest one-third of the waves present—as shown in Table 9–3. Setting the height of the significant waves at 1.00, then we find that the most frequent waves are about one-half as high, and the average waves are about 0.61. The highest 10 percent will be about 1.29 times as high as the significant waves. Thus, given the wave height for part of the wave spectrum, we can predict the other parts of the spectrum.

From on shipboard, the largest waves are the ones which are the most impressive. The largest waves occur in the open ocean, where they are formed by strong winds blowing over large bodies of water. Such waves occur most frequently at stormy latitudes, where the storms tend to come in groups traveling in the same direction, with only short periods separating them.

Thus the waves of one storm often have no chance to decay or travel out of the area before the next storm arrives to add still more energy to the waves, causing them to grow still larger. Typhoons and hurricanes (as the one in Fig. 9–10) do not form exceptionally large waves, because their winds, although very strong, do not blow long enough from one direction.

FIG. 9–10
Hurricane driving waves against the North Bayshore retaining wall at Biscayne Bay, Miami, Florida, Sept. 21, 1964. (Photograph courtesy NOAA)

There are reliable reports of waves 13 to 15 meters high in the North and South Atlantic and the southern Indian Ocean. It appears that these ocean regions rarely produce waves much higher, for several reasons. First, winds rarely blow from one direction long enough to produce waves that are significantly higher, and when the wind changes direction, waves produced under previous wind systems are destroyed or greatly modified. Also, the stormy areas of all oceans experience equally severe storms at one time or another, so that the major difference between ocean areas is the maximum fetch upon which the wind can act. In the North Atlantic the maximum effective fetch is about 1000 kilometers. With such a fetch, a wind blowing about 70 kilometers per hour can produce waves about 11 meters high; with an unlimited fetch, the same wind could produce waves about 15 meters high.

As might be expected from comparing ocean-basin dimensions, the Pacific holds the records for giant waves. The largest deep-water wave for which we have very reliable data was measured in the North Pacific on February 7, 1933. The Navy tanker *U.S.S. Ramapo* encountered a prolonged weather disturbance which had an unobstructed fetch of many thousands of kilometers. The ship, steaming in the direction of wave travel, was relatively stable, and the ship's officers were able to make several measurements of wave height as shown in Fig. 9–11 which showed that one wave was at least 34 meters high. The wave period was clocked at 14.8 seconds and the wave speed

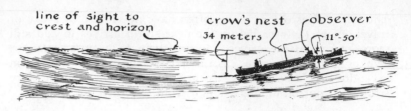

line of sight to crest and horizon   crow's nest   observer
34 meters   11°-50'

FIG. 9–11
Measurement of a wave 34 meters high by the *U.S.S. Ramapo* in the Pacific Ocean, February 7, 1933. This is the largest wave ever measured in a reliable manner.

at 102 kilometers per hour. This was truly a gigantic wave.

An impressive amount of energy is dissipated by breaking waves in the surf. A single wave 1.2 meters high, with a 10-second period, striking the entire West Coast of the United States, is estimated to release 50 million horsepower. (For comparison, Hoover Dam produces about 2 million horsepower per year.) Most of this energy is released as heat, but it is not detectable, because, for one thing, water has a high heat capacity, and, perhaps more important, there is extensive mixing in the *surf zone*, where waves break forming surf, so that the heat is mixed through a large volume of water and carried away from the beach.

WAVES IN SHALLOW WATER

Some wind waves are destroyed when they encounter winds blowing in the opposite direction. Other waves interact and cancel each other. But most wind waves end up in the surf zone as breakers when they encounter the coastline.

Except for the very longest waves—such as seismic sea waves or the tides—most waves move through the deep ocean without experiencing any effects of the bottom. As waves approach the coast, they are increasingly affected by the presence of the bottom, changing gradually from deep-water to shallow-water waves. The wavelength and speed continually decrease, while the wave period remains constant. At the same time, the wave height first decreases slightly, then increases rapidly as the water

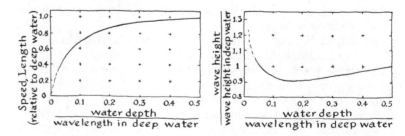

FIG. 9–12
Waves change as they enter shallow water. Speed and wave length decrease as the water becomes more shallow; wave height first decreases, then increases.

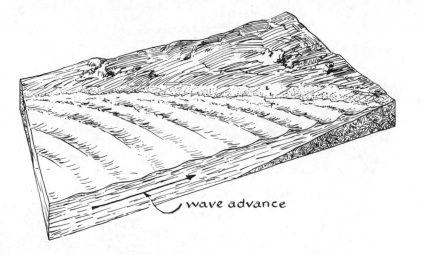

wave advance

FIG. 9–13
Refraction of a uniform wave train advancing
at an angle to a straight coastline over a
gently sloping, uniform bottom. Note the bend
in the crests as the waves approach the
beach. Such waves would cause a longshore
current moving to the right near the beach.

depths decrease to one-tenth the wavelength, and the wave
crests crowd closer together. (This series of events is illustrated
in Fig. 9–12.)

Furthermore, the direction of wave approach changes upon
entering shallower water, so that we commonly see the breakers
nearly parallel to the coastline when they reach the beach, although they may have approached the coast from many different
directions. This process, known as *wave refraction*, occurs because the part of the wave still in deeper water moves faster
than the part that has entered the shallower water. This causes
the crest to rotate into a position parallel to the depth contour
of the bottom in the shallow water near shore, as shown in
Fig. 9–13.

In the simple case just discussed, the ocean bottom was sloping uniformly away from the beach. Obviously, this is not
always the case, and irregularities of the ocean bottom have a
pronounced effect on wave refraction. For example, submarine
ridges and canyons cause wave refraction such that the wave
energy is concentrated on the headlands and spread out over
the bays, as shown in Fig. 9–14. This results in more rapid
erosion of the headlands; the eroded material is often deposited
in adjacent bays, eventually creating a simpler, less rugged
coastline. An example is the refraction of waves by the Hudson
Submarine Canyon off the New York–New Jersey shore, where
the waves of certain periods are focused on the southern shore
of Long Island (illustrated in Fig. 9–15), with much less energy
reaching parts of the New Jersey coast.

As waves encounter very shallow water, wave height increases
and wavelength decreases. Consequently, wave steepness (H/L)
increases, and the wave becomes unstable when the height of
the wave is about 0.8 the water depth and forms a *breaker*. The
belt of nearly continuous breaking waves along the shore or

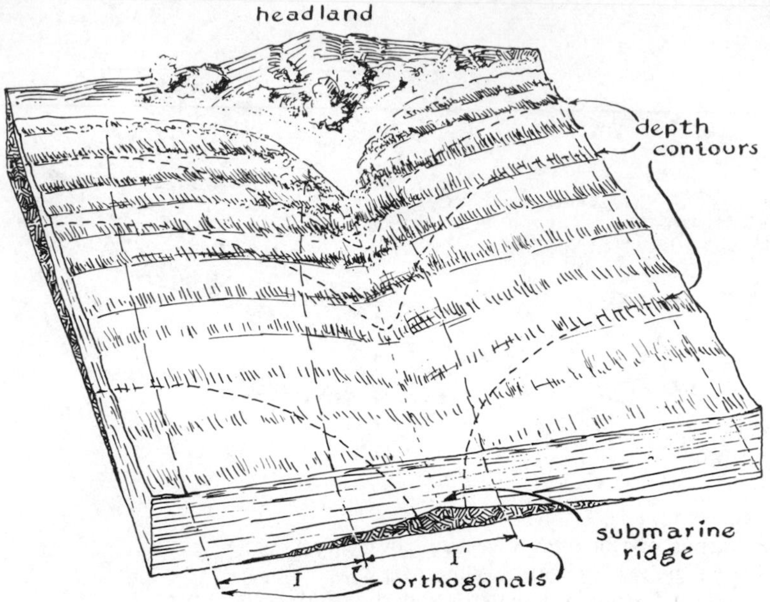

FIG. 9–14

Wave refraction causes energy initially equal at *I* and *I'* to be concentrated on the headland and to be spread out over the adjacent bay.

over a submerged bank or bar is known as the *surf*. These breaking waves are distinctly different from the breaking of oversteepened waves in deeper water, where the tops are blown off by the wind. Remember that in breakers the water moves with the wave, rather than the wave form passing through the water.

Several types of breakers are shown in Fig. 9–16 and classified in Table 9–4. The *spilling* type and the *plunging* type behave differently and form under different circumstances. The spilling breaker is easily visualized as an oversteepened wave where the unstable top spills over the front of the wave as it

Table 9–4

TYPES OF BREAKERS AND BEACH CHARACTERISTICS ASSOCIATED WITH EACH*

Breaker type	Description	Relative Beach Slope	Water Depth / Wave Height
Spilling	Turbulent water and bubbles spill down front of wave; most common type	Flat	1.2
Plunging	Crest curls over large air pocket; smooth splash-up usually follows	Moderately steep	0.9
Collapsing	Breaking occurs over lower half of wave; minimal air pocket and usually no splash-up; bubbles and foam present	Steep	0.8
Surging	Wave slides up and down beach with little or no bubble production	Steep	Near 0

*After Cyril J. Galvin, "Breaker Type Classification on Three Laboratory Beaches," *Journal of Geophysical Research*, 73(12) (1968), 3655.

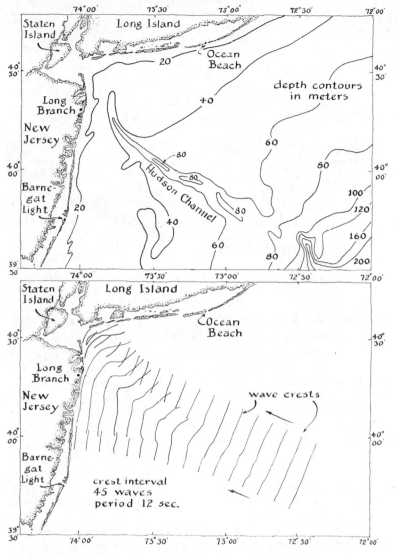

FIG. 9–15
Submarine topography, especially Hudson Channel (a), causes complicated wave-refraction patterns near the entrance to New York Harbor (b). (After W. J. Pierson, G. Neumann, and R. W. James, *Practical Methods for Observing and Forecasting Ocean Waves by Means of Wave Spectra and Statistics*, U.S. Naval Oceanographic Office H.O. Publ. 603, Washington, D.C., 1955)

travels toward the beach. In a spilling breaker, the wave form advances, but wave height (i.e. wave energy) is gradually lost. Spilling breakers are good for surfing.

Plunging breakers are much more spectacular. The wave crest typically curls over, forming a large air pocket. When the wave breaks, there is usually a large splash of water and foam thrown into the air. These waves are even better for surfing.

Plunging breakers tend to form from long, gentle swells ($H/L = 0.005$) over a gently sloping bottom with rocky irregularities. On even steeper bottoms, such waves may break

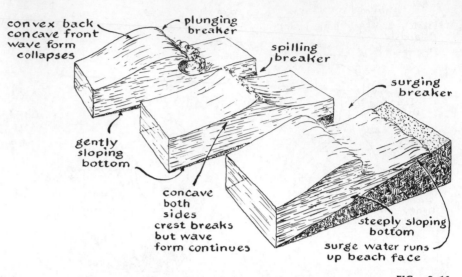

convex back
concave front
wave form
collapses

plunging
breaker

spilling
breaker

surging
breaker

gently
sloping
bottom

concave
both
sides
crest breaks
but wave
form continues

steeply sloping
bottom

surge water runs
up beach face

FIG. 9–16
Types of breakers.

over the lower half of the wave with little upward splash. This is known as a *collapsing* type of breaker, shown in Fig. 9–17.

Usually the surf consists of a mixture of various types of breakers, the result of different type waves coming into the beach and the complex and uneven bottom topography offshore, which is usually changing because of tidal action.

If a wave strikes a barrier, such as a vertical wall, it may be reflected, its energy being transferred to another wave, which travels in a different direction. *Wave reflection* may be seen when small waves in a bathtub are reflected from the sides. In other cases, a wave striking a steeply dipping barrier may form a surging breaker or a turbulent wall of water, which then moves up the barrier as the wave advances and runs back down when the wave retreats.

The height of the surf depends on the height and steepness of the waves offshore, and to a certain extent on the offshore bottom topography. Thus the breakers and surf may be only a few centimeters high on a lake or a protected ocean beach, or many meters high on an open beach. Impressive reports of high surf often come from lighthouses built in exposed positions. For example, Minot's lighthouse, 30 meters tall, on a ledge on the south side of Massachusetts Bay, is often engulfed by spray from breakers, and the glass in the lighthouse at Tillamook Rock, Oregon, 49 meters above the sea, has often been struck by surf.

We have no recorded observations of the waves that cause such surf, but breakers about 14 meters high twice damaged a breakwater at Wick Bay, Scotland, moving blocks weighing as much as 2600 tons. Breakers about 20 meters high have been reported at the entrance to San Francisco Bay, and at the Columbia River estuary on the Pacific coast when onshore gales were blowing. Waves and breakers at a river or harbor are likely to be especially high when the incoming waves encounter a current setting in the opposite direction. Ships may have to wait for days before finding a period when the incoming waves

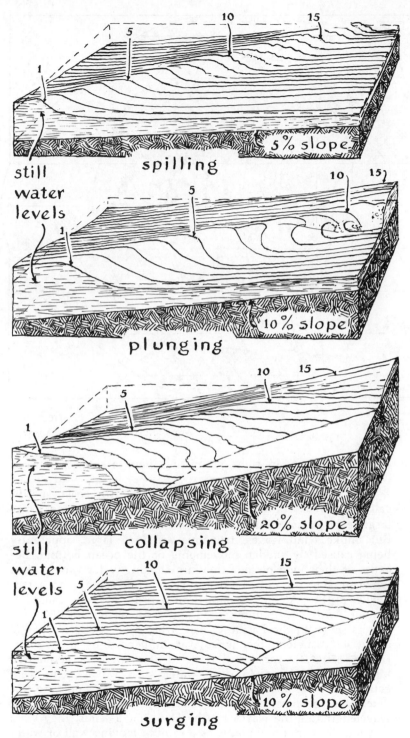

still water levels

5% slope

spilling

10% slope

plunging

still water levels

20% slope

collapsing

still water levels

10% slope

surging

FIG. 9–17
Changes in the water surface through time
(0.06-second intervals) as a breaker advances
toward the beach. (After C. J. Galvin, Breaker
type classification on three laboratory beaches.
Journal of Geophysical Research 72 1968
3651–3659)

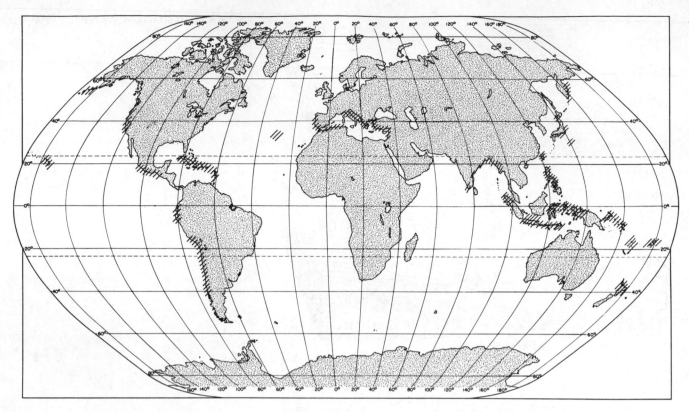

FIG. 9–18
Areas affected by seismic sea waves.

and tidal currents are right, permitting them to enter the harbor with safety.

Like wind waves, *seismic sea waves* strike the coast, specifically the areas shown in Fig. 9–18; such waves often cause large, even catastrophic waves when they enter shallow water. But unlike wind waves, tsunamis are very large, apparently being caused by sudden movements of the ocean bottom. Because of their long periods they are shallow-water waves, even when passing through the deep ocean. For example, an earthquake in the Aleutian Islands on April 1, 1946, caused a tsunami with a 15-minute period and a wavelength of 150 kilometers. Even in the Pacific Ocean, where the average depth is about 4300 meters, this wave was influenced by the bottom ($L/d = 150/4.3$); yet it still traveled at a speed of about 800 kilometers per hour. In deep water such wave crests were estimated to be about a half-meter high, which would be undetectable to ships, especially considering the extremely long wavelength. As the waves hit the Hawaiian Islands, they were driven ashore in a few places as a rapidly moving wall of water up to 6 meters high. These waves also formed enormous breakers that towered up to 16 meters above sea level, where the water was funneled in a valley. More than 150 people were killed in Hawaii, and property damage was extensive. Apparently the wave was highest in the Aleutian Islands, where a reinforced-concrete lighthouse and radio tower 33 meters above sea level were destroyed at Scotch Cap, Alaska. So far as historical records show, Japan alone has been hit by about 150 tsunamis.

Because of their frequent occurrence in areas bordering the Pacific (see Fig. 9–18), a warning net is operated to detect large earthquakes likely to cause tsunamis and to warn those areas likely to be hit. As a result, the tsunami of 1957 killed no one in Hawaii, even though water levels were locally higher than in 1946.

INTERNAL AND STANDING WAVES

So far, we have spoken only about progressive surface waves, but there are other types of waves in the ocean which are not as easily observed, and consequently not as well known. *Internal waves*, as in Fig. 9–19, occur within the ocean rather than at the ocean surface, although in principle they are similar to surface waves. Surface waves occur at an interface between air and water, and internal waves are found at an interface between water layers of different densities—for example, the pycnocline. Since the pycnocline is associated either with the halocline or the thermocline, internal waves can be indirectly observed by studying changes in temperature or salinity at a given depth and fixed location. Movement of the pycnocline due to the passage of internal waves may result in confused or contradictory interpretations of oceanographic data taken at widely separated times from the same area.

Internal waves, it is thought, move as smoothly undulating shallow-water waves. Neglecting the earth's rotation, the speed

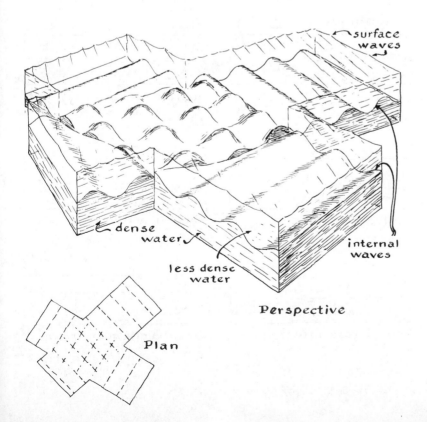

FIG. 9–19

Simple internal waves at interface between water layers of different densities and their interaction to form more complete patterns.

C of an internal wave in meters per second is given by

$$C = \sqrt{g\left(\frac{\rho - \rho'}{\rho}\right)\left(\frac{dd'}{d + d'}\right)}$$

where ρ and ρ' are the densities of the lower and upper layers, respectively; *d* is the depth in meters below the interface; *d'* is the height of the free surface above the interface; and *g* is the acceleration of gravity. Surfaces where internal waves form involve only small density differences between two water layers rather than the much larger density difference between air and water. As a result the height of internal waves can be much greater than surface waves and they generally move much slower.

Standing (or *stationary*) *waves* are yet another type of wave phenomenon in the ocean; standing waves are also important in lakes and play an important role in tidal phenomena. Standing waves can be generated experimentally by first tilting a round-bottomed dish of water and then setting it on a table: the water's surface will appear to tilt first one way and then the other. This type of movement is distinctly different from the progressive waves we have been discussing wherein the wave moves across the body of water—as if we had dropped a pebble in the dish of water.

One may notice that in the standing wave, part of the water surface does not move vertically but acts as a sort of hinge about which the rest of the water surface tilts. This stationary line or point is known as the *node*, and the parts of the water

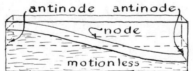

start

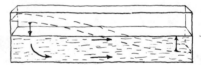

quarter period later

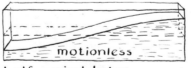

half period later

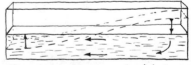

three quarter period later

one period later

one and one quarter periods later

FIG.   9–20
Simple standing wave, with one node—shown at quarter-period intervals.

surface having the greatest vertical movement—known as the *antinodes*—are situated at the walls of the container. More complicated stationary waves may have more than one node and several antinodes in addition to those at the boundaries of the container.

In a stationary wave (as shown in Fig. 9–20), the maximum horizontal water movement occurs at the nodes when the water surface is exactly horizontal. When the water surface is tilted most, there is no water motion. In contrast to the continual orbital motion of the water in progressive waves, in a stationary wave the water flows for a distinct period, stops, and then reverses the flow direction. Also, the wave form alternately appears and disappears, and the water does not move in the circular or nearly circular and continuous orbits associated with progressive waves. Standing waves, also known as *seiches* (pronounced "saysh"), are characteristic of steep-sided basins and are well known in many lakes and in the tidal phenomena of nearly closed basins, such as the Red Sea.

In a simple standing wave (neglecting the effects of the earth's rotation), the period T in seconds can be calculated as

$$T = \left(\frac{1}{n}\right)\left(\frac{2L}{\sqrt{gd}}\right)$$

where n is the number of nodes; L is the length, in meters, of the basin measured in the direction of wave motion; d is the depth, in meters, of the basin; and g is the acceleration of gravity.

Like progressive waves, standing waves can be modified by their surroundings. They are reflected by vertical boundaries and partially absorbed or obliterated by gently sloping bottoms. Standing waves are refracted by moving into depths substantially less than one-half their wave length. Standing waves in large basins such as the Great Lakes are complicated by the influence of the Coriolis effect, and the resulting wave, instead of simply sloshing back and forth, has a rotary motion.

SUMMARY OUTLINE *Wave*—a disturbance of the water surface

Ideal waves

Crest—highest part of wave form

Trough—lowest part of wave form

Wave height—vertical distance from bottom of wave trough to top of wave crest

Wavelength—horizontal distance between successive wave crests (or troughs)

Wave period—in progressive waves: time for successive wave crests or troughs to pass a fixed point; in standing wave: time for water surface to assume initial position

Sine wave—smooth-crested, smooth-troughed ideal wave form used to describe low ocean waves (a mathematical expression)

Trochoid—sharp-crested, flattened trough ideal wave form useful to describe larger waves

Wave steepness—ratio of wave height to wavelength

Deep-water wave—wave unaffected by boundaries, especially ocean bottom, water deeper than $L/2$

Shallow-water wave—wave affected by boundaries, especially ocean bottom, in water less than $L/20$ deep; wave speed $C = \sqrt{gd}$

Waves and their formation

Seismic sea waves—formed by sudden impulsive movement of bottom due to earthquake or sediment slump

Tides—longest waves in ocean

Forced waves—waves formed by disturbance continuously applied: tides are example

Free waves—waves move independently of disturbance; wind waves are example

Capillary waves—ripples, round-crested, V-troughed; wavelength less than 1.7 cm; wave period 0.1 sec; surface tension is dominant restoring force

Gravity wave—most common waves; gravity is dominant restoring force-period 1 sec to 5 min

Formation of sea and swell

Sea—waves under influence of the wind that formed them; short, sharp crests, chaotic, unpredictable

Swell—waves outside generating area; long smooth crests, relatively predictable

Energy for waves usually supplied by wind

Energy transferred to water

Ripples form first, transformed into larger waves, eventually into fully developed seas

Swell forms by gradual transformations

Change to smooth-crested waves

Angular dispersion occurs because of initial differences in direction

Waves separate according to wave length, longer waves travel faster than short ones

Swell can travel tens of thousands of kilometers

Wave height

About 90% of waves less than 6 m

Longest waves: Pacific—34 m; Other oceans 15 m

Waves in shallow water

Refraction—change in wave advance direction upon entering shallow water; energy focused on headlands

Reflection—energy reflected, transferred to newly formed wave moving in direction opposite to original wave

Breakers—unstable waves losing energy; classified as plunging, spilling, or surging

Surf—band of breakers parallel to coast

Internal and standing waves

Internal waves—occur at boundary between water layers

Standing waves (seiches)—occur in nearly closed basins

Entire water surface tilts

Nodes—water level constant; maximum horizontal water movement

Antinode—maximum vertical water movements and changes in water surface level

SELECTED REFERENCES

BASCOM, WILLARD. 1964. *Waves and Beaches: the Dynamics of the Ocean Surface.* Doubleday Anchor Books, Garden City, N.Y. 267 pp. Elementary.

KINSMAN, BLAIR. 1965. *Wind Waves: Their Generation and Propagation on the Ocean Surface.* Prentice-Hall, Inc., Englewood Cliffs, N.J. 676 pp. Advanced mathematical treatment; good descriptions.

RUSSELL, R. C. H., AND D. H. MACMILLAN. 1954. *Waves and Tides.* Hutchinson Scientific and Technical Publications, London. 348 pp. Elementary.

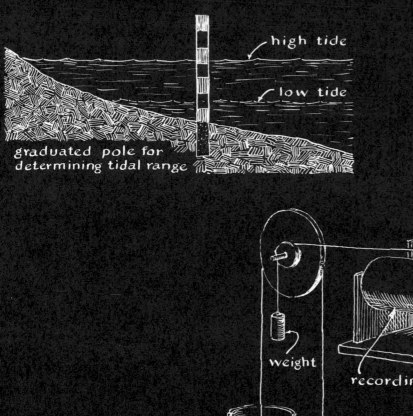

graduated pole for
determining tidal range

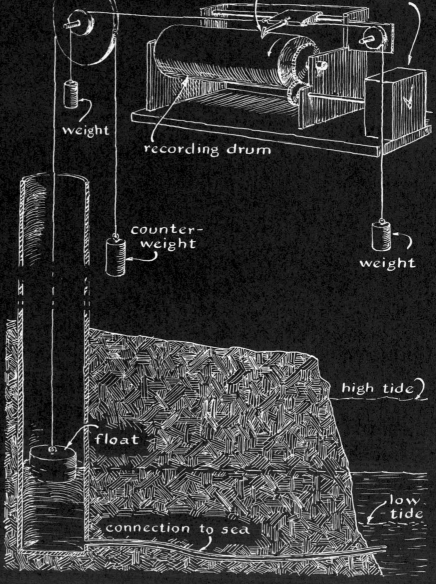

high tide

low tide

recording
pen

clock
motor

weight

recording drum

counter-
weight

weight

high tide

float

low
tide

connection to sea

TIDES AND TIDAL CURRENTS

Graduated pole for determining tidal range.
Simplified diagram of automatic tide guage.

ten

Tides are the pulse of the ocean. Their effects are felt most keenly in the coastal ocean, where periodic rise and fall of the ocean surface, with alternate submersion and exposure of the intertidal zone, profoundly influence plant and animal life. Nor is man oblivious to the tide. Extremely low tides are occasions for digging clams; extremely high tides combined with storms have often caused flooding of low-lying areas and damage to coastal installations. Tidal currents, accompanying the rise and fall of the tide, are by far the strongest currents in the coastal ocean; even large, modern ships prefer to depart a port with an outgoing tide and to enter on an incoming tide, just as sailors did in antiquity.

Like waves, tidal phenomena are easily observed and have been studied since Roman times, at least. Pliny the Elder (A.D. 23–79) correctly attributed tides to the effect of the sun and the moon. The small tides of the Mediterranean Sea could easily be ignored, and most ancient writers did so. Two famous men of action got into trouble by ignoring the tides when their military conquests took them into other ocean areas: Julius Caesar lost part of his fleet and sustained damage to most of his ships during a night of high tide on the coast of England; and Alexander the Great got into similar difficulties at the mouth of the Indus River in the Northern Indian Ocean.

Tides, however, were certainly not strange to those who lived around the North Atlantic. The Venerable Bede (A.D. 673–735) wrote of the extensive tidal observations made by medieval British priests and exhibited a clear understanding of the relationship of tidal phenomena to the moon. He pointed out that separate tidal predictions must be made for each area and that 19 years of observations (corresponding to a complete lunar cycle) were required in order to compile a really accurate tide table.

Among the oldest records of oceanographic observations is the "Flod at London Brigge," a set of tide tables for the London Bridge dating from the late twelfth or early thirteenth century. Probably other seafaring people also had some means of predicting tides, but any such records have been lost, so that we can now only speculate about them. Tables of tides and tidal currents represent one of our most successful efforts at predicting physical processes in the ocean.

TIDAL CURVES

Compared with most oceanographic observations, measurements of the tide can be quite simple, requiring only a pole marked at appropriate intervals and firmly attached to a piling or stuck in the bottom; such a pole is shown in Fig. 10–1. At timed intervals we can record the height of the water surface on the pole—for instance, every hour. When we plot the height of the water surface at each interval, the result is a *tidal curve*.

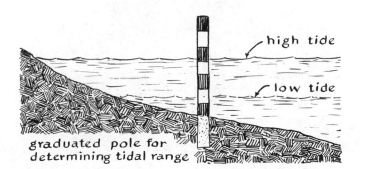

FIG. 10–1 Graduated pole for determining tidal range.

More elaborate installations are needed for the continuous tidal observations made in most major ports. A simplified diagram of such a *tide station* is shown in Fig. 10–2. An enclosed basin is constructed with a narrow connection to the ocean so that the water level in the basin is always equal to the undisturbed sea level outside but is not itself disturbed by normal waves. This avoids some of the uncertainty in tidal observations caused by waves or other disturbances of the water surface. A float on the water surface in the basin is connected to a marker, often a pencil, which draws the tidal curve on a clock-driven, paper-covered drum, indicating the changing sea level. The mechanism operates continuously and requires a minimum of maintenance; normally such a gauge is checked once a day.

Examination of tidal curves from many ports shows that each one has a different tide. Tides are grouped into three types, based on the number of highs and lows per day, the relationship

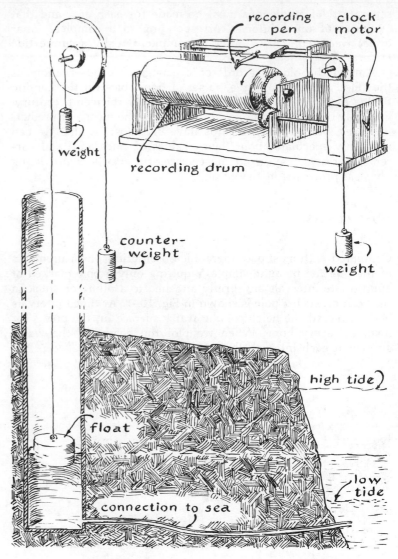

FIG. 10–2
Simplified diagram of automatic tide gauge.

between the heights of successive highs or lows, and the time between corresponding high (or low) stands of sea level. Most tidal curves show two high tides and two low tides per *tidal day* —about 24 hours, 50 minutes. This period corresponds to the time between successive passes of the moon over any point on the earth. The time, either 12 hours and 25 minutes or 24 hours, 50 minutes, between high (or low) tides is known as the *tidal period*.

A few ocean areas, such as parts of the Gulf of Mexico, have only one high tide and one low tide each day (see Fig. 10–3); these are called *daily tides* or *diurnal tides*. Most North Atlantic ports have two high and two low tides that are approximately equal; these are *semidaily* or *semidiurnal* tides. They are relatively easy to predict because high tides tend to occur at a regular time after the moon has crossed the meridian for that port. Tidal predictions for ports with semidaily tides have been

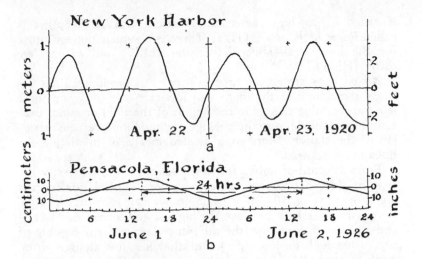

FIG.   10–3

Typical tide curves for (a) semidaily tide, New York Harbor, April, 1920 and (b) daily tide, Pensacola, Florida. (After H. A. Marmer, *The Sea*, D. Appleton, New York, 1930)

made for many centuries, based primarily on an understanding of the lunar cycle.

Tidal curves from U.S. Pacific ports also show two high tides and two low tides per tidal day, but the highs are usually quite different in height, and the low tides differ as well. These *mixed tides* are shown in Fig. 10–4. The higher of the two high tides

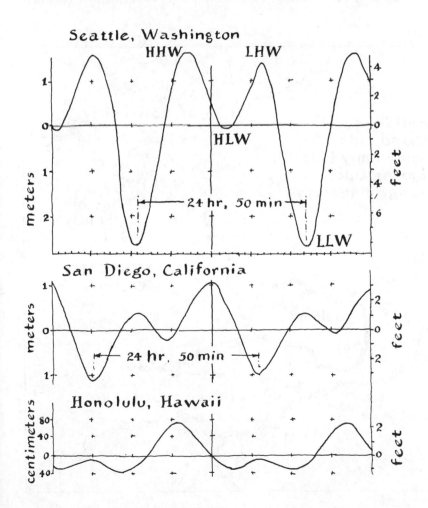

FIG.   10–4

Mixed tides at Pacific ports, April, 1920. (After Marmer, 1930)

is called *higher high water* (abbreviated HHW); the other is called *lower high water* (LHW). There is a similar nomenclature for the low tides—*lower low water* (LLW) and *higher low water* (HLW).

Mixed tides are not as easy to predict as semidaily tides because the timing of the high- and low-tide *stands* (when there is no appreciable change in the height of the tide) does not bear a simple relationship to the passage of the moon over the meridian of the station. More sophisticated means of predicting the tides must be used.

From a record of only a few days' length, we can classify the type of tide characteristic of any harbor. Typical tidal curves are shown in Fig. 10–6. For example, we can measure the *tidal range* (the difference between the highest and lowest tide levels) and the *daily inequality* (the difference between the heights of successive high or low tides). But the tide also changes from week to week. With a record of several weeks' duration, we see a pattern in the changes of tidal range. *Spring tides* occur near the times of full and new moons, and the spring-tidal range is larger than the *mean-tidal range* (the difference between mean high and mean low tides) or the *mean daily range*. During the first and third quarters, the tidal range is least; these are called the *neap tides*. As Fig. 10–6 shows, there is substantial variation

FIG. 10–5
Types of tides and spring-tidal ranges (in meters) on North American coastlines. (Modified after U.S. Naval Oceanographic Office, *Oceanographic Atlas of the North Atlantic Ocean*, Sect. I: *Tides and Currents*. Publ. 700, Washington, D.C. 1968)

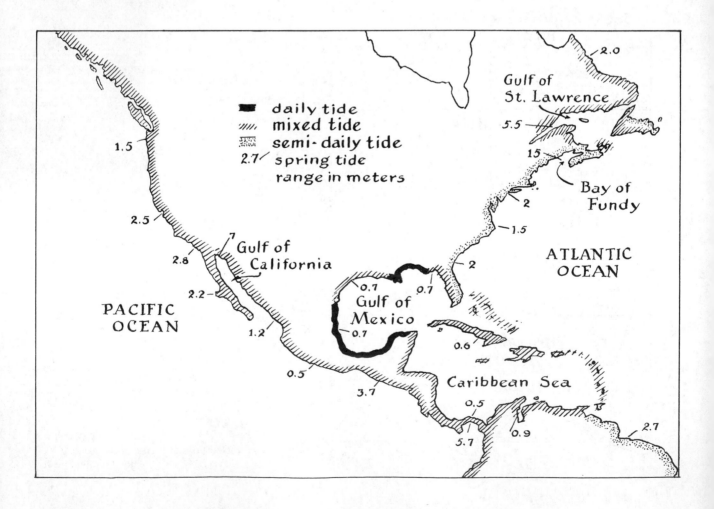

in the tides at the same place during the month. Other, less striking variations occur over periods of several years.

TIDE-GENERATING FORCES
AND THE EQUILIBRIUM TIDE

Although it had long been known that the tides were closely related to the sun and moon, it remained for Sir Isaac Newton (1642–1727) to lay the foundation for understanding the mechanics of the tides. He began by making several simplifying assumptions; his equilibrium theory of the tides assumes a static

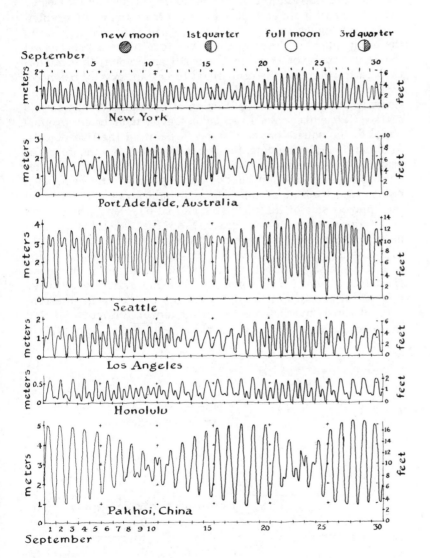

FIG. 10–6
Tidal variations at certain ports during the course of a month. Note the variations in the timing of the spring and neap tides relative to the new and full moon. (After U.S. Naval Oceanographic Office. *American Practical Navigator*, rev. ed., (Bowditch), H.O. Publ. No. 9, Washington, D.C., 1958)

ocean completely covering a nonrotating earth, with no continents. Let us consider the effect of the moon on such a simplified ocean.

Gravitational attraction pulls the earth and moon toward each other, while centrifugal forces, acting in the opposite direction, keep them apart, as illustrated in Fig. 10–7. In many respects, the earth and moon act like twin planets revolving about a common center, which in turn moves around the sun. If the earth and moon were the same size, the center of revolution of the system would be located midway between them. The moon, however, is only about $\frac{1}{82}$ the mass of the earth; consequently the center of revolution of the earth–moon system is located nearer the earth, about 4700 kilometers from the earth's center. This is analogous to an adult and a small child on a see-saw: the adult must sit closer to the pivot to achieve a balance. At the center or pivot between the earth and moon, the gravitational attraction and the centrifugal forces are exactly balanced.

There are, however, small unbalanced forces in the system. Consider the gravitational attraction of the moon on the earth's surface. Remember that the force of gravity is inversely proportional to the square of the distance separating the two objects. In other words, doubling the separation reduces the force of gravity to one-fourth of its former strength. A parcel of water located on the earth's surface is only 59 earth radii away at a point nearest the moon, but 61 earth radii away when it is on the opposite side of the earth (see Fig. 10–7). Hence the moon's gravitational attraction is greatest on the side of the earth nearest the moon, and least on the opposite side of the earth. The centrifugal forces, however, are equal over the earth's surface. On the side nearest the moon, the attraction of the moon exceeds the centrifugal force, so that the water is attracted toward the moon. On the opposite side of the earth, the centrifugal forces overbalance the attraction of the moon so that there, too, a force acts on the water effectively dragging it away

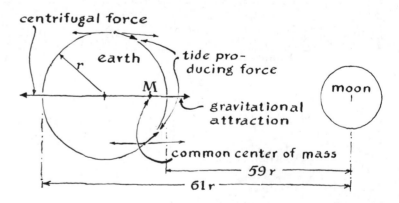

FIG. 10–7
Earth–moon system revolving around their common center *M*. Note that the ocean nearest the moon is about 59 earth radii (*r*) from the moon compared to 61 earth radii for the side farthest from the moon. At the center of the earth, gravitational attraction of the moon is balanced by the centrifugal forces.

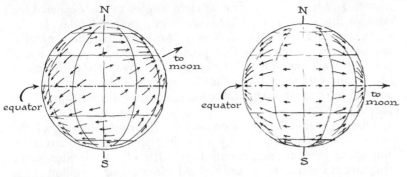

FIG. 10–8
Horizontal component of the tide-producing forces acting on the ocean surface when the moon is in the plane of the earth's equator (a) and when the moon is above the equatorial plane (b). Note the tide-producing forces shift their orientation as the moon's position changes. (After U.S. Naval Oceanographic Office, 1958)

from the earth. These unbalanced forces on the earth's surface, shown in Fig. 10–8, are the tide-generating forces associated with the moon.

To see how the tides are created, let us look at these forces in more detail. Such forces can be represented by vectors (shown in Fig. 10–9 as arrows) which point in the direction in which the forces act. The length of the arrow (the vector) corresponds to the relative strength of the force. Each vector can be resolved

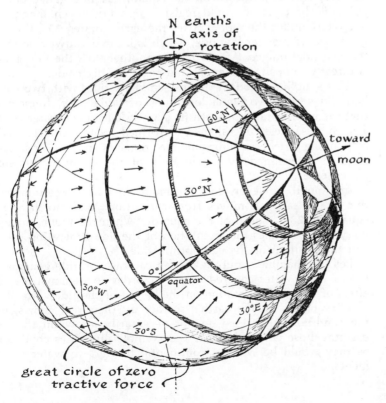

FIG. 10–9
Equilibrium tide (shown by the "fences") when the moon is above the plane of the earth's equator. The height of the tidal bulge is shown by the height of the "fences." Note variations in the tide on any parallel of latitude as the earth rotates. (After W. S. von Arx, *An Introduction to Physical Oceanography*, Addison-Wesley Publishing Company, Inc., Reading, Mass. 1962.)

into two components, acting at right angles to each other. One component acts in a vertical direction, perpendicular to the earth's surface; the other acts in a horizontal direction, parallel to the earth's surface at that point. The vectors point toward the moon on one side of the earth (and away from the moon on the opposite side). The relative strength of the force acting in either a horizontal or vertical sense varies over the earth's surface (see Fig. 10–9).

Directly beneath the moon and on the opposite side of the earth, tide-generating forces act solely in a vertical direction. But these vertical components have little effect on tides, since they are counteracted by gravity, which is about 9 million times stronger. Horizontal components of the tide-generating force are also rather weak, but they are comparable in strength to other forces acting on the ocean's surface and thus are not overwhelmed.

Newton pointed out that these horizontal or tractive forces cause the *equilibrium tide*, which consists of a tidal bulge on the side of the earth nearest the moon and on the side opposite the moon. As a result of these horizontal forces, the water covering the earth is deformed slightly to form an egg-shaped water envelope on our imaginary, nonrotating, water-covered earth. The solid earth itself also responds to these tide-generating forces and deforms slightly, but much less than the ocean waters.

Now if we permit the earth to rotate beneath its deformed, watery covering, we can see how the equilibrium-tide theory explains successive high and low tides. If the moon is in the plane of the earth's equator, the tidal bulges of the equilibrium tide will also be centered on the equator. A tide gauge at any point of the equator would register high tide when that point is directly under the moon. After the earth rotates 90°, the tide gauge would register low tide, which is located midway between the two tidal bulges. After rotating another 90°, the tide gauge is directly opposite the moon and registers high tide.

This simplified model explains semidaily tides with two equal high tides and two equal low tides per tidal day. Remember that the earth rotates beneath its slightly deformed water cover. The more-or-less egg-shaped deformed water surface, corresponding to the equilibrium tide, remains fixed in space, its location determined by the location of the moon in our simplified case.

The moon, however, does not maintain a fixed position relative to the earth. It moves from a position 28.5° north of the equator to 28.5° south of the equator. When the moon changes its position, so does the orientation of the tide-generating forces and the position of the equilibrium tide.

Let us imagine a purely theoretical case of tide-generating forces on a water-covered earth, in order to demonstrate the effect of the earth's rotation on the height of the tides. We shall identify points on the earth's surface by their real names, but the conditions described do not necessarily apply on the real earth at those points, because the example illustrates conditions as they would be without continental barriers to water movements.

If the moon were, for instance, over the site of Miami, there would be a tidal bulge there and also one near the west coast of Australia. These bulges would be particularly well-developed by virtue of being in line with the tide-generating forces. After the earth rotated 180°, these points would again experience a high tide, but a lower one because Miami and Australia would no longer be in line with the moon. Tide-generating forces would now be greatest off northern Peru and in the Bay of Bengal–South China Sea area. (Use of a globe will be helpful in demonstrating this point.) This example shows that the equilibrium theory of the tides not only explains the semidaily tides but the daily inequalities as well.

A similar analysis could be made for the sun–earth system to determine the sun's influence on the tides. There are, however, several important differences. First, the greater distance of the sun from the earth (approximately 23,000 earth radii) is only partially compensated by its greater mass (330,000 earth masses), so that the solar tide-generating forces are only about 46 percent as powerful as those of the moon. Second, solar tides have a period of about 12 hours rather than the 12 hours-25 minute period of the lunar tide. Finally, the sun's position relative to the earth's equator also changes, from 23.5° north to 23.5° south of the equator, but it requires a full year to make the complete cycle, in contrast to the monthly changes of the moon's position from 28.5° north to 28.5° south.

The equilibrium theory also explains variations in tidal range between the spring and neap tides. Spring tides occur every two weeks, usually within a few days of the new and full moons (see Fig. 10–6). Considering their positions in space (shown in Fig. 10–10), we see that during the time of the full and new moons, the solar and lunar tide-generating forces act together, causing large tidal bulges and consequently greater tidal ranges. During the first and third quarters of the moon, tide-generating forces partially counteract each other by acting in different directions, resulting in the lowest tidal range—the neap tides.

There are other, more subtle variations in tides associated with the variable distance of the moon and sun from the earth. When either the sun or moon is closer to the earth, the equilibrium theory predicts that tides will be higher (or lower) than when they are farther from the earth. All these complicated relationships are exactly duplicated every 19 years; all possible positions of the sun and moon relative to the earth occur over that period of time.

Despite the many simplifying assumptions involved in the equilibrium theory, it accounts for many of the features of the tides, and its simplicity helps us to understand such aspects as the daily inequalities. It fails, however, to predict the timing of the tides. According to the equilibrium theory, high tide should occur when the moon crosses the meridian of a given port. This is clearly not the case. If it were true, prediction of the tides could have been done for centuries, whereas in fact it is only in comparatively recent times that tide-predicting computers have been designed to make the necessary complicated calculations.

Newton himself was aware that his equilibrium theory was

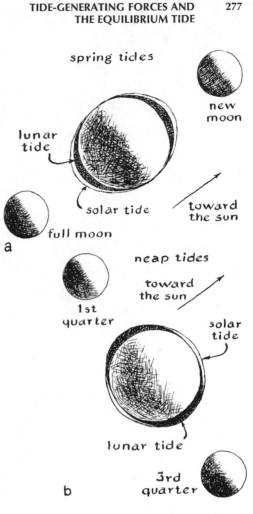

FIG. 10–10, a, b
Relative positions of sun, moon, and earth during spring and neap tides.

not a complete explanation of tidal phenomena. The theory assumed an idealized ocean and dealt only with the tide-generating forces, without taking into consideration the response of the oceans to these forces.

DYNAMICAL THEORY OF THE TIDE

Since Newton's time, mathematicians and physicists have investigated the tides by considering the response of the oceans to the tide-generating forces. This is known as the *dynamical approach*, since it involves a dynamic rather than a static ocean. This approach, like the equilibrium theory, also requires some assumptions in order to deal with the complicated interactions. Among these is the neglect of vertical forces, which is not too serious because they are quite small compared with the earth's gravitational attraction. Perhaps the most important difference between the dynamical treatment and the equilibrium theory is that the former considers the effect of the shape and depth of the basins on the tides. Furthermore, it recognizes that ocean tides do not occur in a static ocean but involve substantial water movements, and thus must take the Coriolis effect into consideration.

The dynamical theory treats the tides as another type of wave phenomenon. As we have seen, tides have many characteristics in common with both progressive and standing waves. Furthermore, treating tides as long-period waves permits the use of powerful mathematical techniques that have proved useful in dealing with waves. Tides can be resolved into several components and each treated separately.

The tide-generating forces may be resolved into *tidal constituents*, the most important being those due to the moon and the sun; some important constituents are shown in Table 10–1.

Table   10–1

SOME IMPORTANT CONSTITUENTS OF THE TIDE-GENERATING FORCES*

	Symbol	Period (hours)	Relative amplitude	Description
Semidaily tides				
	M_2	12.4	100.0	Main lunar constituent
	S_2	12.0	46.6	Main solar constituent
	N_2	12.7	19.2	Lunar constituent due to changing distance between earth and moon
Daily tides				
	K_2	23.9	58.4	Soli-lunar constituent
	O_2	25.8	41.5	Main lunar constituent
	P_2	24.1	19.3	Main solar constituent
Long-period tides (spring and neap tides)				
	M_1	327.9	17.2	Main fortnightly constituent

*After Defant, 1958.

Because of the changing positions of these bodies relative to the earth as many as 62 tidal constituents are used to make tidal predictions, although the four principal ones usually account for about 70 percent of the tidal range.

The response of a bay or harbor—or an entire ocean—to each of the tidal constituents can be considered separately as a *partial tide*. The tide for any location thus consists of the combination of these partial tides (illustrated in Fig. 10–11), just as the complicated waves in a sea can be reconstructed by combining several simple wave trains.

Study of the available tidal curves indicates how the partial tides must be combined for a given port. Then tidal predictions are made by combining partial tides in the appropriate way, normally using some type of computer. Some of the earliest tide-predicting machines were actually simple computers. Development of large, high-speed computers permits more sophisticated tidal models to be studied in greater detail with fewer simplifying assumptions than was previously possible.

Comparison of a tidal record with the profile of a simple wave clearly shows that tides can be considered to be long-period waves. Neglecting the curvature of the earth, the two water bulges are the crests and the intervening low areas are the troughs of a simple wave. Because of their immense size relative to the ocean basins, such waves should behave as shallow-water waves. Their wave length, one-half of the earth's circumference, is about 20,000 kilometers, while the ocean basins have an average depth of only about 4 kilometers, so that $d/L = 4/20,000$. This is much smaller than the limit of $1/20$ for shallow-water waves. If tides were free waves, they would move at a speed of $\sqrt{gd}$ or $\sqrt{9.8 \times 4000}$ meters per second, or about 200 meters per second (720 km/hr). In order to "keep up with the moon," the tides need to move around the earth in 24 hours, 50 minutes, which means that they would have to move 1600 km/hr at the equator.

In order for the tide waves to move fast enough at the equator to keep up with the moon, the ocean would have to be about 22 kilometers deep. Since it is much shallower, the tidal bulges move as forced waves, whose speed is determined by the movements of the moon. The position of the tidal bulges relative to the moon is determined by a balance between the attraction of the sun and moon and frictional effects of the ocean bottom.

Because the world ocean is cut by the north–south-trending continents into various basins, it is not possible for the tide to move east–west across the earth as a forced wave except around Antarctica, where the absence of continents permits the tide to sweep unimpeded through the ocean. Let us see how the tide behaves in the Atlantic Ocean.

After the tide enters the South Atlantic, it moves northward as a free wave. In the South Atlantic, its course is relatively simple, although doubtlessly influenced by the irregular ocean boundaries and probably by the irregular bottom topography as well. As it progresses northward, the wave from the Antarctic tide also interacts with the independent tide of the Atlantic.

Even if the southern end of the Atlantic Ocean were closed off, it would still have a tide of its own. The tide in a basin

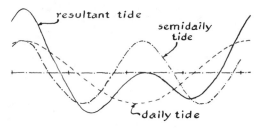

FIG. 10–11
A daily and a semidaily tide when combined produce a mixed tide (After H. A. Marmer, *The Tide*. Appleton-Century-Crofts, New York 1926.)

such as the Atlantic behaves to some degree like a standing wave, as well as having some characteristics of a progressive wave. Every basin has its own natural period for standing waves. If that natural period is approximately 12 hours, the standing-wave component of the tide will be well developed. If the period of the basin is substantially greater than or less than 12 hours, the resemblance of the tide to a standing wave is lessened. Even if a basin lacks a standing-wave tide, it often has a tide that is distinctly different from that in the adjacent ocean.

Tides in the North Atlantic are altered by reflection of the tide wave from the complicated coastline and a standing wave

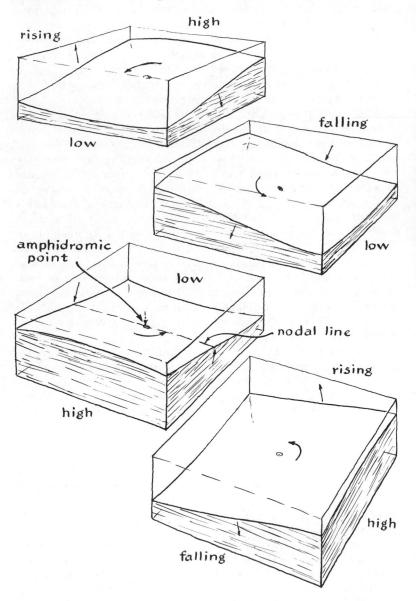

FIG. 10–12
Amphidromic motion of a standing wave tide in a Northern Hemisphere embayment. (After von Arx, 1962)

is set up, but this is not the simple standing wave discussed previously. Since substantial water movements are involved, it is modified by the Coriolis effect. The resulting standing wave is a swirling motion such as might result if we swirled water in a round-bottomed cup. Let us see why such a wave forms and how it behaves.

We shall choose a channel in the Northern Hemisphere and set the water moving northward associated with a standing wave. Because of the Coriolis effect, it will be deflected toward the right, causing a tilted water surface, as shown in Fig. 10–12. When the water flows back again, its surface will be tilted in the opposite direction. The tilted wave surface rotates around the basin, once for each wave period. Near the center of such a system, called an *amphidrome system*, is a point where the water level does not change, the *amphidromic point*.

Figure 10–13 shows the movement of the high water associated with the principal lunar partial tide in the Atlantic. There we see a large amphidrome system in the North Atlantic. Other, smaller amphidrome systems not shown on the map occur in the English Channel and in the North Sea. Points located near an amphidromic point for one partial tide will not be affected by that partial tide, but will likely be affected by

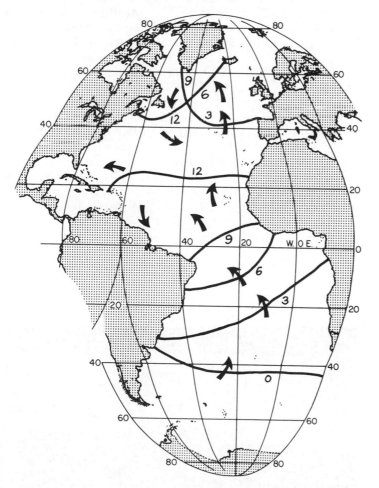

FIG.   10–13
Locations of the high water of the principal lunar partial tide—the major tidal component caused by the moon. The numbers indicate the number of hours since the moon crossed the Greenwich meridian. (After Defant, 1958)

FIG. 10–14

(a) High tide and (b) low tide at Market Slip, Saint John, New Brunswick, on the Bay of Fundy. At the northern end of the bay, on the Petitcodiac River, the profile of the incoming tide surges up the river as a steep wave, known as a tidal bore (c), photographed here at Moncton, New Brunswick. (Photographs courtesy New Brunswick Travel Bureau)

(a)

(b)

another. For example, an area near the amphidromic point for a semidaily tide may well have a daily-type tide.

Because of their dimensions and consequently their natural periods, each ocean basin responds more readily to certain constituents of the tide-generating forces than to others. The Gulf of Mexico appears to have a natural period of about 24 hours; consequently, it responds more to the daily tidal constituents than to the semidaily constituents, and much of the Gulf has a daily tide (see Fig. 10–5). The Atlantic Ocean, on the other hand, responds more readily to the semidaily constituent of the tide-generating forces, and thus tends to have a semidaily type of tide. Because of their size and complex shapes, the Caribbean Sea and the Pacific and Indian Oceans respond to both the daily and semidaily tidal forces and thus have mixed types of tides.

Each marginal sea, or bay, is affected by the tide in much the same way as the Atlantic is affected by the wave from the Antarctic. A tide wave advancing through the bay is reflected by the coast. If the basin has a natural resonance of the appropriate period, a standing wave may be excited as well. The tide in a bay, harbor, or sea is greatly influenced not only by the magnitude of the ocean tide at its mouth, but also by the natural period of its basin and by the cross-section of the opening through which the tide wave must pass. For example, the small tidal range (<0.6 meters) of the Mediterranean Sea can be explained as the result of the small opening through the narrow Strait of Gibraltar into the Atlantic which inhibits exchange of water during each tidal cycle. The small tidal range in several marginal seas of the Pacific can be similarly explained.

Where the natural period of a basin is near the tidal period, it is possible to set up large standing waves, giving rise to exceptional tidal ranges. One famous example is the Bay of Fundy, whose natural period is apparently about 12 hours. Spring tides in the inner part of the bay have ranges of 15 meters (see Fig. 10–5), due to an especially favorable situation. A standing wave is combined with a narrowing valley, which funnels the tide and increases its range. The extreme tidal range and the tidal bore are illustrated in Fig. 10–14. Large tidal ranges (about 13 meters) also occur on the Normandy coast of France and at the head of the Gulf of California (about 7 meters).

(c)

The most successful techniques for studying and predicting these complicated interactions involve harmonic techniques, in which the response of the basin to each of the various tide-generating forces is more-or-less isolated and then combined with all the other responses to other forces, to construct a prediction of the tide for a given location at a given time. With even more sophisticated computers it should eventually be possible to make such predictions based solely on physical principles and using the known characteristics of the oceans and of the particular area for which a prediction is being calculated.

At present, tidal predictions still involve the recombination of information gained from many years of observations for each port.

TIDAL CURRENTS

Tidal currents are horizontal water movements associated with the rise and fall of the sea surface. The relationship between these two types of water movement is not necessarily a direct one. Not every seacoast has tidal currents, and a few areas have tidal currents but no tides.

First, let us look at the currents we would expect to find in a tide consisting only of a simple progressive wave. As we saw in an earlier section, tides must be considered as shallow-water waves because of their extreme length and the relative shallowness of the world ocean. In such a tide, the crest of the wave is high tide and the trough is low tide. Orbital motions of water caused by the tides are ellipses, greatly flattened circles with their long axes parallel to the ocean bottom. In other words, most of the water motions associated with the tides consist of horizontal motions, with little vertical motion involved.

Near coasts, we find the familiar reversing tidal currents in which water periodically flows in one direction for a while, then reverses to flow in the opposite direction. Such tidal currents can

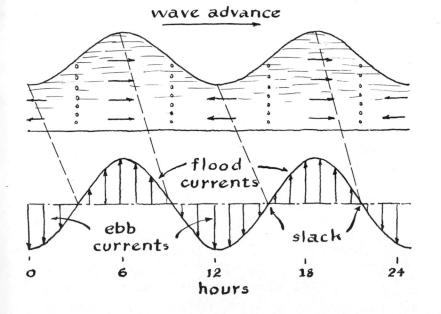

FIG. 10–15

Relationship of tide and tidal currents in an idealized tide consisting of a simple progressive wave. Note that slack water occurs at high and at low tides.

be compared to water movement associated with the passage of a progressive wave as illustrated in Fig. 10–15. As the wave crest moves toward the coast, the water also moves toward the coast; this corresponds to the *flood current*. As the wave trough moves toward the coast, the water moves away from the coast, corresponding to the *ebb current*. Each time the current changes directions, a period of no current, known as *slack water*, intervenes.

In actual fact, tides and tidal currents are rarely that simple. Progressive waves are reflected by the coast, so that the tide observed in most coastal areas consists of several progressive waves moving in different directions, as well as the standing-wave component that we discussed in previous sections. Other complicating factors include friction effects on the waves and nontidal currents such as river currents. As a result, no universally applicable generalization can be made concerning high and low tides and their relation to the times of slack water or maximum currents. Just like the tides, tidal-current predictions depend on the analysis of observations of tidal currents taken over a long period of time.

In contrast to the simple observations of the tides, tidal current studies are difficult and expensive, since they involve measurements of currents. For example, a study of the tidal currents in Long Island Sound in the 1960's required 3 years to make the measurements and involved 160 separate buoy stations, each with one to three current meters operating for periods of 4 to 15 or 29 days, depending on the station location. Because of the

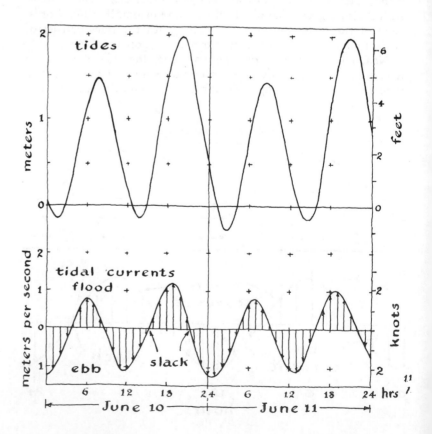

FIG.   10–16
Tides and tidal currents at the Narrows, New York Harbor.

expense involved, our knowledge of tidal currents is usually restricted to coastal areas with much ship-borne traffic (for example, the North Sea) or harbors and bays.

Let us follow the tidal currents through a tidal period in New York Harbor, illustrated in Fig. 10–16. Beginning with slack water before high tide, the flood current increases until it reaches a maximum and then decreases again until it is slack water about an hour after high tide. After this slack the current ebbs as the tide falls. The current again reaches a maximum, then decreases until about 2 hours after low tide, when it is slack water again. Elsewhere in New York Harbor, slightly different tidal-current patterns may be observed. For example, tidal currents near shore usually turn sooner than in the middle of the channel, where the tidal currents tend to be stronger. Tugboat captains often take advantage of these nearshore tidal currents to avoid bucking a strong current in mid-channel.

Tidal currents may be substantially altered by changes in wind or river runoff. For example, river runoff tends to prolong and strengthen the ebb current because more water must move out of the harbor on the ebb than comes in on the flood. Also, different tides will have different tidal currents. In areas with large daily inequalities between successive high (or low) tides, there may be days with continuous ebb (or flood) currents which change in strength during the day.

The strength of a tidal current depends on the volume of water that must flow through an opening and the size of the opening. Thus it is not possible to predict the strength of the tidal current given only the tidal range. For example, the large tidal range in the Gulf of Maine is accompanied by weak tidal currents. Conversely, Nantucket Sound has strong tidal currents but a small tidal range. At a given location, however, the strength of the tidal currents is generally proportional to the relative tidal range for that day. A spring tide, for instance, is usually accompanied by stronger tidal currents than is a neap tide.

Tidal currents are generally the strongest currents in coastal regions. In the English Channel and the Straits of Dover, the tidal currents locally have speeds of 2.5 to 3 meters per second (about 9 to 10 km/hr). Tidal currents of 1.2 meters per second (2.4 knots) occur at the Narrows, the entrance to New York Harbor. At Admiralty Inlet, the entrance to Puget Sound, tidal currents have velocities of 2.4 meters per second (4.7 knots).

In the open ocean, tidal currents are not restricted by the coastline, and they exhibit rotary patterns quite different from the reversing currents observed in coastal areas. Open-ocean tidal currents are continuous and continually change direction. Instead of slack-water periods with no current, such as we find in coastal areas, there are periods in which the current is at a minimum. In the Northern Hemisphere the current usually changes direction in a clockwise sense. Tidal currents in the open ocean are usually weaker, typically about 30 centimeters per second (slightly more than 1 km/hr), than in coastal waters.

A log anchored by an elastic tether near Nantucket Shoals Lightship during a simple semidaily tide would move in a clockwise sense in an elliptical path, returning to its initial point after 12 hours, 25 minutes, as indicated in Fig. 10–17. This again as-

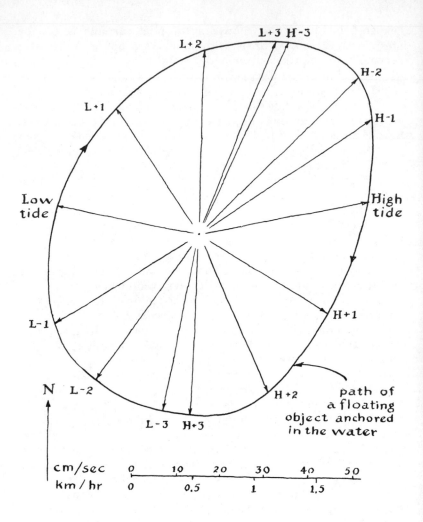

FIG. 10–17
Average tidal currents during one tidal period at Nantucket Shoals Lightship off the Massachusetts coast. The arrows indicate the direction and the speed (shown by the arrow length) of the tidal currents for each hour during a tidal period. The outer ellipse indicates the movements of a log (tied by an elastic line) in the water in the absence of wind or other currents. (After Marmer, 1926)

sumes the absence of wind or nontidal currents. If currents or winds are present, the log moves around the elliptical path at the same time that the whole water body is being moved.

In areas with a mixed tide, where there is a substantial inequality, the current ellipse is more complicated, as shown in Fig. 10–18, because there are two different ellipses. A log placed in such a current pattern moves through two elliptical paths and returns to its original position after 24 hours, 50 minutes, again assuming no wind or nontidal currents.

Since rotary tidal currents constantly change their direction and speed, we use vectors to indicate direction and current strength at hourly intervals (see Fig. 10–17). The pattern formed by the vectors is sometimes called a *current rose*. To understand the current rose, let us imagine that we are on a ship located at the center of the vectors in Fig. 10–17. We throw some dye overboard at a time corresponding to high tide (indicated by H) on a windless day with no other currents. The dye moves generally eastward in the direction shown by the arrow. One hour later (H + 1), we throw more dye overboard. It now moves generally southeastward. If we continue this each hour, we shall find that the dye patches have radiated out from our ship like

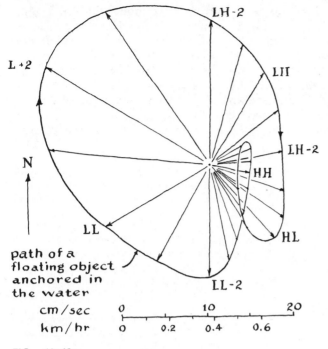

FIG. 10–18
Average tidal currents during one tidal day
at San Francisco Lightship. The radiating arrows
indicate the direction and speed (shown by
arrow length) of the tidal currents for each
hour of a tidal day. The outer ellipse indicates
the movements of an object in the water
(tied by an elastic line) in the absence of wind
or nontidal currents. (After Marmer, 1926).

spokes on a wheel. Each of these dye patches indicates the
direction of the tidal current at the time of release, with the
length of the patch being proportional to the current strength.
After a tidal period (12 hours, 25 minutes) the pattern repeats
itself. Similar results would be obtained in an offshore area with
mixed tides.

A log floating in the water near our ship would also be moved
by tidal currents. Its movements, however, would follow a pattern resulting from a combination of all the instantaneous
currents in that region, including local wind-driven currents and
major geostrophic currents.

TIDAL CURRENTS IN COASTAL AREAS

So far we have considered simple or idealized cases of tidal
currents—reversing currents in harbors and rotary tide currents
in the open ocean. To understand tidal currents found in large
bays or harbors, let us consider Chesapeake Bay, Long Island
Sound, and the North Sea.

Chesapeake Bay on the mid-Atlantic coast of the United
States has relatively simple tidal currents associated with a tide
which behaves primarily like a progressive wave. We shall

follow the changing tidal currents beginning with the time when waters are slack at the entrance (Cape Henry), as shown in Fig. 10–19(a). From just inside the entrance, in the southern part of the bay, up to a second slack-water area, about midway up the bay, tidal currents are ebbing. In the northern part of the bay, beyond the second area of slack water, tidal currents are flooding, and we can see that areas of slack water separate sections with ebb and flood currents.

Two hours later, both areas of slack water have moved into

FIG.   10–19

Tidal currents in Chesapeake Bay: (a) slack water before flood begins at Cape Henry; (b) 2 hours after flood begins at Cape Henry; and (c) 4 hours after flood begins at Cape Henry. Arrows indicate current direction and the numbers give maximum current velocities in kilometers per hour during spring tides. (After U.S. Naval Oceanographic Office Publ. 700, Sect. I, 1968)

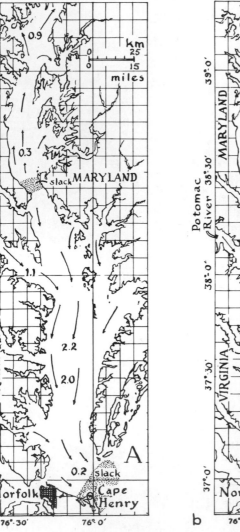

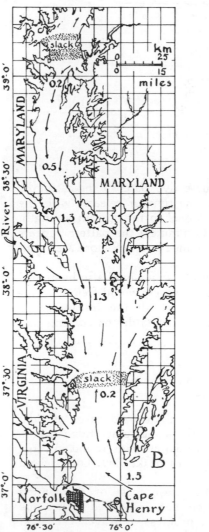

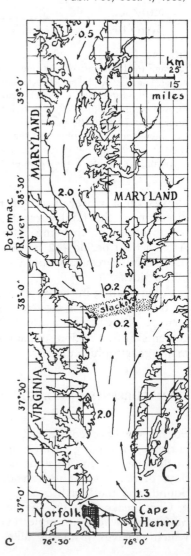

the bay about 120 kilometers, as shown in Fig. 10–19(b). By this time, tidal currents are flooding at the bay entrance and the pattern we observed previously has simply been displaced northward.

Four hours later, the southernmost slack-water area has reached the entrance to the Potomac River, and flood-tidal currents at the mouth have diminished somewhat in strength, as shown in Fig. 10–19(c). A final look at Chesapeake Bay 6 hours after slack water before flood, shows that the area at the bay mouth is again experiencing slack water, but this time it is slack water before ebb. The arrows in the initial picture are now essentially reversed, flood currents being substituted for ebb currents.

Because of its narrowness, there is little tidal flow across the axis of Chesapeake Bay. The reversing tidal currents are therefore relatively uncomplicated, flowing generally along the axis of the bay. In a wider channel, the current pattern would likely be complicated by currents flowing across the channel.

Tidal currents in Long Island Sound do not exhibit the features of a progressive wave as in Chesapeake Bay. Instead, the tidal currents are nearly the same over most of the Sound at any given time. For instance, slack water occurs nearly simultaneously over the entire sound, and flood or ebb currents do likewise, although current strength varies between areas as shown in Fig. 10–20. The area near the western end of the sound is an exception. An area of slack water separates the small area of opposing tidal currents which have advanced into Long Island

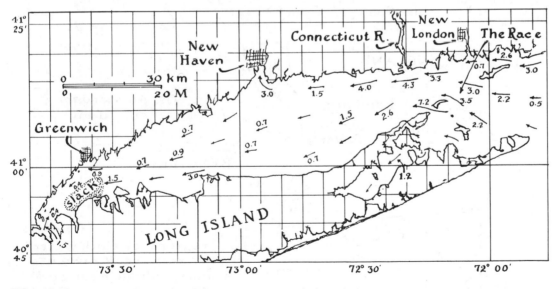

FIG. 10–20

Tidal currents in Long Island Sound after the slack before flood at the Race, the eastern end of the sound. Arrows indicate current direction; the numbers give the current speed in kilometers per hour during spring tides. (After U.S. Coast and Geodetic Survey, *Tidal Current Charts: Long Island Sound and Block Island Sound*, 4th ed., Serial 574, Washington, D.C., 1958)

Sound from the tidal system in New York Harbor, which connects with the sound through several narrow channels.

The situation where simultaneous slack water is followed by ebb or flood currents throughout the water body is typical of a

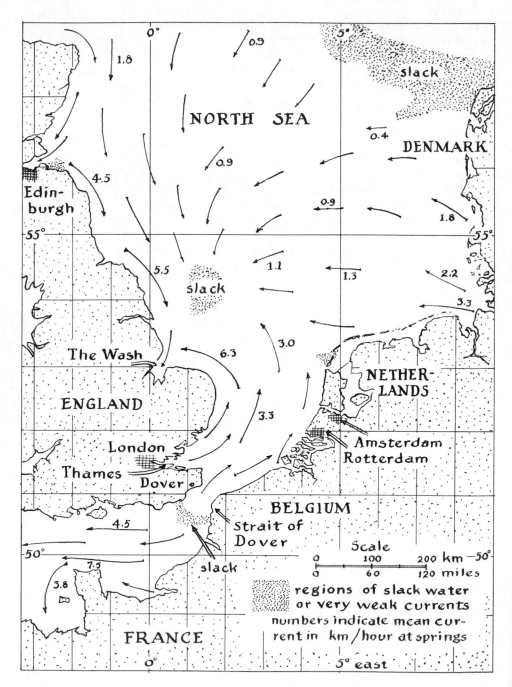

FIG. 10–21
Tidal currents in the North Sea. Arrows indicate current direction; numbers give the current speed in kilometers per hour during spring tides. (After U.S. Naval Oceanographic Office Publ. 700, Sect. I, 1968)

standing-wave type of tide. This type of tide in Long Island Sound results in part from the natural period of the Sound, which is apparently about 6 hours.

For our final example, let us consider tidal currents in the North Sea, illustrated in Fig. 10–21, an area whose tidal currents are relatively well known and rather complicated. This basin is wide enough to permit currents across as well as along the basin axis.

The tide enters the North Sea through the Strait of Dover at the south end and from the large opening at the north. In the English Channel–Strait of Dover, the tide behaves essentially as a progressive wave, but in the North Sea the tide is an amphidromic system. The area of slack water nearest England corresponds to the center of the amphidromic system. Some of the currents parallel the coastline (for example, the coast of Belgium and the Netherlands), as we have discussed in the simple reversing currents; other tidal currents are directed primarily in toward the coast (The Wash, on the east coast of England) or directly off the coast (North Germany–Denmark). (Compare this with the simple rotary tide discussed earlier, in which there is no slack water, only tidal currents whose strength and direction change constantly.)

Each of these three areas discussed above exhibits different aspects of tidal phenomena and illustrates some aspects of the tidal currents associated with them. Collectively, they show some of the complexities resulting from interactions between tide-generating forces and ocean basins.

Tidal currents are affected by many nonperiodic processes in the coastal ocean. In large estuaries, river flow modifies tidal currents, altering their timing so that current prediction is more difficult than predictions of tidal height. Both tides and tidal currents are affected by winds. Times and heights of tides in bays and coastal areas are commonly changed by storms. Hence tide tables often fail to predict the tides as they may be observed under a variety of unusual conditions.

Tides—periodic rise and fall of the ocean surface

Tidal curves—record of changing sea level

    Classification of tides by lengths of the tidal period and relationship between successive highs or lows

    Tidal day—time between successive transits of the moon over a given point, 24 hours 50 minutes

    Tidal period—elapsed time between successive high (or low) tides

    Types of tides:

        Semidaily tide—2 nearly equal high and 2 nearly equal low tides per tidal day

        Daily tide—1 high and 1 low per tidal day

        Mixed tide—2 unequal high tides and 2 unequal low tides per tidal day

    Tidal range—vertical distance between high and low tide

    Spring tide—greatest tidal range

    Neap tide—least tidal range

Tide-generating forces and the equilibrium tide

    Equilibrium tide is formulated for a static ocean completely covering a smooth earth; only the tide-generating forces are considered

    Unbalanced forces (gravitational attraction of the moon and centrifugal force from earth–moon revolution) acting on the earth's surface cause the tides

    Vertical and horizontal components, of which the horizontal is the stronger

    Equilibrium theory explains daily and fortnightly inequalities

    Variable position of the sun, moon and earth causes spring and neap tides

Dynamical theory of the tide

    Includes response of the ocean to the tide-generating forces

    Tidal constituents may be considered separately, as responsible for partial tides

    Partial tides combined for tidal predictions

    Tides as long-period waves, in shallow water, length is one-half earth's circumference

    Natural period of each basin may create elements of a standing wave; form amphidrome system affected by Coriolis force

    Factors controlling tides in nearly enclosed basins:

        Range of tide at ocean entrance

        Natural period of basin

        Size of ocean entrance

Tidal currents—horizontal motions with little vertical component

    Comparable to water movements associated with progressive waves

    Reversing tidal currents near coasts

        Flood current—water moves seaward

        Ebb current—water moves landward

        Slack water—no current

    Complex relationship between tides and tidal currents

    Open-ocean rotary currents flow continuously, changing direction

        Particles move in elliptical paths

        Current rose—formed by vectors indicating direction and speed

Tidal currents in coastal areas

    Chesapeake Bay tidal currents behave like a progressive wave

    Long Island Sound has a standing-wave type of tide

    North Sea has an amphidromic system set up by tidal currents

    Tidal currents may be modified by river flow, winds, storms

DARWIN, G. H. 1962. *The Tides and Kindred Phenomena in the Solar System.* W. H. Freeman, San Francisco. 378 pp. Originally published in 1898; classic, exhaustive treatment of the subject; elementary to intermediate.

DEFANT, ALBERT. 1958. *Ebb and Flow.* University of Michigan Press, Ann Arbor. 121 pp. Descriptive and mathematical; elementary to intermediate.

RUSSELL, R. C. H., and D. H. MACMILLAN. 1954. *Waves and Tides.* Hutchinson, London. 348 pp. Elementary.

Nun Buoy

Can Buoy

ESTUARIES

Floating buoys are used to mark navigation channels or underwater hazards. Various types are characteristically colored and numbered to give definite information about conditions in a harbor or other coastal region.

eleven

An *estuary* is a semienclosed part of the coastal ocean where fresh water from the land mixes with seawater. In many parts of the world, estuaries are the drowned lower portions of rivers. Broad, shallow lagoons on low-lying coasts and fjords in glaciated mountain regions are also estuarine systems.

Centuries ago, when most commercial transport was by water, cities were usually located on large rivers, where ocean-going ships could dock in sheltered waters. Cargo unloaded there could be taken inland by riverboat, or even by canal barge or overland trail. Today, seven of the ten largest U.S. cities are located on estuaries. Virtually every large estuarine system is the site of a major city, originally developed as a port and now the center of a land- and air-transportation network.

The presence of a large city on its shores inevitably changes the estuarine environment, circulation processes, landscape, and quality of the water. Edible fish and shellfish are taken there, and waste materials of every kind are commonly dumped into its waters. The environmental problems of the estuary have become an increasing source of concern in many large coastal cities, to residents and officials alike.

ORIGIN OF ESTUARIES, FJORDS, AND LAGOONS

Many estuaries are the drowned lower reaches of rivers that

thousands of years ago entered the sea much further offshore than they do today. Such coastal-plain estuaries are typically much broader than any part of the river upstream, and may locally be quite deep. The estuary gradually deepens as it progresses toward the sea, except where the river has deposited mud and sand near its mouth. Some estuaries, such as the Hudson at New York City, have an associated submarine canyon extending across the continental shelf.

As we have already learned, the sea stood about 130 meters below its present level during Pleistocene ice ages, because of the large amount of ocean water held as ice in continent-sized glaciers. About 35,000 years ago, when the shoreline was near the edge of the present continental shelf, canyons and valleys were cut by rivers crossing the coastal plain, as the shelf was at that time. Estuaries and lagoons at the edge of the continental shelf were later covered by the rising sea level, and present estuaries formed farther up the ancient river valleys.

Estuaries and lagoons make up 80 to 90 percent of the Atlantic and Gulf coasts, but only about 10 to 20 percent of the Pacific coast, as can be seen in Fig. 11–1. Mountain building on the west coast has left little low-lying coastal plain. The coastline is so mountainous that an estuary can form only in the few places where a river or former glacier has cut through the mountains to reach the sea.

On the Pacific coast of the United States, river-drainage basins are generally small. Large desert areas behind the mountains contribute little water to any river system. Much of the

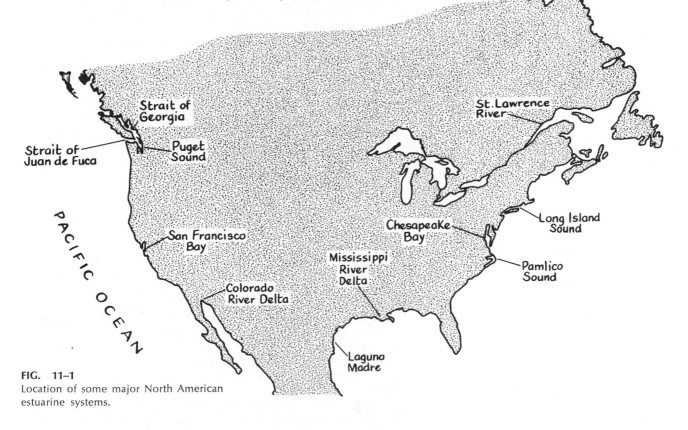

FIG. 11–1
Location of some major North American estuarine systems.

rain that falls in the western half of the United States drains into the Gulf of California (via the Colorado River) or the Gulf of Mexico. The largest estuarine systems (Table 11–1) on the

Table 11–1

CHARACTERISTICS OF SOME NORTH AMERICAN ESTUARINE SYSTEMS

	Estuary area (km²)	Estuary volume (km³)	Mean water depth* (m)	Annual fresh-water discharge (km³/yr)	Land area drained (10³ km²)	Major rivers
Chesapeake Bay System						
(Maryland–Virginia)	11,000	67	6.1	65	110	Susquehanna
Potomac River	1,280	7.3	5.7	12	36	Potomac
James River	650	2.3	3.5	10	26	James
Raritan Bay						
(New York–New Jersey)	230	1.1	4.5	1.8	3.4	Raritan
New York Harbor	159	1.2	7.5	19.4	34.7	Hudson
Long Island Sound						
(New York–Connecticut)	3,180	62	19.4	21	40.7	Connecticut Housatonic
Pamlico–Albemarle Sound						
(North Carolina)	6,630	23.9	3.6	7.8	51	Neuse Pamlico
Strait of Juan de Fuca						
(Washington–British Columbia)	4,370	490	112	nd†	nd†	
Puget Sound						
(Washington)	2,640	185	70	36.5	37.6	Skagit Snohomish
Strait of Georgia						
(British Columbia)	6,900	1,025	156	145	270	Fraser
San Francisco Bay						
(California)	1,190	6.2	5	40	161	Sacramento San Juan
Laguna Madre						
(Texas)	158	1.1	0.9	−0.85‡	nd†	

†nd = no data.
*Mean depth = volume/area.
‡Evaporation exceeds river runoff plus rainfall.

Pacific coast, San Francisco Bay and the Strait of Juan de Fuca system, formed when sections of the continent containing former river valleys sank below sea level because of active mountain building in the region.

From the Strait of Juan de Fuca northward along the Pacific coast and along most of the Canadian coastline, the shoreline has been greatly modified by glacial action. This has created narrow, deep, steep-sided *fjords* and fjordlike estuaries like those in Fig. 11–2, perhaps the most spectacular of all estuarine systems. Fjords occur in other high-latitude coastal areas such as Norway and Tierra del Fuego in South America. Glaciers moving from the mountains and down the river valleys carved gorges hundreds of meters deep, and often deposited boulders and clay at the mouths of the valleys. This glacial debris forms an underwater sill at the entrance to many fjords, restricting movement of bottom waters.

A *lagoon* is a broad, shallow estuarine system. Flow of water between the estuary and the coastal ocean is restricted by the presence of a barrier beach located offshore, generally paralleling the shoreline. Lagoons are invariably shallow, because barrier beaches form only in waters a few meters deep, where wave action suspends sediment and transports it along the coast. The barrier beaches enclosing a lagoon are interrupted at intervals by narrow inlets through which tidal currents move water in and out of the lagoon.

On the United States Gulf Coast and on the Atlantic coast south of New England, long stretches of barrier beach separate lagoons from the coastal ocean. Often a single stretch of beach will isolate several lagoons and estuaries. All parts of such a system communicate with the coastal ocean through the same set of narrow inlets.

In New England and the Pacific Northwest, where the coastline has been scoured by glaciers, lack of sediment moving along the coast prevents formation of extensive bars or spits. On much of the Pacific coast, the continental shelf is too narrow and steep to permit barrier beaches to form.

FIG. 11–2
The rugged topography of a small inlet on Canon Fjord, Ellesmere Island, in the Canadian Arctic was cut by a glacier of which a remnant (the large, white mass) can be seen at the upper left. The icebergs in the water come from such glaciers. Note the small floes of sea ice in the foreground. (Photograph courtesy National Film Board of Canada)

When all inlets to a lagoon are closed by storms or changes in wave or sediment conditions, the lagoon is transformed into a lake. Fresh water flowing into such a lake from local rivers would raise its level until it overflowed its boundaries, if it were not for the fact that sandy beaches are permeable to water. Seepage through barrier beaches keeps lake levels fairly constant.

ESTUARINE CIRCULATION

An estuary acts as a two-way street for water movements. To see how an estuarine circulation develops, let us consider a simple, hypothetical river on a nonrotating earth. River water would flow into the estuary and then, being less dense than seawater, would spread out as a layer over the stationary seawater beyond, as shown in Fig. 11–3. It would move generally seaward in the surface layer over the more dense salt water, and the two layers would be separated by a more-or-less horizontal pycnocline zone. Because of the marked density discontinuity at the pycnocline, there would be very little water flow across it. The only movement of salt or other material across this boundary would be due to the diffusion of individual molecules, which is slow compared to most processes in the ocean.

Such a simplified estuary does not exist in reality, but it illustrates the basic flow pattern found in all estuaries. Actually, friction occurs between the seaward-moving surface layer of fresh water and the seawater below it causing currents, as shown in Fig. 11–4. Because of the frictional drag between the two moving layers, water is also entrained (dragged along) from below and incorporated into the surface layer. The salinity (and therefore the density) of the newly mixed water is less than that of the more saline water below the pycnocline; hence it cannot return through the pycnocline. Mixing of water across

FIG. 11–3

Schematic representation of a simple salt-wedge estuary showing the two-layered structure with a landward flow in the salty subsurface layer and a seaward flow in the less saline surface layer. [Redrawn from D. W. Pritchard, "Estuarine Circulation Patterns," *American Society of Civil Engineers Proceedings*, 81(717) (1955), 11 pp.]

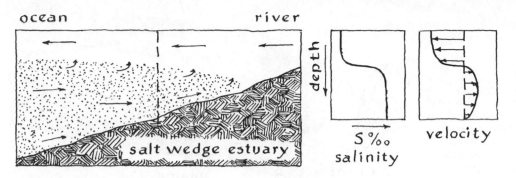

ocean   river

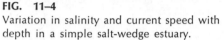

salt wedge estuary

depth

5‰ salinity

velocity

FIG. 11–4
Variation in salinity and current speed with
depth in a simple salt-wedge estuary.

the pycnocline is nearly an irreversible process, although some
fresh water is incorporated in the lower layer.

As a result of this upward movement of salt water into the
surface layer, the salinity of the surface layer increases in a seaward direction. One volume of fresh water ultimately mixes
with several volumes of seawater. At the same time, water in
the subsurface layer moves toward the land to replace seaward-
flowing salt water entrained in the surface layer. This landward
flow along the bottom is much greater in volume than the fresh-
water flow into the estuary.

The subsurface salt water in such an estuary generally forms
a wedge with its thin end pointed upstream; hence in its simplest form, this is called a *salt-wedge estuary*.

If the frictional drag between the two layers is great enough,
internal waves may form on the pycnocline. Such waves at the
boundary between the two layers may actually break, like the
surf seen at the beach. This greatly increases the rate of mixing
between the layers. The more mixing that occurs in the estuary,
the greater is the landward flow in the subsurface layer.

Simplified salt-wedge stratification such as we have described
is rarely observed in estuaries. Certain systems do closely approximate it, however, especially where river flow is large and
tidal range is low. Salt-wedge estuaries are found where a large
river discharges through a relatively narrow channel, such as the
lower passes of the Mississippi River or the Columbia River
during flood stages. Forces associated with large river flows may
be so much stronger than the tides that they dominate the estuarine circulation. This forms a nearly ideal salt-wedge estuary,
one in which relatively little salt water is mixed into the upper
layer. Nearly fresh water is discharged into the coastal ocean,
where it mixes with ocean water.

As river flow decreases in an estuary, the importance of tidal
effects increases. A single estuary may exhibit characteristics of
the salt-wedge estuary during times of flood, but during low-
flow periods of the year it is usually influenced much more by
tides and tidal currents, departing substantially from the pattern
of a simple salt-wedge stratification.

Tidal currents cause mixing between layers, so that the waters
become only moderately stratified, as Fig. 11–5 illustrates. Tidal
flow causes *turbulence* (random movements) throughout the
water column, and this in turn increases mixing so that more

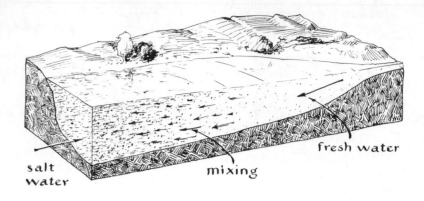

FIG. 11–5
Schematic representation of water movements in a moderately stratified estuary. There is a net landward movement in the subsurface layers and a seaward flow in the surface layers, although the difference between the layers is not as pronounced as in the salt-wedge estuary. (After Pritchard, 1955)

salt water is transferred from the subsurface to the surface layer. Some fresh water from the surface also mixes downward. Consequently, salinity decreases in a landward direction in both the surface and subsurface layers. As in the salt-wedge estuary, however, there is a net landward flow in the subsurface layer, replacing salt water lost from the system, and a net seaward flow in the surface layer removing both fresh water and salt water from the estuary, as Fig. 11–6 indicates.

Let us consider the volume of flow in these two layers. Remember that river water has almost no salt as it enters the head of the estuary. Assume that the seawater moving in along the bottom has a salinity of $35^0/_{00}$. A mixture of equal volumes of fresh and salt water gives a salinity of $17.5^0/_{00}$; mixing 2 volumes of salt water with 1 volume of river water results in a salinity of about $23.4^0/_{00}$ as shown by Fig. 11–7. By the time surface-water salinity has reached $34^0/_{00}$, 1 volume of river water has mixed with 39 volumes of seawater. In other words, at this point, landward flow in the subsurface layer is 39 times the volume of river water discharged at the head of the estuary. The volume of surface water moving seaward in the surface layer will be 40 times the original river discharge.

Where tidal effects are greater, stratification is diminished in the estuary. Increased river flow causes a greater degree of stratification. In an estuary with small river flow but large tides and tidal currents, the waters may be mixed almost completely from top to bottom, as illustrated in Fig. 11–5.

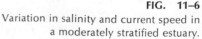

FIG. 11–6
Variation in salinity and current speed in a moderately stratified estuary.

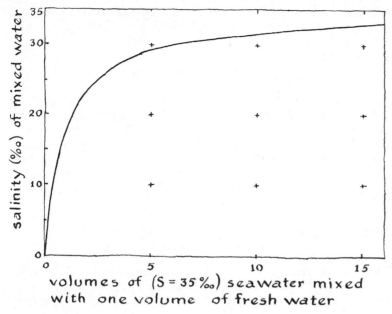

FIG. 11–7
Relationship between salinity of water masses
formed in an estuary and the relative amounts
of seawater (salinity $35^0/_{00}$) and fresh water
(salinity $0^0/_{00}$) involved in the final mixture.

Because of their greater depth, fjord systems have more
complicated circulation systems. Many fjords have a sill at the
entrance which cuts off most of the deeper water from communication with the adjacent ocean, as can be seen in Fig. 11–8.
Alberni Inlet on Vancouver Island, British Columbia, for example, has a sill which comes to within 40 meters of the surface. The basin behind the sill is 330 meters deep, so that nearly
seven-eighths of its volume is partially isolated from communication with the ocean.

Fresh water flowing into the fjord forms a low-salinity surface layer which moves seaward, setting up a typical estuarine

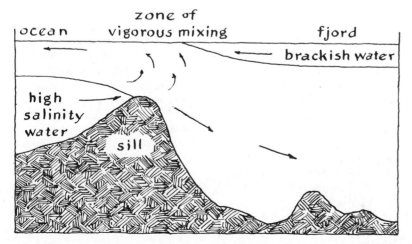

FIG.   11–8
Schematic representation of circulation in a
simple fjordlike estuary. Note the water mass
formed by mixing surface and subsurface
waters while flowing over the sill. [After
M. Waldichuk, "Physical Oceanography of the
Strait of Georgia, British Columbia," *Journal
of the Fisheries Research Board of Canada*
14(3) (1957), 321–86]

circulation. This layer, however, often extends no more than a few tens of meters below the surface and usually involves only the waters above the top of the sill.

The deeper waters of a fjord are almost completely isolated from the surface circulation, but are strongly affected by conditions at the fjord mouth, outside the estuary. Since river flow, tides, and winds have little effect on this deep water, its circulation is controlled primarily by density differences. If the water at or slightly above the sill outside the estuary becomes more dense, it can flow into the fjord and displace the deeper water there, which then moves out slowly. Strong winds can affect the deep waters by setting up *seiches* (standing waves), which cause the waters to "slosh" back and forth.

TIDES IN ESTUARIES

Tides in the coastal ocean also affect estuaries and their tributaries. The tide may enter the estuary as a progressive wave and travel upstream. In many estuaries, the wave is gradually damped (reduced in height) by friction of the channel and opposing river flow. But in some estuaries, the tide wave reaches the end of the estuary and is reflected back. The net result is a standing-wave type of tide, where the tide range at the end of the estuary exceeds the range closer to the estuary entrance. For example, at the entrance to Chesapeake Bay, the tidal range is about 1 meter. This diminishes to about ⅓ meter at the entrance to the Potomac estuary. As the tide progresses up the Potomac, it varies little in range until, near Washington, D.C., the tidal range again increases to about ⅔ meter. A similar situation exists in the Hudson River. As Fig. 11–9 illustrates, the tidal range is about 1.1 meters in New York Harbor, decreasing to slightly less than 1 meter at West Point; at Troy, New York (the uppermost extremity of tidal effect), the tidal range is slightly greater than 1.1 meters.

Tides dominate the circulation in most estuaries. Any current that one sees in an estuary is most likely of tidal origin. To observe the estuarine circulation, it is necessary to make current measurements extending over at least one tidal cycle. During this period, water flows upstream on the flood current a distance nearly equal to its downstream travel on the ebb current. Therefore, when averaged, tidal currents cancel one another, leaving a nontidal current caused by the estuarine circulation. A current meter placed in the surface layers would indicate a net flow seaward, and one in the deeper layers a net flow landward.

Another way to observe the estuarine circulation and its effect on tidal currents is to time the duration of ebb currents as compared to the duration of flood currents. In the open ocean, these are equal in length; in an estuary, the tide must oppose river flow, and the ebb and flood currents are unequal—surface ebb currents are longer than flood currents. For instance, in the St. Lawrence River at Quebec, the rising (flood) tide lasts 5 hours, 2 minutes, and the falling (ebb) lasts 7 hours, 23 minutes.

Another effect of the tide opposing river flow is that the tide

FIG. 11–9

Changes in tidal range on the Hudson River.

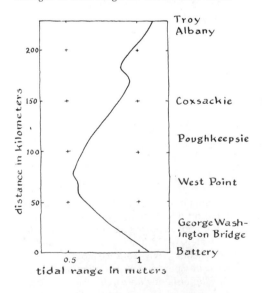

wave becomes markedly asymmetrical as it moves into the river. In some shallow, funnel-shaped estuaries, this asymmetry becomes so marked that a small wall of water moves upstream as the tide comes in. This is known as a *bore*, and it is well developed on the Petitcodiac River at the northern end of the Bay of Fundy, and in the mouth of the Colorado River in the Gulf of California.

BIOLOGICAL PRODUCTIVITY OF ESTUARIES

Estuaries are usually rich in those nutrients (phosphates and nitrates) which support large phytoplankton populations; these in turn provide food for a host of zooplankters, fishes, and benthic organisms. In fact, estuaries play a major role in the productivity of the coastal ocean, serving as home, nursery, and breeding ground for many species.

Estuaries derive nutrients from three sources, the nearby coastal ocean being the predominant one: nutrients in subsurface waters are carried into estuaries. Secondly, rivers supply nutrients leached from soils in the river-drainage basin, or dissolved in rainwater. Finally, nutrient-rich waste products are washed into the estuary from industries, cities, and farmlands along shores and on rivers; this last is an increasingly important nutrient source in estuaries.

Estuarine circulation tends to retain nutrients within the system. While growing, planktonic organisms take up nutrients; after death they sink into subsurface layers and decompose while being carried landward. Nutrients released by decomposition are then returned to the surface layer by entrainment of subsurface waters, where they are again available to phytoplankton. The estuarine circulation also carries to the surface substances, other than phosphates and nitrates, that may be produced on the bottom—for example, vitamin B-12, which is apparently produced by bacteria living on the bottom.

The nutrients which rivers contribute directly to an estuary often have little effect on its productivity. During the summer, growth of fresh-water plants can strip river water of its nutrients. In winter, when river plants are not growing due to low temperatures or lack of light, river waters contribute significant amounts of nutrients to the estuary. However, the abundance of riverborne nutrients has little effect on the estuarine phytoplankton, since this is the period in which its growth is inhibited by lack of light or by low temperature.

River discharge does have a significant—though indirect—influence on phytoplankton growth in the coastal ocean. The estuarine circulation is akin to upwelling. Subsurface waters resupply nutrients to the surface waters and are usually much more important than the contribution of nutrients by undiluted river water. Also, the stability imparted to the surface waters by the low-salinity river waters helps phytoplankton to remain in the sunlit surface layer. This permits an earlier onset of phytoplankton production than might have been possible in the absence of river discharge.

Plankton in surface layers are unaffected by conditions in the deeper waters. Some of these are eaten at the surface, others die and fall to the bottom. In a normal estuary, this deposited organic material is used for food by bottom-dwellers or is consumed by bacteria. Where bottom currents are sluggish, however, as in a fjord, specialized conditions may occur due to depletion of oxygen in sediments and in very deep waters.

When organic matter is digested, dissolved oxygen is used as well. Where oxygenated water is resupplied slowly relative to the rate at which organic matter is being deposited, dissolved oxygen in the bottom water may be markedly depleted. When this happens, oxygen-using organisms, including most bacteria, can no longer live in or on the bottom. Only certain anaerobic bacteria can survive by decomposing sulfate from seawater and releasing hydrogen sulfide as a metabolic by-product. To most organisms, hydrogen sulfide is a poison, and the presence of these bacteria makes the bottom of a fjord uninhabitable to virtually all other forms of life. In the absence of burrowing organisms, delicate layering is preserved in bottom deposits. Annual layers called *varves* record seasonal changes in organisms growing in surface layers and variable sediment contributions from rivers.

CHESAPEAKE BAY

Chesapeake Bay is a classic example of a coastal-plain estuarine system. It is formed from the drowned valleys of the Susquehanna River and its tributaries, including the Potomac, the James, the York, and the Rappahannock. In addition, dozens of smaller tributaries drain into arms, inlets, and bays on all sides, giving Chesapeake Bay the complicated, branching form seen in Fig. 11–10. The southernmost arm of the system, the James River, was probably a separate river flowing into the ocean through its own estuary before the area was flooded by the present sea level.

The seaward boundary of Chesapeake Bay (see Fig. 11–10) is commonly considered to be a line between Cape Charles and Cape Henry. The upper boundary is at the "fall line" on the Susquehanna River. Along the "fall line" waterfalls or rapids in the river mark the boundary between the soft, easily eroded sediments and sedimentary rocks of the coastal plain and the hard rock underlying the rolling hills of the Piedmont. The fall line on Atlantic coast rivers marks the limit of navigation for ocean vessels, and the landward extent of tidal influence.

The total volume of fresh water discharged each year by rivers flowing into Chesapeake Bay approximately equals the total volume of the estuary. The Susquehanna River supplies 49 percent of the annual fresh-water discharge, the Potomac River contributes 18 percent, and the James River about 15 percent. Nearly all the fresh water is discharged into the bay along the western margin, where rivers drain the slopes of the Appalachian Mountains. Very little fresh water comes from the small rivers and swampy areas along the eastern shore of the bay.

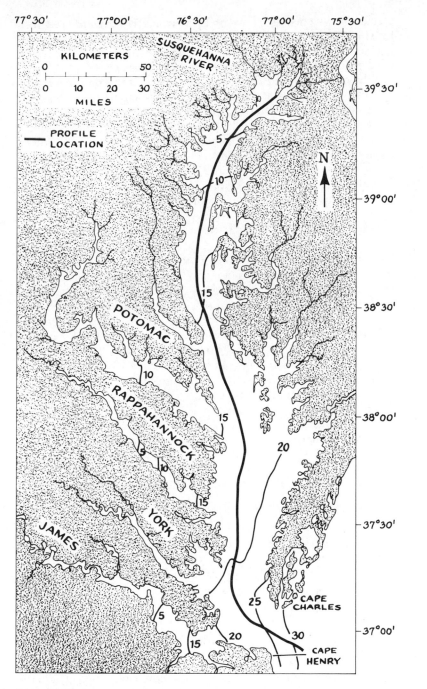

FIG. 11–10
Chesapeake Bay and its tributary estuaries.
Contours show surface salinity in water.
[After D. W. Pritchard, "Salinity Distribution
and Circulation in the Chesapeake Bay
Estuarine System," *Journal of Marine Research*
11(2) (1952), 106–123]

Tidal currents in Chesapeake Bay are mainly the reversing
type common to coastal waters. Depending upon location in the
bay, maximum current speed may be as low as 25 centimeters
per second or as high as 50 centimeters per second. As in all

FIG. 11–11
Typical distribution of surface salinity in Chesapeake Bay in winter. (After Pritchard, 1952)

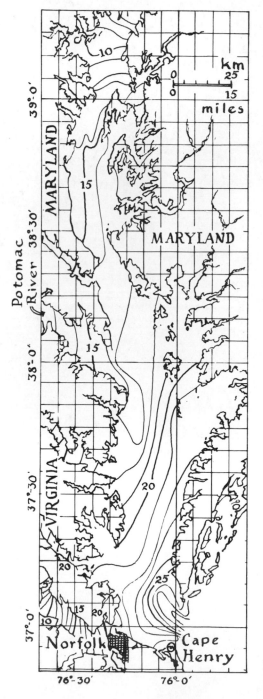

estuarine systems, net flow at the surface is seaward, resulting in ebb currents that begin sooner and persist longer at the surface. Inflow at the bottom is reflected in the early initiation of the flood tides and their greater persistence.

Like most drowned river systems, Chesapeake Bay has a wide mouth, offering little resistance to interchange of water with the coastal ocean. Estuarine waters near the entrance are nearly as saline as seawater. In the upper reaches of the bay, there is little salt in the water, and the system has many of the characteristics of a tidally controlled lake, as indicated in Fig. 11–11.

Salinities are lowest during spring, when high river flow prevails throughout the region. In summer and fall, low river discharge results in maximum salinities. Because of the large discharge of fresh water along the western shore and the deflecting effects of the Coriolis effect, surface-water salinity is generally greater on the eastern side of Chesapeake Bay than along the western side (see Fig. 11–11). Estuarine currents contribute to this effect by moving surface water parallel to the eastern shore. In any given area, there are daily variations which are related to the tidal current and to the discharge of local streams.

Salinity variation with depth is typical of estuarine systems. In January, surface-water salinity in the middle of the bay may be as low as $14^0/_{00}$, while bottom-water salinities are about $18^0/_{00}$, as indicated in Figs. 11–12(a) and (b). In the same area during September, the salinity of the surface layer is higher ($16^0/_{00}$) because of the reduced river flow. Likewise, the salinity of the bottom water is also higher ($22^0/_{00}$), due in part to changes in ocean conditions at the mouth of the estuary. There is also a reduction in the fresh water, which is mixed downward from surface layers owing to the effects of the tidal currents.

Temperature in Chesapeake Bay does not show as clear cut a distribution as does salinity. The system is relatively shallow (see Table 11–1), so that the surface-water temperatures are controlled primarily by local weather conditions (particularly in summer), and by the local discharge of waste heat from along the shores. Steam-powered electrical generating plants use bay waters for cooling; when returned to the bay, these waters are often warmed 10°C or more.

During the winter, water temperature generally decreases toward the head of the bay, as a result of the inflow of warmer oceanic water near the mouth. The cooling of lower-salinity waters at the head of the bay frequently causes ice formation during January and February. Ice forms first in the shallow, protected bays and inlets along the shore, and then gradually extends out toward open water. In especially cold winters, ice covers large areas stretching from shore to shore in the upper estuary.

Strong winds have a pronounced effect on the bay. In addition to the waves they generate, storm winds can also change sea level locally, sometimes causing flooding of low-lying coastal areas. In general, winds only modify currents in Chesapeake Bay; currents set up by river discharge and the tide are usually the predominant influence.

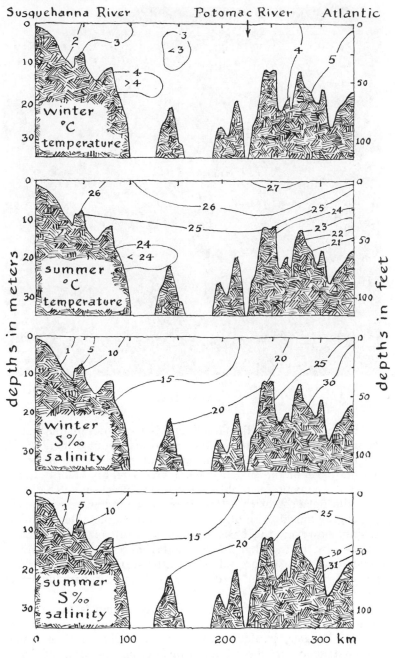

Susquehanna River   Potomac River   Atlantic

winter °C temperature

summer °C temperature

winter S‰ salinity

summer S‰ salinity

depths in meters

depths in feet

FIG. 11–12
Distribution of temperature and salinity with depth in Chesapeake Bay in winter and summer. Profile location is given in Fig. 11–10. (Data from Chesapeake Bay Institute, Johns Hopkins University)

PAMLICO SOUND

Pamlico Sound is a large, bar-built embayment on the North Carolina coast, a shallow estuarine system consisting of a complex of drowned river valleys. It is a broad lagoonlike area, cut off from the ocean by barrier beaches through which several small shifting inlets provide access to the ocean; as illustrated in Fig. 11–13.

FIG. 11–13
Pamlico Sound is formed by barrier islands (the light-colored narrow strips) partially isolating drowned river valleys from the adjacent coastal ocean. Note the plumes of turbid water coming out of the inlets and the cloudlike water masses moving along the coast. The puffy white masses at right are clouds. (Photograph courtesy NASA)

The shallowness of Pamlico Sound is its predominant characteristic. Maximum depths in the sound are about 8 meters; the average depth is about 3.6 meters, and some of the more restricted parts of the system have average depths of only about 1 meter. The wind inevitably has a major effect on so shallow a bay: during most of the year, Pamlico Sound waters are very thoroughly mixed, with only a slight stratification. Despite wind mixing, however, the basic estuarine nature of the system is revealed by the fact that the bottom layers tend, on the average, to be about $0.7^0/_{00}$ more saline than the surface layers. This is a consequence of the landward flow of salt water in the subsurface layers.

Another significant feature of Pamlico Sound is the narrowness of the inlets connecting the bay and the ocean. Because of restricted water circulation through the inlets, there is almost no tidal action in the sound. Considering the flow of water through the inlets during a single tidal change, it can be calculated that the average tidal range is only about 5 centimeters. In contrast, a moderate wind (13 knots) from the southwest (the maximum fetch for the sound) can set up a "wind tide" that raises water levels by 30 centimeters on the downwind side of the bay.

The narrowness of the inlets also strongly affects salinity distributions. Highest salinities (up to $35^0/_{00}$) occur near the inlets, where nearly undiluted ocean water enters the sound. Lowest salinities generally occur near the river mouths. Salinity varies from between 9 and $19^0/_{00}$ near the mouths of the Neuse and Pamlico Rivers, and has comparable values near the entrance to Albermarle Sound (Fig. 11–14). Strong winds cause rapid mixing of the ocean and fresh waters so that there is only a slight difference in salinity or temperature between surface and bottom waters.

The shallowness of the sound also strongly influences water temperatures, which in general closely follow local air temperatures. Waters are warmest (typically about 29°C) in June and July, when insolation is at a maximum. Lowest temperatures (about 6°C) occur during late December and early January, when insolation is at a minimum. Water temperatures can change by 2–3°C in a single day; in fact, the temperature cycle is more typical of continental climates than maritime climates.

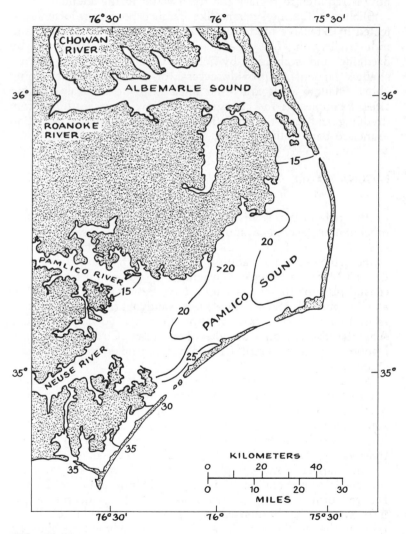

FIG. 11–14
Surface-water salinity in Pamlico Sound, mid-April, 1950. [After E. W. Roelofs, and D. F. Bumpus, "The Hydrography of Pamlico Sound," *Bulletin of Marine Science of the Gulf and Caribbean* 3(3) (1953), 181–205]

The sound is apparently too shallow to have much effect on the region's climate.

Rivers emptying into Pamlico Sound drain an area of about 51,000 square kilometers, nearly 8 times the total area of the sound (which is about 6600 square kilometers). The volume of fresh water discharged by the rivers each year is approximately equal to the volume of Pamlico Sound, so that river discharge has a major effect on water salinity. Salinities are lowest in February through May, when river discharge is high and evaporation low. Salinities are highest in late summer and fall. Considering the average river flow and water salinity in Pamlico Sound, it appears that the fresh water has a residence time of about 5 months. In other words, about 5 months at average river flow is required to replace the fresh water in the sound.

Shallow estuarine systems like Pamlico Sound are often subjected to extensive modification as a result of urbanization and industrialization. The small, shifting inlets are deepened by dredging and stabilized by jetties. Channels are dug across shallow lagoons to provide access for deep-draft ocean-going ships. Shallow areas are often filled in to be used as building sites. Examples of the changes in shallow bays resulting from dredging and land-fill operations can be seen in southern San Francisco Bay, California, and Tampa Bay, Florida.

LAGUNA MADRE

In the two estuaries discussed above, local river runoff and precipitation exceed evaporation; consequently, salinities are lower than those found in the open ocean. One of the few *hypersaline* (above-average salinity) *lagoons* in the United States is Laguna Madre, a series of long, narrow lagoons on the low, flat coastal plain of the Texas Gulf Coast, between the Rio Grande and Corpus Christi (see 11–15). A sandy barrier island (Padre Island) parallels the coast for approximately 160 kilometers and separates the lagoon from the open ocean. Circulation in this lagoon is quite different from other coastal systems we have studied.

At each end of Padre Island, narrow inlets connect with the open ocean. Before these inlets were afforded the protection of man-made jetties, they closed and reopened frequently as a result of storms and the action of longshore currents. The barrier island not only forms a lagoon, but also isolates a small drowned river valley, Baffin Bay, from the Gulf of Mexico.

This barrier-island complex was formed when sea level approached its present position. Large shallow areas in Laguna Madre, formed during a hurricane in 1919, divide the lagoon into two smaller basins, each with its own circulation. Baffin Bay, at the northern end, is the deeper, averaging about 1.6 meters. Much of Laguna Madre is extremely shallow (average depth 0.9 meter); many areas are covered by only a few centimeters of water at high tide.

Inlets at each end are quite narrow, so that there is relatively little tidal interchange with the open ocean. Also, this section of the Gulf of Mexico has a narrow tidal range, less than 0.5 meter.

Consequently, the tidal range in Laguna Madre is only a few centimeters, quite similar to conditions in Pamlico Sound. Wind effects are often more important than the tide in both these shallow lagoons. The wind tides produce changes in water level, predicted to exceed 1 meter, based on observed wind velocities.

Winds also control the currents in Laguna Madre. Winds from the north cause currents to set southward, with flow out the southern inlet. Winds from the south result in essentially the

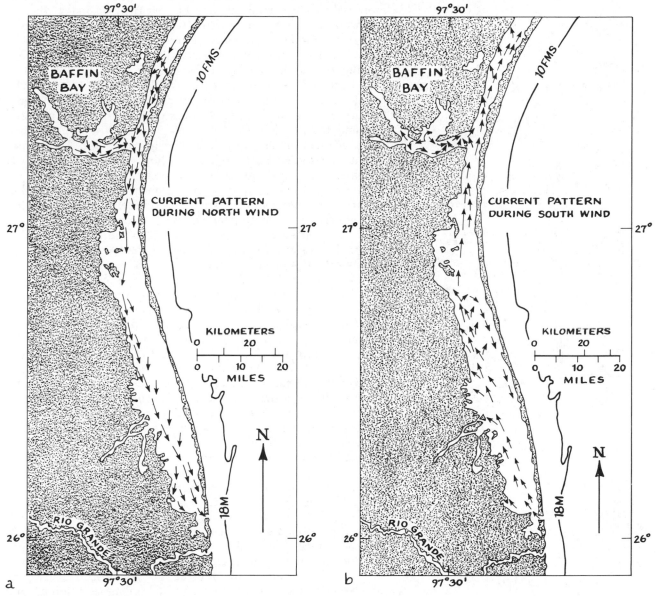

a

b

FIG. 11–15

Postulated current patterns in Laguna Madre during north (a) and south (b) winds. [After G. A. Rusnak, "Sediments of Laguna Madre, Texas, in F. P. Shepard, et al. (eds.), *Recent Sediments, Northwest Gulf of Mexico,* American Association of Petroleum Geologists, Tulsa, Okla., 1960, pp. 82–108]

FIG.   11–16
Surface-water salinity in Laguna Madre,
July, 1957. (After Rusnak, 1960)

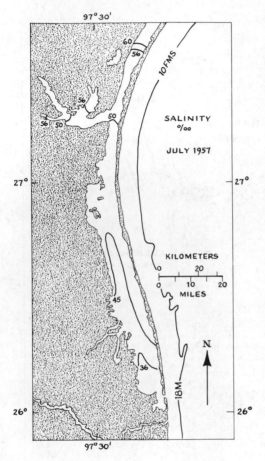

opposite direction of the currents, as indicated in Fig. 11–15. Note that the shallow areas in the center of Laguna Madre greatly influence currents set up by winds from the south. Back-and-forth seichlike water movements are sometimes set up in the lagoon by winds blowing steadily from a single direction. These act to stir up sediments, so that lagoon waters are often turbid.

Laguna Madre is located on a semiarid coast, and evaporation exceeds rainfall for most of the year. Runoff is obtained from only a few small streams or irrigation drainage ditches. Intense local storms (more than 10 centimeters of rain in a few hours) provide large amounts of fresh water, but at infrequent intervals. Between storms, the salinity of the water (often exceeding $80^0/_{00}$) in the lagoon is usually substantially greater than in the adjacent coastal ocean, as Fig. 11–16 indicates. The erratic character of the fresh-water discharge results in extreme salinity variations. The highest salinities recorded are between 110 and $120^0/_{00}$. Surface-water salinities as low as $2^0/_{00}$ have been recorded immediately following a cloudburst. These low-salinity areas persist for a relatively short time before salinity is increased through evaporation and mixing.

In the restricted, shallow environment of Laguna Madre, surface-water temperatures are also extremely variable. During the winter, cold, continental air masses come down from the north, chilling surface waters. Because of the extreme cooling and lack of escape channels for fish in shallow bays, there are mass mortalities among temperature-sensitive fish during spells of unusually cold weather. Substantial fish mortalities ("fish kills") can also occur in summer at times when water temperatures and salinities become intolerable to some species. The dredging of ship channels and inlets has apparently reduced fish mortality by providing escape channels and shelter.

Temperature and salinity data indicate that there is very little stratification within the shallow bay, primarily because of mixing by winds. Near the inlets at either end of the lagoon, mixing is especially vigorous because of tidal currents. When there is a large fresh-water discharge with formation of a low-salinity surface layer, the less saline water flows toward the ocean and is replaced by a landward flow of subsurface ocean water. Normally, as we have seen, excess evaporation in Laguna Madre sets up a circulation system opposite to the typical estuarine pattern. As a rule, the lowest-salinity water occurs at the surface and comes from the adjacent ocean: it flows into the lagoon, where it is subjected to evaporation; as its density increases, it sinks into lower layers. The outflow from the lagoon occurs in the subsurface layer, where salty, denser water flows out into the open ocean, shown schematically in Fig. 11–17.

As in many other small estuarine systems, the dredging of channels and construction of tidal inlets and jetties and other engineering works has modified the circulation and changed the biology of Laguna Madre. The estuary was substantially altered in 1949 when dredging was completed on the Intra-Coastal Canal. This channel increased water exchange between the northern and southern portions of the lagoon, and prevented the near-drying-up of one or the other of these basins,

sea level

dense Laguna Madre water

FIG. 11–17
Schematic profile of Laguna Madre, Texas
before construction of Intracoastal Water Way.
[After J. W. Hedpeth, "The Laguna Madre of
Texas," *Twelfth North American Wildlife
Conference Transactions* (1947), pp. 364–80]

as had happened before construction of the canal. A bridge and
solid-fill causeway at the northern end of the lagoon has greatly
restricted passage of water from one side to the other.

The waters of Laguna Madre are highly productive of fish.
In the early 1940's, slightly more than half of the total fish
production for Texas was taken there. In part, this is a consequence of the deterioration of such nearby areas as Galveston
Bay, once an excellent fishing ground. There is also some evidence that the high salinity of the waters causes certain fishes to
grow larger.

Bay waters receive wastes discharged by urbanized and industrialized areas, as well as runoff from agricultural land,
which brings silt, pesticides, and excess fertilizer into the water.
Declines in Laguna Madre fish production during the 1960's
have been ascribed to the cumulative effects of these waste discharges, especially pesticides.

PUGET SOUND

Puget Sound, illustrated in Fig. 11–18, is a fjordlike estuary,
quite different from the three we have discussed so far. It is
part of an extensive inland waterway that includes the Strait
of Georgia to the north. The Strait of Juan de Fuca connects
both parts of this system with the coastal North Pacific Ocean.

Puget Sound was carved by glaciers from an older river-
valley system. A lobe of a continental ice sheet extended into
the area, greatly modifying the older, river-cut valleys, and cutting deep, steep-sided basins now flooded with seawater. The
ice lobe extended just beyond the southern margin of Puget
Sound, gouging elongated basins as deep as 280 meters, separated by ridges or sills that come to within 45–60 meters of
the surface. Since the ice retreated, there has not been enough
sediment deposited to fill these basins; in fact, many of them
are still floored with glacier-borne sand and gravel. Many small
and a few moderate-sized rivers flow into Puget Sound, but none
apparently carries enough sediment to fill the deep irregular
basin.

Where the mouth of a river is protected from strong tidal
currents and wave attack, small deltas have formed. The Skagit
River is the largest river entering Puget Sound, and it has
formed the largest delta.

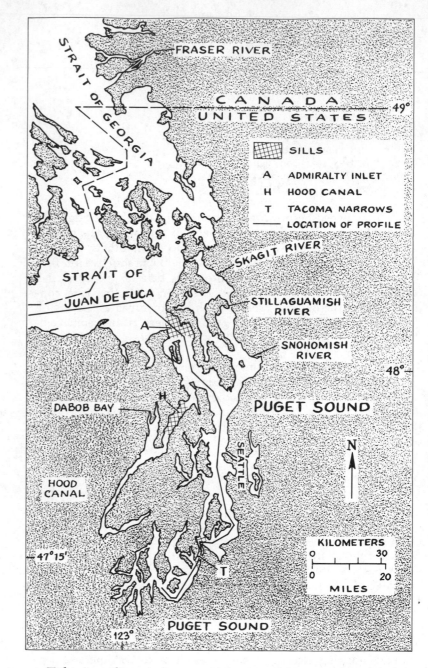

FIG. 11-18
Puget Sound and its adjacent water bodies.
Location of profiles is indicated.

Tides are the primary control on circulation in the Sound. Tidal range increases from the entrance, where the average tidal range is about 1.5 meters, to 3.3 meters at the southern end. About 5 percent of the total volume of the sound is exchanged on each tide. Since there is a mixed type of tide, this amount of water is exchanged twice daily; much of the water taken out on an ebb tide returns on the following flood tide. Still, there is substantial exchange of water between Puget Sound and the Strait of Juan de Fuca.

Tidal currents provide much of the energy involved in mixing the waters of this system. Tidal currents are quite strong, especially over the sills: 2.4 m/sec (4.7 knots) at Admiralty Inlet, 2.6 m/sec (5.1 knots) at Tacoma Narrows. These, combined with irregular topography, cause extensive mixing of surface and deeper waters at the sills as shown schematically in Fig. 11–8. Away from the sills, tidal currents are much weaker, usually less than 50 centimeters per second (1 knot).

Because of the fresh water discharged from rivers and also that falling as rain directly on the surface, Puget Sound waters have relatively low salinities. Subsurface waters entering the Strait of Juan de Fuca typically have salinities of $33.8^0/_{00}$. Passing through the strait they are mixed with waters diluted by river discharge until they reach the sound, where salinities are between 29 and $31^0/_{00}$. Near river mouths, surface layers may have salinities substantially less than this, especially when there is little wind to mix them with the more saline waters beneath.

Puget Sound has an estuarine-type circulation with net seaward flow in surface layers, and a net landward flow along the bottom. Temperature and salinity of the bottom waters are strongly affected by conditions in the Strait of Juan de Fuca, as indicated in Fig. 11–19. For example, temperature and salinity of the deep waters are highest in autumn, as a consequence of upwelling and coastal circulation near the entrance to the strait.

Temperature and salinity of surface layers are modified by local climatic conditions in Puget Sound. Extensive mixing at the sills constantly brings deep waters to the surface and carries the surface waters downward. In winter, this warms the surface layers, which rarely drop below 8°C, except in semi-isolated parts of the sound, where accumulations of low-salinity water may become cold enough to freeze. In summer, surface-water temperatures reach only 13°C. This limited temperature range is the result of the rapid mixing downward of the sun-warmed surface layers. The surface area of Puget Sound is small, so that there is only a limited amount of surface water than can be warmed by the sun. In contrast, there is a large volume of water in the sound that is much too deep to be warmed by the sun.

Since surface temperatures remain nearly constant, the land surrounding Puget Sound also has a limited temperature range, with the cool summers and relatively warm, wet winters typical of maritime climates—in contrast with Pamlico Sound, where the shallow waters have relatively little effect on regional climate.

In addition to restricting the temperature and salinity ranges of surface waters, mixing also keeps the surface layers well supplied with nutrients necessary for phytoplankton growth. Rarely if ever is phytoplankton growth limited by nutrient depletion in Puget Sound, although in winter it is often inhibited by insufficient light for photosynthesis.

These same mixing processes also supply abundant dissolved oxygen to subsurface waters throughout most Puget Sound basins. Some isolated basins (such as Hood Canal) or silled basins (such as Dabob Bay) exhibit seasonal depletion of dissolved oxygen in subsurface waters, especially during late summer. It seems that the productivity of the surface layers is high in these inlets, and the rate of resupply of dissolved oxygen is not rapid enough to make up the loss due to decomposition of organic matter sinking out of the surface layers.

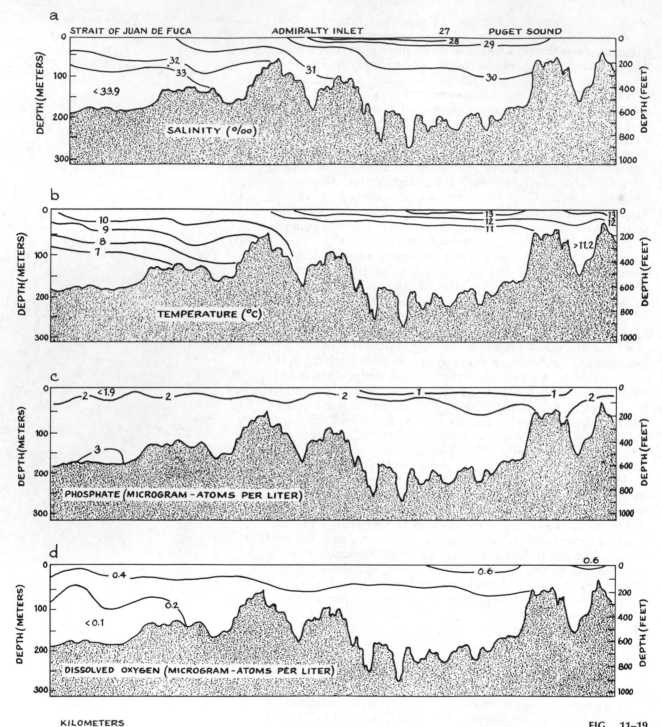

FIG. 11–19
Variation with depth of salinity (a),
temperature (b), phosphate concentration (c),
and dissolved-oxygen concentration (d), during
summer. (After University of Washington
Department of Oceanography, *Puget Sound
and Approaches; a Literature Survey*, Seattle,
1954 3 vols.)

Because of abundant phytoplankton growth in Puget Sound and sediment discharged by rivers, waters are turbid much of the year. Typically, water color varies between brownish and greenish during spring and summer, when phytoplankton production is high and rivers draining glaciers and snowfields are near their maximum discharge. Both river discharge and phytoplankton growth form turbid patches that can readily be seen from the air. During the autumn, surface waters are clearest because productivity and river discharge are low. At this time of year, the waters of the sound more nearly resemble coastal-ocean water offshore.

In contrast to shallow estuaries like Laguna Madre or Pamlico Sound, Puget Sound is little affected by wind. Mixing near river mouths is strongly affected by local winds, but circulation patterns and water levels are influenced only by the largest storms. In part, this is a consequence of the protected location of the sound: it is bordered on three sides by mountains that greatly reduce wind velocities. The only extensive fetches in the sound are along the axes of various arms such as Hood Canal, where the fetch for a north–south wind is about 55 kilometers, and in the main basin, where it is 65 kilometers. In most areas, the largest wind waves observed rarely exceed 1.3 meters, except in Dabob Bay at the northern end of Hood Canal, where waves as high as 2.4 meters have been reported.

LONG ISLAND SOUND

Long Island Sound is an estuary partially cut off from the sea by Long Island, part of New York State. The island itself is a glacial moraine, a pile of sand, gravel, and boulders deposited by a continental ice sheet that covered the area until about 18,000 years ago. This barrier has been only slightly modified by marine processes—wave attack, tidal currents, and longshore movement of sediments.

Long Island parallels the mainland coast, as illustrated in Fig. 11–20, so that the sound possesses certain characteristics of a lagoon as defined earlier. It differs from a typical lagoon, however, in that it is deeper and much larger.

The major rivers enter Long Island Sound from its northern (Connecticut) side (Fig. 11–20). The Connecticut River, the largest river flowing into the sound, enters near its eastern, or open, end. Contrast the resulting surface temperatures and salinities as shown in Fig. 11–21 with Chesapeake Bay in Fig. 11–11, where the Susquehanna River enters at the head.

The western end of the sound is not entirely closed. A tidal strait, the East River, connects it with New York Harbor and the Hudson River estuary, and through it some fresh water flows into the sound from the Hudson River. This fresh-water flow is much smaller than that of the estuarine circulation caused by rivers discharging into the sound proper.

Because of the relatively small amount of fresh water discharged by rivers and vigorous mixing due to tidal currents, Long Island Sound waters are only moderately stratified. During

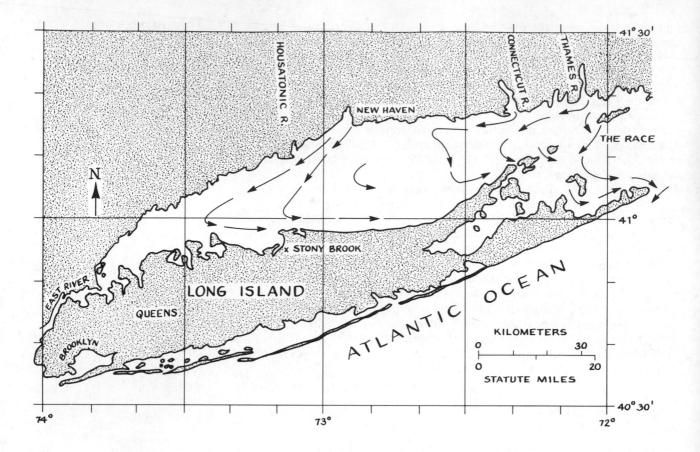

FIG. 11–20
Long Island and Long Island Sound, showing
the postulated net circulation of surface
waters. [After G. A. Riley, "Hydrography of
the Long Island and Block Island Sounds,"
*Bulletin of the Bingham Oceanographic
Collection*, 13(3) (1952), 5–39]

peak river discharge, stratification is moderately well developed
near the mouths of the various streams. Thermal stratification
is most conspicuous between March and August or September,
when surface layers are warmed by the sun, as can be seen
from Fig. 11–22. Temperature differences between the surface
and the colder bottom layers amount to 5°C or less. From
October through February, when waters in the sound are cooling and supplying heat to the surroundings, the layers are mixed
almost completely and there is no temperature stratification
from top to bottom. Temperature of the surface layer is greatly
dependent on local weather conditions; during a hot summer, for
instance, surface layers are warmer than during a relatively cool
summer.

Salinity stratification is always present to some degree, and
depends not on the warming or cooling cycle but upon the
supply of fresh water from rivers. About half of the total annual river discharge occurs during the period from April through
June; river discharge is low during the rest of the year. During
periods of high fresh-water discharge, surface salinity may be
as low as 24 to 25⁰/₀₀; at this time, the bottom salinities range
from 25 to 26⁰/₀₀. During periods of low river discharge, salinity
stratification is reduced. Surface salinity may be as high as
28⁰/₀₀, the salinity of the bottom waters being a few tenths of
1⁰/₀₀ greater on the average.

Mixing of Long Island Sound waters is caused primarily by

tidal currents. The basic estuarine flow pattern is exhibited by differences in the strength of the tidal currents. In the surface layers, ebb currents are stronger and persist longer than in the bottom layers, where the flood currents tend to be stronger and persist longer. For example, on a typical tidal cycle, a parcel of surface water moved 6.5 kilometers eastward; at the same time, the corresponding bottom water moved 2.2 kilometers westward. Thus, in that particular tidal cycle, the separation between and the movement of the surface and bottom waters amounted to 8.7 kilometers. Considering the sound as a whole (see Fig. 11–20), observations at its eastern end indicated a net eastward movement in the surface layers, at speeds ranging from 5 centimeters per second in September to 14 cm/sec in

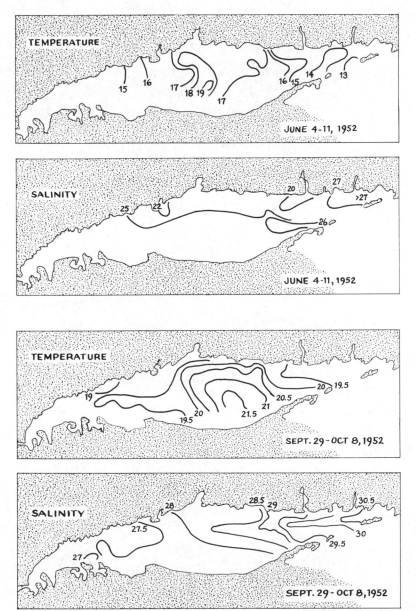

FIG. 11–21
Long Island Sound surface-water temperatures (°C) and salinities ($^0/_{00}$) in late spring (upper) and early autumn (lower). (After Riley, 1952)

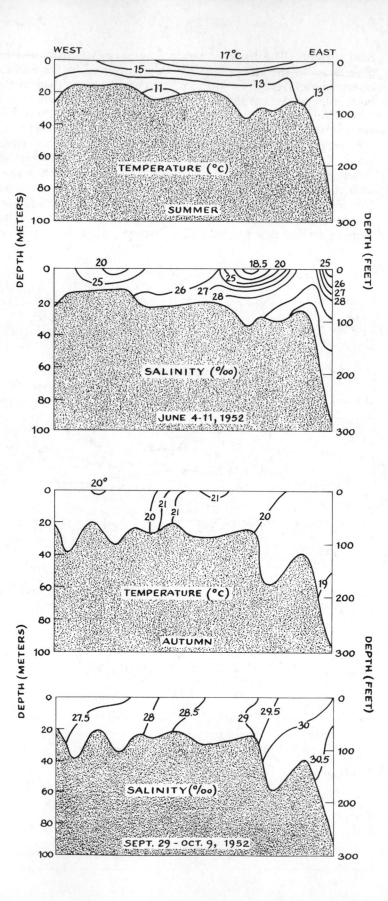

FIG. 11–22
Distribution of temperature and salinity in Long Island Sound in June and September–October, 1952. Profiles run lengthwise through the Sound. (After G. A. Riley, "Oceanography of Long Island Sound, 1952–1954—IX. Production and utilization of organic matter." *Bulletin of the Bingham Oceanographic Collection*, 15 (1956), 324–44)

January. This area is known locally as The Race because of its strong tidal currents. Observations of the bottom waters at the same location suggest that the net movement throughout the year is westward, moving into Long Island Sound.

The amount of fresh water discharged into this estuary amounts to about one-third the volume of the sound in a year's time. Salinity reduction caused by this fresh-water discharge is relatively small, indicating that mixing is vigorous and rapid, and that the volume of seawater brought into the sound through movement of the subsurface layers is much greater than the volume of fresh water discharged by the rivers. On the average, during the year, the seawater inflow along the bottom is about 3.8 times the volume of the sound. This suggests that the residence time for a parcel of water in the sound is on the average about 3 months. Obviously, the time to move a particular body of water out of the sound depends on where and how it entered the estuary. Studies of Long Island Sound suggest that movement of the surface water across a 40-kilometer section requires about 2 months, while the movement of the bottom water, being somewhat more sluggish, requires about 3 months to travel the same route.

Origin of estuaries, fjords, and lagoons

Estuary—a semi-enclosed coastal body of water where fresh water from the land mixes with salt water

Often formed by drowning of river valleys

Fjord—drowned, glacially eroded valley, usually found in mountainous regions in high latitudes

Lagoon—broad, shallow water body separated from ocean by spit or barrier beach

Rise of sea level following retreat of glaciers has had major effect on present estuaries and lagoons

Sea level reached present height about 3000 years ago

Estuarine circulation

Seaward flow of less saline water at surface, formed by mixing fresh water and entrained seawater from below

Landward flow of more saline water at depth, brought in to replace salt moved seaward in the surface layer

Salt-wedge estuary—strong discontinuity between surface and subsurface layers; occurs where river flow predominates over tidal effects

Moderately stratified estuary—more intensive mixing causes less pronounced separation between layers; occurs where tidal effects are more pronounced

Fjords—estuarine circulation above sill depth, sluggish circulation in waters below sill depth; little affected by surface conditions

Tides in estuaries

Progressive waves are damped out in estuaries

Tidal currents dominate currents in estuaries

Biological productivity of estuaries

Estuaries highly productive—nutrients replenished by upward movements of subsurface waters, by river discharge, and by waste discharges

Estuarine circulation tends to trap nutrients in the estuary

Where productivity of surface layers is extremely high and resupply of dissolved oxygen to bottom water is slow, the bottom waters may be depleted in oxygen so that benthic organisms cannot grow; sediments then may form with distinctive structures—varves

Chesapeake Bay—coastal-plain estuary

Formed by drowning of former river-valley system

Tidal effects important

Pamlico Sound—lagoon (bar-built estuary)

Barrier islands and narrow inlets isolate system from coastal ocean

Wind effects predominate over tidal effects

Laguna Madre—lagoon where evaporation exceeds fresh water supply

Isolated from Gulf of Mexico

Circulation involves seaward-flowing subsurface layer and landward-flowing surface layer

Highly variable water temperatures and salinities

Puget Sound—fjordlike estuarine system
    Tidal currents important
    Mixing of waters over sills plays major role in distribution of
    properties
Long Island Sound—a glacial moraine isolated from the ocean
    Moderately stratified
    Tidal effects predominate
    Waters replaced in about 3 to 4 months

LAUFF, GEORGE H. 1967. *Estuaries*. American Association for the Advancement of Science, Washington, D.C. 757 pp. Collection of scientific papers on estuaries; their physics, chemistry, geology and biology; intermediate level, specialized.

SELECTED REFERENCE

THE COASTAL OCEAN

twelve

Eddystone Lighthouse (left). This famous lighthouse is situated on a rocky reef about 24 kilometers south of Plymouth, England. The first structure on this reef was built in 1698. Two subsequent structures succumbed to fire or storm. This sketch represents the 1789 lighthouse, designed by John Smeaton. The shaft of the tower was constructed of granite masonry with a remarkable interlocking construction especially contrived to prevent individual stones from being pulled out by the forces of heavy seas. In 1881, when the foundation stones began showing signs of failure, this structure was removed and a more elaborate one, designed by Sir James N. Douglass, was built on an adjacent rocky shoal.

Trinity Shoals Lighthouse (right). This is one of a pair of all-metal lighthouses at Timbalier Bay in the Gulf of Mexico. The structure rests on nine screw piles, one in the center surrounded by eight at a radius of about 6 meters. The object of the design is to avoid the great forces which heavy seas set up in large areas of masonry. This structure dates from the late eighteenth century.

The deep, open ocean, where processes are unaffected by any boundary, save the air–water interface, includes all but a fraction of the ocean's volume, as well as most of its surface area. It is the coastal ocean, however, such as shown in Fig. 12–1, that we as land-dwellers are most likely to encounter. Boundaries—the shallow bottom and the neighboring continent—have such marked effect on the coastal ocean that it is convenient to consider it as a distinct oceanographic region. In this chapter, we re-examine the dynamic phenomena of the marine environment—waves, tides, currents, wind, temperature effects, evaporation and dilution—but this time in the context of the coastal ocean.

Recall that the coastal ocean lies over the continental shelf. Where this shelf is narrow, as on the United States Pacific coast, coastal-ocean features sometimes extend seaward of the continental-shelf break. Average water depth on the shelf is about 70 meters, in contrast to the open ocean, where depths typically range from 4000 to 6000 meters.

Winds influence coastal circulation in two ways. First, blowing over the surface of the water, winds set surface waters in motion, which may in turn cause storm surges, coastal currents, or upwelling. Secondly, winds generate waves which affect the coastal ocean through their actions on shorelines and mixing of surface waters.

A consistent feature of coastal processes is their compressed time scale as compared with the same processes in the open ocean. This, of course, is a result of the shallowness of coastal waters—conditions change rapidly because relatively little water

FIG.   12–1
The coastal ocean near Houston, Texas, with sunlight reflecting off the water. Galveston Bay, in the center is separated from the coastal waters of the Gulf of Mexico by barrier islands, including Galveston Island. In the lower-right portion of the photograph the twisted course of the Brazos River can be seen. Note the abundance of water in this area, both natural lakes and reservoirs formed by damming streams. The irregular white masses in the lower portion of the picture are clouds. Houston lies between the Brazos River and Galveston Bay. This view was taken from Apollo 7 on October 17, 1968, from an altitude of approximately 190 kilometers. (Photograph courtesy NASA)

is there to be heated, cooled, evaporated, diluted, or moved around. We shall observe this pattern of abrupt change as we discuss general features of coastal oceans. Finally, we shall examine two typical coastal regions—the New York Bight and the Washington–Oregon continental shelf.

TEMPERATURE AND SALINITY IN THE COASTAL OCEAN

Large temperature and salinity changes occur in shallow, partially isolated waters of the coastal ocean. These variations are well beyond anything found in the open ocean. Winds, for instance, affect both water temperatures and salinities to a marked degree. On the United States Atlantic coast, winds coming from the continent are much hotter than the ocean in summer, and much colder in winter. They therefore exert substantial influence on coastal-water temperatures.

Winds from the continent are dry, having lost much of their moisture through precipitation; therefore, such winds usually cause extensive evaporation when they blow across coastal waters. Where winds blow across the ocean toward the coast of a continent, as in the northwestern United States during winter months, they have little effect on the coastal ocean.

Salinity extremes occur in coastal waters, again usually in isolated embayments. Highest surface-water salinities occur in the western Mediterranean and northern Red Sea areas, exceeding average surface-water salinities for their latitude by more than 4 parts per thousand. It is difficult to say where the lowest coastal-ocean salinities occur, although it would certainly be in or near the estuary of a major river.

As was discussed previously, there is net evaporation from nearly all ocean surfaces. Only along the coasts is excess fresh

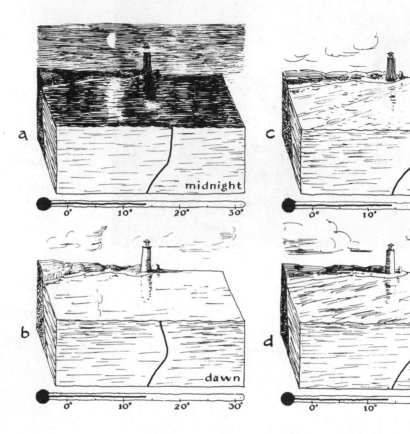

FIG. 12–2

Variation in near-surface water temperatures during the course of a day. At midnight (a) the surface zone is nearly isothermal. Cooling continues throughout the night so that by dawn (b) the surface layer is cooler than the waters beneath. After the sun comes up, the surface layer is warmed, and by noon (c) the surface layer is distinctly warmer than waters immediately below the surface. Warming continues until later afternoon, when the surface temperature is highest. Later, the surface layer begins to cool and is slightly cooler at dusk (d). During the night, the surface zone cools until it is again isothermal around midnight. Daily temperature changes are usually confined to the upper 10 meters or less. [Redrawn from E. C. LaFond, "Factors Affecting Vertical Temperature Gradients in the Upper Layers of the Sea," *Scientific Monthly*, 78(4) (1954), 243–53]

water discharged from the continents to be mixed back into the ocean. Note that isohalines tend to parallel the ocean boundaries, reflecting the fact that the greatest net evaporation occurs near the centers of the ocean.

The ocean's highest and lowest surface temperatures occur in coastal waters. Surface-water temperatures exceed 40°C in the Persian Gulf and in the Red Sea during the summer, whereas the surface of the open ocean rarely (if ever) exceeds 30°C. Lowest surface-water temperatures are controlled by the point of initial freezing of seawater; generally, this is about −2°C, depending on the surface-water salinity. In winter, sea ice forms in many high-latitude coastal areas, particularly in shallow bays and lagoons where cooling is rapid due to the large surface area and small volume of water present.

Warming of surface waters during the day can be detected fairly readily in coastal waters, especially those protected from winds and waves. This variation occurs despite the large heat capacity of water and the ability of the ocean to mix the additional heat through many meters of water beneath the ocean surface. Generally, temperatures of surface waters are highest in mid-afternoon and lowest at dawn; variations are indicated in Fig. 12–2.

Tides cause temperature changes in coastal waters by moving water masses of different temperatures and by causing vertical movements of waters. Tidally induced temperature changes are especially noticeable in shallow areas, such as salt marshes. In a salt marsh, water temperatures are highest on a summer day after the water has flooded the marsh. The water is warmed by contact with the warm marsh deposits and directly by absorption of solar radiation. Therefore, water coming out of a marsh on a summer day is commonly many degrees warmer than waters surrounding the inlet to the marsh. When the tide rises, bringing water into the marsh, it is commonly several degrees cooler than the water that flowed out several hours earlier.

The seasonal temperature cycle in the coastal ocean is most obvious at mid-latitudes, where seasonal changes in air temperature are largest. In many parts of the coastal ocean, waters are thoroughly mixed during winter months, so there is little temperature difference between bottom and surface waters, as Fig. 12–3 demonstrates. During spring the surface waters are warmed by the sun. If strong mixing continues, the warming may be reflected by a gradual warming of a surface layer many meters thick. If mixing is restricted, the warming may be felt only in a shallow layer. During the first part of summer, mixing continues and the surface layer becomes warmer and the thermocline is usually more pronounced. Surface layers are commonly warmest in mid- or late summer, depending on local climatic conditions. During autumn, the surface layer cools by radiation, evaporation, and continued mixing with deeper waters.

COASTAL CURRENTS

In the coastal ocean, currents are strongly affected by the winds.

Because of the shorter time scales and smaller distances, coastal currents appear, disappear, and even change flow directions within relatively short periods of time. During periods of light winds and reduced river flow, coastal currents may become too weak to be readily discernible. Because of its importance, let us consider this process is some detail.

Winds blowing from certain directions move surface waters shoreward. Continuing wind may hold these waters near the coast, depressing the pycnocline and inhibiting upward water movements.

The resulting sloping sea surface causes geostrophic currents, which parallel the coast. These wind-induced currents are strongest when river runoff is large and winds are strong. They show marked variability compared with the slow, steady flow of the deep-ocean currents.

The coastal geostrophic circulation is bounded on the seaward side by the steadily flowing boundary currents that move water around the major ocean basins. Along United States coasts, currents often set in the opposite direction to the boundary current offshore. When winds diminish, or river discharge is low, coastal currents weaken or disappear. Such conditions are common in late summer, when coastal waters tend to form large, sluggish eddies, rather than distinct coastal currents. With the onset of winter storms, these eddies are broken up, the surface layers are mixed, and the coastal circulation is re-established within a few days.

Water discharged by rivers or derived from coastal precipitation must eventually move across the coastal currents to mix

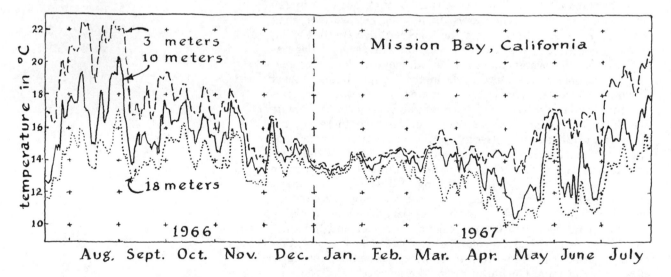

FIG. 12–3
Water temperatures at three different depths at an oceanographic research tower located about 1.6 kilometers offshore from Mission Bay, California, in waters about 18 meters deep. [After J. L. Cairns and K. W. Nelson, "A Description of the Seasonal Thermocline Cycle in Shallow Coastal Water," *Journal of Geophysical Research*, 75(6) (1970), 1127–31]

into the open ocean. Since the circulation tends to parallel the coast, transfer of fresh water across the system can be very slow. (Alternatively we can say that the *residence time* is long.) For example, within the Middle Atlantic Bight, between Nantucket Shoals off Cape Cod, and Cape Hatteras, North Carolina, the fresh water necessary to account for the lowered salinity of the water is equivalent to about 2½ years of discharge of the region's rivers. (By comparison, in the Bay of Fundy it takes about 3 months for the fresh water to be replaced, and about 3 to 4 months in Delaware Bay.)

When prevailing winds move surface waters away from the coast, subsurface waters move upward to replace them, as seen in Fig. 12–4. This process, called *upwelling*, is common along the western sides of the continents, in the eastern boundary-current areas. Coastal upwelling occurs along the California–Oregon coast, the western coast of South America, western Africa,

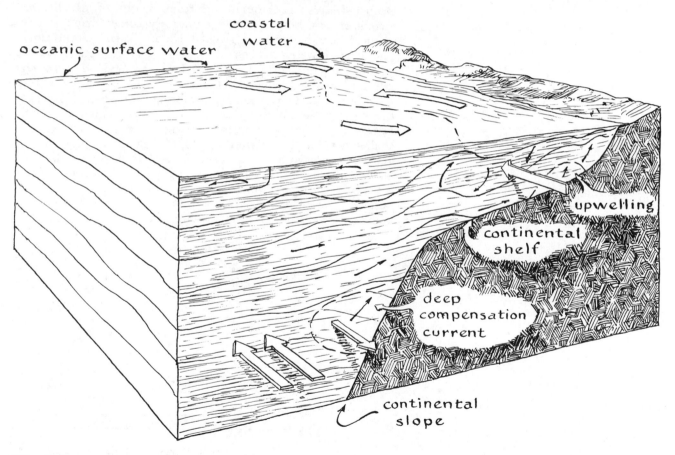

FIG. 12–4
Schematic representation of surface and subsurface circulation in an area of coastal upwelling. (Nearly horizontal lines show equal density surfaces.) [After H. U. Sverdrup, "On the Process of Upwelling," *Journal of Marine Research* 1(2) (1938), 155–64; and T. J. Hart, and R. J. Currie, "The Benguela Current," *Discovery Reports*, 31 (1960), 123–298]

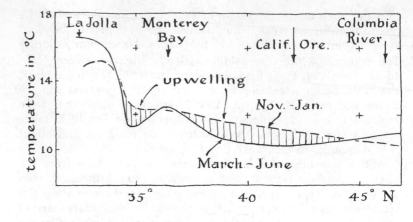

FIG. 12–5

Temperatures along the California–Oregon coast during November–February (no upwelling) and March–June (upwelling active). Note the localized temperature changes caused by upwelling. (After Sverdrup et al., 1942)

northwestern Africa (Morocco), and the western coast of India.

Upwelling has a pronounced effect on the coastal ocean. Deeper waters are brought up from depths of usually 200 meters or less, but they are significantly colder than the sun-warmed surface layers. Upwelled waters usually form small, cold surface-water masses near the coast or in near-surface waters. Both the surface ocean and nearby land areas are significantly cooled when upwelling is active, as Fig. 12–5 indicates.

Another effect of upwelling is to bring nutrient-rich waters to the surface, where they can support phytoplankton growth. Even when subsurface waters do not reach the surface, the shallowness of the thermocline locally enhances productivity, because photosynthesis can occur in waters below the thermocline. Nutrients are constantly resupplied by the continued upward movements of subsurface waters. These waters move vertically about 50 meters per month. Upwelling areas are among the world's richest fishing grounds. Areas of pro-

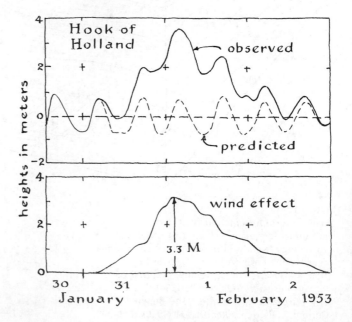

FIG. 12–6

Storm surge of 1953 in the North Sea. Sea-level changes resulting from the surge were estimated by subtracting the predicted level from the observed sea levels. (After Groen, 1967)

nounced upwelling have photosynthetic-carbon production rates from 2 to 10 times the average production in open-ocean waters. In fact, a map of upwelling regions nicely outlines most of the biologically productive ocean areas (which can be seen in Fig. 17–2).

Effects of upwelling are most noticeable along arid coasts, in subtropical areas. The absence of fresh water there prevents the formation of a low-salinity surface layer along the coast which would cover the upwelled waters. Cold, upwelled waters cool the adjacent coast; this effect is especially noticeable along the coast of California. There, too, abundant sunlight insures sufficient light for photosynthesis in the nutrient-rich waters.

In areas with fresh-water discharge, the low-salinity surface layer covers the upwelled waters, and inhibits wind-induced upwelling. Along the Oregon coast, upwelling effects are noticeable when the low-salinity surface layer is moved seaward, exposing the cold upwelled waters beneath. Along the Washington coast, there is some upward movement of subsurface waters, but the thick, low-salinity surface layer (formed by abundant river discharge) prevents upwelled water from reaching the surface. Therefore, effects of this limited upwelling are not readily discernible along the Washington coast.

STORM SURGES

In *storm surges*, strong winds pile up water along the coast, causing changes in sea level. When this happens in a partially enclosed ocean area, it sets up a *seiche*, or standing wave. In the North Sea, for example, a northwest gale (25 meters per second) has a fetch of about 900 kilometers extending from Scotland to the Netherlands. Winds blowing across the North Sea (water depths of about 65 meters) can cause sea level to rise more than 3 meters on the Dutch coast. Such a storm surge occurred on January 31–February 1, 1953, with disastrous flooding on the low-lying Dutch and English coasts. Wind caused sea level to rise about 3.3 meters as seen in the tidal curve shown in Fig. 12–6. This, combined with high tides and strong waves, broke through the dikes and dunes protecting the coast.

It is possible to predict the height of a storm surge caused by a given storm. The actual rise in sea level usually differs from the predicted rise because ocean-basin shapes must usually be simplified to permit calculations to be made. For example, the North Sea has a restricted opening through the Strait of Dover, so that storm surges caused by a northwest wind are somewhat lower owing to the "leak." Other complications include currents and seiches set up by the storm and interactions with astronomical tides. Furthermore, data on storm winds are often inadequate for making accurate predictions.

A second type of storm surge more closely resembles a wave which moves with the storm that caused it. The surge associated with a storm—say, a hurricane—usually consists of three parts. First comes a slow, gradual change in water-level beginning several hours before the storm's arrival. Water-level changes associated with the *forerunner* of the 1953 storm in the North Sea where the water level dropped slightly before the storm arrived. The so-called forerunner is usually felt widely along the coast; apparently it is caused by the regional wind system rather than just the center of the storm itself.

When the hurricane center passes, it causes a sharp rise in water level called the *surge*. This usually lasts about $2\frac{1}{2}$ to 5 hours; rises in sea level of 3 to 4 meters have been observed—usually localized, but slightly offset from the center of the storm. Combined with extremely high waves generated by the storm, hurricane surges can be extremely destructive.

Following the storm, sea level continues to rise and fall as the resurgences or oscillations set up by the storm pass. These are more-or-less free, wavelike motions of the water surface, and have been termed the "wake" of the storm, like the wake left by the passage of a ship through the water. These resurgences can be quite dangerous, particularly as they are often not expected after the storm itself has subsided.

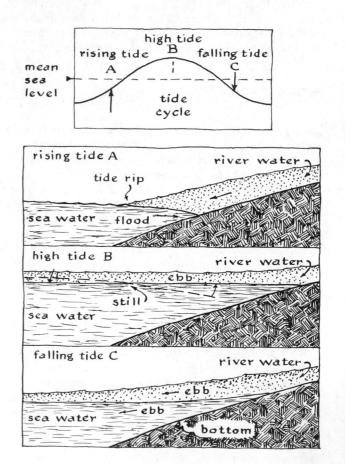

FIG. 12–7
Idealized representation of currents and water masses in a river mouth during a tidal cycle. [After J. P. Tully, and A. J. Dodimead, "Properties of the Water in the Strait of Georgia, British Columbia, and Influencing Factors," *Journal of the Fisheries Research Board of Canada*, 14(3) (1957), 241–319]

Tropical storms and hurricanes frequently cause storm surges along the Gulf Coast of the United States. In 1900, Galveston, Texas was destroyed and about 2000 people were killed by a storm surge resulting from a hurricane. In 1969, the second strongest recorded storm to hit the Gulf Coast caused millions of dollars of damage; even with advance warning, it killed several hundred people. A disastrous storm surge occurred in 1876, on the Bay of Bengal, when 100,000 people were killed. In 1970 a storm surge hit the same area killing an estimated 500,000 persons.

TIDE EFFECT ON RIVER DISCHARGE

Tides are another major driving force in the coastal ocean. Here, rotary tidal currents of the open ocean are typically replaced by reversing tidal currents. Tidally controlled conditions persist for a few hours; for instance, the tide rises for approximately 6 hours, and then falls for 6 hours. To illustrate the influence of the tides on coastal-ocean processes, let us consider the discharge of a large river flowing into the coastal ocean, as illustrated in Fig. 12–7.

As sea level rises ("tide coming in"), the denser seawater flows under the low-salinity water from the river, forming the salt wedge discussed in Chapter 11. For a time, the surface and subsurface currents are in opposite directions: flow is seaward at the surface and landward near the bottom. Because of the rising sea level, flow in the upper level diminishes until it finally equals the speed of the subsurface current flowing in the opposite direction. At this time, a convergence forms (known as a *tide rip*, marking the point where the opposing flows are equal. Note (in Fig. 12–7) that the boundary is inclined. Little or no river water flows past the convergence, since the landward flow is exactly equal to the seaward flow in the surface layer. Rips are easily recognized by the lines of floating debris that collect in them.

As the tide continues to rise, surface flow is eventually reversed, and at this time both surface and subsurface waters flow upstream. Both the salt wedge and the rip move up the estuary. For a time, both fresh water and seawater accumulate in the estuary, and there is no river discharge into the coastal ocean.

Approaching slack tide, but before high water is attained, the flood current diminishes to a point where it is weaker than the river flow. At this time, the subsurface water is still flowing landward, although the rip is being pushed seaward at the surface by river water. Finally, there is no longer any opposing flow at the surface, and seaward flow accelerates as the accumulated water is discharged from the estuary. In the coastal ocean, the momentum of the river flow is quickly dissipated and the newly discharged waters are moved by currents and winds away from the mouth of the estuary.

The rise and fall of the tide thus tends to control the timing of river-water discharge into the coastal ocean. Most fresh water is discharged on the ebb tide and enters the ocean as a series of pulses. River discharge tends to form cloudlike parcels of low-salinity water that move freely and are detectable by their lowered surface salinities. When turbid with suspended sediment, these parcels of recently discharged water are easily recognizable from the air, as Fig. 12–8 demonstrates.

Boundaries around clouds of recently discharged water often remain sharp, resembling the original tidal rip. Because of intensive mixing at the interfaces, the surface is often quite choppy. Sometimes the interfaces can be heard as they pass, because of the turbulence generated by the extensive mixing.

When water parcels are discharged intermittently, each tends to overrun the previous one, forming a complex of irregular structures and a confused arrangement of rips, as shown in Fig. 12–9. As the cloudlike masses move away from the river mouth, mixing continues, boundaries are obscured, and eventually a large lens or plume of low-salinity water forms. Such a plume may be followed for many tens or hundreds of kilometers off the mouths of major rivers. This is especially true where there are no other large rivers discharging in the area, as shown in Fig. 12–10.

When partially mixed, low-salinity water is discharged from an estuary, it continues to mix with adjacent and underlying water in the coastal ocean. Within a few tens of kilometers of the estuary, mixing results from tidal currents—the currents

FIG. 12–8
Small estuaries along the coast of South Carolina discharge turbid waters into the Atlantic. Note the small size of these sediment-filled systems, in contrast to the much larger Chesapeake Bay or New York Harbor, which remain unfilled by sediment. (Photograph courtesy NASA)

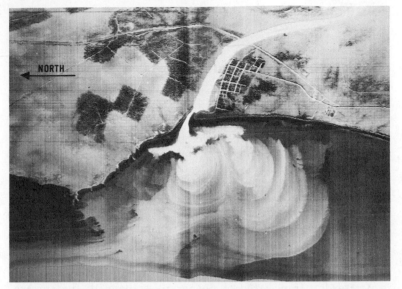

FIG.   12–9
Fresh water from the Quinault River discharges into the northeast Pacific Ocean at Taholah, Washington. Electronic sensors detect differences in water temperature which outline the cloudlike water masses of low-salinity colder water mixing with coastal-ocean waters, which appear dark because they are warmer. (Photograph courtesy NASA)

arising from river discharge—and waves. In this mixing zone, water characteristics are normally extremely variable. Individual water parcels may have relatively sharp boundaries or interfaces, each with extremely active mixing taking place. The newly mixed water formed at boundaries, having intermediate density, flows away from the interface, leaving a plane of active convergence separating the less-saline parcels from adjacent waters.

Beyond the zone of initial mixing, near the river mouth, water parcels are partially or totally mixed and lose their individuality.

FIG.   12–10
Muddy water discharged by the Irrawaddy and nearby rivers moves through the Gulf of Martaban on the eastern side of the Bay of Bengal (Indian Ocean). This photograph was taken in November, 1966, from an altitude of 295 kilometers. (Photograph courtesy NASA)

Here the effluent from the estuary is indistinguishable from the rest of the coastal water.

OCEANIC EFFECTS ON LAND CLIMATES

The ocean moderates climatic conditions on land. Coastal areas typically have a *maritime climate*, where the annual temperature range is generally much narrower than in regions far inland, as shown in Fig. 12–11. Far from the ocean, *continental climates* prevail, with their characteristically wide temperature ranges.

Oceans store large amounts of heat because of water's high heat capacity, permitting heat storage without an accompanying large temperature increase. Furthermore, the ocean (or any large body of water) stores heat through several meters of water due to mixing, and thus retards loss of heat to space by back-radiation at night. In contrast, rocks on land have a low heat capacity, so they become hot during the day. But since this heat is not readily transferred to deeper rocks beneath the surface, most of it is lost at night by back-radiation to space. This may be observed in any desert area, where intense daytime heat is followed by a chilly night.

During summer, continents heat up and warm the atmosphere above them. The relatively less dense continental air mass rises and is replaced by moist air from the ocean. As the oceanic air mass rises, it cools and releases moisture, often as heavy rains. This is the *monsoon* circulation, especially well developed over India as a result of seasonal atmospheric circulation shifts caused by the warming and cooling of the Asian land mass. A similar air-circulation pattern prevails throughout much of the United States midcontinent during summer. Moist air from the Gulf of Mexico supplies large amounts of moisture during the summer, usually as thundershowers.

Mountain ranges block the flow of air from the ocean and thereby affect climatic conditions over large continental regions. North–south mountain ranges of the western United States block the flow of air from the Pacific, creating a rainshadow over interior valleys so that they receive little precipitation. Air cools as it rises to flow over the mountains. Rainfall is sufficient to support a temperate rainforest on the western slopes of the Coast Range of Washington, whereas deserts or near-desert conditions occur east (downwind) of the mountains. The coastal portion of the western United States thus has a maritime climate, characterized by moderate temperatures, winter and summer, and abundant rainfall. Continental climates with wide temperature ranges are common in the interiors of the continent, away from the ocean's moderating influence (see Fig. 12–11).

On the Eurasian continent, by contrast, the Alps, Himalayas, and associated mountains run generally east–west. These ranges have little effect on the flow of maritime air over Europe and Asia, and the maritime climate extends many hundreds of kilometers inland. These east–west ranges do, however, block north–south air movements and very likely cause the climate of the Mediterranean region to be warmer and drier than it might be in their absence.

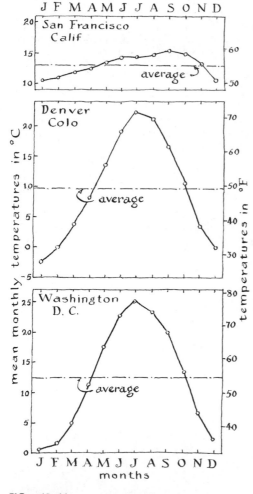

FIG.   12–11

Temperature variation at three locations: San Francisco has a typical maritime climate, with cool summers and relatively warm winters; Denver has a typical continental climate— cold winters, warm summers; Washington, D.C. has a generally continental climate, though tempered by its proximity to the ocean.

Because of the variability of the coastal ocean, it is worthwhile to divide it into distinct regions (as in Fig. 12–12), according to geologic history and environmental characteristics (as set forth in Table 12–1). Coastal zones are analogous to river-drainage basins on land. Several can be identified along the coast of North America.

Let us consider some characteristics of coastal zones, listed in Table 12–2. Surface waters tend to be somewhat restricted (although much less so than river-drainage basins on land) by current systems and to a lesser extent by topography. Little sediment or continental runoff is transported between coastal-ocean regions, but rather tends to remain in the region where it was discharged. Hence a given region will tend to have similar seasonal patterns of surface temperatures and salinities throughout.

Each coastal region has had a similar recent geologic history throughout, so that both land forms and the continental shelf tend to be similar. For example, the recent glaciation of the New England area has caused both a glacially eroded coastline lacking a coastal plain and a glacially eroded continental shelf. Farther south, where there has been no recent glaciation, the

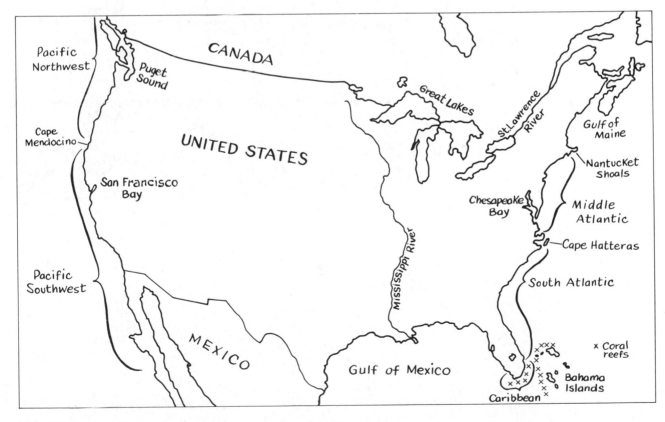

FIG.   12–12
Coastal-oceanic regions of the United States.

Table 12–1

ENVIRONMENTAL ASPECTS OF U.S. COASTAL-OCEANIC REGIONS*

Coastal Region	Shoreline	Continental Shelf	Boundary Current
Gulf of Maine	Rocky, embayed by drowned river valleys; fjords	Rocky, irregular, 250–400 km wide	Labrador
Middle Atlantic	Smooth coast, embayed by drowned river valleys and lagoons	Smoothly sloping, 90–150 km wide	Labrador
South Atlantic	Smooth, low-lying, marshy; many lagoons	Smoothly sloping, 50–100 km wide	Gulf Stream
Caribbean	Irregular; mangrove swamps, coral reefs	Eastside 5–15 km wide; Westside 300 km wide	Gulf Stream
Gulf of Mexico	Smooth, low-lying; barrier islands, marshes, lagoons	Smoothly sloping, 100–240 km wide	Seasonally variable
Pacific Southwest	Mountainous, few embayments	Irregular, rugged, average 15 km wide	California
Pacific Northwest	Mountainous, few embayments, some fjords	Irregular (Oregon) to smoothly sloping (Washington), 15–60 km wide	California
Southeast Alaska	Mountainous, many embayments and fjords	Rocky and irregular in places, 50–250 km wide	Alaska

*Data from U.S. Dept. of Interior, National Estuarine Pollution Study, 1970.

Table 12–2

OCEANOGRAPHIC ASPECTS OF U.S. COASTAL-OCEANIC REGIONS*

Coastal region	Temperature average (range) (°C)	Salinity average (range) ($^0/_{00}$)	Tide†	River discharge Water (km³/yr)	River discharge Sediment (10⁶ tons/yr)
Gulf of Maine	8 (18–0)	30.3 (31–29)	SD	64	?
Middle Atlantic	12 (23–2)	31.5 (31.8–31.3)	SD	95	15.3
South Atlantic	24 (30–10)	35.5 (36.0–32.5)	SD	137	58.1
Caribbean	27 (30–22)	35.7 (36.3–34.9)	SD	10	?
Gulf of Mexico	23 (30–12)	32.3 (37.0–30.3)	D	712	360
Pacific Southwest	15 (20–13)	33.6 (33.8–33.3)	M	21	24
Pacific Northwest	10 (13–7)	30.8 (32.5–28.5)	M	113	127
Southeast Alaska	5 (13 to −1)	31.9 (32.1–31.2)	M	?	?

*Data from U.S. Dept. of Interior, National Estuarine Pollution Study, 1970.
†SD—semidiurnal tide
  M—mixed tide
  D—diurnal tide

coastal plain widens and barrier beaches or islands border the coast. Similar topography extends offshore, on the continental shelf, where submerged lagoons, barrier islands, and beaches make up the present sea bottom.

We shall consider two coastal-ocean areas that typify major aspects of coastal circulation. The circulation from the Gulf of Maine to Cape Hatteras will be covered in our discussion of the New York Bight, which is that part of the coastal ocean lying between the eastern end of Long Island and the entrance to Delaware Bay at the southern end of New Jersey. A West coast region, the Washington–Oregon coast, will be considered in detail, for comparison.

NEW YORK BIGHT

The New York Bight, between Long Island, New York, and Cape May, New Jersey—partly shown in Fig. 12–13—has had a long maritime and urban history. This region, typical of the coastal ocean from Cape Cod, Massachusetts, to Cape Hatteras, North Carolina, includes the Hudson and Raritan River estuaries. These serve as harbors for the New York metropolitan area. Offshore are the Labrador Current and the Gulf Stream system. Conditions in this stretch of coastal ocean resemble those in many parts of the world near the mouths of large, industrialized river basins, where coastal waters serve as transportation routes, waste-disposal areas, and recreation and fishing sites.

In these regions, as in most coastal water, temperatures, salinity, and currents change with the seasons. Surface-water salinities tend to be highest (29 to 32.8$^0/_{00}$) in late winter (February or early March) when river discharge in the area is low because of frozen rivers. At this time, surface-water tem-

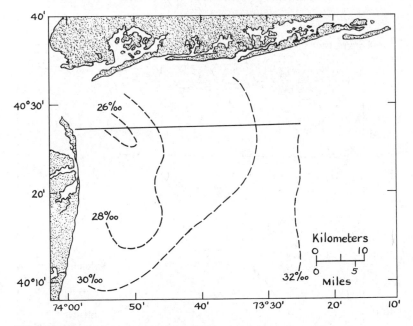

FIG. 12–13

Part of the New York Bight, showing the distribution of surface-water salinities in April, 1948. The line indicates the location of Fig. 12–15. [After B. H. Ketchum, A. C. Redfield, and J. C. Ayers, "The Oceanography of the New York Bight," *Papers in Physical Oceanography and Meteorology* 12(1) (1951), 4–46]

peratures are at their lowest, ranging from −1° to +1°C along the coast. Ice forms in partially isolated shallow waters. Because of the strong cooling and intensive mixing caused by winter storms, coastal waters are nearly homogeneous; there is almost none of the stratification or vertical stability that is typical of summer and autumn conditions. Put another way, in winter, the cold is "felt" all the way to the bottom of the coastal ocean.

Fresh water discharged by rivers (primarily the Hudson River) is quickly mixed with nearby seawater, as Fig. 12–13 indicates. But fresh water added must be removed eventually from the coastal area, otherwise coastal waters would become less and less saline, which in fact does not happen. Surface water moves obliquely away from the coast. Coastal currents transport it parallel to the shore, while there is a net flow seaward as part of a basic estuarinelike circulation. Drift bottles released about 30 kilometers from shore are rarely recovered from New Jersey or Long Island beaches, but are instead often picked up on European beaches, having floated thousands of kilometers across the Atlantic, probably carried in the Gulf Stream system.

River discharge peaks during spring; between late March and the end of May, about 50 percent of the annual total river discharge flows into the ocean. At this time, surface-water salinity off New York Harbor drops to $26^0/_{00}$, and the surface water salinity throughout the region falls to $32^0/_{00}$ or less. A combination of lowered salinity and warming (indicated in Fig. 12–14) forms a low-density surface layer (indicated in Fig. 12–15) so that the water column is now stratified. A shallow

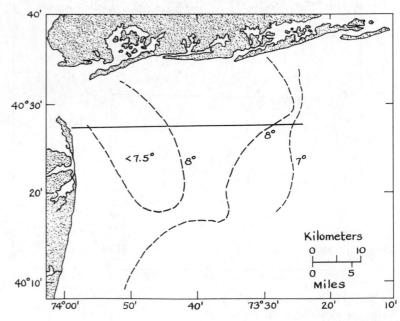

FIG.   12–14
Surface water temperatures off New York Harbor, April, 1948. (After Ketchum et al., 1951)

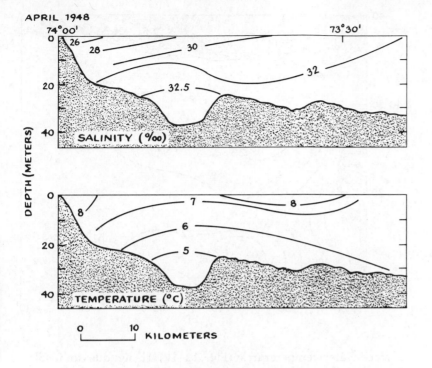

APRIL 1948

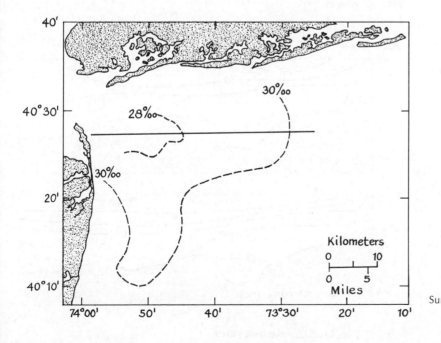

FIG. 12–15
Distribution of temperature and salinity at
depth in the New York Bight, April 1948.
Location of profile is shown in Fig. 12–13.
(After Ketchum et al., 1951)

plume of Hudson River effluent floats generally southward,
parallel to the New Jersey coast but a few kilometers offshore.
It is usually separated from the coast by a belt of more saline
water.

These stratified conditions persist into summer, surface water
becoming slightly less saline (as indicated in Fig. 12–16) and

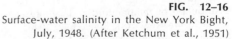

FIG.   12–16
Surface-water salinity in the New York Bight,
July, 1948. (After Ketchum et al., 1951)

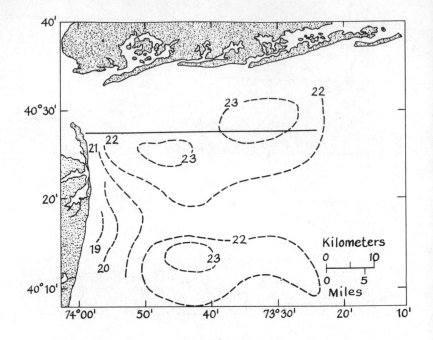

FIG. 12–17
Surface-water temperature in the New York Bight, July, 1948. (After Ketchum et al., 1951)

surface-water temperatures (Fig. 12–17) rising due to insolation. The layer of low-salinity water (Fig. 12–18) is usually about 10 to 20 meters thick, although it may locally contact the bottom, apparently as a result of wind-induced sinking (also known as downwelling) of coastal waters. As river flow diminishes during the summer and wind speed decreases, the plume of water coming from New York Harbor becomes less obvious and forms a series of ill-defined eddies or patches of water which move slowly over the continental shelf. Currents are sluggish at this time, in the absence of large river flow and strong winds.

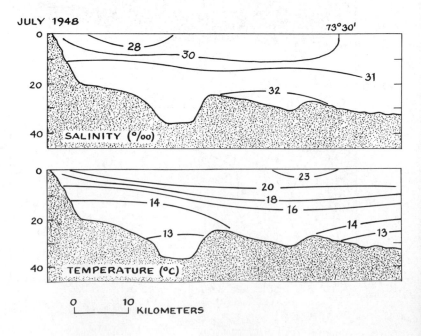

FIG. 12–18
Distribution of temperature and salinity with depth in the New York Bight, July, 1948. Location of profile is shown in Fig. 12–16. (After Ketchum et al., 1951)

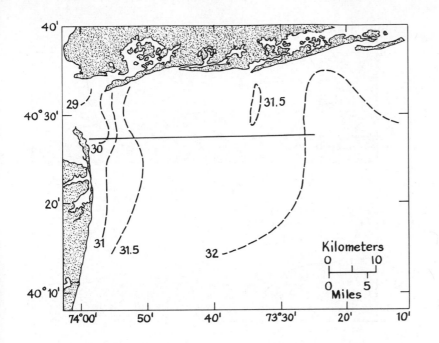

FIG.   12–19
Surface-water salinity in the New York Bight,
October, 1948. (After Ketchum et al., 1951)

During winter, stratification of the coastal water is broken
down as the surface waters are chilled (see Fig. 12–20), and
low salinity surface waters (see Fig. 12–19) are mixed to greater
depths by storm winds and seas. Eventually, isohalines and isotherms become nearly vertical and trend parallel to the coast;
the water becomes warmer and more saline as you go farther
from shore (as shown in Fig. 12–21). Through the winter, cooling is most intensive nearest the coast, where the water is most
shallow. There is some flow of warmer ocean water along the
bottom.

Although about 2½ years of river discharge would be needed

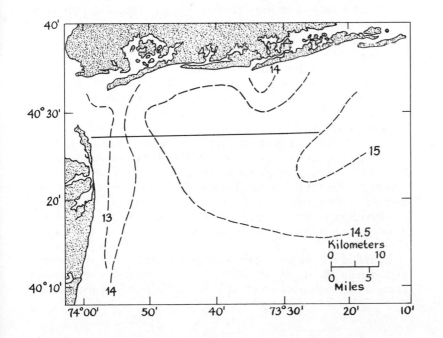

FIG.   12–20
Surface-water temperature in the New York
Bight, October, 1948. (After Ketchum et al.,
1951)

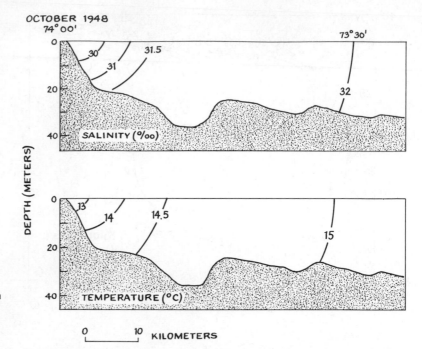

FIG. 12-21
Distribution of temperature and salinity with depth in the New York Bight, October, 1948. Location of profile is shown in Fig. 12-18. (After Ketchum et al., 1951)

to replace all the ocean water in the 2000 square kilometers of coastal ocean nearest New York Harbor, it requires generally only about 6 to 10 days for a tracer solution introduced into the river mouth to move through this part of the ocean. This results from the flow of water involved in the coastal currents. In general, variations in river flow have little demonstrable effect on the average rate of movement through the area; total water flow through this stretch of coastal ocean is anywhere from 50 to more than 200 times the incoming river flow. Wind-driven currents carry water from the region of Nantucket Shoals southward along the Long Island–New Jersey coast.

Within a few kilometers of New York Harbor, the flow near the bottom is generally landward. Elsewhere, the near-bottom flow tends to parallel the coast, moving slightly landward. On the edge of the continental shelf, net flow seems to be slightly seaward. Subsurface currents appear to be little affected by variability in winds or river runoff. Perhaps the overlying water acts as some sort of filter to average out these effects. It may also be that our data are too crude to show short-term variation in subsurface flows.

As the nation's largest city, and one of its oldest, New York and its harbor are subject to the types of contamination problems that plague most coastal areas. For centuries municipal wastes, waste chemicals, and waste heat have been discharged into the harbor, whence they flow into the coastal ocean. Hudson River water can be traced in the ocean by its high dissolved-iron concentrations, resulting in part from discharge of waste chemicals.

Many of the region's sewers were not connected to treatment plants even as late as the 1970's, so that the raw sewage from about 1 million people poured into local rivers and the harbor.

These wastes depleted waters of dissolved oxygen, in addition to contributing undesirable color and odor to the harbor waters. Many beaches and shellfish-producing areas have had to be closed because of high concentrations of sewage-associated bacteria.

In colonial times, these wastes were discharged in small quantities and could be diluted and dispersed by the harbor circulation. After a time the harbor could no longer accommodate the volume of wastes discharged, and the result was substantial deterioration in water quality, becoming quite noticeable early in the twentieth century.

Efforts to clean up the harbor suggest that the trend in water-quality deterioration may be halted, hopefully even reversed permanently. New, improved treatment plants and techniques promise to remove some of the untreated wastes from the harbor. In the past, waste material from the city and its harbor has been transported to the continental shelf a few kilometers outside the harbor entrance. In the mid-1960's, 5 million to 10 million tons of construction debris and dredged material from navigation channels were also deposited on the continental shelf off New York Harbor each year. This is no longer considered a satisfactory technique for management of urban wastes, for in the vicinity of some disposal sites many square kilometers of ocean bottom have been deprived of living organisms.

Even after the harbor's waters had undergone unmistakable deterioration, the volume of waste discharged to the coastal ocean could be accommodated with little or no noticeable change in coastal-water quality. The coastal ocean, however, has a circulation system that tends to retain many of the wastes discharged to it. These are then moved along the coast by currents. The coastal ocean, like the harbor, has a finite capacity to assimilate such materials. Should this capacity be exceeded in the future, parts of the coastal ocean may experience conditions similar to those in the harbor.

Regulation and control of waste discharges is an increasingly important branch of coastal-ocean management. Another aspect is development of techniques for control of large-scale accidents. Modern society uses a bewildering array of liquid fuels and chemicals in vast quantities. To achieve economies in handling and transport, these are commonly moved in large ships. In fact, a dominant trend in ship building since the mid-1950's has been the development of larger ships, especially supertankers, which unfortunately pose serious spillage problems for the coastal ocean. These spills may be chemically dispersed, or they may be concentrated and removed. The choice of method in a particular situation depends on the substance involved, the degree of risk to marine life or possible health hazards, and (of course) costs. Accurate and detailed knowledge of all aspects of the coastal ocean are needed to deal with such problems in a satisfactory way.

NORTHEAST PACIFIC COASTAL OCEAN

To understand the coastal circulation along the Pacific coast of the United States, we begin offshore with the *West Wind Drift*. Primarily controlled by prevailing winds, this current brings water from the Asiatic side of the Pacific toward the North American coast. Since high rainfall and large river runoff make the entire North Pacific Ocean an area of net dilution, surface waters have salinities of about 33–34⁰/₀₀ by the time they reach North America. Because of the high latitude, these surface waters are also relatively cool.

The presence of a low-salinity surface layer in the North Pacific causes an estuarinelike circulation north of about 40°N. As surface water flows toward the center, subsurface waters rise and are incorporated into the upper layers, which are thereby substantially increased in nutrients. Phytoplankton growth is inhibited during much of the year, however, because both the high latitude and extensive cloud cover limit insolation.

After flowing nearly due east across the Pacific, the West Wind Drift is deflected by the North American continent. Its waters flow more-or-less parallel to the continental margins, forming eastern boundary currents—the California Current setting southward, and the Alaska Current setting northward. The point of divergence shifts seasonally: it is about 50°N in summer and about 43°N in winter.

Since current speeds are low and flows rather diffuse, there is no sharp boundary separating the coastal-current system from the eastern boundary current, in contrast to the situation prevailing in the Gulf Stream system along the Atlantic coast. We must, therefore, choose some criterion by which to separate the boundary current from the coastal circulation. Along the Pacific Northwest coast of the United States, the 32.5⁰/₀₀ isohaline is taken to separate the low-salinity coastal waters from the California Current offshore. (The 32.0⁰/₀₀ isohaline is sometimes used instead.)

During winter, locally the rainy season, precipitation along the coast and runoff from numerous coastal rivers causes an almost continuous band of low-salinity water along the coast. This can be traced from the central Oregon coast to the Strait of Juan de Fuca. At this time of year it is difficult, if not impossible, to distinguish the water contributed by a single river.

During summer and autumn, the Columbia River is the largest source of fresh water along the entire west coast of the United States. Relatively little mixing occurs in the estuary of the Columbia, so that its waters are still rather fresh as they enter the ocean. The effluent forms a distinctive water mass— the Columbia River plume, having a salinity of less than 32.5⁰/₀₀. Within 50 kilometers of the river mouth the plume is mixed extensively by tidal currents and waves; farther away, currents and regional winds move it parallel to the coast much as a column of smoke is moved in the air by gentle winds.

After entering the ocean, the plume of low-salinity water has little direct influence on surface circulation. During the winter, it is held near the coast by southwesterly winds, forming a band of northward-moving, low-salinity water derived from all the coastal rivers—the Davidson Current. This can be recognized from northern California to the Strait of Juan de Fuca.

In late summer (July to October), the Columbia River discharge forms a low-salinity plume extending southwestward

FIG. 12–22

Distribution of surface salinity in September, 1963. Note the large area covered by surface water with salinity less than 32⁰/₀₀, derived primarily from the Columbia River. [After T. F. Budinger, L. K. Coachman, and A. C. Barnes, *Columbia River Effluent in the Northeast Pacific Ocean, 1961, 1962: Selected Aspects of Physical Oceanography*, University of Washington, Dept. of Oceanography Tech. Rept. no. 99 (ref. M63-18) (1964), 78 pp.)

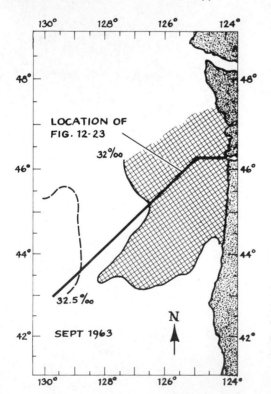

from the river mouth (see Fig. 12–22) as a layer about 30 meters thick (as shown in Fig. 12–23). In October, 1961, for instance, the Columbia River plume covered an area of about 200,000 square kilometers, extending southward about 800 kilometers from its mouth. Following major summer floods, the plume may be even larger.

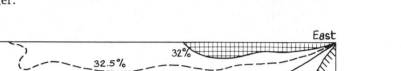

FIG. 12–23
Distribution of salinity with depth near the Columbia River mouth, September, 1963. Location of profile is shown in Fig. 12–22. Waters with salinities less than 32⁰/₀₀ form a relatively thin layer (less than 30 meters thick).

In the coastal eastern North Pacific, winds are generally rather weak during late summer and autumn. Because of the relatively slow mixing of surface layers, the low-salinity plume formed early in the season persists until the first severe storm, usually in October. After that, surface waters are rapidly mixed and the plume can be recognized only near the river mouth (as indicated in Fig. 12–24). The low-salinity water forms a layer (indicated in Fig. 12–25) held near the coast by winds from the southwest.

Over the continental shelf, the landward flow of subhalocline waters (part of the estuarine-like circulation) produces near-bottom currents; these currents are studied using *seabed drifters* —plastic devices (usually brightly colored for ease in spotting) designed to float within a few meters of the ocean bottom. Drifters are dropped into the ocean from ships or low-flying aircraft, in groups of five held together by a "collar" made of rock salt, which dissolves after its weight has dragged the group of drifters to the bottom. Position and time of release of each group is noted, as well as the serial number of each drifter.

Drifters are recovered by beachcombers or by fishermen dragging nets along the bottom. For a small reward, the finder sends the date and location of recovery, with the serial number of the drifter, to the laboratory that released it. The apparent direction and distance traveled by the drifter is calculated by comparing the release and recovery points. The apparent speed is calculated by dividing the apparent distance by the length of elapsed time between release and recovery. This does not tell us the details of the path followed, nor the true speed of drifter movement; it indicates only the net movement of near-bottom currents in which the drifter floated.

Recovery of seabed drifters released on the Washington–Oregon coast continental shelf indicates a net northerly movement of near-bottom waters at depths exceeding 50 meters. In waters less than 50 meters deep, near-bottom currents move toward the coast. This appears to be a result of near-bottom

FIG. 12–24
Distribution of surface salinity in the Pacific Northwest, January–February, 1962. Note that low-salinity surface waters form a narrow band along the coast. [After A. C. Duxbury, B. A. Morse, and N. McGary, *The Columbia River Effluent and its Distribution at Sea, 1961–1963*, University of Washington, Dept. of Oceanography Tech. Rept. no. 156 (ref. M66-31) (1966), 105 pp.]

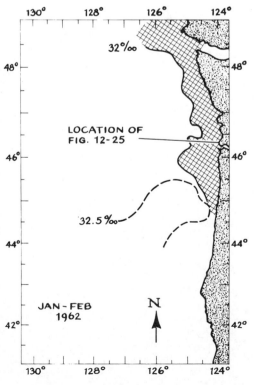

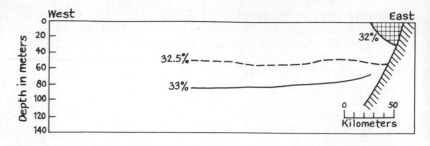

FIG. 12–25
Distribution of salinity with depth near the mouth of the Columbia River, January–February, 1962. Note that the low-salinity waters form a wedge rather than a layer. Location of the profile is shown in Fig. 12–23. (After Duxbury et al., 1966).

water movements and coastal wind-induced upwelling in the summer (shown schematically in Fig. 12–26).

Within about 10 kilometers of the Columbia River there is a net movement of near-bottom waters toward the river mouth, caused by the estuarine circulation. A similar near-bottom circulation pattern is observed within 70 kilometers of the Strait of Juan de Fuca. The fresh-water discharge into the latter system is somewhat less than the discharge of the Columbia River, but the flow of subhalocline water into the Strait of Juan de Fuca is from 5 to 20 times greater than the corresponding inflow at the Columbia River mouth. This is because there is more mixing in the large, silled Strait of Juan de Fuca system. More seawater is carried out in the surface layers for each volume of fresh water entering the estuary, and therefore more seawater must come in to replace it. Strong currents are set up by the flow of large amounts of water through the strait, and the associated near-bottom currents affect a large part of the continental shelf.

In the Columbia River system, the relatively large amounts of fresh water discharged through the narrow, deep river mouth preclude much mixing in the estuary itself. Most of the initial mixing occurs over a relatively large area outside the estuary, setting up a rather ill-defined estuarine circulation. Near-bottom currents associated with this weak circulation are thus observed within only about 10 kilometers of the river mouth.

Columbia River discharge affects phytoplankton production near the river mouth. In the surface waters offshore, depletion of nutrients and insufficient light during part of the year limit primary productivity to about 60 grams of carbon per square meter per year. Near the river mouth, however, primary productivity of carbon in surface waters increases to about 90 g/m²/yr, owing to locally increased supplies of nutrient-rich subhalocline water. Near the Strait of Juan de Fuca, on the other hand, productivity shows no dramatic rise, because nutrient-rich subhalocline waters come to the surface within the estuary itself rather than at the entrance to the Strait.

Despite low population densities along the Washington–Oregon coast, the harbors and the coastal ocean itself have been

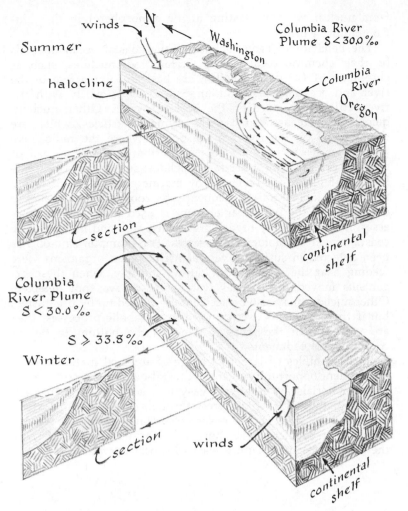

winds N
Summer   Washington   Columbia River Plume S < 30.0‰
halocline   Columbia River
Oregon
section
continental shelf

Columbia River Plume S < 30.0‰
S ⩾ 33.8‰
Winter
section   winds
continental shelf

FIG. 12–26
Block diagrams showing generalized surface and subsurface circulation in the Pacific Northwest region, near the mouth of the Columbia River in winter and summer. Compare the surface and subsurface currents for the two seasons. Also note the effects of upwelling along the Oregon coast in summer and the downwelling in winter.

subject to contamination. Waste chemicals and paper mill wastes in various estuaries cause locally severe pollution problems. Indeed, marine life in some of the harbors has been nearly destroyed, for short periods, by large accidental releases of wastes into estuarine waters.

This particular coastal area is unique because of the large volume of radioactive wastes that it has received. From the 1940's to the early 1970's, plutonium-producing reactors on the Columbia River about 600 kilometers upstream from the river mouth released large volumes of slightly radioactive waters to the river. These radionuclides escape primarily through corrosion in the pipes used to cool the reactors. The warm and radioactive cooling waters are discharged to the river. (Modern reactors recirculate the cooling waters, using the heat for electrical-power generation.)

Large volumes of radioactive wastes have thus been carried by the river to be discharged into the coastal ocean. In 1957, these waste discharges accounted for more than 95 percent of all radioactivity released to the environment (excluding fallout from nuclear weapons testing in the atmosphere) in the non-Communist world.

The behavior of radionuclides in the coastal ocean depends on their chemical composition. Certain radionuclides, such as chromium-51 (28-day half-life), are basically unreactive in the river and ocean; thus chromium-51 moves primarily with the river effluent as part of the surface water. Other nuclides quickly become associated with sediment particles, which are then carried by the river to the ocean, where the radioactive particles move in response to near-bottom currents instead of surface currents. Still other radionuclides, such as zinc-65 (half-life of 245 days), are taken up by marine organisms—probably first by phytoplankton, subsequently by zooplankton and higher organisms. If the particular element is used in metabolic processes, it is retained by organisms at each step and moves in the coastal ocean by biological processes. For example, zinc-65 has been detected in surface layers at night, when organisms were feeding near the surface; but during the day, when these organisms moved into deeper water, zinc-65 was not detected. Other nuclides may also be accidentally picked up by organisms; but if these elements are not used in metabolic processes, they are voided with the feces, sinking to the bottom to be incorporated in sediments.

Such examples of large discharge of unusual materials offer an excellent opportunity to study coastal-ocean processes: radioactive materials are relatively easy to trace, and we can learn much by studying their progress through ocean systems. Many more studies of how the ocean assimilates materials will be necessary in predicting the short- and long-term effects of using the coastal ocean for future waste-disposal operations.

SUMMARY OUTLINE

Coastal ocean—lies over continental shelf, average 70 m deep
    Influenced by boundaries—shallow bottom and continents
    Winds and tides have major effect
    Processes act over shorter time and distances than in open ocean

Temperature and salinity
    Greatest variations occur in coastal ocean
    Salinity changed by river runoff and by evaporation caused by winds blowing from continents
    General estuarine-circulation system in areas of excess runoff
    Seasonal temperature cycle greatest in mid-latitudes

Coastal currents—generally parallel coastline
    Strongly affected by winds, seasonal variations common
    River runoff also influences coastal circulation
    Winds blow surface waters away from coast causing upwelling
        Cold, nutrient-rich water from about 200 m brought to surface

Storm surges—strong winds move waters toward coast
    Creates standing wave in some basins—e.g., North Sea
    Surge is preceded by forerunner and followed by pronounced oscillations of water level

Tide effect on river discharge—discharge of coastal river controlled by tides

Minimum river discharge on rising tide, flow reversed on flood tide

Maximum river discharge on falling tide, ebb-tidal current

Tide rip forms at point where water masses meet

Forms sharply bounded water parcels, which move as cloudlike bodies through coastal-ocean areas

Continued mixing blurs outlines of water parcels

Oceanic effects on land climates

Maritime climates—restricted temperature range, relatively wet climate

Continental climate—wide temperature range, relatively dry climate

Coastal-oceanic regions—characterized by similar topographic and environmental conditions

New York Bight

Part of coastal ocean lying between Long Island and the entrance to Delaware Bay

Seasonally variable currents, strongest when winds and river discharge are greatest

Currents most sluggish during late summer under light winds and little river discharge

Waters strongly stratified in summer, thoroughly mixed in winter

River discharge forms low-salinity layer a few tens of meters thick

Northeast Pacific Coastal Ocean

West Wind Drift diverges at coast forming southward-setting California Current and northward-setting Alaska Current

Indistinct boundary separates California Current from coastal circulation

Area of excess precipitation, therefore surface-water salinities are low, less than $34^0/_{00}$

Coastal circulation most readily delineated by $32.5^0/_{00}$ isohaline

Columbia River discharge can be readily identified in late summer as a plume of low-salinity water moving southwestward

Near-bottom currents set northward along continental shelf

Upwelling occurs as result of winds blowing from north—influences concentrations of nutrients and temperatures of surface waters

SELECTED REFERENCES

IPPEN, A. T., ed. 1966. *Estuary and Coastline Hydrodynamics*. McGraw-Hill, New York. 744 pp. Advanced mathematical treatment of estuarine and coastal-ocean processes.

LAUFF, G. H., ed. 1967. *Estuaries*. American Association for the Advancement of Science, Washington, D.C. Publ. 83. 757 pp. Collection of technical papers on many aspects of estuaries and coastal-ocean processes, emphasizing biology.

PICKARD, G. L. 1964. *Descriptive Physical Oceanography: an Introduction*. Pergamon Press, Elmsford, New York. 199 pp. Elementary treatment of general oceanography, including coastal processes (Chapter 8).

U.S. DEPARTMENT OF THE INTERIOR, FISH AND WILDLIFE SERVICE. 1970. *The National Estuary Study*. U.S. Government Printing Office, Washington, D.C. 7 volumes. Comprehensive study of U.S. estuaries.

U.S. DEPARTMENT OF THE INTERIOR, OFFICE OF THE SECRETARY. 1970. *The National Estuarine Pollution Survey*. U.S. Government Printing Office, Washington, D.C. Senate document 91-58, 91st Congress, 2nd session. 633 pp. Survey of pollution problems in U.S. estuaries.

WIEGEL, R. L. 1964. *Oceanographical Engineering*. Prentice-Hall, Inc., Englewood Cliffs, N.J. 532 pp. Advanced treatment of engineering practice in coastal areas.

SHORELINES AND SHORELINE PROCESSES

Plunging breaker at Fire Island National Seashore, New York.

thirteen

The *shoreline*—the line where land, air, and sea meet—is the most dynamic part of the coastal zone. Here the continents respond to the same forces that affect the coastal ocean: tides, winds, waves, changing sea level, and man.

The *shore*, as typically shown in Fig. 13–1, extends from the lowest tide level to the highest point on land reached by wave-transported sand. It is affected by the aforementioned forces each day. The broad zone (known as the *coast* or *coastal zone*), directly landward from the shore is also strongly affected by the processes that shape the shore. Usually these forces act on the coastal zone only under extreme conditions, such as storms or exceptionally high tides. Violent storms or exceptional tides can be expected to strike most coasts every 10 to 100 years. For example, southern New England has been hit by destructive hurricanes in 1635, 1815, 1938, and 1954. The intervals involved are short periods in the history of a continent, and storms of this intensity have a major effect on the coastal zone.

Processes affecting shores act on time scales ranging from a few seconds to tens of thousands of years. Let us examine some of the more obvious forces, beginning with waves which break, rush up the beach face, and then retreat in a matter of seconds. Along open coasts there are few days without some wave action to move materials on the beach face, dissipating the energy originally imparted by winds at sea.

The daily rise and fall of sea level under the influence of the tides submerges or exposes parts of the beach every 6 or 12 hours. To a great extent this controls the type and vertical extent of the beach, as well as the types of plants and animals that live there. Tidal currents are important in inlets and coastal embayments, but have relatively little influence on open, straight coasts.

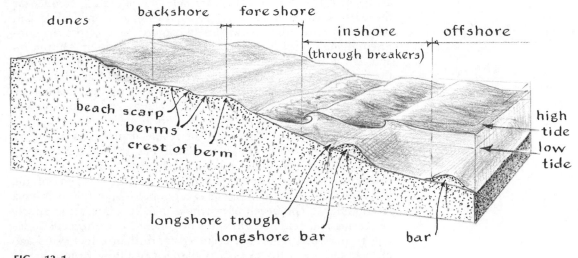

FIG. 13–1
Profile of a typical beach and adjacent coastal zone

Storm surges—relatively sudden large changes in sea level resulting from strong winds blowing across a shallow and partially enclosed body of water—also cause large changes in the shoreline and the coastal zone landward because of the flooding of normally protected areas. Although major storm surges may come only every few 10 or 100 years, the changes they cause often remain visible for decades. For instance, some inlets on the south shore of Long Island broke through barrier islands during hurricanes in the 1930's changing previously isolated and brackish water ponds and lakes into lagoons and estuaries which are now strongly affected by the tides. The inlets are now kept open at the present time by man-made jetties and continued dredging, in order to prevent accumulations of sediment from filling it in and forming solid barrier beach once more.

Other processes controlling the shoreline act so slowly that their effects are difficult to detect over a human lifetime, or even in historical records. Such changes occur over periods ranging from 100 to 10,000 years; even areas with the longest historical records have little reliable information about coasts and their features that predate the classical historians, who wrote about 2000 years ago. To detect and interpret long-term changes it is necessary to rely on the record preserved, often indistinctly, in elevated beaches such as those occurring on the west coast of the United States, an area which is generally rising because of continuing mountain building, or in Scandinavia, where the continent is still rising because of the unloading of the crust caused by melting of the continental ice sheets over the past 20,000 years. In parts of the Mediterranean area, some ancient Roman port cities are now submerged as a consequence of the sinking of the crust, also under the influence of large-scale crustal movements; others have been elevated.

Perhaps the most profound influence on coasts as we now see them was the repeated advance and retreat of the continental glaciers and resulting sea-level changes. When sea level stood at its lowest, about 20,000 years ago, the shoreline stood near the present continental-shelf break. As the glaciers waned and water was released, the shoreline advanced back across the continental shelf until it reached its present level about 3000 years ago. If the remaining ice sheets (Antarctica, Greenland) melt in the future, the sea surface may eventually stand as high as 50 meters above its present level. The shoreline we now see will be submerged, and a complex of beach ridges and lagoon or estuarine deposits will mark its former location.

Thus the present shoreline is the result of processes acting on different time scales. The beach itself has the shortest "memory," recording distinctly only what happened when the most recent waves rushed up and then retreated, or the level of the last high tide; perhaps a recent exceptionally high tide will have left its mark. The beach as a whole usually records the effects of storms and seasonal changes. In general, it is difficult to detect records of events that occurred more than a few years ago on a beach; even the extensive changes resulting from a storm or storm surge in winter are usually quickly obliterated during the following summer unless man intrudes by dredging inlets or building seawalls, groins, or jetties in order to inhibit erosion and reduce sand movements.

COASTLINES

Geomorphologists—those who study land forms—have classified coastlines according to certain common features and have attempted to determine the dominant forces that have created them. Of the coastlines of the world, shown in Fig. 13–2, most exhibit signs of the recent rise of sea level, modified by wave action and by deposition of sand and gravel to form beaches, or mud to form deltas and marshes.

One major type of coastline, called a *secondary coast*, is formed primarily by terrestrial forces; the ocean has not been at its present level long enough to reshape the land features along its margin. An example of such a coastline is seen in the drowned river valleys of the United States' east coast, where the former valley of the Delaware River is now Delaware Bay, and the former valley of the Susquehanna River and its lower tributaries forms Chesapeake Bay. The original shapes of these valleys are virtually intact, although largely underwater and somewhat modified by recent sediment deposition. In glacially carved regions, often mountainous, movements of the great ice sheets created deep valleys with U-shaped cross-sections. Now filled by seawater, the U-shaped bottoms are concealed, but the valleys cutting back into mountains form the typical Norwegian coastal fjords. Fjords also occur in Canada (see Fig. 11–2), Alaska, and southern Chile.

Not all processes acting on the coastline have had the effect of cutting away land surface; other forces were active in leaving behind deposits which now dominate the coastline. Long Island, in New York State, is part of the terminal moraines formed in two periods of Pleistocene glaciation. The moraine is a deposit of sand and gravel originally scoured from the continental surface, carried toward lower latitudes, and left behind at the terminus of the glacier when it melted and retreated. The delta at the mouth of a sediment-laden river is another example of coastline modification caused by deposition of material eroded from the land.

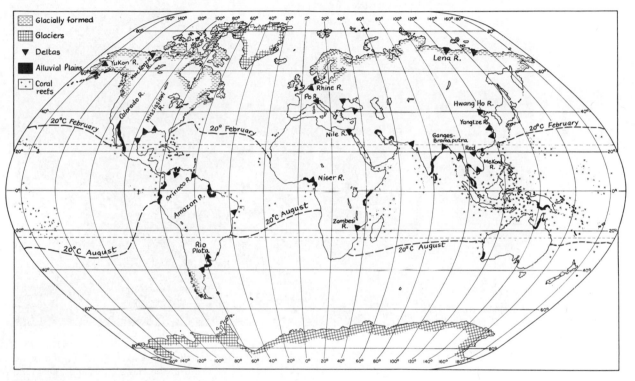

FIG. 13–2
Coastlines of the world [After J. T. McGill, "Map of Coastal Landforms of the World," *Geographical Review*, 48:402 (1958)]

Spectacular but rather rare coastline features are formed by volcanoes. Good examples are found in the main Hawaiian islands where, as shown in Fig. 13–3, volcanic cones and lavas come directly down to sea level. In other areas, the volcanoes have been worn down to sea level, and in still other areas the sea has broken into the top of the volcano, producing a round bay, usually steep-sided and deep, making an excellent harbor. The harbor at Pago Pago (American Samoa) is thought to have such an origin.

A *primary coast* is shaped by marine processes or by marine organisms. Such coastlines are often cut in rocks or sediments soft enough so that even the limited time of the present sea-level stand has been sufficient to cut back the coast. Where bluffs of unconsolidated sand and gravel rise above the shore, they are often cut back by waves to provide a wavestraightened coastline, as illustrated in Fig. 13–4. The materials derived from the erosion of these bluffs are deposited near shore or form small beaches or *baymouth bars* across bays and inlets near the bluffs. Sand moves along the coast by the action of longshore currents, forming *barrier islands* and *spits* that separate such inlets and bays from the open ocean.

FIG.   13–3
The coastline at Waikiki Beach on the island of Oahu in the Hawaiian Islands. Diamond Head in the background is a remnant of an extinct volcano. The original volcanic features of this island have been altered by erosion and by man as demonstrated in the large amount of dredging visible in the yacht basins in the foreground. (Photograph courtesy Hawaii Visitors Bureau)

DELTAS

The sediment load carried downstream by most rivers and streams is either deposited in the estuary or at the river mouth. For most streams, the only sign of their sediment load near the river mouth is a small shifting sand bar which is moved by the tides. The bulk of this sediment is eventually resuspended by wave action and moved along the coast by longshore or tidal currents, to be carried onto nearby beaches or deposited on the continental shelf.

Some rivers, however, carry far more sediment than can be

dispersed along the adjacent coast. In this case, sediment usually settles at the river mouth, forming a delta. This is a fairly common occurrence along coastlines; well-known examples include the Mississippi (as shown in Fig. 13–5), Rio Grande, Frazer, Nile, Tigris–Euphrates, and Rhine deltas. Delta lands are flat, well-watered, and often exceedingly fertile because sediment deposition during flood periods provides a continuous supply of rich soil. Many primitive agricultural societies have developed and flourished on river deltas, the Egyptian culture on the Nile being a familiar example. Lower Mesopotamia, site of the early Sumerian civilization, was an ancient delta, and all over the world deltas are still valued as farmlands. (Their low elevation above sea level makes them vulnerable to flooding by storm surges.)

The relative capacity of the coastal ocean to transport sediment is determined by the tidal range and associated strength of tidal currents, the wave energy dissipated on the coast, and the strength of the longshore currents. The strong tidal currents set up by a large tidal range cause scouring of recently deposited sediment. Waves act to resuspend sediment in the surf zone; they also cause *longshore currents*, which move sediment along the coast in a direction determined by prevailing wave approach.

The Columbia River on the west coast of the United States, discharging between 10 and 30 million tons of sediment per year into the ocean, lacks a delta due to its relatively large tidal range and strong waves at its mouth. On the other hand, the nearby Frazer River, with approximately the same sediment load, has formed a delta in the protected waters of the Strait of Georgia (British Columbia) because strong wave action is inhibited there by the limited fetch for wave generation. The Mississippi River has formed the largest delta in the United States as a consequence of its large sediment load (about 300 million tons of sediment per year) and the low tidal range in the Gulf of Mexico.

Some river systems supply so much sediment that a delta forms in spite of a large tidal range and extensive wave action. The region where the Ganges–Brahmaputra Rivers in India discharge at the head of the Bay of Bengal, shown in Fig. 13–6, is an example. These two rivers, which reach the ocean through the same delta complex, carry about 700 million tons of sediment each year; a large delta has formed as a result of its deposition.

The first step in delta formation, illustrated in Fig. 13–7, is for a stream to fill its estuary with sediment. The Mississippi, for instance, probably filled its estuary soon after sea level reached its present position. Rivers with large estuaries and relatively moderate sediment loads, such as the Amazon River, are still filling their estuaries. Until the estuary is completely filled, little or no sediment escapes to be deposited at the river mouth or to form a delta.

As a delta forms, the river builds channels across it called *distributaries*, as shown in Fig. 13–8, through which the water flows on its way to the ocean. There are often a series of these channels, complexly interconnected; they extend across the delta as long, radiating, and often branching fingers. An active distributary continually builds its mouth further seaward until the distance to the sea is so great that the river can no longer main-

FIG. 13–4
The coastline on the northern shore of Long Island has been modified by wave erosion and sediment transport. Sand and gravel eroded from bluffs at the points (sharp bends in the coastline) move along the coast and smooth original irregularities. In the left center, a small inlet connects a salt marsh (known as Flax Pond) to Long Island Sound. Sand and gravel beaches separate the marsh from the sound. On the right side of the photograph, a relatively large, shallow harbor has been created by a long beach which separates submerged former valleys from the sound. (Official U.S. Coast and Geodetic Survey photograph)

tain flow through that channel. At this time the river shifts course, often during flood, so that the flow proceeds through a different set of distributaries to reach the ocean, and the whole process begins again. The Mississippi River Delta reveals a series of abandoned distributaries, each with its own subdelta, forming a complex *lobate delta* (see Fig. 13–5). Distributary abandonment is not always sudden, but may occur gradually as one channel becomes too shallow to carry a large amount of water and another gradually receives more of the flow so that it becomes enlarged.

Many deltas occur in areas where the land is subsiding. For

FIG. 13–5

Mississippi River Delta and its growth between 1874 and 1940. [After F. P. Shepard, *Delta-front Valleys Bordering the Mississippi Distributaries,* Geological Society of America Bulletin 66(12) (1955), 1489–98]

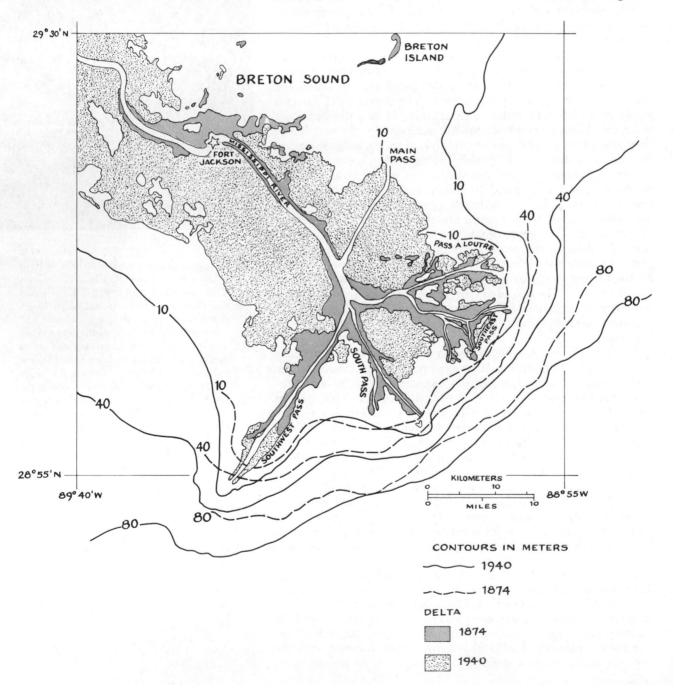

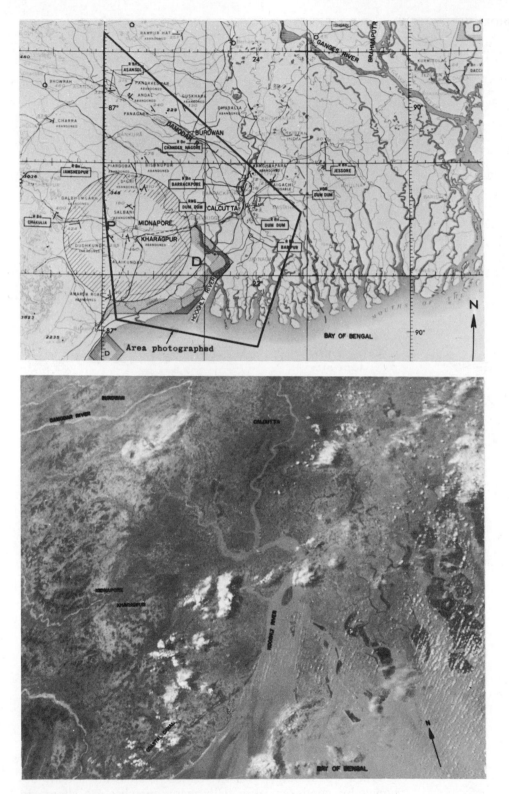

FIG. 13–6
A portion of the delta at the mouth of
Ganges–Brahmaputra Rivers as photographed
from a satellite. (Photograph courtesy NASA)

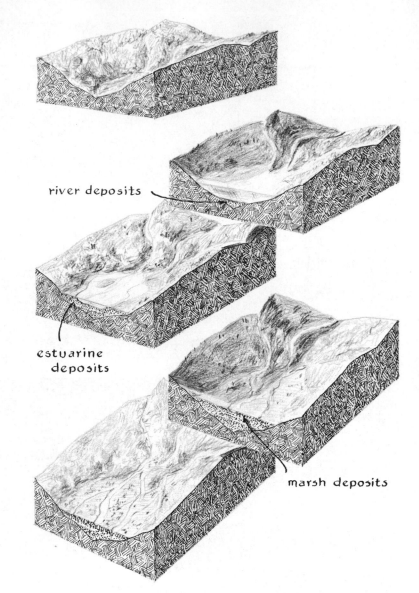

river deposits

estuarine
deposits

marsh deposits

FIG. 13–7
Stages in the development and filling of an
estuary by sediment deposits, converting a
river valley (top) to a salt marsh (bottom).

example, the Mississippi Delta region as a whole is subsiding at
a rate of about 1 to 4 centimeters per year. Furthermore, the
sediment beneath the delta continually compacts and expels
water from the sediment. Consequently, the delta surface is continually subsiding, except in those areas currently receiving
sediment.

When a distributary is active, adjacent areas are usually occupied by ponds or lakes with many marshes, as shown in
Fig. 13–9. Areas between distributaries receive less sediment,
and the deposits there tend to be finer-grained and contain

bulky plant debris (peat). These deposits compact (and subside) more than the coarse-grained material along the distributary banks. The banks, therefore, subside less than the adjacent marsh and eventually stand higher, forming natural levees on which towns and roads are built.

As the marsh subsides below sea level, it is transformed into

FIG. 13–8
A portion of the modern delta of the Mississippi River is pictured here looking south from the "Head of Passes," where one distributary separates into three. The left-hand pass is a natural artery; the other two are kept open for navigation by dredging, and by construction of jetties to direct the river flow—thus increasing its velocity so that less sediment is deposited in the channels. (Photograph courtesy U.S. Army Corps of Engineers, New Orleans District)

a shallow bay. Seawater in these bays is diluted by ground water discharged through the delta, so that the bay is usually brackish. These areas, even though their appearance to an outsider may be unprepossessing, are highly productive of marine life. About 90 percent of commercially important coastal fish and game fish are dependent on these bays or similar marsh areas during some critical stage of their life cycle.

Deltas are rarely formed by rivers draining areas that were covered by ice during the last glacial period. Lakes, gouged out by ice, are common in these areas, and act as traps to prevent the escape of river-borne sediment to the sea. An example is the

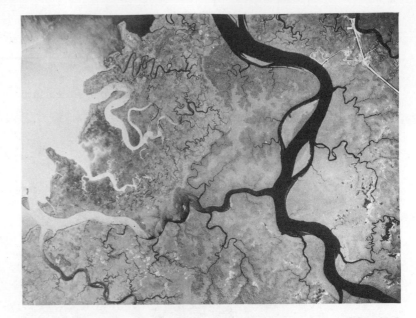

FIG.   13–9
Distributaries of the Suwannee River, Florida, showing the tidal marsh and cypress swamps lying between them. Note the meandering tidal creeks in the marsh. A road and dredged channel are conspicuous in the upper-right portion of the photograph. (Official U.S. Geological Survey photograph)

St. Lawrence River, which drains an area where the Great Lakes act as effective sediment traps. After lakes in the river's drainage basin and the estuary at the river's mouth are filled with sediment, the stream may then be able to transport enough sediment to the ocean to build a delta.

The shape of a delta is controlled in a complex way by the balance between erosion and sediment transport along the coast, on the one hand, and the rate of sediment supply on the other. Where coastal processes predominate, the margin of the delta tends to be rounded, as in the Niger Delta, as shown in Fig. 13–10. Where the sediment supply from the river exceeds the capacity of the coastal ocean to transport it, the delta shape is primarily controlled by river processes rather than coastal processes. The Mississippi Delta's *"bird-foot"* shape results from each distributary having built seaward, with flooding between the distributaries, as previously discussed.

BEACHES

Beaches are probably the most familiar shoreline features. Sand beaches, barrier islands, and bays border the Atlantic coast of the United States from Long Island, New York, to Key West, Florida; in the Gulf Coast region, barrier islands and lagoons are also a characteristic feature. On mountainous coasts, beaches are usually less extensive, being generally restricted to low-

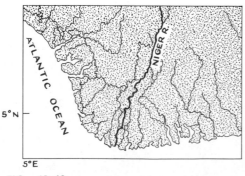

FIG.   13–10
Delta of the Niger River on the western coast of Africa.

lying areas between rocky headlands (as illustrated in Fig. 13–11).

In a very real sense, a *beach* is a sediment deposit in motion. At any time the motion of grains in the surf may be obvious, while the rest of the beach appears quite stable. But a visit to a beach following a major storm will show substantial changes, demonstrating that large segments of the beach move fairly frequently. Higher or more protected parts may move only during exceptionally powerful storms; nonetheless, the whole beach does move, and has justly been called a "river of sand."

To most of us, the word "beach" brings to mind a sand beach composed of grains with diameters between 0.062 and 2 millimeters. On mid-latitude coasts, beach sands are usually derived from chemical alteration and mechanical breakdown of silicate rocks. The mineral quartz is a common constituent of such beaches. In fact, some beach sands are so pure that they are mined to obtain nearly pure quartz sand for making glass. In tropical and subtropical areas where silicate rocks are rare or absent, beach sands come from broken carbonate shells and skeletons of marine organisms. Such beaches are often white or slightly pink—as in the Bermuda Islands, where the broken shells of a red *foraminiferan (Homotrema rubrum)* color the sands. In the Hawaiian Islands, all of the famous black-sand beaches are formed from recently erupted lava (volcanic glass) that make up the islands.

Not all beaches are composed of sand. Where wave and current action is especially vigorous, sand may be washed away

FIG. 13–11
Trunk Bay, a sheltered cove on St. John, Virgin Islands, a beach between rocky headlands. (Photograph courtesy U.S. Virgin Islands)

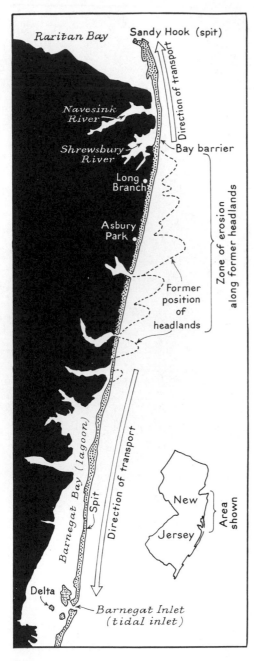

FIG. 13–12
Several types of barrier beaches occur along the New Jersey coast. The beaches are formed by sand moving north and sand from an area of headlands in the vicinity of Long Branch and Asbury Park. [After D. L. Leet and S. Judson, *Physical geology*, 4th ed., Prentice-Hall, Inc., Englewood Cliffs, N.J. (1971), 704 pp.]

faster than it is being brought in, leaving behind gravel or even *cobbles*. Still other beaches consist of mixtures of gravel and sand where wave action is not strong enough to completely remove the sand. Sand–gravel beaches occur on the north shore of Long Island, in New England, and on many Pacific coast beaches.

In general, beaches are accumulations of materials that are locally abundant and which are not immediately removed by waves, tidal currents, or winds. In areas like Long Island, beach sands and gravels are derived from erosion of glacial deposits, originally containing unsorted gravels, sands, and clays. Only the gravels and sands remain on the beaches; silt and clay-sized particles are usually washed out of beach areas by even weak waves or tidal currents. Fine-grained sediments tend to accumulate in areas with little wave action or tidal currents, either on the continental shelf at depths below about 30 meters or in lagoons, bays, or tidal marshes.

Most beaches consist of fine sand and tend to slope gently, with a hard-packed *foreshore* (the zone between high and low water). As the average grain size of the material forming the beach increases, the slope of the beach also increases. Beaches composed of fine sand ($\frac{1}{8}$ to $\frac{1}{4}$ millimeter) have average beach-face slopes of about 3°. Pebble (4 to 64 mm) beaches typically slope about 15° and cobble (64 to 256 mm) beaches typically slope about 24°. Even on a single beach, one can see the slope of the land increase as the grain diameter increases. Changes are easily observed as you go from areas with weak currents, where fine sands can accumulate, to areas where stronger currents remove fine sands, leaving gravels and a more steeply sloping beach. Fine-sand beaches are usually hard-packed, permitting easy walking and even the driving of cars on them. Coarse-sand or gravel beaches are much looser, because the larger grains do not pack as compactly as fine sands.

Beaches typically form near a major sediment source—at the base of a cliff or near a river mouth. Sediment is moved onto beaches by waves and currents, replacing materials either moved out into deeper water or transported along the coast. On the west coast, the Columbia River is a conspicuous example; near its mouth are some of the largest beaches and sand dunes on the Washington–Oregon coast. In most other coastal areas, including much of southern California, beaches tend to be small and located near the rivers which supply their sand.

On the North Atlantic coast of the United States, virtually no river-borne sediment escapes the many large estuaries to enter the ocean. Hence many Atlantic coast beaches are formed either from erosion of nearby cliffs or from sands deposited offshore during a time of lower sea level. In many cases a beach forms where transport of sand along the coast is interrupted by an obstruction such as a headland downstream.

Not all beaches are located on coastal plains. If the land along a coast is low-lying and slopes gently toward the ocean, sand deposits will probably form parallel to the coast and a short distance offshore. If submerged these are called *longshore bars*; but where large enough they form *barrier islands* which typically have a shallow lagoon or bay between the island and the mainland. Such *barrier beaches*, formed by the onshore movements of sands and the longshore movements of currents, are probably the most common type of beach occurring along the low-lying coastlines of the world whose locations are shown in Fig. 13–12. When connected to the mainland, usually at some *headland* (a point of land that juts out from the coast), they are known as *barrier spits*. At small indentations a barrier of sediment may build completely across the mouth of a bay, in which case the barrier is known as a *baymouth bar*. Along rocky coasts, offshore rocks and stacks are sometimes connected to the mainland by beach complexes called *tombolos*. A wide variety of beaches can form within even a single bay.

Two theories have been advanced to explain why barrier beaches form where they do. One is that they develop where there is a base of ancient, submerged sediment. The other is that sand moving past headlands tends to be deposited rather than being moved further downcurrent, forming a series of spits that bridge the mouths of bays in series along a coast, as shown in Fig. 13–13. In any case, barrier bars result from a dynamic balance between longshore currents, wave attack, sediment

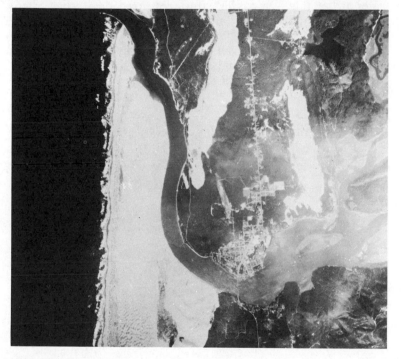

FIG. 13–13
A baymouth bar formed by longshore currents moving sediment northward along the Oregon coast at the mouth of the Siuslaw River. The light-colored spit is an area of active dune migration with no vegetation. Darker areas on the spit are depressions between dunes; some have small lakes in them. Dark-colored areas landward of the spit are heavily vegetated, and sand there is completely stabilized and not readily moved by wind. (Official U.S. Geological Survey photograph)

supply, and bottom topography. They tend to be long and straight, broken at intervals by inlets through which water is borne in and out of the bay or lagoon by tidal currents. The force of these currents keeps the inlets open; thus they act in opposition to the longshore currents, which tend to deposit sediment across the channel.

The barrier beach, like the beach on a coastal plain, is normally in a state of being washed away and replaced at a fairly constant rate. It is extremely sensitive to changes in the force with which waves break on it, or changes in the amount of sand carried toward it by those waves. In a few hours, a major storm can move large quantities of sand to form new inlets or to close old ones.

Coming toward the beach from the ocean, an echo sounder on a small boat would reveal one or more submerged low sand ridges on the ocean bottom, called *longshore bars*, parallel to the shore, and generally situated in a few meters of water (see Fig. 13–1). At extreme low tides, the tops of these bars are often exposed. Separating the longshore bar from the beach proper is the *longshore trough*, which usually remains filled with water at low tide. On coasts with a restricted tidal range, there may be several such bars and troughs developed. Coasts with large tidal ranges usually have only one set of bars, at or near the low-tide mark. Bars can usually be identified, even when submerged, by the fact that waves break on them; on many coasts, several sets of bars can be spotted from the lines of breakers offshore.

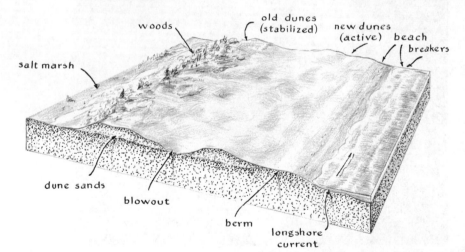

FIG. 13–14

Typical sand beach with active dunes, stabilized dunes and salt marshes. Such a beach–dune complex is found on the south shore of Long Island and at many points along the U.S. Atlantic coast. [Redrawn from F. D. Larsen, "Eolian Sand Transport on Plum Island, Massachusetts," *Coastal Environments of Northeastern Massachusetts and New Hampshire*, University of Massachusetts Coastal Research Group, Amherst, Mass. (1969), pp. 356–67]

Leaving the offshore portion of the beach and coming onto the beach proper, there is a sandy area dipping gently seaward, the *low-tide terrace*. This part of the beach is completely bare at low tide and submerged at high tide. A small scarp or vertical face often occurs at or near the upper limit of the low-tide terrace. This is commonly left by a recent cycle of more intensive wave action which has caused erosion into the beach profile formed during a preceding cycle. The seaward-dipping portion of the beach, collectively known as the *foreshore*, leads up to the *berm crest* or *berm*, the highest part of the beach. There may be several berms present on a beach at any time. Berm crests usually form during storms, and represent the effective upper limit of wave action during that storm. Usually the highest berm on a given beach is formed during winter storms and is often referred to as the "winter berm."

The berm slopes gently downward toward the base of the cliffs or dunes behind the beach. These sands are usually not moved often, as indicated by the large trees that grow there (and often by the accumulations of beer cans and other human artifacts). Where wind action is especially strong, the sands may be winnowed, removing the lighter particles and leaving behind the heavier mineral grains. On some beaches, the heavy minerals remaining give the surface a deep red or purple color, especially noticeable near the base of sand dunes.

Where a beach is not backed by cliffs, dunes are often formed from beach sands blown by the prevailing onshore winds, as illustrated in Fig. 13–14. Where dunes are not actively gaining or losing sediment, they may be colonized by various salt-tolerant plants or trees, to become stabilized through time; this process is shown in Fig. 13–15. Dunes protect the low-lying sand behind them; for example, in the Netherlands the dunes form a

FIG. 13–15
Beach grasses stabilize the dunes at Lighthouse Point near Gainsville, Florida. (Photography courtesy Florida News Bureau, Dept. of Commerce)

vital part of the defense against flooding from the North Sea. Frequently, access to the dunes is limited or prohibited in beach park areas to preserve the slow-growing plants which protect dunes against active erosion by storm winds.

On the Atlantic coast of the United States, extensive areas of dunes occur near Cape Kennedy, Florida, and near Province-town, Massachusetts. On the west coast, dunes are associated with the beaches near river mouths (see Fig. 13–13) and also occur on isolated beaches along the rest of the coast. Since they are controlled by the winds, dunes can form obliquely to the coast. Several intersecting sets of dunes may occur in the same region.

BEACH PROCESSES

Wind-generated waves dominate beach processes. Currents and turbulence generated by waves stir up sediment, and the currents associated with the waves and tidal currents transport sediment parallel to the coast. Transport usually takes place between the upper limit of wave advance on the beach and depths of about 15 meters. Large amounts of sand are transported in suspension; relatively little is transported along the bottom.

Beaches undergo definite seasonal cycles. During seasons of low, long-period swell, sand is moved back onto the beach, usually causing increases in height and width. Longshore bars migrate shoreward, filling in the troughs, and a new berm forms, usually at a level lower than the one preceding.

During periods with high, choppy waves (usually in winter), beaches are commonly cut back. The beach foreshore becomes more gently sloping and a beach scarp may occur as erosion proceeds. Strong longshore currents caused by the waves develop deep channels. Bars develop because of the offshore movement of sand from areas seaward of the breakers. Most of the sand removed from the beach is deposited nearby in the offshore zone to be moved back onto the beach during the next period of smaller waves.

Waves commonly approach the beach obliquely—rarely at right angles to the beach. Even though they are refracted upon entering shallow water so that crests are more nearly parallel to the coast, the process is rarely complete, and most waves approach the shore obliquely (see Fig. 13–17). Wave energy acting parallel to the coast causes longshore currents to move generally in the same direction as the waves when they approached the coast. The current is strongest in the band between the surf zone and the beach. The strongest currents are predicted to occur when the waves approach the shore from a 45° angle. This rarely happens; crests of most waves usually deviate less than 20° from being parallel to the beach when they strike the shoreline.

Along some trade wind coasts, the amount of sand moved by longshore currents exceeds 4 million cubic meters per year. On the south shore of Long Island, littoral currents move about 500,000 tons of sediment westward every year. A comparable quantity of sand moves along New Jersey beaches.

Each wave hitting the beach causes an uprush of a relatively thin sheet of water or *swash* onto the beach face, as Fig. 13–16 illustrates. The water rises until all the energy of the oncoming wave is dissipated or until the water moved by the wave percolates downward into the sand. Any water remaining on the surface runs back down the slope of the beach face.

Since waves rarely strike the beach head-on (see Fig. 13–17), but usually strike it at some angle, the swash rises obliquely across the beach face. When the water with its entrained sediment runs back, it goes directly down the slope of the beach. Sand moving along the beach as a result of wave effects is called *littoral drift*. Its direction can change during a single day or over a season; this is a small-scale phenomenon associated with waves hitting the beach. *Long-shore currents* are large-scale phenomena which also result from waves striking the coast.

Speeds of wave-induced sediment movement can be surprisingly high—up to 25 meters per hour and up to 1 kilometer per day. A more typical rate would be 5 to 10 meters per day on the average. Direction of net movement along the beach is determined by the direction of the strongest and longest acting waves. On most beaches this is the direction from which storm winds come.

Within an individual wave, there is substantial movement of water as it breaks. Water at the surface and along the bottom moves toward the beach, carrying with it materials floating or dragged along the bottom. This accounts for the fact that a beach acts as a convergence zone, collecting all sorts of debris

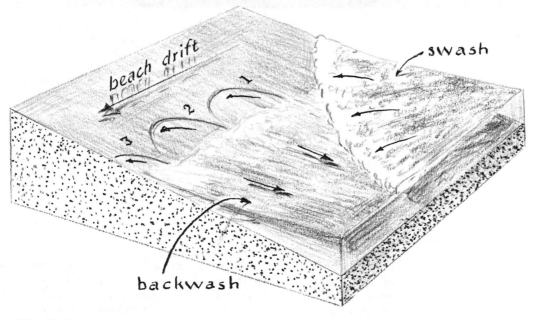

FIG. 13–16
Waves obliquely approaching a beach move diagonally up the beach face, and then the water moves directly down the beach. Consequently sand and gravel move along the beach in a series of arched points.

FIG. 13–17
Waves strike the beach north of Oceanside,
California; their approach at an angle causes
beach drift and longshore currents. [Photograph
from R. L. Wiegel, *Oceanographical
Engineering*, Prentice-Hall, Inc., Englewood
Cliffs, N.J. (1964), p. 372]

as well as sediment. Return flow of the water occurs at mid-depth within the water column as illustrated in Fig. 13–18. In general, this flow occurs through several meters of water and is not terribly strong.

Rip currents are another manifestation of the movement of water toward the beach in the surf and its return flow. After moving toward the beach, water tends to flow parallel to the beach for short distances until it enters a highly localized stream of return flow through the breaker zone—the rip current. In rip currents, the most rapid flow is relatively narrow, and has speeds of up to 1 meter per second until it reaches a distance of perhaps 300 meters from the coast. In this section of the current, it often forms a channel deeper than the surrounding ocean and is therefore recognizable from above by the deeper color of the water.

Seaward of the breaker zone, the current becomes more diffuse and spreads out, forming a "head" to the current. The

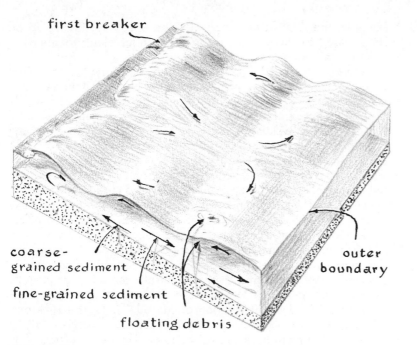

first breaker

coarse-
grained sediment

fine-grained sediment

floating debris

outer
boundary

FIG. 13–18
Schematic representation of breaking waves
and water movement at and below the water
surface. Water depth at the outer boundary is
approximately half the wave length for storm
waves. [Redrawn from R. L. Miller, and
J. M. Zeigler, "A Study of Sediment Distribution
in the Zone of Shoaling Waves over
Complicated Bottom Topography, in R. L.
Miller, (ed.), *Papers in Marine Geology*,
Shepard commemorative volume, Macmillan,
New York (1964), pp. 133–53]

water is caught up at this point in the general flow toward the
beach. Rip currents and their associated water movements form
a cell-like, nearshore circulation system within the breaker zone.

As an example of the processes that control beach development and sediment movement, let us consider the New Jersey
coast (see Fig. 13–12). This 200-kilometer stretch of coast consists of a series of barrier islands separated from the mainland
by bays, lagoons, and tidal marshes. Along this coast, beaches
are interrupted by three rocky headlands and ten major inlets
leading to lagoons or bays. Tides here are about 1.5 meters;
northeasterly storms are common, and there are occasional hurricanes.

At the northern end of the coast, near Sandy Hook, the net
littoral drift is northerly because nearby Long Island shelters
this stretch from waves approaching the coast from the north
and northeast. Over a period of about 100 years, sand accumulation on Sandy Hook amounted to about 400,000 cubic meters
per year. Near Cape May, at the southern end of this stretch of
coast, the net littoral drift is directed southerly and amounts to
about 150,000 cubic meters per year. The total amount of sand
in transit is substantially greater than the net littoral drift. For

example, the total littoral drift at Cape May is estimated to be about 900,000 cubic meters per year, equivalent to about 1 million tons per year.

Between the northern and southern portions of the New Jersey beaches, there is a nodal point—located about 60 kilometers south of Sandy Hook—where the net littoral drift is zero. Sand accumulating at Sandy Hook comes from erosion of this section of headland, as there are no major rivers in the region. Between 1838 and 1953, this section of coast was cut back 150 meters at a rate of about 1.5 meters per year. The materials eroded were about two-thirds sand, which was added to local beaches; about one-third was silt and clay-sized material, which moved seaward and was lost from the coastal area. Extensive construction of seawalls and groins in this section effectively reduced the littoral drift by only about 12 percent.

Inlets and lagoons behind barrier islands act as traps for sediment moving along the coast. Sand is stirred up and put in suspension by waves and then moved into the inlets or lagoons by flood-tide currents. When these slacken, or when sediment encounters the dense vegetation of the tidal flats, it settles out. Without resuspension by waves, the ebb-tide currents fail to move this sediment back out of the inlet. Thus each of the New Jersey inlets accumulates about 200,000 cubic meters—equivalent to 250,000 tons—per year, filling in the navigation channels and necessitating dredging.

MINOR BEACH FEATURES

As a wave breaks and runs up on the beach as a thin film of water (the swash), it carries with it various floating debris, sea foam, and small shells. When the wave reaches its highest point on the beach, the water percolates into the permeable sand, leaving a line of foam known as a *swash mark*. If the next wave runs higher up the beach, it incorporates the preceding swash mark and moves it higher. Where seaweed, shells, driftwood, straw from marshes, or man-made trash are thrown up by storm waves, the line of debris may be quite substantial. If the tide is falling, swash marks are left successively lower, and an entire series may be preserved on the low-tide terrace until the next high tide erases them. Several of the minor beach features are illustrated in Fig. 13–19.

As sea level drops during the tidal cycle, it leaves the sands or gravels of the beach saturated with seawater, which gradually drains out. This dewatering of the beach usually manifests itself in the wetness of the lower part of the low-tide terrace. Where larger volumes of water are discharged, a series of small *rills* (miniature channels) may be cut by the water as it runs down the beach.

Rills cutting deeply enough into the beach or a trench dug perpendicular to the trend of the beach will usually reveal a series of buried layers known as *laminations*. These layers are generally formed by different types of materials deposited as thin sheets. For example, during periods of little wave activity,

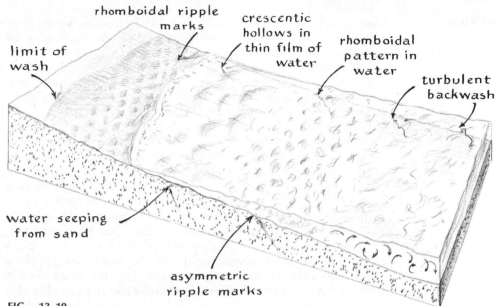

rhomboidal ripple marks

crescentic hollows in thin film of water

rhomboidal pattern in water

turbulent backwash

limit of wash

water seeping from sand

asymmetric ripple marks

FIG. 13–19

Examples of some small-scale features commonly observed on sand beaches. [After A. Guilcher, *Coastal and Submarine Morphology*, John Wiley, New York, (1958), 274 pp.]

mica flakes may be deposited instead of being removed, as they usually are during strong wave activity. When the beach is dry, as during periods of strong wind, some quartz grains are often blown away (as previously described), leaving behind a layer of dark-colored, heavier minerals as in Fig. 13–20. In general, these layers parallel the present beach surface, dipping seaward on the foreshore. There may also be disturbances of older beach surfaces recorded; for instance, air can be entrapped in the

FIG. 13–20

A scarp cut by waves. The boardwalk has been undermined, and snow fences have been placed in an effort to trap wind-blown sand. The dark patches and streaks at lower left are heavy mineral grains left behind by the waves. Some lamination is visible along the scarp. (Photograph courtesy Fire Island National Seashore, National Park Service)

sand, which then migrates to form a dome structure which later collapses, disrupting the layering of the beach. Burrowing of organisms also frequently disrupts the regular layering. On public beaches, it is sometimes possible to find buried automobile tracks that were left on a hard-packed surface and later covered by sand. The depth to which beach-cleaning machines have sifted the sand can often be clearly seen.

Some beaches have well-developed *beach cusps*: rather uniformly spaced tapering ridges, with rounded embayments between them, as illustrated in Fig. 13–21. The regular spacing ranges from less than 1 meter to several tens of meters, and seems to be related to wave height: higher waves are associated with wider spacing of cusps. These cusps seem to form and reform rather quickly, especially on fine-sand beaches. Their origin has long puzzled those who study beaches, but there seems to be general agreement that they are related to the to-and-fro motions of the water running up the beach. When waves strike a beach obliquely, causing longshore currents, the cusps do not form and are usually obliterated. Apparently, the rounded embayments between the points are channels through which water brought onto the beach by waves returns seaward.

FIG.   13–21
Beach cusps at El Segundo, California, photographed from the air. (From Wiegel, 1964, p. 347)

Ripple marks on the sand are another common feature. There are generally two kinds, both formed by water movements. Oscillation ripple marks tend to be symmetrically sloping on both sides, with pointed crests. They form roughly perpendicular to the back-and-forth motion of the water. The other kind of ripple mark is formed by currents; these are assymetrical, with a gently sloping side facing upcurrent, the sharp side of the ripple mark facing downcurrent. This type is formed by unidirectional currents, although both types are commonly found together.

SALT MARSHES

Salt marshes are low-lying portions of coastline which are submerged by high tides but protected from direct wave attack. Their surfaces are nearly flat and generally overgrown by a salt-tolerant vegetation, adapted to periodic submergence.

Marshes form where sediment is available from rivers or from resuspension in waves or strong tidal currents. In some areas, sediment depositing in marshes comes from local sources such as erosion of nearby headlands; this is the source of sediment for many New England marshes, as illustrated in Fig. 13–22. Size and shape are determined by the general outline of the depression in which the marsh forms; vertical extent is controlled by the tidal range. The upper limit of a marsh is generally controlled by the spring tides in an area, which is the

FIG. 13–22
Small salt marshes in nearly filled bays on the north shore of Long Island (Stony Brook Harbor and West Meadow Creek). Both marsh areas have been modified by dredging. (Official U.S. Coast and Geodetic Survey photograph).

highest level to which ocean water can periodically transport sediment.

Most of the marsh area consists of nearly flat-topped banks consisting of sand or mixtures of sand and silt. The tops of the banks, known as *tidal flats*, are commonly exposed at low tide and submerged at high tide. Where a marsh is well protected, the flats are usually covered by dense growths of marsh grass. Where the marsh is big enough and open enough for wind waves to be set up across its surface, tidal flats may be completely barren of plants.

Cutting through the tidal flats are many channels, through which seawater enters and drains from the marsh in response to the tidal cycle. The largest and deepest of these channels contain water even at low tide, and plants do not usually grow in them. The bottom material is generally shifting sand or gravel because strong tidal currents resuspend the finer-grained materials, to be deposited on the flats. These large channels connect with smaller branching and meandering channels which cut back into the marsh and are often exposed at low tide.

Sediment is transported into the marsh by strong currents in the tidal channels. As rising water moves out into the smaller channels and then onto the tidal flats, current velocities decrease. At some point, the current velocity is too low to keep sediment in suspension, so that particles settle out and are eventually deposited. When the tide goes out, current velocity is inadequate to resuspend the sediment grains, which therefore remain where they settled out of the water. Plants on tidal flats also tend to retain sediment, and marshes thus act as effective traps for fine-grained sediment moving in the coastal ocean. Many marshes mark the locations of former small, shallow estuaries.

Some of the largest and best-studied marshes are those lying behind the Friesian Islands on the Dutch North Sea coast and the adjacent North German coast along the edge of the Rhine Delta. In the Netherlands much of the marshland that formerly bordered that delta has been reclaimed by building dikes to exclude ocean waters, and then draining the salt water to make fields suitable for agriculture.

Salt marshes, or *wetlands*, are important biological features of the coastline. In temperate climates, salt-tolerant plants and grasses produce an abundance of food for the large and small animals. Much plant debris escapes to be deposited with the sediment accumulating on the nearby ocean bottom.

In tropical climates, grasses play a less conspicuous role, and *mangroves* such as those in Fig. 13–23 dominate marshes along the estuarine border. These large, treelike plants have extensive root systems. The roots from dense thickets at the water level which provide shelter for both marine and land animals—a zone truly intermediate between land and water. Many forms of life are specially adapted to survive in this environment; the mangrove oyster, for instance, attaches itself to roots and branches that are exposed at low tide, presenting the spectacle of oysters growing on trees!

Mangrove roots trap sediment and organic matter, and eventually the swamp is filled in to be replaced by a low-lying

FIG. 13–23
Mangroves growing out into the water at
Ten Thousand Islands, Florida. (Photograph
courtesy Florida State News Bureau)

tropical forest. Over a 30- to 40-year period, 1500 acres of new
land were created in Biscayne Bay and Florida Bay by colonization of shallow-water areas by mangrove seedlings.

MODIFICATION OF COASTAL REGIONS

Coastal areas have been substantially modified by man's efforts
to adapt them to his own needs—especially near urban centers,
where population pressures are greatest. Improvement of shipping facilities in estuaries and harbors, filling-in of marshes for
industries, airports, and housing, prevention of beach erosion,
and use of coastal areas for waste disposal probably account for
the bulk of coastal modifications in the United States. Because
of their importance to shipping, estuaries have perhaps been
more extensively modified than any other coastal feature.

The dredging of deeper (and usually straighter) channels for
safer passage of ships is perhaps the oldest form of coastal alteration commonly undertaken in this country. Most estuaries
have bars—shallow sand deposits—at their mouths, whose
greatest depth in channels would be around 3 to 5 meters below
sea level. These shallow channels are kept open by river flow
and tidal currents, but they are far too shallow to accommodate
modern vessels; furthermore, they tend to shift location frequently. Dredging them to depths of up to 15 meters has improved navigation in large harbors, but it has also introduced a
host of other problems to the estuary.

Deeper channels act as sediment traps for sands moving along
the coast as well as for river-borne sediment, so that a con-

tinuous program of dredging is required to keep the channels
open. Alteration of normal circulation patterns affects salinity
distributions; for instance, enlargement of inlets may permit
larger amounts of seawater to enter, and water through man-
made channels may flow toward remote areas that would normally be relatively fresh. Fish and bottom life are affected by
such changes.

Disposal of wastes—municipal and industrial, treated and
untreated—into harbors has left large areas of harbor bottom
covered by the solids from these wastes. As a result, many
marine organisms can no longer exist in these areas, and water
quality is often degraded too severely for recreational uses.
Efforts to remedy such situations have been underway for many
years in United States coastal areas, with much remaining to
be accomplished.

While less altered than most harbors, beaches have often
been subjected to conflicting uses and have been greatly altered
in the process. In many areas, beaches are exploited directly as
sources of sand for construction and land-fill operations, making
them unsuitable for recreational use. In other cases, changes in
the local regime of sand transport has caused erosion of beaches.
When a harbor or inlet, for instance, is dredged for improvement of navigation facilities, the resulting deep channels or
basins may act as traps for sediment moving down the coast.
This cuts off the sand supply to beaches downcurrent.

Efforts to prevent accumulation of sediment in inlets have
met with limited success. Seawalls have been built to stem erosion, and groins to inhibit sediment movement along beaches.
Other efforts might include construction of jetties to keep sand
out of the inlets, perhaps combined with pumping of sand past
the inlet. Sand may also be dredged from bays and used to
replenish beaches. These approaches are expensive, and some
are undesirable because of side-effects—such as destruction of
large and valuable marsh areas by dredging.

Once lost, a beach is not easily restored. Several approaches
have been used to protect or rebuild beaches. One of the most
direct approaches has been to construct *groins*—low stone walls,
such as those in Fig. 13–24, built at regular intervals along a
beach to retard sand movement. On the side from which sand
is moving, the beach is widened; but the beach downcurrent is
usually narrowed unless particular care is taken in construction
and placing of the groins. In some instances where groins have
been improperly placed, erosion has actually increased. On the
New Jersey coast, where groins have been used extensively, it
is estimated that they have only reduced the rate of sand movement by about 12 percent.

The other approach to beach stabilization is *replenishment*, a
procedure whereby large volumes of sand are brought in and
put on the beach. In some instances, the sand is dredged from
shallow, protected bays behind the beach. The effects are noticeable but transitory, because the sand moves downcurrent as part
of the regional movement of sediment along the beach. Often,
a single storm will remove the newly added sand.

Salt-marsh areas were commonly bypassed in the early development of coastal areas in the United States. Unsuited for

FIG. 13–24
Groins have been constructed along the ocean
beach (left side of the picture) at Sandy Hook,
N.J., in an attempt to halt the flow of sand
along the beach.

FIG. 13–25
Former marshland on the western side of
Newark Bay, N.J., has been developed by
dredging channels and filling low areas to build
terminals for handling containerized freight
(center of photograph) and Newark
International Airport (left side). Relatively
undisturbed marsh is visible in the background.
(Photograph courtesy Port of New York
Authority)

most agricultural purposes and difficult to build on, they were
often used initially for pasture, then as waste-disposal sites and
only later for building purposes. About 20 percent of Manhattan
Island in New York City is built on "reclaimed" marsh and
shallow harbor areas. Continued population growth has often
resulted in accelerated use of salt-marsh areas as they come to
represent the only available open land. LaGuardia and Kennedy
Airports in New York City are built on filled marsh areas, as
are San Francisco's International, Boston's Logan, and Washington, D.C.'s National Airport; Fig. 13–25 illustrates how this was
done for Newark International Airport. Once dredged and filled,
salt marshes are attractive sites for industrial development because of their convenient proximity to water transport and often
relatively unrestricted waste disposal. Housing developments
have also taken their toll of salt-marsh areas.

In addition, coastal marshes are commonly dredged and filled to form numerous narrow peninsulas and boat channels, offering access to waterways for small-boat operators. In the process, the marshes are altered or destroyed, and circulation within any remaining wetlands is greatly modified. Frequently, the dredged basins and channels provide poor circulation, and the resultant accumulations of wastes—including untreated sewage from housing developments and boats—have caused severe local problems including odors, and fish kills.

Even where marshes are not developed for housing or other purposes, they are often modified by extensive ditching operations as part of mosquito-control measures. Ditches are laid out and dug on regular patterns to improve drainage in the interior of the marsh and thereby eliminate mosquito-breeding areas. Part of the original marsh surface is destroyed and circulation locally altered in the process. Erosion may be accelerated owing to improper location of drainage channels.

Through the combination of all these alterations, it was estimated, in 1970 about 23 percent of estuarine systems (including salt marshes) in the United States had been severely modified, and about 50 percent moderately altered. In 1968 the United States government alone spent more than $870 million on coastal-channel and harbor improvement projects. Many millions more were spent by state and local governments and by private concerns.

In the past, planning and engineering of large-scale coastal modification projects depended primarily on experience accumulated over many years of building (and sometimes rebuilding) coastal installations. During the Middle Ages, for instance, Dutch engineers were already constructing dikes to reclaim shallow areas of the North Sea for agriculture and industry.

Modern coastal engineering depends increasingly on various types of models to predict the effect of structures on the coastal ocean. Where processes are well understood and computation facilities adequate, it is possible to construct mathematical models—expressions describing in precise mathematical terms the nature and effects of physical processes. Using computers, it is then possible to predict the effects of projected structures. Designs can be altered if necessary to achieve the desired objective at minimum cost, or to meet other criteria.

Although the use of mathematical models has become increasingly common since the development of computers, engineers have been successfully calculating the effects of tides and winds on coastal structures for the past several decades. The great Dutch Afsluitdijk ("enclosure dike"), for instance, has kept seawater out of the former Zuider Zee estuary since 1932. Extensive studies of tide and current forces and of the probable strength and frequency of storms preceded its construction. Much valuable agricultural land has been reclaimed from shallow coastal areas behind the dike by first pumping out the salt water, then permitting rainfall to flush away the salt remaining in the soil.

Many problems cannot be handled by mathematical techniques. Either the processes are too poorly understood to be described mathematically or too complicated to be solved by existing computers. For these problems, *hydraulic models* such as the one shown in Fig. 13–26 have proven extremely useful. The model simulates an area on a greatly reduced scale so that physical processes can be modeled and studied relatively inexpensively. Furthermore, the model permits the collection, in a short time, of data that would require perhaps years to collect in the area itself.

FIG. 13–26
Hydraulic model of Galveston Bay showing the inlet in the barrier beach. The model was constructed on a scale of 1:100 vertically and 1:3000 horizontally. Tides and tidal currents are reproduced in the model by a tide generator located in the Gulf of Mexico portion of the model, on the right-hand side. The strips in the model on the bay side are necessary to adjust flow in the model to make it reproduce conditions observed in Galveston Bay. (Photograph courtesy U.S. Army Corps of Engineers Waterways Experiment Station)

Hydraulic models are carefully constructed to duplicate the original feature. Then careful observations are made in the field, and the model is adjusted to reproduce physical processes accurately. Hydraulic-model studies of estuarine systems have been especially useful in studying such diverse phenomena as the effects of breakwaters on waves in harbors, movement of wastes in estuaries, and sediment movement in harbors.

SUMMARY OUTLINE

Shore—extends from low-tide to high-tide levels

Coast—broad zone extending landward from shore

Processes affecting shores are effective over long as well as relatively short times

Storm surges—relatively sudden sea level changes, usually caused by storm winds

Coastlines
> Secondary coast—formed by terrestrial forces: drowned river valleys, volcanoes
> Primary coast—shaped primarily by marine processes or organisms; barrier beaches, coral reefs

Deltas—large deposits of river-borne sediment at the river mouth
> Modified or prevented by tidal currents or waves
> Steps in delta formation: filling of estuary, formation of distributaries, shifting of distributaries
> Shape controlled by river processes and coastal processes

Beaches—deposits of loose sedimentary material moved by waves
> Usually sand (sometimes gravel) derived from local sources, such as rivers or bluffs; nature of material depends on coastal processes

Barrier beaches—islands or spits built by action of waves and currents

Beach coastlines—generally smooth outlines of pre-existing topography
> Offshore—sand bar and trough
> Onshore—low-tide terrace, beach face, berm crest (high point, formed by storms)
> Dunes—commonly behind large sand beaches

Beach processes
> High, choppy waves erode beaches, usually in winter
> Long-period, low swell builds beaches, usually in summer
> Waves approaching beach obliquely cause longshore currents and beach drift
> Water movement within breaking wave is toward land at surface and along bottom, seaward at mid-depths
> Rip currents—return flow of water moved toward beach by breakers

Minor beach features
> Swash—thin film of water from wave moving up beach carries debris
> Rills—minute channels formed by water draining out of beach
> Laminations—layers of colored minerals left behind during periods of strong wind erosion
> Beach cusps—uniformly spaced tapering ridges with rounded embayments between them
> Ripple marks—formed by current movements

Salt marshes—low-lying area covered by vegetation, usually submerged only at highest tides
> Tidal flats—covered by high tide, bare at low tide; may or may not have vegetation cover
> Channels tidal-creek—flooded at all times, avenues for drainage
> Sediment moves into marsh during flood current, not readily eroded by ebb current
> Vegetation also traps sediment—e.g., mangrove swamps

Modification of coastal regions
> Estuarine modification
> Includes dredged navigation channels, stabilization of harbor entrances with jetties, waste disposal
>> Alters estuarine circulation patterns and sediment flow

Salt marshes
   Used for waste disposal, filled for construction of houses
   and airports
   Navigation-channel dredging and mosquito-control drainage
   alters circulation

BASCOM, WILLARD. 1964. *Waves and Beaches: the Dynamics of the Ocean Surface*. Doubleday Anchor Books, Garden City, N.Y. 267 pp. Elementary, well written.

BIRD, E. C. F. 1969. *Coasts*. M.I.T. Press, Cambridge, Mass. 246 pp. Wave, current, and wind effects on the structure of coastlines; elementary.

JOHNSON, D. W. 1965. *Shore Processes and Shoreline Development*. Hafner, New York. 584 pp. A classic study, first published in 1919.

STEERS, J. A. 1969. *Coasts and Beaches*. Oliver & Boyd, Edinburgh. 136 pp. Elementary discussion of beaches and beach processes, examples primarily from Great Britain.

VAN VEEN, JOHAN. 1962. *Dredge, Drain, Reclaim: The Art of a Nation*. Fifth Ed. Martinus Nijhoff, The Hague., 200 pp. Elementary discussion of Dutch reclamation work.

ZENKOVICH, V. P. 1967. *Processes of Coastal Development*. Interscience, New York. 738 pp. Thorough treatment of shore processes, with emphasis on Russia and recent Russian research.

SELECTED REFERENCES

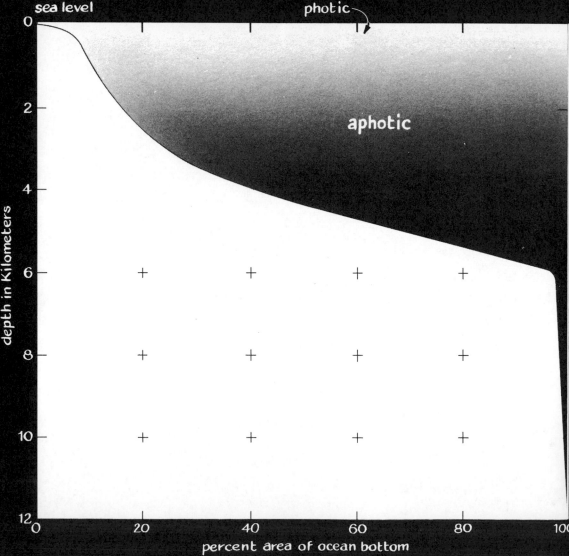

LIFE AND THE OCEAN

FIG. 14–1
Simplified diagram of the marine environment. The deep ocean is
cold, dark, and sparsely populated in comparison with the shallow
coastal zone and sunlit surface layer. All marine organisms ultimately
depend for food on organic carbon produced in the photic zone.

fourteen

Previous chapters have described physical conditions in the ocean, but viewed the ocean as essentially devoid of life. In the remaining chapters, we shall consider the ocean as a complex series of *habitats* for organisms of all types— drifting, swimming, creeping, and attached.

Plants or animals live in every part of the ocean, both in the water and on the bottom, at all depths and at every latitude. We may think of the *habitat* as an identifying address, by which the conditions of life are defined. Each habitat supports a community of organisms whose members are structurally and functionally adapted to survive and reproduce under conditions existing there.

Marine ecologists include biological communities in *ecosystems*—functional units in which living and nonliving components interact dynamically with exchange of materials. We have already learned that organisms influence and control marine processes in many ways; the chemistry of ocean waters, involving both salts and dissolved gases, and also the composition of marine sediments are modified by biologically significant chemical and physical variables. In this chapter, we shall see how some physical and chemical variables in the ocean affect distribution of marine organisms. In the following chapters, we shall show how some typical organisms are adapted to live in the ocean.

Membership in an ecosystem implies dynamic exchange of matter and energy, with the nonliving environment and with other organisms. Each species occupies a unique position or *ecological niche* relative to the others. This niche represents a functional status in the economy of the community; it is not an address (which is the *habitat*) but a profession.

In each ecosystem, organisms occupying different niches are mutually interdependent. This interdependence is most obvious in a *food web* (a complex group of food chains) where organic material originally produced by plants is eaten by animals, which in turn eat each other. When we discuss specific marine communities, we shall see many less direct ways in which organisms depend on one another for food, shelter, and defense against enemies in the deep ocean and along coasts.

Significant changes in the physical or biological conditions of an ecosystem cause corresponding changes in the relationships of its members to one another and to the physical environment. Certain species may disappear, and members of other species may be able to fill newly created niches in the system. So long as biological or physical conditions remain unstable—as when, for instance, production of food exceeds consumption (or vice versa) —the numbers or kinds of organisms in the community continue to change in a process known as *ecological succession;* if and when, a stable ecosystem is established, a so-called *climax community* is the result.

An example of succession would be the changes resulting from dredging New York Harbor. Bottom-dwelling organisms are removed, leaving barren sand. Two kinds of organisms re-populate the area—those that move rapidly, such as crabs and snails, and those which release abundant larvae capable of wide dispersal, such as certain worms. These first communities commonly contain few species, although each one may quickly produce large numbers of individuals.

Figure 14–2 is a highly simplified picture of the flow of energy and nutrients through an ecosystem. Nutrients from seawater, atmospheric gases, and radiant energy from the sun are combined in the surface ocean by *photosynthesis* of *phytoplankton* (microscopic floating plants); these plants are primary *producers.* Radiant energy is thus converted into *glucose* (a simple sugar) or other energy-rich compounds, some of which are stored as potential (chemical) energy. The plants are eaten by small animals (*primary consumers*) and they in turn by larger ones, or *secondary consumers.* Each consumer uses some stored chemical energy to carry out its life functions, and the energy stored in its own body is passed along to the next so-called *trophic level* when it is eaten by another consumer. Some of the nutrients released by animals in metabolic wastes return directly to seawater to be reutilized by plants. Detritus (non-living organic matter) is food for scavengers; in certain ecosystems, scavengers and the organisms that prey on them are considered as forming a detrital food chain, a part of the ocean's food web. In the ocean, it is often difficult to delineate a separate detrital food chain because many organisms, such as shellfish, eat live and dead materials indiscriminately. Bacteria (*decomposers*) also utilize nonliving organic matter and return its nutrients to seawater; they utilize chemical energy but store relatively little of it.

Organic matter sinks from the surface (where it formed) to deeper waters transporting energy and body-building materials

FIG. 14–2

Flow of matter and energy through the marine food web. Nutrient substances are returned to solution in seawater to be reused; potential energy is dissipated, largely as heat.

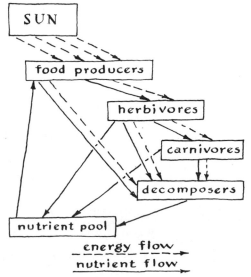

to the ocean depths. Ocean-circulation processes continually carry deep water toward the surface, where nutrients dissolved in it can be reutilized by growing plants. In this way nutrients are constantly recycled in the ocean. Energy flows only one way through the food web—it is not recycled. Without sunlight, there would be no source of energy to sustain life. Throughout our study of marine ecology, we shall be seeing examples of how energy and matter flow through marine food chains in a variety of ecosystems. In the light of what we have learned, we shall try, in Chapter 17, to evaluate present and potential food production in the ocean as it may be applied to human needs.

DENSITY

Most land organisms live surrounded by air. If we compare conditions in our "ocean of gases" with those in an ocean of seawater, we may note first that on earth we live at the bottom of our "ocean" rather than all through it. Some spores and seeds, as well as bacteria and viruses, do drift about, and certain spiders can float in the air attached to a fine thread. In addition, birds and flying insects can often make use of updrafts to support themselves for a while, but all flying creatures must alight from time to time.

The layer of atmosphere actually inhabited by living organisms is relatively shallow. The tops of our tallest trees are only about 70 meters above the earth's surface, and few living creatures penetrate more than a few meters into the soil.

In the sea, while many animals and plants spend at least part of their lives attached to a solid surface, most marine organisms are adapted for a floating or swimming existence in an interconnected, relatively homogeneous medium. Life exists at all depths—from the sunlit surface layer to the dark, cold, deep ocean floor.

One obvious difference between air and seawater atmospheres concerns their relative densities. The specific gravity of seawater, like that of living protoplasm, is around 1.025; the specific gravity of air is around 0.0012. Thus the effective weight of animals and plants in the sea is reduced to a negligible fraction of what it would be on land. This fact has profound implications for the physical structure and mode of life in marine organisms.

In the first place, problems of support for the organs and appendages of the body assume quite different dimensions. The heavy internal skeletons of land animals, developed to hold us upright and to facilitate motion and leverage against the pull of gravity, are not necessary in the ocean; neither are rigid cellulose frameworks characteristic of the larger land plants. The relative masses of individuals are in part determined by support problems. The experiment of great size in dinosaurs was eventually a failure on land, but whales attain lengths of up to 34 meters and a bulk of over 135,000 kilograms (150 tons) without apparent hazard to survival of their species on that account.

Also related to weight in the ocean is the problem of sinking, which has no counterpart on land. Most marine organisms are adapted for survival in a particular layer of the ocean, having specific requirements for light, pressure, and temperature. An individual whose specific gravity is lower than that of seawater will rise to the surface and float partially exposed to the air. Most marine plants and animals are slightly heavier than seawater, and must thus counteract a tendency to sink into ever-deeper water layers.

Flotation is achieved in the ocean in a variety of ways, depending on the size and general structure of the individual. Self-propelled organisms can swim upward from time to time. Very small organisms have an advantage in this respect, because frictional resistance to falling increases as the ratio of

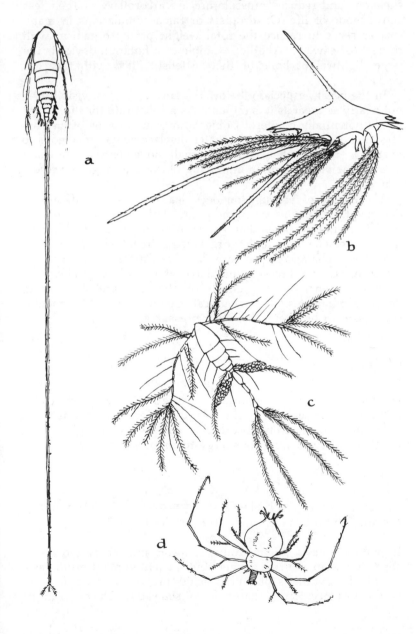

FIG. 14–3
These tiny, floating crustacean animals, related to land insects, have large surface area and relatively small mass (weight). In this figure, (a) and (c) are adult copepods, (b) is the larva of a barnacle and (d) is a lobster larva. (Redrawn from Sverdrup et al., 1942)

surface to mass becomes greater, and most marine organisms are so small that sinking is not a major problem. Many very small plants and animals, such as those in Fig. 14–3, have thin, wispy filaments or appendages on their bodies which increase their ratio of surface to mass; this seems to be an especially common feature in warm water, which offers less resistance to motion (including sinking) than cold water.

A flotation aid common among both large and small marine organisms is the possession of lightweight gas or fluid bladders. Fishes, whose skeleton and muscle tissue is generally more dense than seawater, often have a gas-filled swimbladder located beneath the backbone. The amount of gas in the bladder can be regulated to achieve the appropriate density. Most fishes without a swimbladder are deep-water species, whose lightweight skeletons and reduced musculature are a corollary of their less active mode of life. Oil droplets or fat accumulations in many species serve to reduce the total weight of the organism with respect to seawater. Another example of a flotation device is the large, lightweight bone of the cuttlefish, filled with tiny gas pockets.

In the ocean, especially below the surface layers, visibility is not nearly as great as it is on land. As a substitute for the ability to see potentially dangerous objects, marine animals often possess an extreme sensitivity to water displacements (analogous to hearing), enabling them to detect small movements in the water or even the presence of an unmoving object that might represent danger.

The effects of pressure on deep-sea animals are difficult to document. Organisms are found living at all depths in the ocean; thus it is not obvious that pressure is anywhere limiting to marine life. The more highly organized animals seem to be more sensitive to changes in pressure than are simple organisms—for instance, bacteria. Pressure may have some physiological effects, including such factors as effects on the structure of water in deep-sea organisms. Some chemical reactions, such as fermentation, seem to be altered by great pressures.

LIGHT

For the biologist, the importance of light in the ocean relates primarily to its use by marine plants. Chlorophyll pigments in living plants selectively absorb that part of the sunlight reaching them that provides the energy for *photosynthesis:*

$$\text{Sunlight} \atop {674{,}000 \text{ cal} + 6CO_2 + 6H_2O \underset{\text{(respiration)}}{\overset{\overset{\text{(assimilation)}}{\text{Chlorophyll}}}{\rightleftharpoons}} C_6H_{12}O_6 + 6O_2} \atop \text{Metabolic energy}$$

In photosynthesis, a plant converts water and carbon dioxide to sugar (glucose) and free oxygen. Subsequently, plant protoplasm is built by synthesis of amino acids, proteins, fats, and other vital substances from materials in seawater. Glucose can be easily changed back into its components in the presence of free oxygen by the process of respiration. In this way, the stored energy is released, providing energy for life processes.

In addition to this very important reaction, light is apparently necessary for the production of certain essential trace substances, such as certain vitamins. Synthesized by organisms in the sunlit, surface zone, these substances pass through food webs and eventually become available to animals below the lighted area.

Of every 1 million photons (units of solar radiation) that reach the earth's surface, only about 50 are used for photosynthesis of land plants, and about 40 by marine plants. It is from this radiant energy that all the food and free oxygen on earth are derived. As Fig. 14–4 indicates, the light absorbed most efficiently by chlorophyll is in the blue range—that is, wavelength around 0.45 micron; red light is used to a lesser extent. Some marine plants contain carotenoid pigments, which permit them to utilize a broader spectrum of light for photosynthesis than is usable by chlorophyll alone.

Biologists generally consider the open ocean as being divided vertically into three zones (see Fig. 14–1), based on penetration of sunlight. The *photic zone*, where light is sufficient for plant growth, may extend to 80 meters or more below the surface. The fading or *disphotic zone*, where light may penetrate dimly under optimal conditions, seldom extends deeper than 600 meters. In clearest subtropical ocean waters, the light which penetrates farthest is in the blue part of the spectrum (0.465 micron). This is the wavelength of light most readily absorbed by photosynthesizing plants. In the *aphotic zone*, all life depends on the supply of food from higher layers.

There is a pronounced decrease in the amount of light going from the equator toward the poles, as well as downward from the water surface. These gradients change daily and seasonally in an extremely precise way, as the earth moves relative to the sun—much more predictably than do temperature gradients, for instance. In polar regions, light is adequate for photosynthesis only during the three months or so when high latitudes are most fully illuminated (see Fig. 7–3); in the tropics, it is adequate throughout the year. In between are winters of varying lengths, during which plant food is not effectively produced in the ocean.

Animals, as well as plants, are affected by light. Many animals which live within about 750 meters of the surface and migrate vertically (a characteristic discussed in Chapter 15) are often associated with zones of specific light intensity. Light sources or changes in color of light often guide marine animals toward food. Strong light intensities, however, such as are found at the surface in direct sunlight, may inhibit normal reactions in or actually damage certain plants and animals.

Among animal species, both marine and terrestrial, there is widespread dependance on changes in light level for regulating cyclic physiological processes. Changes in the quality or quantity of sunlight, or even moonlight, can cause an animal to moult, to eat, to produce eggs, or to migrate. Many marine animals, including some bacteria, produce their own light. This may serve

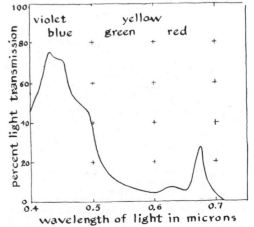

FIG. 14–4

Percentage of light in the visible spectrum absorbed by a phytoplankton population taken from the vicinity of Woods Hole, Mass. [Modified from C. S. Yentsch, "The Influence of Phytoplankton Pigments on the Color of Sea Water," *Deep-Sea Research*, 7 (1960), 1–9]

a variety of functions—as a badge of identity, as camouflage, as a lure for prey, or to illuminate potential food sources.

BIOGEOCHEMICAL CYCLES

Table 14–1 shows the major chemical elements in plankton, in order of abundance in the samples analyzed, which were largely diatoms, the dominant plant of northern oceans. Their elementary composition is comparable to that of other kinds of

Table 14–1

DISTRIBUTION OF PHYSIOLOGICALLY IMPORTANT ELEMENTS IN THE OCEAN

Element	Content: g/100 g Plankton† (dry wt)*	Content: g/m³ Seawater $(35^0/_{00})$*	
	P	S	S/P
O	44.0	8.57×10^5	19,600
C	22.5	28	1.25
Si	20	3	0.15
K	5‡	380	72
H	4.6	1.08×10^5	23,500
N	3.8	0.5	0.13
Ca	0.8	400	500
Na	0.6	1.05×10^4	17,500
S	0.6	900	1,500
Cl	0.5‡	1.9×10^4	38,000
P	0.4	0.7	0.17
Mg	0.32	1.3×10^3	4,100
Fe	0.035	10^{-2}	0.29
Zn	0.026	10^{-2}	0.39
B	0.01‡	5	500
I	0.003‡	6×10^{-2}	20
Cu	0.002	3×10^{-3}	1.5
Mn	0.00075	2×10^{-3}	2.7
Co	0.00005	3×10^{-4}	6
V	0.00005	2×10^{-3}	40

*After H. J. M. Bowen, *Trace Elements in Biochemistry*, Academic Press, London (1966).
†Mainly diatoms.
‡Concentrations poorly known.

organisms (with a few exceptions). Silicon, for example, is a major constituent of diatom shells. Sodium and chlorine are present in relatively small amounts, perhaps because plankton do not contain the large amounts of salty body fluid characteristic of larger animals.

The last column in Table 14–1, S/P, shows the ratio of supply to demand for each element, a figure which varies widely. Elements for which S/P is smallest are nitrogen, phosphorus, and silicon, the first elements to be removed completely from

FIG. 14–5

The phosphorus cycle. (Modified from Harvey, 1960)

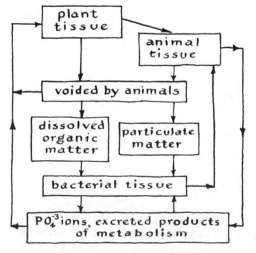

seawater during periods of rapid plant growth. When the supply of these nutrients has been exhausted, plant growth ceases, even if all other necessary elements are present in abundance. In seawater, nitrogen, phosphorus, and silicon are available for use by plants as compounds of oxygen (NO_3^-, PO_4^{3-}, and SiO_2).

The most common skeletal minerals among marine organisms are calcium carbonate ($CaCO_3$) and silica (SiO_2). Seawater is usually supersaturated with $CaCO_3$, and so the supply to organisms is plentiful. Its biological precipitation, however, is inhibited by low temperatures and high pressures; consequently, deep-water and high-latitude forms tend to have siliceous skeletons, or else poorly developed, fragile structures made of $CaCO_3$.

Table 14–2 indicates the remarkably constant atomic ratios of carbon, nitrogen, and phosphorus in plant plankton. Plants

Table 14–2

ATOMIC RATIOS OF CARBON, NITROGEN, AND PHOSPHORUS
IN THE ELEMENTARY COMPOSITION OF PLANKTON*

	C	N	P
Zooplankton	103	.16.5	1
Phytoplankton	108	15.5	1
Average	106	16	1

*After A. C. Redfield, B. H. Ketchum, and F. A. Richards, "The Influence of Organisms on the Composition of Sea-water," in M. N. Hill (ed.) *The Sea*, Interscience, New York (1963), Vol. II, pp. 26–77.

extract inorganic phosphates and nitrates from surface waters; some can also utilize dissolved organic compounds. Plants flourish in water where nutrients and sunlight are present in adequate amounts, and animal populations usually thrive on the abundant food supply. Nutrients from plants are concentrated in animal protein and skeletal materials, whereas carbon–hydrogen–oxygen compounds (carbohydrates) are "burned" (combined with oxygen) to provide energy, or stored as potential energy.

Plant nutrients not converted into animal tissue or stored as oils are voided as fecal matter, or as undigested, unaltered plant matter. Nutrients often dissolve directly out of fecal matter, thereby returning directly to the water. Dead plants and animals, as well as fecal matter, are often consumed by scavengers. Eventually, particulate organic matter is decomposed by bacterial action. Nutrient elements are returned to the oceans to be recycled through another generation of plants and animals. A very small fraction is lost by incorporation in sediment. The biological cycle for phosphorus is shown in Fig. 14–5, and the cycle for nitrogen in Fig. 14–6.

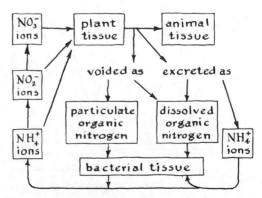

FIG. 14–6
The nitrogen cycle (Modified from Harvey, 1960)

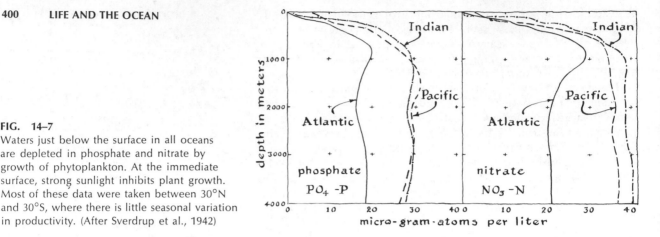

FIG. 14–7
Waters just below the surface in all oceans are depleted in phosphate and nitrate by growth of phytoplankton. At the immediate surface, strong sunlight inhibits plant growth. Most of these data were taken between 30°N and 30°S, where there is little seasonal variation in productivity. (After Sverdrup et al., 1942)

FIG. 14–8
Production of nitrogen compounds from decomposition of phytoplankton in aerated seawater in the absence of light. (After Horne, 1969)

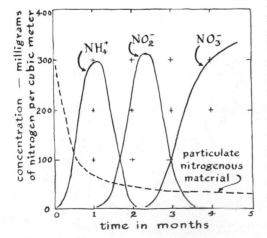

Some organic material sinks to the ocean floor; often there also exists a relatively dense layer of nonliving matter (*detritus*) near the bottom of the photosynthetic zone. Many kinds of bacteria, living on solid surfaces such as bottom sediments and floating debris, utilize phosphates and nitrates from nonliving organic matter, and sometimes from inorganic compounds, to synthesize proteins. In order to do this, they require energy sources, such as carbohydrate-rich plant detritus and dissolved organic compounds. A wide variety of bacterial "flora" act to decompose organic matter in the oceans; their chemistry is complex and as yet not completely understood.

Bacteria are a major food source for many very small animals in bottom communities, and to some extent for animal plankton as well. Bacteria release CO_2 to the environment but store little carbohydrate, and thus do not serve as an important energy source for great numbers of large animals. The total weight (biomass) of bacteria at any given time is small relative to that of other organisms in an ecosystem, but they put nutrient-rich, high-protein materials directly into animal food chains. Bacterial activities cause release of nitrates and phosphates to the water and thus play a major role in the ocean's nutrient cycles.

Phosphate compounds are fundamental to photosynthesis. Measurements of chlorophyll content in seawater usually correlate closely with measurements of phosphorus in the dissolved and particulate matter. As marine plants are increasingly grazed by herbivores, the amount of phosphorus contained in the total mass of plant matter declines and that contained in animal matter rises. As phosphate is utilized in photosynthesis, oxygen is released to the environment. These two constituents are thus inversely related in the production–decomposition cycle of plants.

Bacteria play a comparably important but more complex role in the nitrogen cycle. Nitrate is utilized by phytoplankton in the same ratio as is phosphate (see Table 14–2), and the ratio of nitrate to phosphate concentration is quite uniform in all seas and at all depths (see Fig. 14–7). Most of the nitrogen released from the decomposition of organic matter is in the form of ammonia (NH_3); this is oxidized to nitrite (NO_2^-), and finally to nitrate (NO_3^-), by bacterial activity. This process may take up to a few months, as shown in Fig. 14–8. Nitrogenous wastes, excreted by animals, are similarly oxidized. Some kinds of bacteria reduce NO_3^- and NO_2^- to NH_4^+ (ammonium ion) or to free nitrogen (N_2), and some are able to incorporate (fix) free nitrogen into organic materials.

Figure 14–9 indicates the seasonal variation in amounts of limiting nutrients present in the English Channel over a 12-month period. The amounts present at different seasons reflect utilization of the nutrients at times of rapid plant production, especially of diatom *blooms*, a high concentration of diatoms in the water. In spring and again in late summer, an explosive bloom depletes the nutrients; the curve for silicon (in Fig. 14–9) shows this especially well. During winter, when a lack of sunlight and a mixing of surface waters below the photic zone prevent any substantial growth of plants, nutrient levels in surface waters are at a maximum.

Under conditions of nitrate or phosphate depletion, marine plants can sometimes survive with greatly lowered internal concentrations of these nutrients, at least for short periods. Phytoplankton are able to extract phosphate from seawater very rapidly, and to "hoard" much greater amounts than they actually need for growth. For this reason, they may continue to grow for several generations after the supply in the water has been depleted.

Nitrates and phosphates are steadily lost from sunlit waters as organic particulate matter sinks below the photic zone. Decomposition of organic matter in nonphotosynthetic zones results in nutrient enrichment of deeper waters (see Fig. 14–7). Plant life would soon cease if there were no means for returning these essential ions to the surface. Wind-induced mixing accomplishes this at mid- and high latitudes, where winter storms break down the thermocline and permit deeper water to circulate to the surface. Upwelling of deep waters along coasts and at the equatorial divergence also brings nutrients to the photic zone. At low latitudes, there is a steady upward migration of subsurface waters through the thermocline.

Oxygen and carbon dioxide cycles in the ocean are also governed by biological activity. Oxygen in surface waters is nearly at equilibrium with atmospheric gases. It is made by photosynthesis in the photic zone (usually less than 100 meters in depth) and it is used in the respiration of plants and animals and in the decomposition of organic matter, probably by bacteria. Oxygen supplied at or near the surface diminishes steadily with depth in the top several hundred meters of ocean water, where most biological activity takes place.

In the deep ocean, bottom currents transport water that has been saturated with oxygen at high latitudes—for example, off Greenland, as indicated in Fig. 14–10, which presents figures for various latitudes. Oxygen demand at great depths is limited. Low temperatures reduce the metabolic rate of marine organisms, so that they use relatively less oxygen than do warm-water animals. Scarcity of food keeps deep-water populations small. Decomposition of detrital material is also minimal, since most of the supply is consumed by bacteria or other organisms

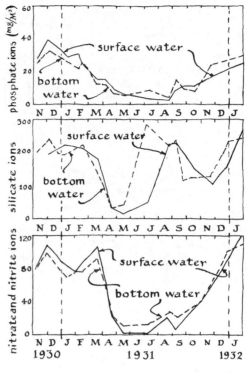

FIG. 14–9

Seasonal variation in nitrate, silicate, and phosphate in the English Channel. The curve for silicate reflects the fact that, in the channel, diatoms (yellow-green algae with siliceous shells) bloom profusely in spring and again in late summer. Their shells are voided by animals rather than being incorporated into animal tissue. Channel waters are so shallow and well mixed that there is little difference in nutrient concentration throughout the water column at any time. [After J. E. G. Raymont, *Plankton and Productivity in the Ocean*, Macmillan, New York (1963), 625 pp.]

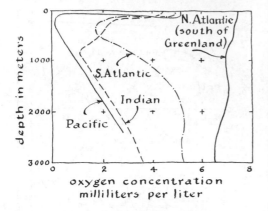

FIG. 14–10

Characteristic oxygen profiles of major ocean basins. Positions: 12°N, 137°W; 0°N, 80°E; 9°S, 5.5°W; 55°N, 45°W. (After Dietrich, 1963)

near the surface. Consequently, deep-ocean waters contain much of the oxygen with which they were saturated when they sank below the surface.

As has been previously noted, deep waters slowly move upward at mid- and low latitudes. The relatively oxygen-rich profile of the Atlantic Ocean (see Fig. 14–10) reflects the fact that surface waters are introduced at both poles, whereas the northern boundaries of the Pacific and Indian Oceans are blocked by land barriers. The time elapsed between the sinking of a water parcel and its re-emergence at the surface is therefore relatively shorter in the Atlantic than in the other two major ocean basins. South of Greenland, where the North Atlantic Deep Water sinks throughout the year, there is little variation in dissolved-oxygen profile throughout the water column.

As deep water rises in mid- and low latitudes, its dissolved-oxygen content is gradually reduced due to utilization by marine organisms. At these latitudes, in warm and temperate waters, there is a zone, varying in depth from 150 to 1000 meters, where dissolved oxygen concentrations are at a minimum (see Fig. 14–10). Little or no dissolved oxygen from the surface is mixed downward to this layer, and little oxygen remains in the water that has risen from great depths.

In certain areas of restricted circulation the deep waters are devoid of dissolved oxygen (i.e., are *anoxic*)—among these basins are the Black Sea, the Carioco Trench (Caribbean Sea), certain deep-ocean basins, and many fjords and inlets off Norway and British Columbia. In summer some relatively shallow estuaries may also have no oxygen below the layer of no net motion. In the absence of oxygen, *anaerobic* organisms (those which can live in the absence of free oxygen) break down nitrates (NO_3^-) or sulfates (SO_4^{2-}) to obtain oxygen for metabolic processes, giving off NH_4^+, N_2, and H_2S (hydrogen sulfide) as by-products. Decomposition of organic matter proceeds at greatly reduced rates under anoxic conditions. Sediments accumulating in these basins are usually rich in carbon as a result of this.

Carbon dioxide dissolves in water, forming carbonic acid, as shown in Fig. 14–11. This acid ionizes slightly, yielding carbonate and bicarbonate in equilibrium. As carbonate is used

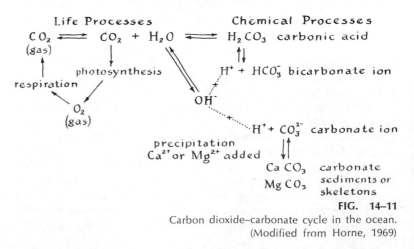

FIG. 14–11

Carbon dioxide–carbonate cycle in the ocean. (Modified from Horne, 1969)

in plant production, the equilibrium shifts to the right in Fig. 14–11 and causes more bicarbonate to dissociate. Carbonate precipitation in skeletons represents a net CO_2 loss to the cycle.

Another biologically significant aspect of CO_2 in the ocean is that the weak acid (H_2CO_3) acts as a buffer. Addition of acidic substances (H^+) causes the equilibrium in Fig. 14–11 to shift to the left, creating more bicarbonate. This ionizes very little, so that the pH (a measure of the relative alkalinity or acidity of a solution) remains fairly constant. Thus respiration and decomposition processes producing CO_2 hardly affect the pH of seawater, nor does the removal of CO_2 in photosynthesis. Surface layers, which experience a great deal of photosynthesis, respiration, and decomposition, vary more in pH than does deep water. Below the photic zone, the tendency is toward lowered pH (greater acidity) due to respiration of animals and bacteria, which releases CO_2.

Marine organisms extract from the water or their food those elements required for life processes. Table 14–3 shows concentrations of fundamental elements such as carbon, nitrogen, phosphorus, and silicon in typical marine organisms. A variety of trace elements are also selectively abstracted by many organisms as indicated in Table 14–4; some of these elements are thought to be unessential to life processes. Silver (Ag), cadmium (Cd), chromium (Cr), and lead (Pb), for instance, have no known func-

Table   14–3

RELATIVE COMPOSITION OF MARINE ANIMALS* (Na = 100)

Element	*Calanus* (Copepod)	Fish (Average)	*Archidoris* (Nudibranch)	Seawater	Concentration Factors Copepod	Fish	Nudibranch
Cl	194	—	180	180	1.1	—	1.0
Na	100	100	100	100	1.0	1.0	1.0
Mg	5.6	36	156	12.1	0.46	3.0	12.9
S	25.9	259	7.1	8.4	3.1	31	0.85
Ca	7.4	52	262	3.8	1.9	13.7	69
K	53.7	383	20	3.6	15	109	5.5
Br	1.7	—	—	0.6	—	—	3
C	1113	c. 4100	c. 480	0.26	4.3×10^3	15.8×10^3	1850
Sr	—	—	11	0.12	—	—	92
Si	1.3	—	—	0.001	13×10^3	—	—
F	—	—	69	0.01	—	—	6.9×10^3
N	280	1276	107	0.001	28×10^4	1276×10^3	107×10^3
P	24.1	256	6	0.0001	24×10^4	256×10^4	6×10^4
I	0.04	—	—	0.0005	80	—	—
Fe	1.3	1.3	0.23	0.0002	6×10^3	6×10^3	1000
Mn	—	0.0008	—	0.0001	—	8	—
Cu	—	0.008	0.43	0.0001	—	80	4.3×10^3

*After Sverdrup et al., 1942.

tion in biological systems, and in high concentrations some of these trace elements are damaging to organisms. This situation has serious implications in view of man's increasing use of the oceans as a repository for trace-element by-products and waste materials. In the absence of any mechanism for excretion of such substances, they accumulate in plant and animal fats and proteins. Such materials tend to be conserved in ecosystems, because they are passed along in food chains. Some may substitute for elements essential to metabolic processes; cadmium, for example, substitutes for zinc in humans, resulting in impairment of fat metabolism. Elements incorporated in animal or plant skeletons are likely to have shorter lifetimes in an ecosystem and to be more quickly incorporated in sediments instead.

Table 14–4

ELEMENT-ENRICHMENT FACTORS IN SHELLFISH*

| Element | Enrichment factors | | |
	Scallop	Oyster	Mussel
Ag	2,300	18,700	330
Cd	2,260,000	318,000	100,000
Cr	200,000	60,000	320,000
Cu	3,000	13,700	3,000
Fe	291,500	68,200	196,000
Mn	55,500	4,000	13,500
Mo	90	30	60
Ni	12,000	4,000	14,000
Pb	5,300	3,300	4,000
V	4,500	1,500	2,500
Zn	28,000	110,300	9,100

*After Horne, 1969.

DISSOLVED AND SUSPENDED ORGANIC MATTER

The ocean contains an enormous amount of matter in suspension, ranging in sizes from ocean-going ships down to microscopic particles (which are essentially dissolved in the seawater). Concentrations may vary widely throughout the world ocean; however, they are generally the lowest (less than 1 milligram per liter) in very deep water and in mid-ocean regions. Near coastlines, concentrations of inorganic and organic matter—both particulate and dissolved—rise to several parts per million or more, causing the yellow-green or brownish cast of turbid coastal waters (see Fig. 7–2).

Suspended and dissolved organic matter may be in excess of suspended sediment in a given area. Where less lithogenous material is present, the organic contribution is proportionally greater. Organic detritus may consist of unorganized or loosely aggregated protein, lipids, and carbohydrate molecules, together with plant and animal fragments, fecal matter, bacteria and microscopic plankton, and human wastes. A much greater amount of organic matter is present in seawater in the dissolved state, as Fig. 14–12 shows, than in any other state. It should

FIG. 14–12
Distribution of organic material in the photic zone. (Redrawn from Horne, 1969)

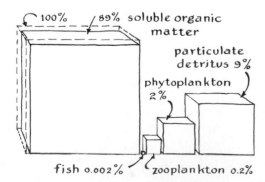

be noted, however, that sometimes the relative proportions of dissolved and suspended matter are more dependant on differences in filtering methods used in measurements than on absolutely reproducible differences in relative quantities.

Figure 14–13 shows diagramatically the relationships between living and nonliving matter in the ocean. Bacteria can utilize large dissolved organic molecules such as polysaccharides, proteins, and long-chain peptides only when they are adsorbed on or incorporated into some solid or coagulated substrate. Bacteria may, however, utilize solutions of simple sugars and amino acids derived from the hydrolysis (decomposition in water) of more complex organic molecules.

Most of the material suspended in seawater eventually settles out. Colloidal particles, however, are a constituent of seawater almost in the same sense that dissolved matter is. Random motion of water molecules provides sufficient upward force to keep these very small particles from sinking for some time; theoretically, they could remain dispersed indefinitely unless somehow aggregated to form particles that can settle out.

Much dissolved and suspended material is surface-active, meaning that, once having come in contact with any kind of surface or interface between dissimilar substances, it will tend to stick there. Water alone does not effectively wash it off the surface—that is, back into the dissolved or suspended state.

The surface-active character of dispersed organic material has far-reaching biological significance in the ocean. It results in a high concentration of material that would otherwise be extremely dilute, and thus makes it available to marine organisms in a form they can utilize as food. Many bottom-dwelling and drifting animals are adapted to filter this fine material from seawater or to scrape it off solid surfaces.

Seafarers and boat owners have long been aware of the fact that anything remaining beneath the ocean surface for a period of days or even hours becomes coated with a slimy film. Following this will be a gradual encrustation of phytoplankton, bacteria, barnacles, seaweeds, and other attached organisms. Piers, ships' hulls, in fact, any solid object placed in the water will soon become covered with these *fouling organisms*, as they are called. Some of these are simply seeking a firm surface for attachment (one is illustrated in Fig. 14–14), and some are attracted by the adsorbed organic material that collects on virtually every surface in the ocean.

Certain substances exhibit chemical or physical properties that especially favor adsorption of surface-active materials. Bottom clays and silts, for example, commonly collect organic material in concentrations up to 10^5 times the amount found in a dissolved or suspended state. Beach sand can retain perhaps one-tenth as much on a weight basis. This affects distribution of marine species; for instance, organisms adapted for scraping food substances from a solid surface (e.g., many kinds of snails) can flourish where such material is abundant.

Surface-active molecules tend to collect at air–water interfaces (see section on surface tension in Chapter 6). In calm lake waters, enough material may accumulate at the surface to support whole communities of tiny insects. In the ocean, wind

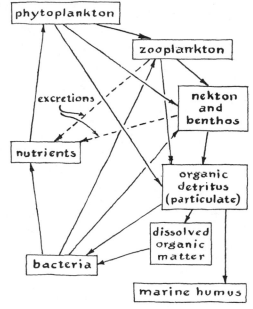

FIG. 14–13

Relationship of organic detritus and dissolved matter to living matter in marine food chains. (Revised from Raymont, 1963)

FIG. 14–14

The tunicate *Molgula*, a common fouling organism in Chesapeake Bay. These were found on the side of a tank. (Photograph courtesy Michael J. Reber)

and wave action generate smooth, glassy patches or *slicks* of organic material in bays and coastal areas. When winds exceed about 7 knots, the slicks break up into regularly spaced *windrows* (streaks), with their long axes oriented generally parallel to the wind direction. Windrows are often caused by Langmuir circulation, with alternate left- and right-handed helical vortices, which is set up by the wind in the surface-water layer; this phenomenon is illustrated in Fig. 14–15.

Floating particles respond to this cellular circulation (shown in Fig. 14–15). Swept toward the lines of convergence of the surface current, slicks, phytoplankton, and debris accumulate in rows, adsorbed on bubbles of foam. Because of the interaction between density differences and the velocity of water circulation, the entire mass resists being drawn down into the convergence, although a certain amount of material continues to circulate with the water.

Zooplankters, the tiny floating animals who eat plant plankton, may accumulate in areas of downwelling. Often, when air temperature is significantly different from water temperature, a thermal gradient exists at the convergence. Water coming across the top of circulation cells has been warmed (or cooled) by contact with the air; tiny marine animals may be sensitive to this difference in temperature. They may also be able to detect the changed color of the light coming through the aggregated material at the surface. Possibly, too, some means of chemical reception, or smell, indicates the presence of food. In any case, concentrations of zooplankton beneath zones of convergence may be up to 100 times greater than those in adjacent waters.

Another interesting example of the behavior of concentrated organic matter is the formation of "marine snow"—flakes of fragile, amorphous material, broken up by common sampling methods, which have been described by observers in submersibles operating in both deep and shallow waters. The particles range in size from 5 microns to several millimeters. They usually contain organic material, including plankton and bacteria, as well as inorganic components.

Tiny bubbles near the surface, or suspended particles, may

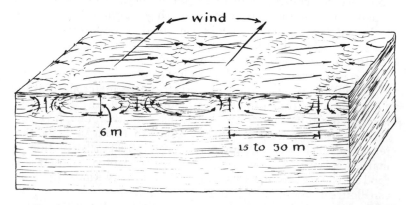

FIG. 14–15
Oily surface films and debris form windrows along lines of convergence in the Langmuir circulation set up by a steady wind.

serve as substrate for adsorption of dissolved and suspended hydrophobic compounds. These aggregations appear as amorphous, spherical aggregates in surface layers, high in carbohydrate but low in protein. The subsequent coalescence and additional adsorption of dissolved and particulate matter give rise to the clouds of firmer, plate-shaped flakes visible in deeper water. The flakes seem to have a higher bacterial content and to be high in protein relative to the amorphous aggregates seen near the surface. The large amounts of organic material organized in this way are a significant substrate for bacterial growth, and are also important food for the many animals who feed by filtering particles from seawater.

SALINITY

An animal or plant existing in seawater may be thought of as a container of salty water whose ionic composition is controlled within narrow limits. The organism is suspended or floating in a medium of salty water whose composition is different in slight but significant ways. The most critical difference as far as the organism is concerned is the proportion of water in the two solutions. This is because the cell wall (the "container") is semipermeable—that is, penetrable by water, but not by certain ions. In the process known as *osmosis*, when a concentration gradient exists between the solution inside and the one outside a container, water molecules generally pass through the semipermeable membrane toward the more concentrated solution, while many fewer molecules move in the opposite direction.

Present-day bony fishes, whose body fluids are less saline than seawater, are believed to have evolved in fresh water or in estuaries. Some ancient fishes were similar to the modern shark family, characterized by soft, cartilaginous skeletons and body fluids *isotonic* (equal in osmotic pressure) with seawater. Presumably, these entered fresh water via estuaries, gradually acquiring a tolerance for increasingly low salinities. There they developed bony skeletons and low-salinity body fluids. Eventually, some of them re-entered the ocean, and many lost their capacity to tolerate fresh water; but still others, such as the mullet, can pass from areas of low to high salinity without difficulty. Other species (shad, salmon) spend certain stages of their life cycle in fresh or brackish water, as we shall see when we discuss estuaries as nursery grounds in Chapter 17.

If an organism lives in a medium less dilute than its internal fluids, it constantly loses water and requires a mechanism for water retention. This is especially critical in small tide pools, where seaweeds, mussels, worms, and sea anemones may be isolated in a puddle of water every time the tide goes out. Heavy rain pouring fresh water into the pool dilutes the seawater so that the body fluids of marine organisms are distinctly concentrated with respect to the water around them. On a warm day much of the water in a tide pool may evaporate, leaving its inhabitants in a highly saline solution until they are submerged by the rising tide. An osmotic gradient is thus set up which may cause extra water to accumulate in body tissues; failure to eliminate the extra water can cause bursting of delicate cell walls.

In many estuarine and brackish-water forms, water content of body fluids is regulated by release of water through the gut. In others, the excretory or respiratory systems are modified for that purpose. In sharklike species, for instance, the urea content in their blood is maintained at quite a high level when they are living in the ocean. When the animals enter an estuary, they excrete urea to maintain osmotic balance with the surrounding water.

Some animals give off salts through their respiratory membranes (gills) to maintain osmotic balance if there is a sudden dilution in the environment. The intertidal green alga *Chaetomorpha linum* accumulates or extrudes KCl to compensate for changes in osmotic potential of the environment. The amount of fluid in the plant cells thus remains constant when heavy rains dilute the shallow coastal waters in which it grows.

Many animals selectively retain essential ions, at the same time adjusting to the osmotic pressure outside. The estuarine crab *Hemigrapsus*, for instance, has a specialized gland which acts to maintain the magnesium-ion concentration of its blood at about one-third that of the surrounding medium, regardless of salinity conditions. On the other hand, sodium-, potassium-, and calcium-ion concentrations are regulated by a mechanism in the gills so that they remain fairly constant whether the animal is in the ocean or in an estuary. These ions are more concentrated in the blood than in the environment of an estuary, but are comparable to the levels of seawater in a high-salinity environment.

The body fluids of most marine invertebrates are essentially isotonic with seawater. Such animals are limited to areas of the ocean where salinity is constant. Many can tolerate some change in the composition of their internal fluids; often, a sudden dilution of the environment causes the metabolic rate to increase sharply during a period of adjustment to new conditions. To survive in a coastal or estuarine environment, organisms must possess either specific osmoregulatory mechanisms or a high degree of tolerance to change.

Not all types of accommodation to problems of dilution and concentration are osmotic. Most marine fishes have skin, scales, and frequently a mucous coating to minimize penetration by water or salts that would cause osmotic imbalance. Some worms and mollusks can produce protective coats of slime. Shelled invertebrates in estuaries (clams, mussels) often seal their shells tightly against a hostile environment, and wait for better conditions. Some worms and clams retreat into burrows, out of contact with the surrounding water. Soft-bodied animals such as hydroids and worms can often contract sensitive organs or even their entire bodies temporarily to reduce exposure of surface areas. Usually there is a limit to the amount of time that can safely be spent in such a protected condition. For example, a long period of abnormally high river discharge is sometimes fatal to an oyster population, owing to the prolonged lowered salinity.

Salinity changes limit species distribution in certain marine environments. In general (as with most variable factors), the reproductive and juvenile stages are most sensitive. Some estuaries invertebrates, such as the Gulf shrimp, are tolerant to varying salinities in adult life, but return to the constant salinities of the open ocean to spawn. Others, such as oysters, spawn only in low-salinity coastal areas—a particularly favorable adaptation in that such predators as starfish and oyster-drill snails, which feed on oysters offshore, are inhibited by low salinities.

Most marine species are adapted to living in high salinity waters, in the open ocean or shallow seas. In coastal and estuarine areas where strong salinity gradients exist, the total number of species normally declines as salinity is reduced, reaches a minimum in brackish-water areas ($S \approx 5^0/_{00}$) and then increases in fresh water areas. Most animals living near the mouth of an estuary are marine species which have developed a tolerance for reduced or variable salinities. Some of these may penetrate the waters far upstream; a smaller number of inhabitants are fresh-water species with a tolerance for brackish conditions. A few organisms are uniquely adapted to estuarine conditions and can survive either in fresh water or in seawater. Dominant species in an estuary are often transients who spend part of their lives there and the rest in the more saline waters offshore.

There is generally a minimum number of species where salinity variation is greatest. Marine organisms penetrate farther upstream, and fresh-water species farther down, where tides are small and the salinity gradient is stable. Figure 14–16 shows the distribution of marine, brackish-water, and fresh-water organisms in a section of the Tees River (England) estuarine sys-

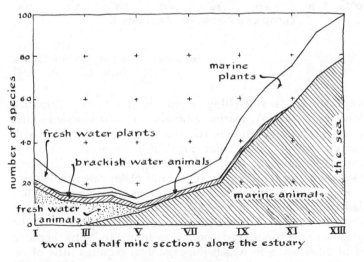

FIG. 14–16
Distribution of plants and animals according to salinity tolerance in the Tees River estuary, on the northeast coast of England. (After W. A. Alexander and others, *Survey of the River Tees.* Water Pollution Research Technical Paper No. 5. (London: Her Majesty's Stationery Office, 1935), pp. 66)

tem. In addition to those organisms whose location is more or less constant, many motile species come and go with the tides, thereby remaining in a zone of optimum (usually relatively high) salinity.

TEMPERATURE EFFECTS ON MARINE ORGANISMS

Except for coastal or estuarine areas, where salinity is a significant variable, temperature is perhaps the most important factor governing distribution of marine organisms. Of course, when we refer to temperature variations in the ocean, we are talking about small changes relative to those on land. Many marine organisms are adapted to living within a narrow temperature range; a few can live at almost any temperature, and no part of the ocean is entirely devoid of life.

Life styles of plants and animals along the shores, in the shallow seas, and in the surface layers of the oceans vary in large measure according to water temperature; thus we may designate *polar*, *temperate*, and *tropical* populations, as defined in Fig. 14–17. Species peculiar to each ocean region have physiological optima (most efficient metabolic processes) at temperatures typical of their normal environment.

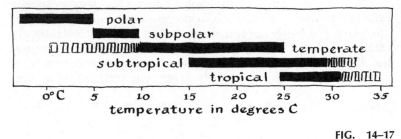

FIG. 14–17
Temperature limits of biogeographic zones.
(After Hedgpeth, 1957)

Except for the sea-dwelling mammals (porpoises, dolphins, sea lions, and whales), marine animals are cold-blooded—that is, their body temperature is for the most part uniform with the temperature of the environment. Within the acceptable range for a species, processes such as breathing rate (O_2 utilization), photosynthesis, general energy level, and often rate of development accelerate as temperature increases. In general, cold-blooded organisms can more easily tolerate great temperature changes toward low extremes than toward the high end of the tolerable range. Pronounced cooling is likely to bring about a period of quiescence, in which the organism requires little or no food and less oxygen than normal. An obvious parallel among land animals is the hibernation of bears, moles, and other mammals who "sleep" at lowered body temperatures while their normal food is in short supply.

The optimum temperature for a particular species is not necessarily the same for all stages and functions of life. Maximum and minimum levels of tolerance may vary with the stage of the life cycle; eggs and larvae commonly have a narrower temperature tolerance than mature adults. Individuals which drift in currents out of the temperature zone where spawning would normally occur may be unable to reproduce successfully. Organisms living in areas where there is wide seasonal variation in temperature are often limited to breeding during certain times of the year.

There is little temperature barrier to species distribution below about 2000 meters, where temperatures are uniformly low as previously shown in Fig. 7–20. Floating and swimming organisms native to this region are widely distributed. Certain species, tolerant of variations in depth and salinity, occur at high latitudes both north and south, as well as in the cold deep waters of mid- and low latitudes. Animals of the deep trenches are more likely to be distributed in patches—they tend to be trapped by the topography.

Temperature sensitivity determines how the composition of populations will change in a given area during the year. Particularly in temperate latitudes, where there is marked seasonal variation, species common to subarctic and subtropical climates may breed in the same area but at different times of the year. The seed crop that multiplies when the temperature is right is sometimes brought by currents from another area. In other cases, small stocks live at the breeding ground all year; these may enter a dormant or resting phase that persists until temperatures are again suitable for active growth and reproduction. Many fishes react to seasonal temperature changes by migrating in a regular pattern, their life cycles adapting accordingly.

Evidence of the dominant influence of temperature, rather than pressure or salinity, in distribution of species, is seen in temperate zones, where coastal waters undergo considerable seasonal temperature change. Here, coastal populations are often quite different from deep-sea populations at the same latitude. By contrast, in the Arctic, where the range of temperature between deep and shallow water is much smaller, species found in the deep ocean also thrive in shallow coastal waters.

Where industrial operations such as nuclear power stations have been designed to utilize water from a river or coastal region for waste-heat removal, communities of organisms may be strongly affected. Very large volumes of water are utilized for such purposes as cooling power plants and many organisms are killed in intake structures or by sudden heating within the plant. During a temperate summer, migrant organisms from lower latitudes may be attracted to the detritus-rich plume of warm water discharging from the industrial plant. They may remain there long after they would normally have migrated out of the region, and if operations are shut down, even briefly, organisms may be killed by thermal shock. Also, breeding cycles of native organisms may be altered or inhibited by the attendant abnormal temperature conditions.

Additional hazards to marine life are often associated with waste-heat disposal operations. Chlorine added to the intake water to prevent fouling within the cooling system eventually enters the discharge area. In addition, since waterpipes for these plants are often made of toxic metals (such as copper or nickel) to discourage fouling organisms, traces of these substances may be discharged to the environment. Oysters in such areas, for example, may accumulate copper in concentrations many times higher than in the surrounding water, the oyster "meat" sometimes taking on a bright green color as a result. When fuel oil is used for power in waterfront industries, spills present another hazard.

Several interesting relationships apparently depend on temperature. For instance, communities living in zones of high and constant temperature in the sea as well as on land are characterized by a large number of species, each containing a relatively small number of individuals. Conversely, Arctic and subarctic communities usually consist of many individuals of a few species, as indicated in Fig. 14–18. Temperature also affects the size of individuals within a species: cold-water flora and fauna tend to grow larger than related warm-water species, and individuals of the same species may be larger at maturity at high than at low latitudes.

The relationships between temperature, size of organism, number of individuals in a particular population, and diversity of species in a habitat or a biological community seem in some cases easily explained, in others not. Of first importance is the fact that for each 10°C rise in temperature, the metabolic rate of a cold-blooded animal approximately doubles. This is generally measured in terms of rate of oxygen consumption. On the average, sexual maturity occurs at an earlier age in warm climates than in colder ones, while life span is shorter.

In other words, developing and aging seem to be accelerated by increased metabolic rate. This means less growing time before reproduction, a smaller size at maturity, and more rapid successions of generations for organisms residing in warmer environments. It may also mean that the changes in genetic material which produce mutations (variations in form of physiology, inheritable by subsequent generations) are much more likely to occur as temperature increases—in other words, a more rapid succession of generations may promote an accelerated rate of genetic change within a population. This could be one explanation for the extreme diversity (large number of species, each containing few individuals) of tropical communities.

Naturalists have sometimes theorized that life is "easier" in the perpetual summertime of tropical waters than in either the changeable temperate zone or the near-freezing temperatures of high latitudes. A great variety of plants and animals inhabits the tropical ocean, especially along the shallow, sunlit margins of islands and atolls. This would seem to bear out the theory that warmer waters are somehow more hospitable. On the other hand, most observers have concluded that the total mass of living organisms (biomass) is larger on the average in subpolar areas than it is in the tropics.

Most of the great food fisheries and whaling grounds of the world are in very cold waters. Extensive upwelling at these latitudes results in high phytoplankton productivity and consequent abundance of food at every level of the food web. In addition, large numbers of fish of a single species are typical of subpolar areas, making them easier to catch and to process than the varied tropical and subtropical species.

FIG. 14–18
Diversity gradient for species of calanoid copepods (planktonic crustaceans) from the upper 50 meters indicates that number of species decreases with decreasing temperatures. Numbers of individuals are greatest at subpolar latitudes. [After A. G. Fischer, "Latitudinal Variations in Organic Diversity," *Evolution.* 14 (1960), 64–81]

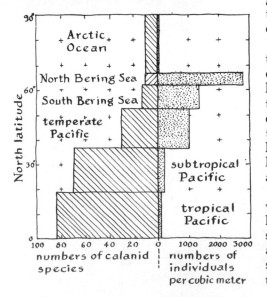

In tropical climates, temperatures are relatively uniform throughout the year. Organisms with a wide range of structural and functional variation can survive there, including some of the variant species created by mutation. In areas where temperatures fluctuate sharply during the year, or where they approach the limits of temperature tolerance (as in polar waters), a smaller fraction of the new species may survive. The tropical environment is said, therefore, to provide a greater degree of "receptivity" than any other temperature zone.

The pioneer work of Charles Darwin in this area showed that evolution of species by natural selection implies competition with other species for occupancy of a finite number of ecological niches. Normally, each niche is occupied by one species only, in a given community. If two species compete for occupancy of a niche, the superior competitor usually crowds out the other.

Where intense competition exists among species in a biological community, each species tends to occupy a narrow niche where living conditions are optimal. In this way, individuals can avoid competing with members of other species. Where few competing species are present, competition tends to concentrate between members of the same species. Individuals then minimize competition by adopting other life styles. This may mean accepting different foods, depositing eggs in a less sheltered spot, or attaching oneself to a rock where there is greater temperature variation than is optimal for that organism.

In tropical areas, where a relatively large number of species coexist, each is restricted to a narrow, highly specialized ecological niche, capable of supporting only a few individuals. If tropical conditions also provide a larger number of niches than elsewhere, as has been suggested, species diversification may progress more rapidly there than in other regions, with a lower rate of extinction.

To imagine how biological communities may evolve in an area previously devoid of life, we can refer to the simple model in Fig. 14–19. Through mutation and natural selection, a few immigrant species evolve into a diversified community capable of filling available ecological niches. Changes in the environment or in the structure of the community may result in the extinction of some species. As the community "matures," the rate of extinction of old species is likely to approach the rate of establishment of new species within the community, but does not necessarily reach or exceed it.

There is one particularly significant characteristic of tropical oceans not yet mentioned in this context—their relative constancy in temperature through geologic time. Large-scale changes in temperature have a drastic effect on biological communities (see Fig. 14–20), and we know that the tropical climate has been among the earth's most constant feature. There, communities have been evolving for longer times than most elsewhere, probably with a minimum of catastrophic disturbance. In polar and temperate zones, populations have been periodically extinguished by such major changes as the effects of ice ages over the last 2 to 3 million years. Evolution beyond 30°N and 30°S latitudes has received such serious setbacks that communities are relatively "immature," and hence much less diverse.

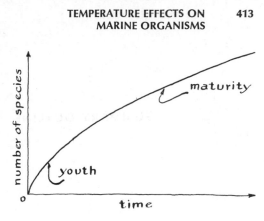

FIG. 14–19
Increasing diversity of species in a maturing biological community. (After Fischer, 1960)

FIG. 14–20
Setbacks to species diversification (extinction of species) in an area subject to fluctuations in climate over millions of years. (After Fischer, 1960)

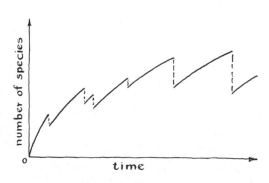

Since there are fewer species in less mature communities, individual organisms of these species may be present in great numbers.

SUMMARY OUTLINE

Marine habitats define conditions of life for organisms that live in them

Ecosystem—a community of interdependent organisms interacting dynamically with each other and with the physical environment

    Ecological niche—a functional status in the ecosystem

    Food chain—sequence of organisms in which each is food for another

    Food web—the most obvious example of interdependence among organisms, a group of interrelated food chains

    Ecological succession—change in species composition of an ecosystem resulting from changed conditions; final result is climax community

Nutrients and energy flow through an ecosystem

    Producers and consumers are different trophic levels in food web

    Nutrients are recycled and returned to surface seawater after release by decomposition of organic matter

    Energy is not recycled

Density

    Physical structure and mode of marine life different from terrestrial environment

    Sinking problem necessitates buoyancy mechanisms

Light

    Light essential in photosynthesis

    Light penetrates only surface layer in intensity adequate for photosynthesis

    Poles receive sufficient light for photosynthesis only 3 months a year; tropics all year

    Many animals respond to light stimuli, e.g. to migrate, to molt

Biogeochemical cycles

    Principal skeletal minerals in marine organisms are calcium carbonate and silica

    Carbon, nitrogen, and phosphorus are extracted by plants from seawater in constant ratio; nutrients are concentrated in animals; carbohydrates are converted into energy

    Bacteria—important as decomposers and protein synthesizers

    Phosphates important in photosynthesis—lack of phosphate in seawater often limits photosynthetic activity

    Nitrogen cycle—bacteria oxidize nitrogenous wastes in complex cycle

    Amounts of nutrients in seawater vary, depending on growth of plants and animals throughout the year; nutrients recycled from deep water to surface by oceanic circulation

    Oxygen profile in the ocean a measure of biological activity in absence of exposure to atmosphere at the surface

    Carbon dioxide in equilibrium with carbonate compounds in seawater acts as a buffer against acidity changes

    Marine organisms tend to extract trace elements from seawater; these often become concentrated and conserved in ecosystems

Dissolved and suspended organic matter

    Dissolved and suspended organic matter in seawater exceeds living organic fraction

    Surface-active nature of this material influences its distribution e.g., fouling, surface slicks, marine snow

Many marine organisms have body fluids of different salinity from seawater

Water moves through semipermeable membranes by osmosis in organisms

Regulatory mechanisms permit the organism to be more-or-less independent of salinity

Some organisms selectively retain or excrete certain ions

Many have nonpermeable shells, scales, or burrows

Reproductive and juvenile stages most affected by variable salinities

In estuaries, number of marine species declines as salinity decreases; fewer organisms are adapted for brackish water

Effect of temperature on marine organisms

Temperature is the most important factor governing distribution of species on a worldwide basis

Within acceptable range, cooling is better tolerated (by means of metabolic slowdown) than heating for cold-blooded organisms

Coastal populations at temperate latitudes must tolerate great temperature changes

Waste heat may affect local populations in a variety of ways

Communities at high, constant temperatures (e.g., tropics) are characterized by great species diversity; polar regions have least diversity

Accelerated metabolism, shorter life cycle, smaller-size typical of high-temperature species

Greatest total mass (biomass) of organisms often found in subpolar waters

Where great numbers of species coexist (warm climates), specialization of function is greatest, to minimize competition from other species

Evolution of species at low latitudes has suffered fewer setbacks due to critical climatic changes; therefore, more time has been available for biological accommodation; therefore, more diversity

SELECTED REFERENCES

BATES, MARSTON. 1960. *The Forest and the Sea*. Random House, New York. 277 pp. Essays on ecology and natural history, in a lyrical vein, by a well-known biologist.

COKER, R. E. 1962. *The Great and Wide Sea*. Harper, New York. 325 pp. General survey of oceanography and marine biology.

HARVEY, H. W. 1960. *The Chemistry and Fertility of Sea Waters*. Harvard University Press, Cambridge. 240 pp. Technical but clear and concise introduction to biogeochemical relationships in seawater.

HEDGPETH, JOEL W. (ed.) 1967. *Treatise on Marine Ecology and Paleoecology*, Vol. I: *Ecology*. The Geological Society of America, New York. 1296 pp. Detailed, scholarly monographs on habitats and biological conditions in the ocean, with emphasis on biological communities; extensive bibliographies.

HORNE, R. A. 1969. *Marine Chemistry: The Structure of Water and the Chemistry of The Hydrosphere*. Wiley–Interscience, New York. 588 pp. Advanced.

ISAACS, JOHN D. 1969. "The Nature of Oceanic Life," *Scientific American* (Sept. 1969). 221(3):146–62. Excellent summary of conditions that affect food relationships in the marine environment.

ODUM, EUGENE P. 1969. *Fundamentals of Ecology*, 2nd ed. W. B. Saunders, Philadelphia. 546 pp. Classic elaboration of basic ecological principles.

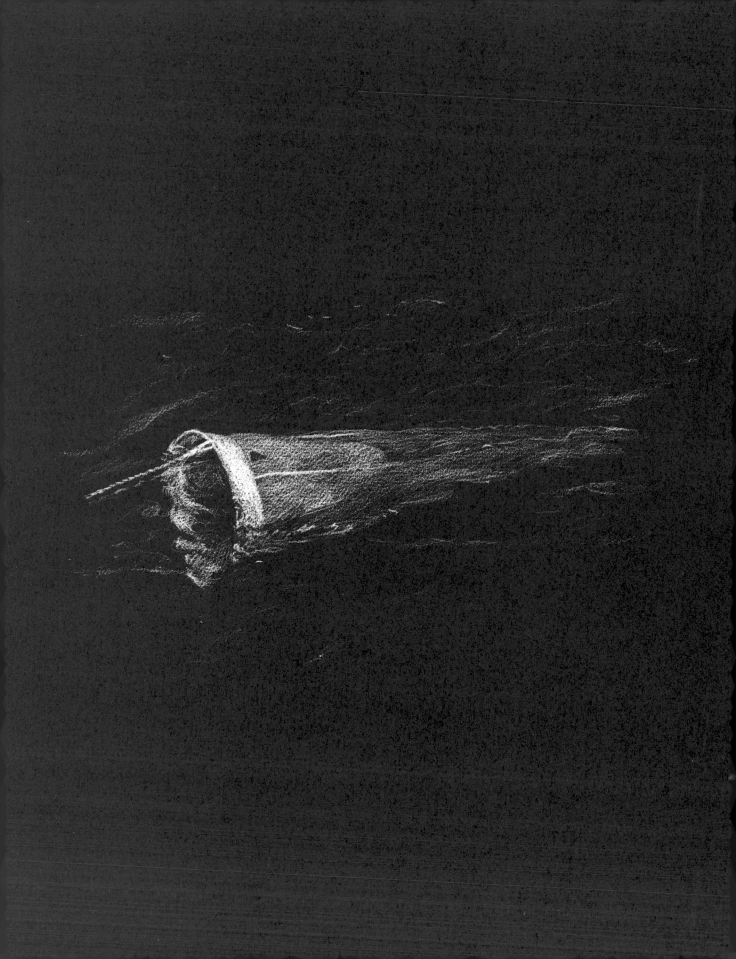

MARINE PLANKTON

Small plankton net.

fifteen

P*lanktors* are plants and animals which drift passively in the water because their swimming ability is limited or non-existent. Many can swim actively, but they are so small that they usually cannot move faster than the waters are flowing. Most large-scale migrations of plankton are vertical—from near-surface to deeper waters and vice versa. In most cases their horizontal movements are controlled by water currents.

Plankton are often classified by size, according to the openings in the nylon or silk mesh nets used to collect them. The largest plankton (*macroplankton*) range from 1 millimeter in diameter to 1 meter or more across (large jellyfish); with very few exceptions (some large diatoms), all within this size range are animals. A net of smaller mesh will serve to trap organisms between 0.5 and 1 millimeter across; these are mainly animals also, but plants of this size are likely to occur in coastal areas. Fine mesh nets are used to collect smaller plants, and some young animals and eggs as well, in the range 0.06 to 0.5 millimeter (*microplankton*). The tiniest plants (*nannoplankton*) and bacteria may be less than 0.005 millimeter across; these are not taken in nets, but filtered, allowed to settle out, or centrifuged from samples of seawater, or even removed using special filter paper.

PHYTOPLANKTON

The plant plankton, or *phytoplankton*, is the basis for life in the deep ocean, because it is the original source of food for marine animals. In general, phytoplankters are a better source of protein and fat than land plants, containing about 25 percent or more of each, depending on the species. Dominant phytoplankton types are described in Table 15–1. Figure 15–1 shows some ex-

Table   15–1

DOMINANT FORMS OF MARINE PHYTOPLANKTON

Type and characteristics	Location	Color and appearance	Method of reproduction
Blue-Green Algae: small, relatively simple cellular structure; cell wall of chitin	Mainly inshore, warmer surface waters—tropics	Blue-green or red; rafts of mottled filaments, or a colored "bloom" in the water	Simple fission (division of each cell into two); alternating growth and fission
Diatoms: silica and pectin "pillbox" cell wall, sculptured designs; the most important phytoplankton for productivity in the coastal ocean; has floating and attached forms	Everywhere in the surface ocean, especially in colder waters; can live in and around polar ice; some assimilate preformed organic matter below photic zone; some form a "resting spore" under adverse conditions	Yellow-green or brownish; single cells or chains of cells; radial or bilateral symmetry; many have conspicuous spines and other flotation processes	Division with splitting of nuclear material; average reduction of one cell-wall thickness at each division (see Fig. 15–2); when limiting size is reached, cell contents escape, grow in size, and form a new, larger cell
Dinoflagellates: large, diverse group; second to diatoms in productivity; many live as animals, ingest particulate food; some have cellulose "armor"	Found in all seas, & below photic zone; some are parasitic; great diversity among warm-water species; some can enter a resting stage when necessary for survival	Usually brownish, one celled; possess 2 flagellae (whiplike locomotive structures); many are luminescent	Simple, longitudinal or oblique divisions— daughter cells achieve the same size as parent before dividing
Coccolithophores very small—down to 0.005 mm; covered with calcareous plates, embedded in gelatinous sheath; important source of food for filter-feeding animals	Mainly in open seas, tropical and semi- tropical; sometimes proliferates near coasts; some animal-like forms found at depths to 3000 meters	Often flagellated, often round or football- shaped single cells; when present in great numbers, they give the water a milky look	Individuals form cysts from which spores arise to develop into new individuals
Silicoflagellates: very small, have silica skeleton; some nonphotosynthesizing forms	Widespread, but not numerous in colder seas worldwide	Single-celled, 1 or 2 flagellae; starlike, open shells	

amples of important types within each broad classification.

Distribution of phytoplankton in the ocean varies geographically, seasonally, and annually. Each ocean region supports characteristic dominant groups. Population densities are extremely variable, depending primarily on light intensity and nutrient concentrations, but also on rate of grazing by hebrivorous animals and rate of sinking below the sunlit zone. Factors governing productivity of phytoplankton in the ocean will be

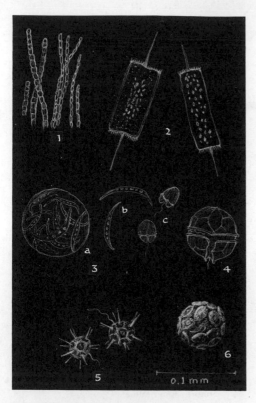

FIG. 15–1

Characteristic marine phytoplankton: (1) chains of the blue-green algae, *Oscillatoria;* (2) a diatom, *Ditylum,* (3) a dinoflagellate *Gymnodinium,* showing reproduction by formation of a cyst (a) in which secondary cysts (b) develop; these become free-swimming individuals (c); (4) a dinoflagellate, *Goniaulax,* (5) a silicoflagellate, *Distephanus;* (6) a coccolithophore, *Coccosphaera.* Note the protective covering of coccoliths—tiny calcareous plates.

discussed in detail when we study marine food resources in Chapter 17.

A major difference between plant production in the sea and that on land is that most marine plants are tiny cells, widely dispersed in a relatively dense fluid medium. Comparing the amount of plant food produced in a grassy meadow to that in a moderately rich area of the ocean, the land plants are hundreds of thousands of times more concentrated in relative mass. In fact, marine conditions limit plant life to such microscopic dimensions and meager density.

The evolution of nonattached plants in the ocean favored development of very small forms, because compounds which plants require are distributed more or less uniformly in seawater, in extremely dilute form. On land, plants live at the interface between solid earth and atmosphere. Roots extend downward a short distance to reach nutritive compounds left there by previous generations of plants. Leaves grow in the air where CO_2 and sunlight are plentiful. In the ocean, however, such small extensions cannot reach richer sources of light or nutrients. Much of the most nutrient-rich water in the sea lies below the depth of light penetration.

On land, air moves around us, removing gaseous wastes and bringing new supplies of needed gases. In the sea, a floating object does not move at all with respect to the water medium. Each water parcel, together with whatever may be suspended in it, moves as a unit. Thus a free-floating plant depends to a great extent on molecular diffusion in the water for nutrient intake and waste excretion. Small size (large ratio of surface to mass) facilitates exchange of ions between the plant and the water medium. For this reason, very few large plants are known to grow and reproduce successfully while floating in the open ocean.

Attached seaweeds are analogous to terrestrial plants: they remain fixed while water circulates around them, so that there is relative motion between plant and surrounding water. As essential gases and nutrients are depleted, fresh supplies are constantly being brought to the plant. The most successful floating plants, in terms of productivity in the ocean, have evolved the ability to move relative to the water medium. The flagellated phytoplankton (see Table 15–1) generally propel themselves upward, toward the light, with lashing motions of their whiplike flagellae. Planktonic diatoms, on the other hand, tend to sink slowly through the water. Turbulence in the surface layer recycles enough of these tiny plants toward the surface to prevent all individuals from sinking out of the sunlit photic layer. Those individuals which sink below the surface zone and are unable to return upward serve as food for organisms in deeper waters. (Diatom reproduction is illustrated in Fig. 15–2.)

The mechanisms which govern distributions of phytoplankton species are only partly understood. Selective grazing by zooplankton may sometimes favor survival of plant cells of a particular size in a certain area. Chemical by-products of one species may create conditions toxic to some other species. Upwelling areas are favorable to diatom populations, not only because nutrients are constantly being supplied from below the euphotic zone, but also because the upward flow of water helps diatoms remain near the surface. Dinoflagellates, on the other hand, are relatively more abundant in areas of slow downwelling—for instance, off some continental coasts in summertime; presumably, their swimming abilities are adequate to permit them to remain in the photic zone.

Armored flagellates and certain algae produce poisonous or growth-inhibiting by-products. The presence of these substances in the water may prevent normal development of succeeding phytoplankton populations.

In warm coastal waters, sudden "blooms" of phytoplankton may occur, causing the water to take on a red, brownish, or other color; this phenomenon is called "red tide" or "red water." These blooms may or may not be poisonous to fish and shellfish; sometimes they cause catastrophic kills. Toxicity may be due to the release of poisons, or to secondary conditions such as oxygen deficiency. Blooms of this type are favored by high nutrient concentrations, a stable water column, suitable temperatures, the presence of some individuals that have survived from previous generations to act as a seed crop, and certain essential micronutrients in the water—such as the B-vitamins. Experimentally obtained evidence has indicated that certain ions or radicals, such as dissolved iron compounds, may also trigger red tides.

ZOOPLANKTON

Marine plants need sunlight, and are thus restricted to the photic zone. Although marine animals are not limited in this way, many of them live by grazing on the phytoplankton and thus also have to live in surface waters.

Nearly all groups of marine animals have representatives among the *zooplankton*, minute floating animals. Besides those organisms whose whole life is spent drifting or swimming haphazardly, a great many bottom-dwelling species have planktonic larval stages. Adult organisms, living on sand or rock beneath the shallow waters of the continental shelf, release eggs or live young to the water. After a period of time the eggs develop into larvae. One or several *metamorphoses* (marked and abrupt change in body form) transform larvae into adults, and these return to the sea floor, usually for the remainder of their lives. The reproduction and metamorphosis of the common barnacle is shown in Fig. 15–3. The development from egg to adult may take anywhere from a few hours to a few weeks, depending on the species. Within a species, temperature often affects the length of planktonic embryonic or larval life.

Having planktonic larvae serves to disperse bottom-dwelling species, especially those that are attached to one spot but have young which drift far from where they hatched as eggs. On the other hand, drifting larvae may fail to find a suitable spot for attachment and so perish before metamorphosis to the adult stage can be completed.

A few animals can accomodate to either a drifting or a benthic mode of life, depending on water conditions. Still others,

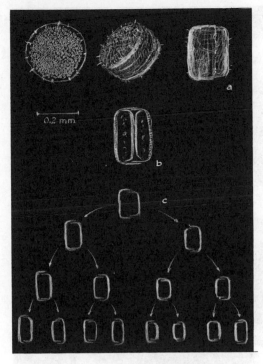

FIG. 15–2
Reproduction in diatoms. There is a "pillbox" shell (a), perforated by pores for exchange of metabolic products. When a diatom grows large enough to divide, the "lid" and "box" separate (b), and each gets one-half of the cell contents. A new "box" is then secreted over the exposed protoplasm, so that the original shell-half becomes the new "lid." Progressively, the daughter cells become smaller (c) in this type of division.

such as the *Obelia* in Fig. 15–4, are at first fixed to the bottom in an asexual stage, and later give rise by budding to a free-floating, jellyfish, sexual stage. Sexually produced larvae return to the bottom to repeat the cycle of alternate generations.

Most of the *nekton* (swimming animals such as fishes) also have planktonic eggs and young, which make an important contribution to the zooplankton and serve as food for larger animals.

Much of the food in the ocean occurs as extremely small, widely dispersed particles. An obvious method of gathering such food is to filter large quantities of water. The finer the filter, the smaller the particles that can be trapped, but the more energy that is required to pump water through it. In order to avoid having to filter large quantities of water, zooplankton show many adaptations for trapping food selectively. Elaborate combination trap-and-filter devices, many with sticky mucous coatings, have been developed by many animal groups.

There are many more kinds of marine zooplankton than we can deal with here, and our selection is admittedly somewhat arbitrary. An effort has been made, in the following discussion, to include representatives of the most important animal groups in the ocean, as well as a few that have unusually interesting adaptations for occupancy of specialized niches in the marine environment. The organisms are introduced in approximate order of increasing anatomical and physiological complexity.

Protozoa are microscopic animals, often one-celled. Most of

FIG. 15–3

Developmental stages of the acorn barnacle *Balanus*. The larval "nauplius" (upper left) swims freely, metamorphosing into the secondary "cypris" stage (upper right and center, showing inner and outer views). Having found an appropriate location for its stationary adult life, the animal forms a hard shell which can be opened for feeding or closed for protection. Cutaway views (bottom) show the animal inside its shell, closed and open.

FIG. 15–4

The coelenterate *Obelia* has feeding polyps into which waving appendages sweep food particles. It also has reproductive polyps which give rise to tiny, free-swimming jellyfish (*medusae*). These reproduce sexually to form a new, sessile animal.

them remain permanently among the plankton, although there are some bottom-dwelling forms. Most float freely, but some attach themselves to larger animals. Two important members of the protozoa are *Foraminifera* and *Radiolaria*.

Foraminifera illustrated in Fig. 15–5, are related to the amoeba, an inhabitant of fresh-water ponds well known to students of biology. The marine animal consists of a simple, amorphous speck of protoplasmic jelly which moves about by pushing out temporary "arms" and then pouring the rest of its cell contents into them. It ingests food by surrounding a particle with the formless *pseudopodia* (*pseudo*: "false"; *podia*: "feet") and dissolving or breaking the cell wall to take the food particle into the cell. Wastes are excreted in a similar manner—that is, directly through the cell wall. *Foraminifera* are surrounded by a calcareous outer shell, with perforations for the pseudopodia. As the animal grows, successive pseudopodia are covered with a layer of the shell material, giving it a spiny outline. Other forms secrete a chambered shell, like that of a miniature spiral snail.

Radiolarian protozoa have an internal capsule of chitin and a siliceous skeleton. The skeleton may be arranged as a sphere or as concentric spheres, with radiating pieces sculptured and branching into literally thousands of different intricate and beautiful designs, as shown in Fig. 15–6. These animals feed on other small animals and plants, often including detritus and bacteria. Some radiolarians live closely associated with phytoplankton, the two exchanging food materials to mutual benefit; others, including many species which live below the photic zone, are carnivorous. Individuals range in size from 0.1 millimeter to over 10 millimeters. *Foraminifera* and *Radiolaria* are most common in the warmer waters of the world, where their shells make important contributions to ocean sediment, as discussed in Chapter 5.

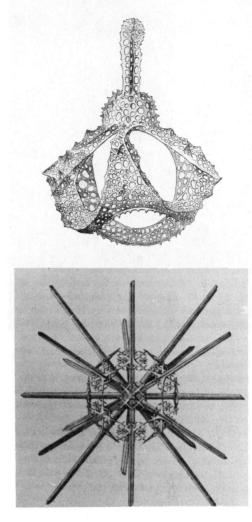

FIG. 15–5
Foraminifera: these marine protozoans *Hastigerina* (a) and *Globigerina* (b) have spiny skeletons that increase flotation and after death contribute to deep-ocean sediments. (From H. R. Brady, Report on the Foraminifera dredged by H.M.S. *Challenger* during the years, 1873–1876. Report of Scientific Results, H.M.S. *Challenger*, Zoology, vol. 9, 814 pp. 1884.)

FIG. 15–6
Radiolarians: these delicate, intricately formed skeletons are among the most beautiful organisms in the ocean. (From E. Haeckel. Report on the Radiolaria collected by H.M.S. *Challenger* during the years 1873–1876. Report of Scientific Results, H.M.S. *Challenger*, Zoology, vol. 18, 1803 pp. 1887.)

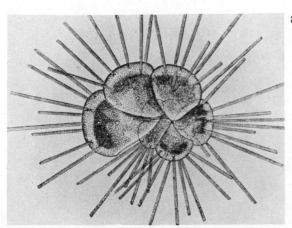

a

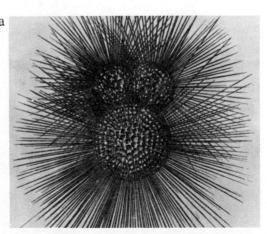

b

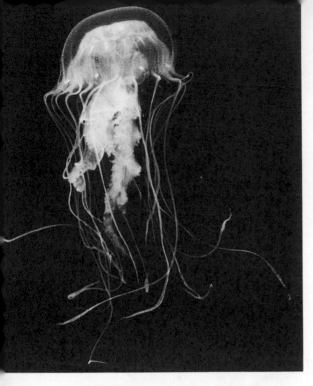

FIG. 15–7

The adult jellyfish *Chrysaora* measures about 10 cm across the bell and is found worldwide. The unpleasant sting of its tentacles, a mechanism for paralyzing prey, at times drives all but the most daring swimmers from parts of Chesapeake Bay. (Photograph courtesy of Michael J. Reber)

There are many very small protozoans, important consumers of the nannoplankton. *Tintinnids* ("bell-animalcules") are common throughout the world ocean. Their small bodies (a few tenths of a millimeter) are enclosed in goblet-shaped or tubular shells, with a circle of hairlike *cilia* around the mouth. The abundance of animals in a given area may vary widely, depending on seasonal environmental changes. In the open ocean, ciliate protozoans of simple structure and extremely small size (5 microns or so in diameter) are often abundant in plankton samples. (The tintinnid, as well as many other forms discussed in this chapter, is illustrated in the frontispiece associated with Chapter 17.)

Coelenterates are tubelike, relatively simple animals having a continuous body wall around a digestive cavity that has only one opening. This opening is surrounded by tentacles for capturing swimming or drifting animals. Many planktonic members of the group are attached during part of their life cycle, an example being the previously mentioned hydroid *Obelia* (see Fig. 15–4). Notice in Fig. 15–4 that the jellyfish or *medusa* stage is actually a modification of the same basic tubular structure seen in the feeding polyps of the attached, or *hydroid, stage*. Coelenterate medusae swim by rhythmic contractions of the "bell."

This group includes common jellyfish, such as the one in Fig. 15–7, as well as the spectacular colonial medusae (*siphonophores*) of the open sea (for example, the Portuguese man-o'-war illustrated in Fig. 15–8). Both jellyfish and siphonophores paralyze their prey with stinging cells that consist of a barb attached to a sack of poison. Dangling tentacles entangle the

FIG. 15–8

The Portuguese man-o'-war (*Physalia*) has a purple, air-filled bladder that floats at the surface. This animal is colonial in the sense that beneath the float hang clusters of polyp "persons," some of which entrap, paralyze, and engulf their victims, digesting them and absorbing their juices on the spot. Other polyps serve a reproductive function, producing sexual medusae, similar to those of *Obelia* (Fig. 15–4). The man-o'-war fish, immune to *Physalia*'s poison, feeds on discarded scraps and acts as a lure to attract victims. (Photograph courtesy Wometco Miami Seaquarium)

food and sweep it toward the mouth, inside the bell. Some forms swim upward, then sink with bell and tentacles extended, trapping prey beneath. A siphonophore is not a single animal, but a colony of medusae and polyp "persons" living together and functioning as one individual. Some species (e.g., *Physalia*) have a gas-filled float, but most have one or more swimming bells. Prey-catching, reproductive, and digestive polyps trail below like a fisherman's drift-net.

Although somewhat similar in appearance to coelenterate medusae, the beautiful and delicate *Ctenophora* (such as those shown in Fig. 15–9) are considered to be a separate group of animals. Small and jellylike, they are variously known as "sea-walnuts," "sea-gooseberries," or "comb-jellies." Some have trailing tentacles for capturing prey. Voraciously carnivorous, ctenophores often occur in great numbers, and may substantially reduce populations of crustaceans and young food fishes.

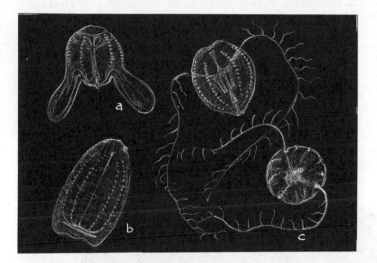

FIG.   15–9
Sea-walnuts (ctenophores) include various forms, three of which—(a) *Bolinopsis*, (b) *Meroë* and (c) *Pleurobrachia*—are shown here. They swim by motions of the fused cilia, of which there are typically eight rows on the body. Medusae, on the other hand, swim by pulsations of the bell.

Planktonic worms belong to several major animal groups, among the most important being the "glass-worms" or "arrow-worms," *Chaetognatha*, such as those in Fig. 15–10 (the name means "bristle-jawed"). Mature individuals, 2 to 8 centimeters in length, are active and abundant from the surface to great depths in all seas. They are transparent, and possess chitinous spikes embedded in muscle around the mouth. These spikes can open out like a fan and then turn inward as seizing jaws to consume their prey. By flexing their long bands of longitudinal muscles, arrow-worms can dart rapidly through the water in pursuit of the great quantities of small zooplankton their diet demands. There is no larval stage; small worms hatch directly from fertilized eggs.

FIG.   15–10
These chaetognath worms include several species of *Sagitta* common to North Atlantic waters. These are sometimes used as tracers for water masses, since different species are found in waters having different nutrient compositions. [Redrawn from G. E. Newell and R. C. Newell, *Marine Plankton: a Practical Guide*, Hutchinson Educational, London (1967), 221 pp.]

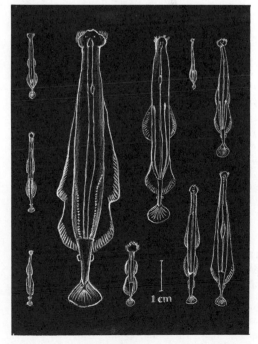

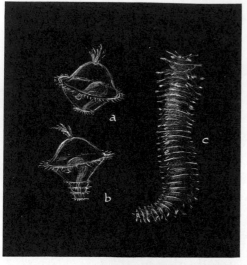

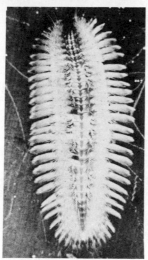

Common marine *segmented worms* include the *Polychaeta* (bristle-worms), named for the rows of chitinous bristles along their sides. Most have planktonic larvae; adults (shown in Fig. 15–11) live on the ocean floor, where they burrow in soft sediment or inhabit fibrous tubes. Some species, like *Tomopteris* (see Frontispiece, Chapter 17) are permanently planktonic. At first the young worm is toplike or rounded, as in Fig. 15–11(a). The mouth is at the "equator," below a ciliated girdle, with the anus at the bottom. Waving cilia send the animal spinning through the water, and currents set up by the spinning motion move food particles toward the mouth. This type of larva is called a *trochophore*. As development continues, other ciliated rings and segments appear below the mouth, as shown in Fig. 15–11(b). In some forms the trochophore becomes the head of the adult worm, in others the girdle and other larval structures are suddenly sloughed off when the animal settles on the ocean floor and completes metamorphosis, as shown in Fig. 15–11(c).

The *Mollusca* include snails and slugs (*gastropods*), clams and oysters (*bivalves*), squid and octopus (*cephalopods*), as well as a few less important forms. Most of this group are benthic as adults, with planktonic larvae. The development of a gastropod provides a good example of molluscan larval adaptations for feeding and locomotion, as well as illustrating the typical metamorphosis. The first stage is a trochophore, which usually remains in the original egg capsule, shown in Figs. 15–12(a) and (b). In the later, free-swimming stage, the ciliated girdle is extended at the sides to form large lobes—shown in Fig. 15–12(c)—which not only support the increased weight of the larva (now

FIG. 15–11
Larval stages of generalized polychaete worm. The trochophore stage (a) spins through the water like a top by waving its tuft of cilia; the gut can be seen within. In (b) and (c), we can see how the larva metamorphoses to adult form. (Redrawn from Hardy, 1965) Photograph of an adult benthic bristleworm. (Photograph Wometco Miami Seaquarium)

FIG. 15–12
Metamorphosis of a marine snail: (b), (d), and (g), which diagram the internal structure of (a), (c), and (f), respectively, show that the gut, originally straight, twists into a loop so that mouth and anus are both at the bottom of the shell, where the opening is in the adult form. (Redrawn in part from Hardy, 1965)

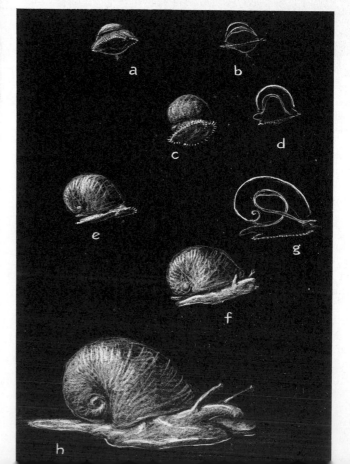

known as a *veliger*), but also prevent the typical spinning motion of the trochophore, so that the animal swims straight ahead. The lobes extend out beyond the body of the larva, and the internal organs twist into a loop, as shown in Fig. 15–12(g). Mouth and anus use the same shell aperture; when the head and foot are withdrawn for safety inside the shell, a hard plate (the *operculum*) attached to the foot covers that opening tightly. As the shell develops, the snail sinks to the ocean floor to take up benthic life. The development of the oyster (illustrated in Fig. 15–13), which does not have to twist its body to fit into a single shell, is a variation on the same basic theme.

The *pteropod* group of gastropods are permanent members of the plankton. They occur in dense swarms in all seas, but especially in warmer waters—where their shells may make an important contribution to sediments. In pteropods, the characteristic foot of the snail family is modified into fins, as illustrated in Fig. 15–14. Rows of beating cilia set up currents that drive small food particles against a mucous secretion, and then bear the sticky mass toward the mouth. In surface waters pteropods eat phytoplankton, in deeper currents they are carnivorous.

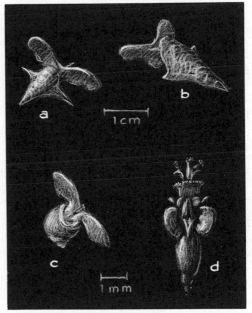

FIG. 15–14
Marine pteropods: *Clio*—(a) and (b)—found throughout the world ocean, *Spirialis* (c), and *Dexiobranchaea* (d). (Redrawn in part from Hardy, 1965)

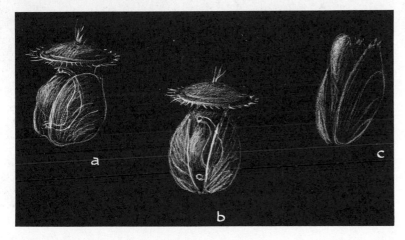

FIG. 15–13
Metamorphosis of a generalized bivalve, showing the trochophorelike larva (a) and (b) with a shell forming at the sides. It has been drawn as if transparent, to show formation of the gut. Nearing the adult state (c), the foot is visible where the shell is partially open.

Some cephalopods are permanent members of the plankton, but most are large and active swimmers or crawlers. There is no larval form; the drifting or floating young have the form of tiny adults. Their long arms and tentacles with suckers are adaptations for preying on swimming animals.

Arthropods are perhaps the most important group of zooplankters. We are all familiar with shrimp, lobsters, and crabs, and even perhaps with barnacles, but it is the planktonic forms which are significant in marine food chains. Arthropods are divided into two extremely important classes, plus a few minor groups. On land, *Insecta* are the most numerous of all animals; in the ocean, *Crustacea* (shown in Fig. 15–15) occupy a com-

FIG.   15–15

Planktonic crustacea. Clockwise, beginning at the bottom:

Amphipods hatch in brood-pouches, emerging as tiny adults. Some groups are locally very abundant in the plankton. The *Hyperiid* type shown here has characteristically huge compound eyes. The species illustrated is *Themisto*, an important food of the North Sea herring.

Cumaceans are more common in shallow coastal waters than in the deep ocean. They often remain on the ocean floor during the day, rising to surface layers at night to feed.

Mysids are a shrimplike form. Many species live near continental margins, but a few are found at depths of 4000 meters or more. Metamorphosis, from egg through nauplius stages to adult form, takes place in a brood-pouch under the front legs of the female.

Copepods (meaning "oar-footed") occur at all depths, in all parts of the ocean. They have been termed the "insects of the sea" because of their abundance and avriety. The eggs hatch into free-swimming naupli, which often moult many times before attaining adult form.

Euphausids are relatives of the edible shrimp, which is a bottom-dweller. All euphausids are planktonic, and are found everywhere in the ocean. Many are filter-feeders, especially in high-latitude waters; others have grinding jaws to accommodate larger food particles.

Many *Isopod* species have migrated from the ocean and are now land-dwellers. The species *Gnathia*, of which the larva (left) and adult male are shown here, lives as a fish parasite during the larval stages, but is free-swimming in adult life.

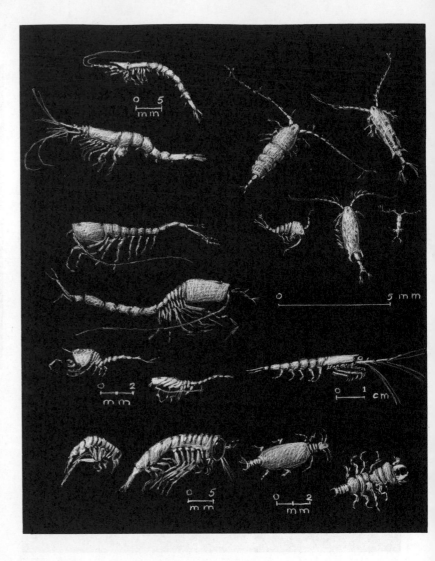

parable position, constituting 70 percent or more of the zooplankton both by bulk and by numbers.

Like all arthropods, crustaceans are segmented and have several pairs of appendages, each pair usually specialized for a particular function—feeding, movement, sensation, reproduction, or other behavior. The stiff outer shell typically contains calcium. There is a characteristic free-swimming *nauplius* larval stage, which commonly molts many times, each time with some change of form (as illustrated in Fig. 15–16), before assuming the adult shape.

Two permanently planktonic species are of particular importance. The first, the *copepods* (some are shown in Fig. 15–17), are present in all seas at all seasons. They are among the most numerous animals in the sea, possibly even in the world, and

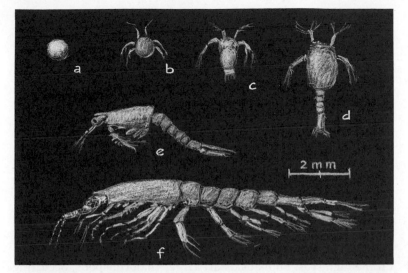

FIG.   15–16
Life cycles of *Euphausia superba:* Egg (a),
early and late nauplius stages—(b), (c), (d)—
the intermediate *mysis* stage, and mature adult
are shown.

FIG.   15–17
Calanoid copepods: (a) *Calanus finmarchicus*,
(b) *Rhincalanus*, (c) *Pseudocalanus*, (d) *Para-*
calanus, (e) *Microcalanus*, (f) *Euchaeia*,
(g) *Temora*, (h) *Eurytemora*, (i) *Metridia*,
(j) *Centropages*, (k) *Isias*—all found in the
vicinity of Great Britain. (Redrawn from
Newell, 1967). (l) and (m) are tropical varieties
of *Calocalanus*. [Redrawn from A. Hardy, *Great*
Waters, Harper & Row, New York (1967) 541 pp.]

FIG. 15–18

Echinoderm larvae: (a) larva of sea star *Luidia* (front and side view); (b) pluteus (easel-shaped) larva of heart urchin *Spatangus*; (c) two stages of larval development in the sea star *Asterias rubens*; (d) pluteus larva of brittle-star *Ophiothrix*; (e) three species of sea urchin, pluteus larvae (left to right)—*Echinocardium*, *Psammechinus*, and *Echinocyamus*.

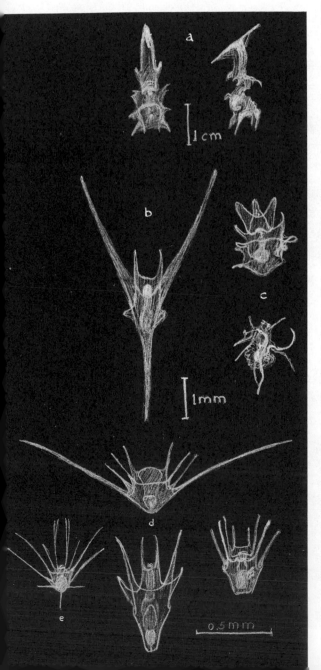

among the foremost of plant consumers. Many of their 6000 species are benthic and some are parasitic, but the greatest number of individuals are planktonic, ranging in length from about 0.3 millimeter to about 8 millimeters. Copepods have many feathery, curved bristles which form a filter chamber behind the mouth. By constant movement of the appendages near the head, two opposing spiral currents are set up. One moves the animal forward, the other forces a stream of water into the filter chamber. Tiny plants and fine particles thus trapped are passed along to the mouth. Copepods in northern seas, their bodies rich in proteins and fats, are eaten in great numbers by herring, whales, and many of the other organisms that are important as a source of food for man.

Shrimplike *euphausids*, commonly known as "krill," are another important crustacean group. Dense swarms of these animals feed on diatoms, and themselves comprise the chief food of the large, filter-feeding whales, as well as of many fishes on the high seas. Euphausids are larger than copepods, occurring up to 5 centimeters long, and they are found on or near the bottom as well as in surface waters. The group includes herbivorous and, especially in warm waters, carnivorous species.

There are many other important types of marine crustacea, of which we shall mention only a few. The *cladocerans* ("water-fleas"), *Evadne* and *Podon* (see Frontispiece, Chapter 17), sometimes occur in tremendous concentrations in coastal oceans (e.g., the North Sea) and in estuaries (e.g., Chesapeake Bay). They are small—around 1 millimeter long—and possess a characteristic single large eye formed by fusion of the two larval eyes.

The marine insect *Halobates*, though not a crustacean, is a unique and interesting member of the marine arthropod plankton. An air-breather, *Halobates* lives on top of the water rather than in it, apparently supported by surface tension. Individuals have been captured thousands of kilometers from shore in the Atlantic and Pacific Oceans, but the details of its feeding mechanisms and reproductive cycle are as yet unknown.

The last major invertebrate group, *echinoderms*, occur in the plankton only as larvae and are less important than the others we have mentioned. Benthic forms include mobile sea stars (starfish), sea cucumbers, and sea urchins, and attached sea lilies. Adults are radially symmetrical, a body plan common among attached and slow-moving organisms, but their typical *pluteus* larval stages are bilaterally symmetrical as seen in Fig. 15–18.

Chordates, with notochords and gill slits, have a relatively recent evolutionary history. A primitive chordate division, which lacks a true backbone, is represented in the plankton by the *tunicates* (sea-squirts). Among the most interesting of these animals is the tadpolelike *Oikopleura*, shown in Fig. 15–19, which surrounds itself with a delicate, gelatinous "house." Water, containing food particles, enters the "house" through a mesh which filters out large objects. The water is then pumped through fine tubes which permit passage to only the very smallest plankton, such as coccolithophores and dinoflagellates. Whenever the filter mechanism becomes clogged, the animal creates a

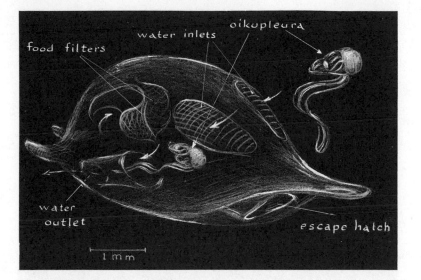

food filters

water inlets

oikopleura

water outlet

escape hatch

1 mm

FIG.   15–19

Oikopleura is shown here in the gelatinous,
balloonlike structure which it secretes around
itself in order to filter plankton from the
water. Undulations of the animal's tail drive
a stream of water through the delicate structure,
where microscopic food particles are trapped
on filters, to be eaten later. Oikopleura can
swim out of this "house" if danger seems
imminent, and secrete a new one at leisure.

new one from its own secretions. *Oikopleura* and related forms
are important constituents in the diet of some food fishes.

Transparent, barrel-shaped *salps* and *doliolids*, illustrated in
Fig. 15–20, occur in significant numbers among warm-water
plankton. These tunicates also feed on tiny plankton by pumping water through their bodies. They seem not to be widely
eaten by fishes and thus may represent more or less of a dead
end in marine food chains. They float singly or in chains, and
are sometimes seen in great masses at the surface in warm, quiet
waters.

Vertebrates, including the bony and cartilaginous fishes and
mammals of the sea, make up the rest of the chordates. Fish
eggs, larvae, and juveniles (shown in Fig. 15–21) are an important source of food for carnivores in the ocean. Some fishes,
such as herring and sand eel, attach their eggs to rocks or
vegetation, some lay them in "nests" near the shore, and some
lay them in gelatinous masses; but most fish eggs are released
and fertilized in deep water near the continental margins. These
drift with the plankton, and when sufficiently developed they
hatch and begin to feed. Depending on temperature and species,
this may be as soon as 1 or 2 days after release. They may then
remain in the planktonic stage for weeks. Some larvae upon
hatching are still virtually embryonic.

Most fishes discharge their eggs directly into the ocean where
they are particularly likely to be eaten before they reach matu-

FIG. 15–20

Tunicates (sea-squirts): these animals pump water through their bodies by contraction of the muscular bands encircling them. *Doliolum* (a) has a life cycle of alternative sexual and asexual generations—the asexual is illustrated here. Reproduction is by budding— little groups of cells are formed on the underside of the mature individual (upper left), whence they migrate along the body of the "parent." They then attach themselves in rows to an appendage at the front end, the *cadophore*, shown enlarged at lower left. As the cadophore grows longer, the buds grow to maturity. Some remain attached as a colony; others break off to live as individuals. (Redrawn from Hardy, 1965).
Salpa (b) also exhibits alternation of generations—the asexual stage is shown here, in solitary and aggregate forms (the individual at lower right is an aggregating type). At the top of the group is a "nurse" salp with a cluster of buds on its underside that will break off to form sexually reproducing individuals. (Redrawn from Hardy, 1965)
(c) At top—solitary (right) and aggregate forms of *Salpa democratica*; center—solitary (left) and aggregate forms of *Salpa asymmetrica*; below—solitary (bottom) and aggregate forms of *Salpa zonaria*. (Redrawn from Newell, 1967)

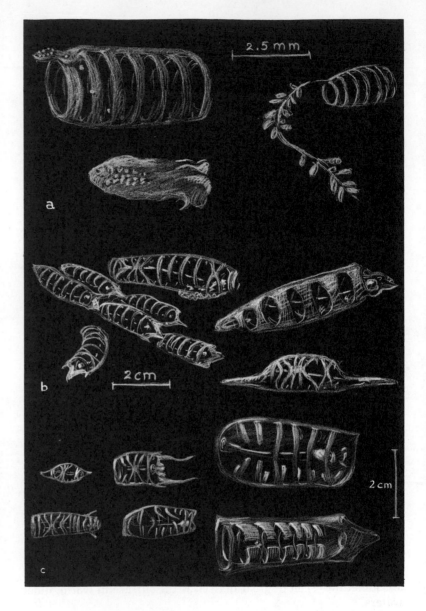

rity. Consequently, such fishes usually produce relatively more eggs than those species which spawn in more sheltered areas or protect their eggs in some other way. While herring, for instance, lay tens of thousands of eggs in a single spawning period, haddock and plaice release hundreds of thousands, and the figure for cod is in the millions. Some fishes bear their young live (Pacific surf perches, some sharks); some bury their eggs in gravel or sand (Pacific salmon, grunion, some smelts). The eggs of some fishes develop in a protective, horny capsule (some sharks and skates); these are not planktonic, and therefore they are much less abundant. Some small adult fishes, mostly deep-water species, are planktonic.

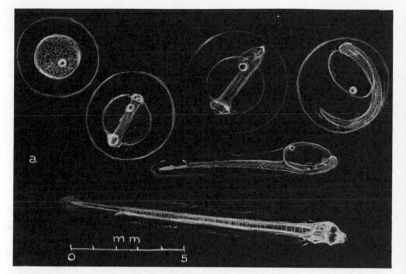
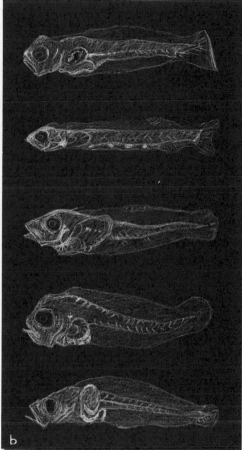

FIG. 15–21
(a) Development of the Atlantic fish *Sardina pilchardus*; the tiny fish is at first seen developing inside the egg. (b) Larvae of (from the top) mackerel, anchovy, rockfish, croaker, halibut.

FACTORS INFLUENCING ZOOPLANKTON DISTRIBUTION

Zooplankton fall into two main groups—coastal plankton living in continental-shelf waters, and oceanic forms characteristic of open seas, with overlapping among certain species. Coastal plankton are usually larger, whereas open-ocean populations contain great numbers of extremely small animals. Continental-shelf waters contain a high proportion of eggs and larvae of benthic organisms and fishes, including nearshore and estuarine organisms tolerant of low-salinity waters. One must also consider the influence of temperature on the distribution of plankton in the coastal- and open-ocean areas. Recall that temperature provinces may be visualized as parallel bands circling the earth (discussed in Chapter 14).

Coastal regions have greater diversity of species than one finds in the open ocean. Variations in temperature, salinity, and turbidity, and differences in the character of the bottom, together with wave, current, and tides, create variety in the environment, with corresponding diversity among species and communities. In addition, coastal waters are rich in nutrients because coastal-circulation regimes rather quickly recycle them back into surface waters.

In the open ocean, conditions are often uniform over large areas. The main differences are based on change of temperature with latitude. In general, warm waters support the most diverse populations. Biologists have identified many hundreds of open-ocean, warm-water zooplankton species, but only about one-tenth as many are native to polar areas. Of course, most of the world ocean lies within the tropic-subtropic belt, which stretches around the world with relatively little barrier to communication. Thus, many warm-waters species are found in all three major oceans.

Abundance of polar and subpolar plankters varies sharply according to the time of year. There is a single, short growing season for plants (1 to 1.5 months), mainly diatoms. It is a rich bloom when it does occur, for then days are long and the waters within the photic zone are abundantly supplied with nitrates and phosphates. After the diatom bloom is underway, the seed crop of zooplankton, having spent the winter in a dormant state in deep waters, rises to the surface to breed extensively while food supplies last. Their numbers drop off sharply with the approach of winter; only a small percentage survives until the next growing season.

A few polar species are found in the high-latitude surface waters of both hemispheres. They apparently move under the equatorial belt of warm surface water, or were able to do so relatively recently. An even smaller number, mainly a dozen or so species of copepod, are found in every part of the world ocean.

In addition to the surface communities, hundreds of species of crustaceans, coelenterates, and chaetognaths inhabit the deep sea, from 400 meters depth to the bottom, where cold and dark prevail year-round. All of these are carnivores or detritus feeders.

There is great variety in depth of maximum zooplankton populations from one area to another. In the coastal ocean, numbers of species and individuals increase toward intermediate depths. In the open ocean of the Northern Hemisphere, maximum numbers of individuals occur in waters above 500 meters, but the maximum number of species is found between 1000 and 3000 meters depth. In the southernmost Atlantic, the maximum number of species has been found to occur at 600 to 700 meters, and studies in the Indian Ocean have found little zooplankton below 1500 meters.

Where food is scarce, zooplankters reproduce less frequently, live longer, and utilize available food more efficiently. Where food is abundant, generations may succeed each other every 10 days or so. Under such conditions, food ingested in large quantities may pass through the animals virtually undigested.

Distribution of temporary plankton (larval stages only) is particularly dependent on temperature conditions. At low latitudes, where water temperatures are virtually uniform throughout the year, spawning can occur at any time, and benthic organisms may prolong their larval stages indefinitely. Most tropical benthic organisms have a free-swimming stage in the life cycle.

In temperate zones, breeding and larval stages are restricted to a certain time of the year, depending on the species. Immigrants from the subtropics breed in summer, those from subarctic regions in early spring or fall. In polar waters, the short summer season limits the number of temporary plankton species to those forms that can tolerate a very brief season of abundant plant food; only about 5 percent of marine invertebrates are able to complete their metamorphosis within a few weeks at low temperatures, so most polar benthos (bottom-dwelling organ-

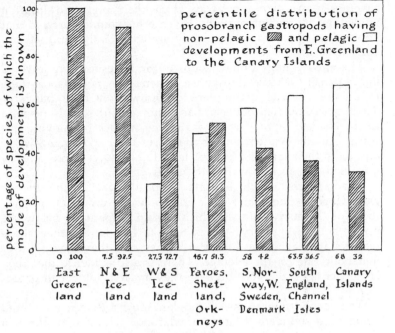

percentile distribution of prosobranch gastropods having non-pelagic ▨ and pelagic ☐ developments from E. Greenland to the Canary Islands

0 100	7.5 92.5	27.3 72.7	48.7 51.3	58 42	63.5 36.5	68 32
East Green- land	N & E Ice- land	W & S Ice- land	Faroes, Shet- land, Ork- neys	S. Nor- way, W. Sweden, Denmark	South England, Channel Isles	Canary Islands

y-axis: percentage of species of which the mode of development is known

FIG. 15–22

Moving south from polar to mid-latitude waters, the percentage of species of snails having pelagic (free-swimming) larvae increases with increasing temperatures. (After Raymont, 1963)

isms) have no free-swimming stage. (This is also the case for benthos of the deep-ocean floor.) Figure 15–22 shows results of a study made in the North Atlantic, documenting the effects of a temperature gradient on mode of larval development among gastropods.

Thus far we have discussed organisms which regularly inhabit surface or coastal waters. In the open ocean, however, particularly at mid- and low latitudes, some planktonic larvae occupy deeper layers, especially those from organisms that spend their adult lives at depths of 500 to 1500 meters. Most mid-level fishes, as well as many crustaceans and mollusks, have larvae that mature near the surface, where, feeding on nanno-plankton and tiny zooplankters, they have a better chance for survival than in the relatively barren waters below.

ZOOPLANKTON MIGRATION

Most zooplankters migrate vertically each day in order to remain at a preferred level of light intensity. At sunset they rise toward the surface to feed on phytoplankton, and at dawn they swim downward to a midday maximum depth as far as 800 meters below the surface. A copepod 2 millimeters long can swim upward at a speed of 15 meters per hour, and downward 3 times as fast; a larger euphausid can climb at a rate of 100 meters per hour, and descend even more rapidly.

There are several advantages to this behavior: first, the dark provides relative safety from predators. Second, cooler water in deep layers lowers metabolic rate, so that less energy is expended during the resting period. Third, in some areas, surface currents move plankton away from their preferred location while they are feeding at night. By entering countercurrents that flow below the surface layer, they are effortlessly returned up-current by day.

The upward migration of zooplankton in response to lowered light intensity is a useful response. When currents transport them to an area of rich phytoplankton growth or to a patch of detritus accumulated between Langmuir cells, they may respond to the shadow of the food mass at the surface by moving upward. This takes them out of the lower layer of moving water and halts their lateral drift. At nightfall, they can rise directly into food-rich surface waters to graze.

Zooplankters which migrate vertically enrich the deeper water layers. They, and their metabolic by-products, provide food for the many animals which remain perpetually below the photic zone. Migratory species tend to live longer than those which stay near the surface at all times, and they have a lower metabolic rate on the average. There is also evidence that they lay fewer eggs, possibly because they are less subject to predation.

The vertical migrations of zooplankton have been monitored using echo sounders. Masses of plankton, along with the fish who follow them in their daily cycle, often form a sound-reflective layer at mid-depth, known as the "deep scattering layer," detected in every part of the world ocean. (Certain organisms—for instance, fish with swimbladders—give back a good echo, whereas jellyfish are poor sound reflectors.) The presence of this layer, detectable at less than 200 meters at night and as deep as 700 meters by day, mystified scientists for years until an understanding of vertical migration provided an explanation.

Those plankton lacking a pattern of daily vertical migration exhibit a constant tendency to be carried out of their breeding range by surface currents. Therefore, mechanisms must exist to ensure that a seed crop or breeding population survives until the next breeding season. Complex cycles of seasonal migration have been described for zooplankton populations in coastal regions, where organisms carried away from the area by currents at one season are brought back at another. Breeding cycles are often found to correlate with such seasonal migrations. In Antarctic waters, for instance, large populations of euphausids drift northward in the upper 200 meters of water during the summer months, while food is plentiful, and then sink to between 500 and 1000 meters depth, depending on species. Deeper currents carry them southward in winter; when summer comes again they rise, breed, and repeat the cycle.

Distances traveled in this way can amount to hundreds of kilometers. One species of euphausid shrimp has a 2-year cycle of development, adapted to Antarctic current patterns. Spawning starts in November or December, when adults are close to the Antarctic Convergence. Eggs, released at a depth of about 200 meters, sink lower and are carried southward as larvae for 9 months by deep subsurface currents. During this period, the larvae have some vertical diurnal migration. They rise in the Antarctic Divergence and drift northward in surface currents for 12 months as adolescents, during which time they do not migrate vertically. When they return to the original breeding site, the cycle is repeated.

Dense irregular concentrations of phytoplankton and zooplankton are often found over large areas of the world ocean; descriptions of such noncontinuous distributions often refer to it as "plankton patchiness." Patchiness may sometimes be related to the cyclic nature of the food demands of animals (e.g. zooplankton). Sometimes dense swarms of zooplankters occur alternately with areas having abundant phytoplankters. If the animals seem not to be attracted by the plants nearby, it may be that they are in a nonfeeding stage of their life cycle. Alternatively, the plants may be of a type that the zooplankters are unable to eat. Patchiness can also be caused by physical processes such as currents or Langmuir cells as previously noted.

The origins of water masses can sometimes be determined by analyzing their plankton populations, for a water mass having particular characteristics—such as a certain temperature and salinity—sustains a characteristic type of zooplankton, which in turn can breed successfully only under those conditions. As the mass is carried by currents from one part of the world ocean toward another, water of different types is entrained. Gradually, the physical and chemical properties of the original mass change until they can no longer be used to define its point of origin. Thus "indicator species," which can survive as adults under the new conditions but are incapable of breeding can be useful in tracing water movements over long distances.

Medusae of *Obelia*, for instance, if collected in the open ocean, would indicate that the water in which they were found had coastal origins. Deep-living organisms taken at the surface identify water from areas of upwelling. Organisms that can breed only at tropical temperatures sometimes survive as individual adults far into a temperate region, and thus serve as a tag for an equatorial water mass. By keeping track of an organism with a wide range of tolerance for change and of those following species suited only to the new conditions, fairly wide dispersals can be traced. Phytoplankton are less valuable for studies of this type because they usually do not live long enough.

Two species of the arrow-worm *Sagitta* have been used for many years to predict the productivity of fishing grounds in the North Sea. *Sagitta setosa* is found in a low-phosphate water mass and *Sagitta elegans* in another mass having different biological properties and a much richer average plankton crop. The relative position of these masses varies from year to year, and fishermen prefer the *elegans* water mass because the chances of a good catch are greater there than in *setosa* water. Indicator species used to study water-mass movements include salps off East Australia, and oceanic fish to determine how much ocean water enters the Strait of Georgia, in western Canada.

COLOR AND BIOLUMINESCENCE

Small animals in the ocean are for the most part colorless and transparent (particularly those in the upper layers), which renders them less visible to predators. In warm seas and clear, offshore waters, surface plankton are more likely to be blue than any other color. Below the illuminated zone, from about 500 meters to 3000 or 4000 meters depth, large numbers of animals are predominantly red or dark brown. Some of these migrate hundreds of meters upward at night. Such colorations have obvious survival value—in surface waters, for instance, blue animals are protectively colored, and in dimly lit layers, where red light rays have all been absorbed, reddish animals who migrate upward from great depths appear black or are virtually invisible.

Many marine animals have the power to generate light chemically (*bioluminescence*), without heat. This characteristic is particularly noticeable at night, when sailors see the wake from their ship glowing with swarms of luminous flagellates or jellyfish. Disturbance of the water surface excites the animals and causes them to emit light.

Many marine bacteria which possess bioluminescence shine constantly and impart a glow to the many animals whose whole bodies or specialized organs serve as hosts for luminescent microorganisms. Larger animals often have light organs which they can control—turning them off and on and regulating the light's intensity. Among zooplankton, jellyfish are particularly light-productive, and many small crustaceans give off luminous clouds when startled. Most deep-sea crustaceans, cephalopods, and fishes are equiped with light organs, frequently concentrated near the eyes and along the sides. Maximum production is usually in the blue-green part of the spectrum—around 0.47 micron; this is near the point of maximum light transmission

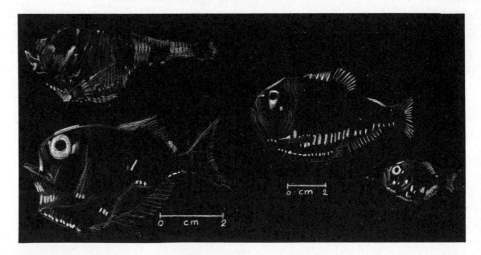

FIG. 15–23

Four species of hatchet-fishes, all small— the largest species is *Argyropelacus gigas,* at top. These fishes are common at depths of 100–500 meters, although some have been taken at depths up to 2000 meters. They are characterized by patterns of luminous organs and by their distinctive "hatchet" shape.

for seawater, and the point of greatest visual acuity for many deep-water organisms.

Bioluminescence may serve a variety of functions in the ocean. Many squids and deep-sea shrimps emit a cloud of luminous ink when threatened in order to confuse the enemy. The lantern-fishes (see Appendix IV) and hatchet-fishes (illustrated in Fig. 15–23) are noted for the many different patterns of their light organs, each characteristic of a particular species. Often the patterns are different for males and females of the same species; presumably, these enable potential mates to recognize each other. Regularly emitted flashes of light or distinctive arrangements of light organs may also permit members of a given species to find one another and remain together, forming schools.

In the deep ocean, fish often possess a shining "lure" in or near the mouth (see Vignette VI, Chapter 17) for attracting organisms who are themselves seeking food. Many kinds of marine fishes, crustaceans, and squids, including luminous species, are readily attracted to a moving light. Below the sunlit zone, such "living light" serves not only to attract food organisms but perhaps also may help to illuminate the prey.

Movements of planktonic plants and animals are dominated by water movements

Phytoplankton

Ultimate source of food for nearly all marine animals

Conditions of life for marine plants favor microscopic size

Nutrients very dilute in ocean waters

Water medium does not move relative to the plant plankton; therefore, transfer of materials between the plant and the environment takes place by molecular diffusion

Distribution of organisms governed by oceanographic conditions

Zooplankton

Feed mainly in surface waters

Many benthic forms have planktonic larvae

Filter-feeding a common adaptation

Protozoa—mainly single-celled or noncellular—planktonic throughout life

Include *Foraminifera, Radiolaria*, ciliates

Open-ocean forms often extremely small

Coelenterates—many alternate generations

Include medusae and colonial forms

Ctenophores—voracious carnivores

Planktonic worms

Include chaetognaths, active and carnivorous

Polychaete larvae develop from trophophore structure

Mollusca

Gastropods—some benthic, have veliger stage after trochophore

Pteropods—planktonic as adults

Crustacea—correspond to insects on land

Nauplius is typical larva

Copepods are the most numerous, of primary importance in food chains

Euphausids important in high-latitude seas

Echinoderms—occur mainly as larvae

Chordates—recently evolved

Sea-squirts (*Oikopleura*) have intricate feeding mechanism

Salps and doliolids important as consumers in warm waters, have few enemies themselves

Vertebrates—fish larvae: lay many eggs; most larvae are eaten by predators

Factors influencing zooplankton

Coastal species

Many larval benthos

Diversity of species, related to diversity of environments

Polar species—little diversity, abundance varies seasonally

Maximum numbers are found in surface layers, maximum number of species at intermediate depths

Life cycles are shorter where food is abundant and temperatures are high

Distribution of larval plankton varies seasonally in temperate zones

Many deep-ocean species have surface-dwelling larvae

Zooplankton migration

Daily vertical migration typical—feeding at surface during the night, resting in darker, deep water by day

Maintenance of population at appropriate breeding location depends on taking advantage of current direction changes at different water levels, to return to breeding area if carried away on surface currents

Patchiness of plankton caused by complexities of life cycles, grazing requirements
Water masses can often be identified by plankton populations
Color and bioluminescence
Absence of color in surface layers a protective adaptation
Bioluminescence a common adaptation; also in very deep water, to confuse predators, and to attract and illuminate food organisms

SELECTED REFERENCES

FRASER, JAMES. 1962. *Nature Adrift: the Story of Marine Plankton.* G. T. Foulis, London. 178 pp. Concise, readable survey of marine-plankton forms, their behavior and ecology.

GRAHAM, MICHAEL. 1956. *Sea Fisheries: Their Investigations in the United Kingdom.* Edward Arnold (Publishers) Ltd. London. 487 pp. Good discussion of relationships between plankton and production of fishes and shellfish. Well illustrated sections on fishing techniques and equipment.

HARDY, ALISTER. 1965. *The Open Sea: Its Natural History—Part 1: The World of Plankton.* One-volume ed. Houghton Mifflin, Boston. 335 pp. A classic in the field, by a well-known British naturalist; nicely illustrated.

IDYLL, C. P. 1964. *Abyss: The Deep Sea and The Creatures That Live In It.* Thomas Y. Crowell Company, New York. 396 pp. Elementary discussion of marine life with particular emphasis on deep-ocean organisms; also discusses various curiosities such as sea monsters.

RAYMONT, J. E. G. 1963. *Plankton and Productivity in the Oceans.* Macmillan, Pergamon Press, New York. 625 pp. Comprehensive survey of plankton research literature, including aspects of physical oceanography as they relate to marine biology.

RUSSELL-HUNTER, W. D. 1970. *Aquatic Productivity: An Introduction to Some Basic Aspects of Biological Oceanography and Limnology.* The Macmillan Company, London. 306 pp. Modern treatment of plankton and productivity in lakes and in the ocean, emphasizing North Atlantic forms. Good sections on plankton and world food production.

STEELE, J. H. (EDITOR) 1970. *Marine Food Chains.* University of California Press, Berkeley and Los Angeles. 552 pp. Collection of papers dealing with relationships between plankton, productivity, and energy transfers in the ocean and estuaries. Intermediate to advanced in difficulty.

WIMPENNY, R. S. 1966. *The Plankton of the Sea.* American Elsevier, New York. 426 pp. Emphasis on the taxonomy of animal plankton; well illustrated.

THE BENTHOS

FIG. 16–1
On this rugged stretch of Maine coast, brown algae grow thickly along a steep rock face, then cease abruptly except near the bottom, where growth extends further out toward the water.

sixteen

Benthic organisms live in a two-dimensional world, in contrast to the free movement in three dimensions experienced by swimming and floating animals. There are three main strategies of benthic life: attachment to a firm surface, free movement on the bottom, or concealment below the surface of the ocean floor. These correspond roughly to three basic ways of obtaining food: filtering it from seawater, catching it by active predation, or scavenging bottom detritus. We are familiar with filter-feeding and predation from our previous discussion of plankton, but deposit-feeding is an adaptation peculiar to bottom-dwellers.

Many animals use some combination of these life styles. A crab or a worm for instance, may live in a sand or rock burrow, but emerge to hunt actively for prey. Animals with heavy armor for protection, such as snails or sea urchins, have limited mobility due to the weight and bulk of the shell, so they often scavenge for detritus on the bottom rather than seek live victims. Fixed sea anemones, on the other hand, are often predators.

For attached plants, the marine environment offers severe limitations. Very few kinds of attached plants live in the sea, in contrast to the great variety of animal life. The reverse is true on land, where there are many kinds of plants and relatively less variety among animals. Only 2 percent of the ocean is shallow enough for photosynthesis to be carried on by attached plants—that is, approximately in the region above the 30-meter depth contour. Most plants cannot maintain a firm attachment to mud or sand; they require a substrate such as a reef or rock surface (as in Fig. 16–1) that can be gripped firmly enough to resist wave and current action. Fixed seaweeds thus make little contribution to total food production in the oceans, although they are a factor in the high productivity of coastal areas.

In very cold, shallow water, surface-living bottom life (*epifauna*) is relatively sparse. A short growing season for plants and extreme cold, together with sharp salinity changes as ice melts and freezes, make the polar coastal ocean a rigorous climate. In warmer water, the coastal, and especially the intertidal zones, are much richer in bottom life, and far more diversified. Where light is adequate for phytoplankton production during a large part of the year and salinity variation is moderate, coastal waters support substantial epifaunal communities. Rocks, reefs, holes and crannies, tide pools, and muds rich in organic detritus all offer a variety of exposures to light, air, and wave action and create diversity in intertidal habitats. Farther offshore the bottom is generally sandy; and out farther yet, fine-grained, soft muds predominate. In the tropics, where water is warm and temperature constant, making seasonal adjustment unnecessary, the potential for diversification of coastal species is enormous.

Diversity of life styles among benthic organisms is related to diversity of benthic habitats. Slight variations in composition along a section of the continental shelf, for instance, may favor the growth of quite different populations of animals. On a rocky headland with different weather exposures, there is often a different assortment of seaweeds, crustaceans, and so forth at each location.

Many free-swimming larvae of benthic animals have some ability to postpone metamorphosis while seeking a preferred substrate. They settle to explore the bottom, then swim away if it does not suit. Possession of this ability greatly improves an animal's chances of finding a site where it can survive and reproduce. Attraction to an established colony of the same species often seems to be a factor in the selection process. Oyster larvae, for instance, prefer a bed of oyster shells (even empty ones) to any other substrate. On the other hand, rejection of previously colonized areas can provide incentive for migration of the species into fresh territory.

Large numbers of recently metamorphosed bivalves, barnacles, or polychaetes may suddenly settle in a new area; but mortality is high among these young animals, and wholesale disappearance of such young colonies is very common. Especially among groups that breed rapidly and do not live long, there is often wide variation in the size of a brood from year to year. Rate of growth can also affect the standing crop; fast-growing organisms can quickly take over an area, squeezing out previous residents. The abundance of five species of clam from shallow waters along the coast of Norway, indicated in Fig. 16–2, illustrates these yearly fluctuations.

Among relatively long-lived species, where many generations are present at all times, communities tend to be less subject to fluctuation. Nonpelagic larval development also contributes to stability of benthic populations, because planktonic larvae may

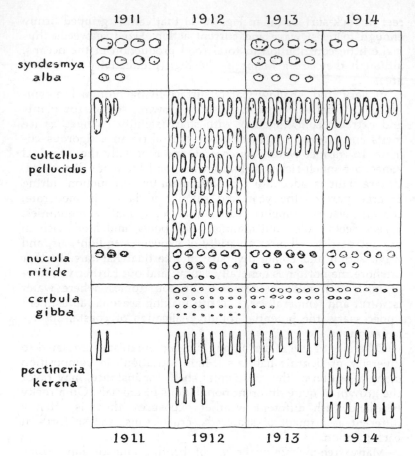

FIG. 16–2
Annual variation in abundance of five clam species studied over a 4-year period. (After Raymont, J. E. G. 1963. Plankton and Productivity in the Ocean. Macmillan, Pergamon Press, New York, 625 pp.

drift away from the colony so that older individuals are not replaced as they die.

ROCKY BEACHES

Intertidal areas provide a vertical sequence of habitats, known as *zones*, for benthic organisms. Each zone or horizontal band is occupied by a distinctly different assemblage of plants and animals. This is most easily seen on a rocky beach, where many plants and animals are attached to the rocks; on sandy or muddy shores, zonation is also present, but most of the zones here occur below the sediment surface. Even on rock beaches it is not altogether obvious what is living where, since many creatures hide in moist, protected crevices when the tide goes out.

Conditions in intertidal areas range from nearly always dry (highest-high-tide level) to nearly always submerged (lowest-low-tide level). Population zones are often separated by a sharp line rather than there being a gradation of one population into another. This is in keeping with the intense competition for living space in intertidal regions. At the line where the two populations meet, neither enjoys an advantage over the other; above and below the line populations are determined by differences in ability to endure exposure to air, and often also by the ability of their predators and diseases to survive the same conditions. Such zonation among barnacles on the Maine coast is shown in Fig. 16–3.

Where winds, sun, or waves create a severe climate, or where the rock face is very steep, attached seaweeds may be sparse or absent. In more protected areas, a variety of plants may cling to bare rocks along the seashore. Above the high-tide mark, seawater only washes over that area at higher spring tides and during storms, but salt spray is a regular feature. Resistance to drying is the prime requirement in this region. Several varieties of blue-green algae, lichens (an alga growing symbiotically among the filaments of a fungus) and certain small snails are common. The filamentous green alga known as sea-hair sometimes grows in moist, protected spots.

Between the high-tide and low-tide marks, especially where heavy surf is a factor or where tidal scouring is intense, the ability to cling tightly to the rock is a necessity. Barnacles such as those illustrated in Fig. 16–3 often dominate the upper, more exposed regions, where the rock is covered by water less than half of each tidal day. In shady, protected areas, the barnacle zone extends farther up the rock face than it does on dry, sunny surfaces. These animals feed while covered by water at high tide, filtering small particles from the water with their feathery "feet" (see Fig. 15–3). The shell has a complex, four-part lid which shuts tightly during exposure to air because of the need to conserve moisture during the long period of air exposure.

Flat, cone-shaped snail called limpets also commonly reside in rocky intertidal areas. These animals also feed underwater, by passing slowly over the rock and scraping off accumulated detritus. During low tide, or when subjected to wave action, a limpet returns to his own depression on the rock face, hollowed out by years of limpet occupancy so that it is a perfect fit for the animal's rather flat, smooth-rimmed shell. The seal is so good that a partial vacuum is created beneath: it is nearly impossible to dislodge a limpet from his rock by hand. Another favored spot for attachment is on top of another limpet.

Farther down the rock face, tube-worms and sea anemones may compete with the barnacles; closer to the low-tide mark, dense communities of mussels are quite common, and various kinds of attached algae as seen along the Maine coast in Fig. 16–4. Among plants, preference for a specific light intensity govern distribution of species. Green algae may be found in areas ranging from somewhat above the low-tide mark to a depth of perhaps 10 meters in temperate and tropical zones. Varieties range from lush-looking bright green sea-lettuces, up to 1 meter in length, to delicate, mossy types only a few centimeters long. Red algae grow worldwide, but most abundantly in temperate and tropical seas. They prefer the dim light of very deep water or well-shaded pools. Brown algae flourish best

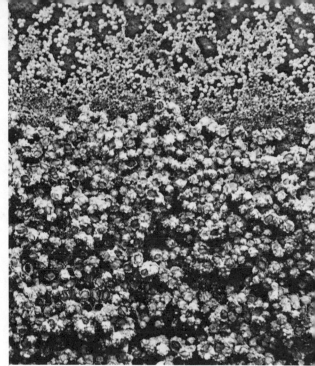

FIG. 16–3
Zonation among barnacles, Maine coast. A small species dominates the upper layer, separated by a sharply defined boundary from the larger species below.

FIG. 16–4
A rocky intertidal area in Maine shows zonation of attached animals and plants: barnacles above, brown algae below, and at the bottom, underwater even at low tide, green and red algae grow in a shady pool. Snails, worms, sea stars, and tiny crustaceans live among the plants.

in colder water, although some species are found on rocky coasts throughout the world. An abundant form in North America is the brown rockweed, with its clusters of berrylike floats. Marine algae have no true roots, but absorb nutrients from seawater directly into their fronds. Some benthic algae are attached to the bottom by root-like structures called "hold-fasts."

Where rocks rise irregularly from the sea, are broken into boulders, or are arranged as a series of "stairstep" levels, tide pools contain a more-or-less specialized group of plants and animals. A protected orientation in crevices of the rock often permits more delicate organisms to live in these pools, and a large variety of species may be present in a small area, as shown in Fig. 16–5. On Southern California beaches, colorful sea anemones such as those in Fig. 16–6 are abundant; these are relatives of the corals and hydroids, having tentacles which bear stinging cells for capturing small animals. Tiny, shrimplike crustaceans, swimming worms, and many kinds of snails are common tide-pool residents, especially where deep crevices or beds of seaweed retain an appreciable amount of water during low tide. Evaporation, overheating, and oxygen depletion occur on warm, sunny days, whereas heavy rains cause dilution within a matter of minutes. Success for a tide pool organism thus depends on a tolerance for sudden changes in temperature, salinity, and dissolved-oxygen content of the water.

FIG. 16–5
Many varieties of red and green algae shade one another in this tide pool on the Maine coast, providing shelter for small shellfish and worms. The area photographed is about 1 meter across.

The most populous zone on a rocky beach tends to be located around and below low-tide level. Starfish and crabs are common, usually hidden in rocky crevices. Small scavenging snails are found where protected niches contain stagnant water and decaying debris brought in by the tide. Hermit crabs have a

FIG. 16–6
Warm-water sea anemones with sea stars at
the New York Aquarium. The animals at left
center and far right are open; the mouth
opening is at the center of the "flower."
When startled, they close instantly, giving the
appearance of a wrinkled stump (right center).

particularly successful mode of life (shown in Fig. 16–7) where
the waves are not too violent. These lively, shrimplike creatures
have no protective covering on the rear parts of their bodies, so
they appropriate empty snail shells thrown up by the waves.
When the animal outgrows a shell, he abandons it and finds a
larger one.

Sea anemones (shown in Fig. 16–8), sea urchins, and sea-
cucumbers may be locally abundant below the low-water mark.
Colonies of delicate hydroids (such as those in Fig. 16–9) grow

FIG. 16–7
Symbiotic (mutually rewarding) relationship
between a hermitcrab and a sea anemone.
The stinging tentacles of the anemone afford
protection to the crab, who in turn provides
transportation in search of fresh food sources.
(Photograph courtesy Marineland of Florida)

FIG. 16–8
Sea anemones grow thickly among rocks below
the low-tide mark, to depths of about 15
meters in the vicinity of New York Harbor.
These were photographed at the New York
Aquarium.

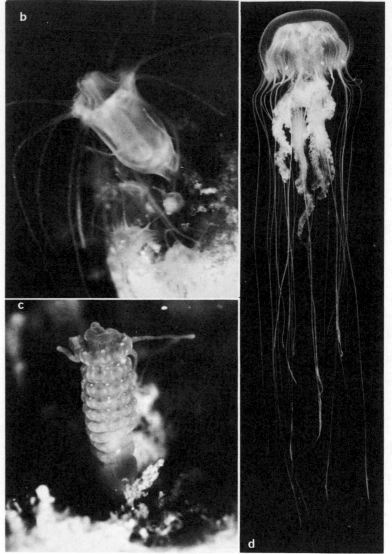

FIG. 16–9

The coelenterate *Chrysaora quinquecirrha*, a native of Chesapeake Bay: (a) hydroid polyps, growing on an oyster shell; (b) detail of a typical feeding polyp, about 1 millimeter tall —the mouth, surrounded by sixteen tentacles, is at the top of the polyp; (c) reproductive polyp shows budding of *ephyrae*, which when released grow into full-grown medusae in 2 to 4 weeks; polyp may give rise to two to twelve ephyrae, and may produce a new stack up to four times in one summer (Photographs courtesy of Chesapeake Biological Laboratory, Michael J. Reber, photographer); (d) full-grown medusa of *Chrysaora*, about 10 centimeters across the bell; their unpleasant sting sometimes closes the bay to swimmers. (Photograph courtesy Michael J. Reber)

FIG. 16–10

The nudibranch *Craten pilata*, length about 20 millimeters, which feeds actively on hydroid polyps in Chesapeake Bay. This shellless snail breathes through the gill-like projections along the sides. Because of their noxious taste, they have few natural enemies. (Photograph courtesy Chesapeake Biological Laboratory, Michael J. Reber, photographer)

where waters are quiet but not stagnant, as do the nudibranchs (such as the one in Fig. 16–10) which feed on them. Mussels often cling to rock surfaces above and below the low-tide level; this common bivalve secretes tough threads which firmly anchor it to the rock surface and to other individuals in a colony of mussels.

Lobster boats are commonly seen offshore from rocky beaches. Lobsters (as in Fig. 16–11) and crayfish ("spiny lobsters," shown in Fig. 16–12) scavenge on hard bottoms, both near shore and far out on the continental shelf, around most of the North American continent. They walk about at night, searching for worms, mollusks, and organic debris, and usually seek shelter during daylight under rocks or among seaweed. In autumn, Maine lobsters move offshore to breed, probably seeking the more stable climate found in deeper waters.

The octopus (shown in Fig. 16–13) is another fairly common

FIG. 16–11
American lobster, which lives off the East coast of the United States. (Photograph of a model in the U.S. National Museum, Washington, D.C.)

FIG. 16–12
Spiny lobster (crayfish), similar to a true lobster but lacking the large claws, is found in the warm waters of the West and Gulf coasts. (Photograph courtesy Marineland of Florida)

resident of hardbottom continental shelves. Being shy, it seeks the protection of cavelike rock formations, and usually only emerges at night to hunt for food.

BENTHOS OF COASTAL SEDIMENTS

Infauna—animals which live buried in sediment—are much less conspicuous than are the members of surface-dwelling plant and animal populations (*epifauna*). Some are *filter-feeders*, which pump large quantities of water across a filtering device, using mechanisms similar to those we have encountered in the plankton. Others are selective *deposit-feeders* (as in Fig. 16–14), which lift or suck organic particles out of the mud. Many infauna feed unselectively on sediment deposits, eating their way through the substrate and absorbing its food content through the gut; these include some nudibranchs, sea-cucumbers, and many worms (such as those in Fig. 16–15).

Distribution of infauna is largely governed by the nature of the substrate. Animals which feed by filtering plankton and detritus from seawater predominate in sandy deposits below the low-tide level. Muds and clays, however, are more suitable for those which feed unselectively on sediment; this is because the smaller particles in fine-grained deposits have a relatively higher ratio of surface to volume than larger sand grains, and generally adsorb more dissolved and suspended organic matter. Thus detritus supports a population of bacteria and very small animals which are an important source of nourishment for deposit-feeders. For filter-feeders, high concentrations of very fine particles suspended in the water are a disadvantage. Turbidity of near-bottom water resulting from sediment scoured by tidal

FIG. 16–13
Octopus. The eyes are on top of the head, below is a siphon through which water is ejected after passing over the gills. There are eight tapered tentacles, or arms, all equipped with double rows of suckers. Maximum size is around 50 kilograms with a 9-meter armspread, but many common species are much smaller. (Photograph courtesy Marineland of Florida)

FIG. 16–15
Sipunculid worms can retract the tentacle-bearing head into the body. Many live in soft ocean sediments and feed by swallowing great quantities of mud and sand, from which the organic component is extracted in the gut. (Photograph of a model in the U.S. National Museum, Washington, D.C.)

FIG. 16–14
Benthic bivalves: the cockle (left) is a filter-feeder; the clam *Tellina* (right) sucks edible particles from the sediment surface. (After Hedgpeth, 1957)

currents (or by bottom currents in nontidal areas) can clog the delicate mucus-coated filtration devices with which these animals collect food particles.

Even for deposit-feeders, the most desirable sediments are not necessarily entirely silt-or clay-sized particles, especially where waters contain large amounts of dissolved or suspended organic matter. Excess organic material in the sediment may result in a degree of oxygen depletion intolerable to most benthic animals—for instance, where an estuary receives nutrient-enriched runoff from the land. Studies in eastern coastal waters of the United States indicate that about 3 percent organic matter is optimal for deposit-feeding bivalves.

Mud-burrowing animals require specialized breathing structures that will not become clogged with fine silt. Breathing through the skin—as sea stars do, for example—is a poor adaptation for a muddy environment. One widely used mechanism involves setting up a flow of water through the burrow, or in and out of the shell of a bivalve. In this way, oxygen is supplied for respiration, fine particles are swept off delicate gill membranes, and a constant food supply is maintained.

For filter-feeders, fine sands are more suitable than coarse. Particle size does not affect the animal directly, but where grain size is large, bottom currents are generally too strong. Fine sands occur where currents are gentle, steady, and free of suspended sediment. In intertidal areas, deposit-feeders generally predominate, whatever the substrate, because an unstable sediment surface prevents the accumulation of phytoplankton or other particulate food utilized by filter-feeders.

Many deposit-feeders compact their fecal residues into pellets that may be extremely durable on the bottom. This tends to reduce turbidity at the sediment–water interface, and is thus beneficial to certain benthic organisms. In deep, quiet water, the bottom may be nearly covered by these fecal pellets, which often

FIG. 16–17
Fiddler crab and his burrow: (a) fiddler posed
on the sand; (b) about to dart into his
burrow, fighting claw held protectively over
his head; (c) a freshly made burrow surrounded
by lumps of compacted sediment excavated
by the crab and deposited outside.

FIG. 16–16
Tidal marsh at high tide (a) and at low
tide (b), when the surface is wet mud
rather than water.

exhibit a form or texture characteristic of the species which
made them.

In the absence of strong wave or current action, intertidal
sediment deposits may be stabilized by marsh grasses. Food
production is extremely high in salt marshes, and sediments
there are rich in detritus. Walking through the area at low tide,
one might think nothing lives there, but the ground is full of
little holes which testify to an active, if temporarily hidden,
population. Mussels live buried in the wet mud, and so do burrowing worms. A tidal marsh on Long Island is shown in Fig.
16–16.

Around high tide, fiddler crabs (shown in Fig. 16–17) dig
their watertight burrows. These burrows may be up to 1 meter
in length, but they are dug obliquely so as to be as shallow as
possible in order to allow the crab to remain dry. While the tide
is low, a fiddler picks through the mud, selecting bits of plant
or animals detritus. He enters his burrow when flooding is imminent, sealing the doorway against the rising tide, and remains
inside until the water subsides. Fiddlers require only enough
contact with the sea to keep their gill chambers moist, although

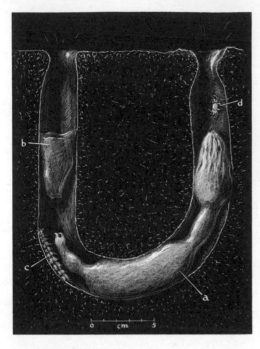

FIG. 16–18
The "innkeeper worm" and his guests in the burrow: (a) *Urechis caupo*; (b) slime net for trapping food particles; (c) scale-worm *Hesperonoë*; (d) pea crab *Scleroplax*.

experiments have shown that they can survive for weeks immersed in seawater. They seem to be evolving in the direction of a purely terrestrial habitat.

The burrowing land crab *Cardisoma* and the beach scavenger *Ocypode* (ghost crab) seem to maintain even less contact with seawater than does the fiddler. These species have enlarged gill-chamber linings which function as lungs. Temperate to tropical environments seem to favor this adaptation to a semiterrestrial way of life; relatively higher humidities at these latitudes retard dessication of the animal's gill membranes between its periodic returns to the wave line. In the tropics, certain crabs spend virtually their entire adult lives on land, sometimes even in deserts, migrating to the sea only to lay their eggs. Isopods such as the wood-boring "gribble," possessing a brood pouch where the eggs can hatch in a moist environment, have abandoned the marine environment completely. These adaptations indicate that shorelines have provided a route to the land for certain crustaceans, and others may be evolving in the same direction.

Where an intertidal deposit is scoured by tidal currents, there may be no covering of grass, and the area is a *mud flat* instead of a marsh. A common inhabitant of Pacific mud flats at or below the low-tide mark is the pink echiuroid worm *Urechis*, shaped like a lumpy banana. It constructs a deep, semipermanent, U-shaped burrow with narrow openings, not by eating the mud but by scraping it away with bristles at each end of its body. Inside, *Urechis*, (shown in Fig. 16–18), secretes a funnel-shaped, fine-meshed mucus net. The narrow end of the net fits over the worm's neck like a collar. As *Urechis* pumps water through its body and through the burrow, food particles are trapped in the net. When the funnel clogs up, the worm eats it and constructs a fresh one farther along the burrow. Wastes and debris are ejected with a vigorous blast of water from the pumping gut.

Urechis eats only the finest particles caught in the filter and discards the rest, a habit which supports some other inhabitants of the burrow. A certain species of worm (*Hesperonoë*) lives squeezed in behind the host, one or more tiny pea crabs scurry around behind him, and small fish called gobies sometimes shelter in the burrow.

A deep burrow has several advantages where flat, muddy substrate is subject to exposure at low tide. Mud flats drain slowly and incompletely, so that water continually seeps into any hole or tunnel beneath the surface, even when the opening is above the tide. The environment within the burrow thus remains fairly constant while changes are taking place outside.

Dense stands of a seed-bearing marine plant known as *eel-grass* grow in shallow-water sediments, as shown in Fig. 16–19, along both coasts of the United States. Eel-grass has an underground stem and ribbonlike leaves as much as 1 meter long. Specialized communities of plants and animals find food and shelter in eel-grass beds; its leaves become densely coated with *epiphytic* (growing attached to another plant) diatoms, blue-green algae, protozoans, bacteria, organic detritus, and small hydroids. Tube-building crustaceans and worms attach themselves to the leaves as well, and small grazing shrimps and

FIG. 16–19

Model of an eel-grass community, Woods Hole, Mass. Scallops, shrimps, and crabs are shown in the eel-grass; several kinds of worms live in the soft bottom. The burrow at left contains the polychaete *Chaetopterus*, which maintains a current through the tunnel by flapping leaflike *parapods* ("side-feet"), seen along the middle of its body. The burrow lining is a tough, parchmentlike material, shown in the W-shaped burrow at center. At right, the tube-dwelling polychaete *Diopatra*, with gills emerging from the mouth of the tube. (Photograph courtesy American Museum of Natural History)

snails scrape off accumulated detritus. This regular cleaning of the eel-grass by animals is essential to the continued growth of the epiphytes, especially the diatoms, for if they become coated with detritus they die for lack of sunlight. Other common residents of eel-grass communities include tunicates and scallops.

The species composition of an eel-grass community depends in part on local conditions such as temperature, depth, and current speed. Where currents are strong, for instance, small crustaceans living in tubes (which they themselves secrete and cement to the blades of eel-grass) are likely to survive, while heavier, crawling species would be swept away by the water. Infauna grow profusely in the sediments, because the grass has a stabilizing effect and turbidity is reduced. The eel-grass and epiphytes are a food source for many animals in the community, and detritus for bottom-dwellers is abundant. In addition, sediments are usually firm enough to permit construction of permanent burrows; holes dug in coarse sands tend to collapse, although some worms and other animals are able to build tubes which prevent such accidents.

Below low-tide level, in the muddy sediments of the continental shelf, there are usually large numbers of deposit-feeding bivalves (such as the clam in Fig. 16–20). Where water is too deep for algae and eel-grass, sediments are often soft and easily suspended, due in part to constant reworking by burrowing organisms, such as polychaetes and mollusks. A few tens of clams per square meter can rework all newly deposited sediment once or twice each year, to a depth of perhaps 2 centimeters. Attached animals are unable to grip the semiliquid substrate, and filter-feeders are choked by an excess of turbidity.

The deposit-feeding clam *Yoldia* is equiped to take advantage of this habitat. Its low-density body is often swept up by strong bottom currents, and suspended along with clouds of silt from the soft bottom. Heavier species, such as an occasional filter-feeding bivalve that tries to settle there, tend to be buried and suffocated by these turbid flows. If *Yoldia* happens to be buried by a load of fresh sediment, it quickly tunnels upward toward the new mud–water interface, where oxygen is available. If it is brought to the surface by wave action it just as quickly digs in again.

Where deposit-feeders are abundant, filter-feeders are scarce. This may happen where the activities of the former make the area less easily habitable by the latter. In this case, one organism forces another out of the area—not by competing for food or even for living space, but by the creation of an undesirable habitat. On the other hand, oxygen introduced by biological reworking of sediment permits bacteria, protozoans, and other small benthic animals to live deeper below the sediment–water interface than is possible in compacted, oxygen-deficient muds.

MICROORGANISMS IN MARINE SEDIMENTS

Microscopic organisms play a vital role in marine food chains. Figure 16–21 shows a generalized scheme of the dynamic interrelationships among pelagic and benthic organisms, including the very small consumers and decomposers that live at and near the surface of sediment deposits.

Marine sediments contain large numbers of small benthic animals called *meiobenthos* (from the Greek *meion*, meaning "smaller"). These may include various kinds of *Foraminifera*, worms, bivalves, copepods, and other small crustaceans, as well as many other tiny creatures which do not fit easily into the major animals groups that we have already discussed. In general, shallow-water and intertidal muds support a richer population of animals in this size range than do purely sandy deposits or the deeper, finer sediments offshore. Meiobenthic organisms are adapted to crawl, climb, and swim between sand grains feeding on protozoans, phytoplankton, and bacteria, or on fine particles coated with an organic film. They tend to penetrate more deeply into sandy substrate near the beach than they do offshore, although the top 1 or 2 centimeters are usually the most densely populated. In intertidal sediments, many organisms move up and down continuously in correspondence with the tidal rhythm, so as to remain near the air–water interface in flooded sand. Hundreds of thousands of individuals may be present beneath 1 square meter of intertidal beach.

In intertidal and shallow-water food chains, meiobenthos are important all out of proportion to both their size as individuals and their abundance. The mass of meiobenthos may be on the order of 1 gram or more per square meter. Energy flow through the population is very large with respect to its total mass; respiration rate is high, and new generations are produced perhaps twice a day. Numerically, they may be 100 times as abundant as the macrobenthos (animals longer than 500 microns), but by weight they make up only about 1 percent of the total benthos. They are eaten in great numbers by larger animals.

Microbenthos (among which are bacteria and fungi, ciliate

FIG. 16–20

Feeding orientation of *Yoldia limatula*. Half-buried in mud, the clam uses its feeding palps to collect sediment and bring it into the mantle cavity. The large inedible fraction is forcibly ejected as a cloud of loose sediment, creating mounds of laminated, reworked sediment as the clam feeds. When resting, *Yoldia* commonly burrows a few centimeters below the surface. [After Donald C. Rhoads, "Rates of Sediment Reworking by *Yoldia limatula* in Buzzards Bay, Massachusetts, and Long Island Sound," *Journal of Sedimentary Petrology*, 33(3) (1963), 723–27]

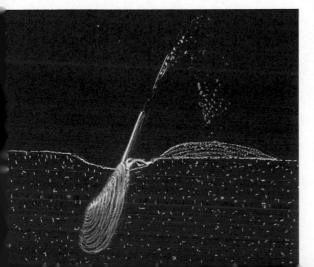

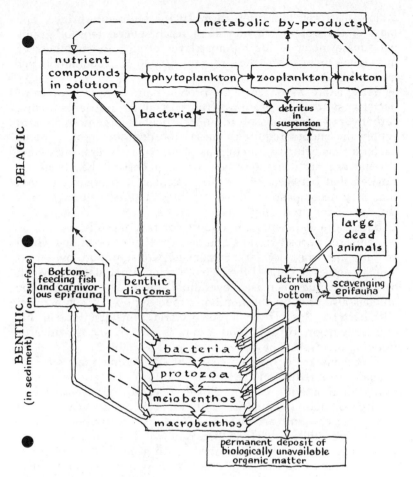

FIG. 16–21
Diagram showing the marine food cycle in a
simplified form. The upper portion represents
the pelagic component, below that the
epifauna and bottom-feeding nekton, and at
the bottom of the diagram the infauna and
marine humus. Arrows indicate the direction
of flow of matter and energy. (After Raymont,
1963)

and flagellate protozoans, amoebae, benthic diatoms, and filamentous algae) are smaller than 100 microns and range in weight
from 10^{-4} to 10^{-9} milligrams. Autotrophic organisms live at the
surface and heterotrophs mainly within the sediment. In nearshore deposits, the microbenthos often constitute an even
smaller mass than do the meiobenthos, but at the same time a
hundredfold larger number of individuals. Most of their food
is used up in respiration and reproduction; relatively little contributes to increased biomass. Generally speaking, in large
animals there is a greater biomass relative to total energy flow,
and a relatively slow rate of aging and reproduction. Very small
animals tend to reproduce rapidly, to consume large amounts of
food relative to their own body mass, and to have a high reproduction rate.

When large quantities of organic matter are present, however, a dense bacterial film often forms at the sediment surface. The mass of the population may then reach several tens of grams per square meter in the topmost layer of silt, and numbers of individuals may reach the order of 10,000 to 100 million per gram of bottom mud.

Bacteria are a concentrated protein source. Many kinds of bacteria synthesize proteins from dissolved nutrients, using carbohydrates from dissolved and particulate organic matter to supply the energy required. These so-called *chemo-autotrophic* bacteria thus actually function as primary producers, increasing the net organic content of marine sediments. It has been estimated that between 25 and 40 percent of the organic carbon that they decompose is converted into bacterial cell substance, the remainder producing carbon dioxide.

Bacteria in sediments are a source of fats and oils, as well as proteins, for deposit-feeders and also for the meiobenthos which in turn are eaten by larger animals. Consumption by larger animals serves to control the size of bacterial populations, which otherwise could increase severalfold in a day. Meiobenthic animals also feed on the nonliving food supply, in competition with bacteria; the relationship of bacteria to meiobenthos in this detrital portion of the food chain is shown by downward-trending arrows in the lower portion of Fig. 16–21.

The chemistry of marine sediments, including their state of oxidation (or reduction) and their acidity, is largely controlled by bacterial activity. Carbon dioxide released by decomposition of organic matter combines with the calcium in seawater to precipitate calcium carbonate ($CaCO_3$). Where supplies of free oxygen are limited, bacteria break down the abundant sulfate (SO_4^{2-}) in seawater to obtain the oxygen needed for respiration. The remaining (reduced) sulfur combines with hydrogen to form hydrogen sulfide gas (H_2S) or the iron sulfide that gives subsurface muds their characteristic black color. Breakdown of organic phosphates (PO_4^{2-}) results in precipitation of inorganic calcium and iron phosphates in sediments. Many bacteria and some algae use elemental nitrogen to form amino acids or ammonia.

Several types of marine bacteria are active and abundant at very low temperatures, even below 0°C. Their activities alter chemical, physical, and biological characteristics of bottom deposits in the deep ocean. Aerobic bacteria (which must have dissolved oxygen) live in the top 5 to 10 centimeters, while anaerobic types (which obtain oxygen by breakdown of sulfates or other compounds) penetrate to 40 or more centimeters below the surface; both types, however, are most abundant in the top few centimeters and decrease in numbers with depth. The principal determinant of abundance seems to be the amount of organic material present.

A certain amount of the organic material that reaches the ocean bottom is extremely resistant to decomposition and forms a more-or-less permanent *marine humus* (see Fig. 16–21). Bacteria in sediments break down many resistant organic substances, making them available to deposit-feeders, but much organic material resists even bacterial decomposition. Fecal material from deposit-feeders, for instance, resists physical and chemical degradation. Sediment below the zone of bacterial action seems to contain as much organic material as do surface layers in deep-ocean deposits. The oxygen deficit that generally occurs below the surface may be responsible for this, in that anaerobic decomposition processes seem to be less efficient than aerobic ones. Chitin, for example, contains a large proportion of the nitrogen in marine sediments, but this is attacked slowly in its absence. Fermentation of cellulose, a rather resistant organic substance common in plant fibers, is known to occur in deep-ocean sediments where turbidity currents have deposited materials that originated on the continents.

REEF COMMUNITIES

Many assemblages of benthic organisms apparently evolve because the members all require about the same set of physical conditions. Such assemblages are *habitat groupings*, rather than ecosystems based on mutual dependence among members. A classic example of an ecosystem, one in which a dominant organism actually creates the environment, is found in a *reef community*.

Reef building is an ancient tactic of benthic animals, as evidenced by remains of wave-resistant reefs that can be identied in rocks 300 to 400 million years old (Silurian period). Several different kinds of organisms can grow rapidly enough to build and maintain reefs. The essential element is secretion of a hard skeletal or shell structure whose parts are cemented together to form a rigid framework, resistant to waves and currents. In this section we shall briefly discuss oyster reefs, coral reefs, and reefs built by the tube-dwelling worm *Sabella*.

Oysters are sessile bivalves, abundant in flat, tidal areas worldwide. Their life cycle, and the ecology of the shallow-water communities which they dominate, have been thoroughly studied for commercial reasons. Oysters have been an important product in many regions (see Fig. 16–22).

Each mature female (shown in Fig. 16–23) produces many millions of eggs in a year. Most of these are fertilized, but only a small fraction of the larvae survive the planktonic stage (which lasts a few weeks, more or less, depending on latitude) and settle to the bottom to develop as adults. Larvae select a suitable spot, preferring oyster shells to any other substrate, and apparently favoring live oysters over dead shells. One to 5 years' growth is required for production of a fully mature oyster, during which time most of those which settled together as larvae have been killed by crowding, competition for food, or predation.

Oyster beds consist of enormous numbers of individuals growing in heaps, their shells cemented to rocks and to one another. Certain species flourish on the flats of partially enclosed bays and river mouths, especially where moving water brings fresh supplies of plankton and oxygen and there is a minimum of silt. (An exception to this is the Virginia oyster *Crassostria virginica*, which has a great tolerance to silt and is in fact

FIG. 16–22
Oyster dredging in Chesapeake Bay. (Photo courtesy Michael J. Reber)

FIG.   16–23
Oysters (male on left, female on right) puffing
out sperm and eggs. (Michael J. Reber)

adapted to life in a silty environment.) Under particularly advantageous conditions, as in places along the Gulf Coast, extensive oyster reefs are formed.

Living where rivers meet the sea, oysters are exposed to whatever urban or agricultural wastes may be flowing out of the estuary. This is a serious problem on heavily populated, industrialized coasts such as the United States Atlantic seaboard. The colony may be suffocated by excessive siltage, or poisoned by sewage and chemicals (including pesticides). If not actually killed, oysters may be seriously weakened and thus less able to resist disease and predation by natural enemies. Or they may retain and transmit such human pathogens as hepatitis or typhus.

Oyster beds are biological communities, with various ecological niches being filled by many different consumers. Oysters pump many gallons of water past their feeding and respiratory structures each hour. Plankton or other food particles are caught on a mucous net which is moved steadily toward the mouth by ciliary action. A certain amount of the food concentrated in this way is selectively swept out of the shell again, before the oyster has a chance to consume it. Other filter-feeding animals are attracted by the presence of a concentrated food supply.

Attached and free-swimming worms, barnacles, crabs, mussels, fixed coelenterates, and sponges live in oyster beds. Those which secrete cementing materials, such as tube-worms (shown in Fig. 16–24), add to the bonding which holds the oysters together in heavy clusters, thus increasing the stability of the structure. Crevices between the shells provide homes for other, more delicate filter-feeders. Microorganisms, flatworms, and tiny crabs may feed parasitically on the fixed animals, while certain fishes (gobies and killifish, for example) are especially well adapted for life in the shallow tidal waters over the beds. The latter prey on small crustaceans and soft-bodied animals among the rocks and crevices. Some animals, like the pea crab, even live inside the shells of live oysters.

Sea stars, snails, crabs, and various fishes prey directly upon oysters. The notorious oyster-drill snail chemically dissolves and physically bores a hole through the heavy shell to extract its contents. Sea stars exert tremendous suction with their pneumatically powered tube-feet to pull open the valves, then they decompose the soft meat by everting their stomachs to surround the oyster tissues.

On the Atlantic and Gulf coasts, oysters thrive in various different habitats. In the Chesapeake Bay area, for example, there are important natural beds in the intermediate salinities of the estuary (7–18⁰/oo), where oysters can survive but their principal enemies and diseases cannot. Experiments have indicated that, in the absence of predation, oysters grow faster and more prolifically near the mouth of the estuary than they do upstream. In the high-salinity waters of the seaside along the eastern shore, they can survive only in the intertidal zone.

Sabellarian polychaete worms (shown in Fig. 16–25) are found in rocky coastal areas of the eastern and western United States. Members of the group build a tube of mucus which hardens into a parchmentlike substance. These worms have beautiful, feathery gills, often brightly colored, which are the only part of the worm that ever emerges from the end of the tube. The gills, which absorb oxygen and also trap plankton and detritus in a mucous coating, have eyespots which are sensitive to changes in light intensity: the passing of a shadow causes them to vanish at once into the tube.

Individual worms are common residents of North Pacific tide pools, where they cement themselves firmly between rocks. Some colonial sabellarians live in the surf zone where the shoreline is rocky enough to support their tubes. They depend on the breakers to supply filterable food particles and sand grains, which they bond with mucous secretion to make strong, wave-resistant structures. As a sabellarian colony grows, new worms settle around and over old ones, and a porous reef is formed which may extend for several meters along a beach or breakwater.

Barnacles, algae, mussels, and other sessile organisms colonize on any fixed structure in the intertidal zone. The process, termed *"fouling"* because it reduces the speed of passing ships, begins with an accumulation of bacterial slime. Benthic diatoms and protozoans appear next and multiply rapidly, utilizing adsorbed organic compounds and products of bacterial decomposition. Hydroids and multicellular algae follow, and then come the planktonic larvae of barnacles, mussels, and snails. Eventually, the ecosystem reaches a balanced state where all ecological niches are filled and immigrant organisms can no longer readily establish colonies, so that ecological succession ceases.

Very often, a single organism dominates this final system, or *climax community*, as in the oyster reef based on oysters, or the sabellarian reef on *Sabella*. The "dominant" organism modifies the environment and consequently influences the species composition of the entire community.

FIG. 16–24
Calcareous tube-worm *Hychoides dianthus*, commonly found encrusting oysters or other mollusks. (Photo courtesy Michael J. Reber)

FIG. 16–25
This photograph of sabellarian "feather-duster" worms was made at the Wometco Miami Seaquarium.

FIG. 16–26
Encrusting coral growing under experimental
conditions at the Chesapeake Biological
Laboratory. (Photo courtesy Michael J. Reber)

FIG. 16–27
Fish swimming over a coral reef, Midway
Island. (Photo courtesy U.S. Geological Survey)

Reefs modify their environment physically in that they
change the current patterns and the temperature and salinity
distributions of the water that flows around them. Oyster rèefs
and worm reefs modify currents relatively little; in fact, they
build themselves up by encrustation on an existing substrate
in order to take better advantage of currents and waves in the
first place. Coral reefs (especially atolls), on the other hand,
actually create a complex environment where none would
otherwise exist (see Coral Reef section in Chapter 3). Coral
animals and the algae associated with them build not only a
reef but a lagoon where water-circulation patterns trap nutrients
and thus modify the environment over a large region. Flow of
water around the lagoon also tends to retain nutrients, creating
an oasis of life in a normally barren ocean area.

An individual coral polyp resembles a small sea anemone,
except that the coral animal builds itself a cup-shaped, calcareous skeleton, as shown in Fig. 16–26. In tropical waters, corals
grow in large colonies, by new individuals budding from the
side of a parent animal. As individuals die, consumed by predators or scavengers, subsequent generations build over and
around the disused skeletons so that a reef may continue to
grow for millions of years.

Reef-building corals must remain below the surface of the
water, except perhaps for a short period at low tide, but they
cannot survive below about 100 meters depth. They grow best
at temperatures (23–25°C) and salinities (27–40^0/$_{00}$) which permit $CaCO_3$ to precipitate easily from seawater.

One-celled algae, *zooxanthellae*, are incorporated in the tissues of reef-building corals. In this permanent, symbiotic relationship, the algae make their photosynthetic products directly available to the host animal, and in turn receive nutrients and CO_2. *Zooxanthella* photosynthesis apparently alters CO_2 concentrations in coral animal tissue, greatly increasing the animal's ability to extract $CaCO_3$ from seawater for production of the limestone skeleton. The fact that active reef building is limited to the photic zone is apparently related to the light requirements of these algae.

The biological communities that flourish in and around reefs on Pacific atolls are some of the most productive in the ocean, supporting extensive benthic communities and abundant fish populations as seen in Fig. 16–27. The amount of living plant tissue on a reef may greatly exceed the animal material. Coralline algae add encrustations of calcareous material to the reef, and filamentous green algae are embedded all over the reef surface. This continuous sheet of algae manufactures food during the day, using nutrients released by animals and bacteria. At night, coral polyps extend their tentacles (shown in Fig. 16–28) and wave them to and fro, creating gentle movement of water over the reef from which plankton is filtered.

Thousands of kinds of exotic-looking invertebrates and fishes live on or around the reef, but often there are relatively few individuals of any one kind. Some, like the sea star in Fig. 16–29, consume the live coral animals and algae, and others feed on

FIG. 16–28
Branching whip coral, *Leptogorgia:* (a) branches of coral with a redbeard sponge in the background; (b) close-up of a feeding polyp showing tentacles extended. (Photo courtesy Michael J. Reber)

FIG.   16–29
Crown-of-thorns sea star (*Acanthaster planci*) feeds on coral reef at Guam. In the lower photo, the same area of coral is shown as it appeared a few months later. The coral is dead and its skeleton has been overgrown by algae. (Photo courtesy Westinghouse)

detritus. Parrot-fish and other browsers eat away at the reef itself to digest food materials embedded in the limestone, thereby releasing clouds of calcareous silt. Shells of dead benthic organisms are incorporated into the newly building outer layer of the reef structure, or they contribute to the large amounts of calcareous sediment that accumulate in the lagoon and in the cavernous interstices of the reef itself. Benthic foraminiferans, mollusks, and calcareous algae are major sediment contributors.

There are countless adaptations for occupancy of specialized niches in such a highly diversified, competitive community. It has been estimated that upward of 3000 different animal species may coexist on a large reef tract. One of these, the moray eel, shown in Fig. 16–30(a) will serve as an example.

The moray is a fish well adapted for the reef environment. During the day it generally remains out of sight in a cave or crevice. When hunting, it puts its head out of the hole and weaves back and forth like a snake, trying to catch the scent of a victim. After dark, a moray may swim about in the open, still relying on smell rather than vision to locate food. The animal's bite is vicious, and typically it never lets go. Interestingly enough, however, it does not try to harm the neon goby swimming nearby [see Fig. 16–30(a)].

The relationship between the moray and the goby illustrates one of the specialized niches typical of a diversified community. This is known as a "cleansing symbiosis," observed among certain small fishes and crustaceans. These "cleaners" pick parasites and debris from the body of a larger animal, such as a fish, sea anemone [as in Fig. 16–30(b)] or echinoderm. The cleaning organism also acts as a fishing lure, attracting predators which are then devoured by the host. In return for these services, the "guest" gains protection from natural enemies and an opportunity to utilize an otherwise unavailable food source.

Unique mechanisms for defense and for camouflage abound on the reef. Many fishes and invertebrates have poisonous barbs, spines, or stingers, and many have coloration patterns for protection or to present a threatening appearance. Bizarre forms and colors are often attributed to the need for individuals of the same species to recognize each other amid the enormous diversity, in order to mate or to school together for mutual defense.

The various shapes and colors of coralline deposits on a reef depend in part on how budding takes place in a particular species. Pigmentation of the reef-building forms is often a mechanism for optimal absorption of light. Position on the reef with respect to waves or currents often determines the form of algal or coral structure in a particular location.

DEEP-OCEAN BENTHOS

On the deep ocean floor, where soft, flat-lying deposits are the characteristic substrate and environmental conditions are essentially constant, the relatively few species are widely distributed. Patchy distributions occur mainly around seamounts, where epifaunal animals tend to congregate on the relatively firmer substrate and where turbulence probably favors recycling of food materials. Uniformity of structure and function reach a maximum among infauna of the deep trenches, where competition is largely between members of the same species.

There are two possible explanations for this apparent lack of diversity in the deep ocean. First, environmental conditions are very similar worldwide and year-round, and second, relatively few species are adapted for this specialized mode of life. It seems likely that, before the glacial period, deep-water temperatures were 3–8°C higher than at present. The sudden cooling due to freezing of polar ice was very likely catastrophic to organisms living on the bottom. The present fauna, at least at mid-

FIG. 16–30

Symbiotic relationships between reef-dwelling animals: (a) a neon goby picks food from around the teeth of a moray eel, to mutual benefit (photograph courtesy Wometco Miami Seaquarium); (b) the clownfish is not attacked by the sea anemone, for which it acts as a fishing lure; the "flower" snaps shut over anything that touches it, with the exception of the clownfish, which it seems to ignore. (Photograph made at the New York Aquarium)

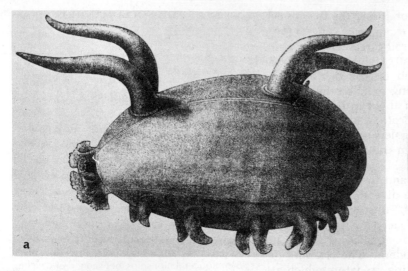

FIG. 16–31
Deep-ocean sea-cucumbers: (a) *Scotoplanes;*
(b) *Psychropotes* (drawn from the underside,
as if crawling on a sheet of glass). These
cylindrical, muscular echinoderms creep along
the bottom by contractions of the body. At the
anterior end the mouth sweeps up sediment,
detritus, and small animals. (From H. Theel.
Report on the Holothurioidea of the *Challenger
Expedition.* Report of the Scientific Results
H.M.S. *Challenger.* Zoology vol. 14 pt. 1
176 pp. 1882; pt. 2, 290 pp. 1886.

FIG. 16–32
Sea lilies have cup-shaped bodies and long
arms. The skeleton is made of calcareous
plates, absent in the vicinity of the mouth.
(Photograph of a model in the U.S. National
Museum, Washington, D.C.)

and low latitudes, may thus represent a group able to tolerate
lowered temperatures. Many deep-ocean species are also found
at intermediate depths, where conditions are similar to those in
the deepest parts of the ocean.

Due to the uncompacted nature of the substrate, deposit-
feeders such as sea-cucumbers (shown in Fig. 16–31) and worms
often predominate. Filter-feeding animals (for example, the
crinoid in Fig. 16–32) are equipped with long stalks that hold
them well above turbid, easily eroded bottom oozes. Predatory
forms—for instance, the abundant brittle-stars (shown in Fig.
16–33)—walk on long, spidery legs that hold their bodies above
the sediment surface. Sponges (as in Fig. 16–34), coelenterates
(as in Fig. 16–35), segmented and unsegmented worms (as in
Fig. 16–36), mollusks and crustaceans—in fact, all the major
groups we have encountered among the shallow-water benthos
—have some deep-water species. Uniform coloration (gray or
black among the fishes, often reddish in crustaceans) and
delicacy of structure are typical in these quiet, dark waters.

The density of living organisms on the deep ocean floor is
on the average very low. It has been estimated that the biomass
of ocean-bottom benthos per unit area is only 1 percent of that
in shallow coastal areas. All food available to these organisms
is ultimately derived from the photic zone. Many larvae of
benthic forms hatch near the surface and mature there before
descending of the bottom. Heterotrophic bacteria are a major
source of food. The deposit- and filter-feeders that live on the

FIG. 16–33
Brittle-star and sea-cucumber, photographed at a depth of about 1500 meters in Hydrographer's Canyon, western North Atlantic. (Photograph courtesy Woods Hole Oceanographic Institution, David Owen, photographer)

FIG. 16–34
Deep-ocean sponges: (a)*Toegeria;* (b) *Lefroyella.* (From John Murray, and A. F. Renard. Report on deep-sea deposits based on the specimens collected during the voyage of H.M.S. *Challenger* in the years 1872 to 1876. Report of the Scientific Results H.M.S. *Challenger* vol. 5. 525 pp. 1891.

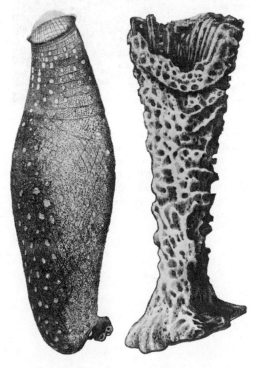

bacteria and compete with them for detritus are eaten by predatory benthos and fishes in the complex food chains of the deep ocean floor, as shown in Fig. 16–21.

Large fishes, sharks, or whales may fall to the ocean floor after death. Such rich food deposits attract large numbers of scavengers; thus the ability to sense food from a distance is an important adaptation in a dark, relatively barren environment. Below about the 4°C isotherm—that is, deeper than one to two kilometers in most of the world ocean—there is evidence that little consumption or decomposition of organic matter takes place while it is falling toward the bottom. The region is vast and sparsely inhabited, and it seems probable that no indigenous bacterial population exists where there are no fixed surfaces. Experimental data also suggest that bacteria descending from above may be inhibited by high pressure.

Near the continents, where the flow of sediment-laden currents through abyssal channels brings plant material to the deep ocean, a specialized population exists. A certain bivalve, for example, bores holes in the coconut husks, bamboo stems, and parts of trees that are carried to the deep ocean by turbidity currents. Abandoned holes may then be occupied by small mussels, worms, or amphipod crustaceans. Plant debris is eventually decomposed by bacteria, which then make an important contribution to the diet of deposit-feeders.

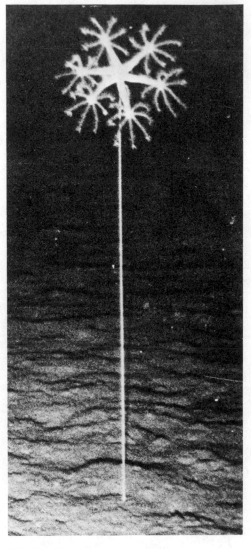

FIG. 16–35

The stalked coelenterate *Umbellula*, whose pencil-thick stem appears to be about 1 meter long, was photographed at a depth of more than 5000 meters off the Atlantic coast of Africa by a Naval Oceanographic Office crew aboard the research vessel *Kane*. (Photograph courtesy U.S. Navy)

SUMMARY OUTLINE

Benthic world: two-dimensional

Three strategies of benthic life:

    Attachment to a firm surface—mainly filter-feeders

    Free movement on the bottom—many are active predators

    Burrowing in sediments—mainly deposit-feeders

Attached plants are relatively few

Most diversity in warm, coastal areas

Many planktonic larvae or benthic animals can select a favorable location

Rocky beaches

    Life occurs in zones, dependent on tolerance for exposure

    Intertidal areas require ability to hold fast to the substrate

    Tide pools offer a specialized environment to a wide variety of plants and animals

    Around and below low-tide level is the most highly populated and diverse area

    Mobile crustaceans inhabit rocky bottoms on the continental shelf

Benthos of coastal sediments

    Infauna are filter-feeders, selective or unselective deposit-feeders

    Distribution governed by the nature of the sediment

        Filter-feeders prefer coarser, less turbid substrate

        Deposit-feeders prefer muds, which contain more organic matter

        Too much organic matter may cause oxygen depletion in sediment

        Burrowing in fine silt requires specialized, self-cleaning breathing mechanism

    Fecal residues of burrowing deposit-feeders stabilize sediments

    Marshes are rich in plants and detritus—support a varied community

    Tide flats favor deep-burrowing organisms tolerant of silt and requiring constant moisture

    Shallow-water eel-grass beds create a stable bottom, offer shelter and food to many organisms

    Soft, fine muds on the continental shelf support specialized infauna, tolerant of turbidity and able to avoid burial by later deposits

Microorganisms in marine sediments

    Meiobenthos (100–500 microns long) flourish in intertidal and shallow-water muds—numerous, high energy flow, short lives, small biomass

    Microbenthos (bacteria, protozoa, diatoms)—may have even smaller biomass

        Bacteria may be dense if detritus is abundant

        Bacteria synthesize protein, utilize carbohydrate, are an important source of protein in benthic food chains

    Chemistry of sediments governed by bacterial activity—in the absence of free oxygen H_2S produced by decomposition of SO_4^{2-}

    Deep-ocean bacteria attack marine humus slowly, anaerobically below the surface—fermentation of cellulose

Reef communities

    Many are ecological communities where organisms are mutually interdependent

Oyster community
  Characteristic of flat, coastal regions with firm substrate
  Food lost from filter-feeding apparatus attracts other attached and motile animals
  Oyster beds offer shelter to small, defenseless organisms
  Predators are fewer farther up the estuary, where waters are brackish
Sabellarian worm community forms in path of breaking waves, other animals settle on the reef for food and protection; community develops by process of fouling and ecological succession
Coral reef or atoll greatly modifies the environment, in terms of current flow in lagoon and around atoll
  One-celled algae live symbiotically in the coral animal
  Highly productive regions—shells of many other animals contribute to the sediment accumulation and to reef structure itself
  Many specializations and adaptations result from intense competition among the many species present

Deep-ocean benthos
  Substrate soft, flat; relative lack of diversity characteristic among animals
  Biomass per unit area only about 1 percent of coastal fauna biomass
  Bacteria an important source of food—all other food sinks from above
  Specialized benthic population lives near continents, receives food in turbidity currents

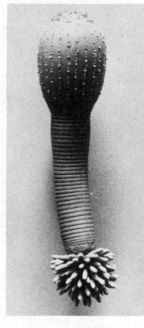

FIG.   16–36
The unsegmented worm *Priapus* swallows its prey whole. It is found in bottom sediments of polar regions in both hemispheres. (Photograph of a model at the U.S. National Museum, Washington, D.C.)

SELECTED REFERENCES

AMOS, WILLIAM H. 1966. *The Life of the Seashore.* McGraw-Hill (in cooperation with World Book Encyclopedia), New York. 123 pp. Nontechnical but authoritative; good for browsing; spectacular color photography.

ARNOLD, AUGUSTA FOOTE. 1968. *The Sea-beach at Ebb Tide.* Dover, New York. 490 pp. A standard field guide, classical in approach with emphasis on taxonomy, first published in 1901.

DIETZ, R. S. 1962. The sea's deep scattering layers. *Scientific American* 207(2):44–64.

HEDGEPTH, JOEL W. (ed.) 1957. *Treatise on Marine Ecology and Paleoecology, Vol. I: Ecology.* Geological Society of America, New York. 1296 pp. Detailed, scholarly monographs on habitats and biological conditions in the ocean, with emphasis on biological communities; extensive bibliography.

MCELROY, W. D. and H. H. SELIGER. 1962. Biological luminescence. *Scientific American.* 207(6):76–91.

NICOL, J. A. COLIN. 1960. *The Biology of Marine Animals.* Interscience, New York. 707 pp. A standard reference text, with emphasis on physiology.

RICKETTS, EDWARD F., and JACK CALVIN. 1962. *Between Pacific Tides,* 3rd ed. Stanford University Press, Stanford. 516 pp. Excellent field guide and reference book, well illustrated and including detailed bibliography.

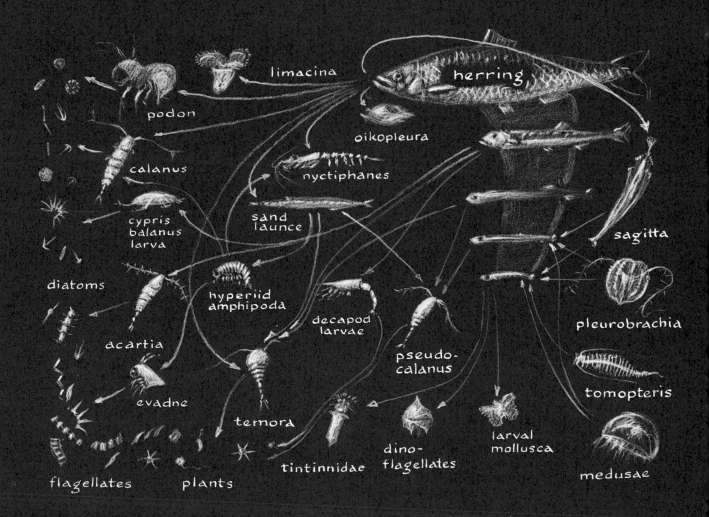

podon

limacina

herring

calanus

oikopleura

nyctiphanes

cypris
balanus
larva

sand
launce

sagitta

diatoms

hyperiid
amphipoda

decapod
larvae

pseudo-
calanus

pleurobrachia

acartia

tomopteris

evadne

temora

tintinnidae

dino-
flagellates

larval
mollusca

medusae

flagellates

plants

PRODUCTIVITY OF THE OCEAN

Food relationships of
the Atlantic herring
(see page 492).

seventeen

We have previously considered the ocean as a physical environment and discussed two of the important modes of life there—planktonic and benthic. In this chapter we shall consider the productivity of the ocean as a whole and draw some conclusions as to how the web of marine life is affected by ocean processes.

Although our approach will be somewhat different from that used when studying the budget of heat or water, the intent is similar: we want to show how energy flows through this complex. Furthermore, we shall take up one aspect of this flow in greater detail—how much food can the ocean supply for man's use? The amount of living material produced in the ocean sets an upper limit on the amount available for harvesting. Man competes directly with other organisms for this food supply, so that large-scale harvesting reduces the amount of food available for other creatures, many of them considered highly desirable.

Most food harvested from the ocean comes from large swimming animals—fishes and whales are two familiar examples. These represent a marine life style that we have not previously considered, the actively swimming *nekton*. Fishes and similar organisms usually can swim faster than the currents. Many of them can migrate long distances and take advantage of seasonal availability of food in different areas. In addition, different portions of their life cycles are often passed in several different ocean areas. Spawning may take place in an area that is particularly favorable for survival and adult life my be spent in a different part of the ocean, where food is more plentiful.

The mobility of the nekton contrasts notably with the limited range of movement characteristic of benthic organisms.

PRIMARY PRODUCTIVITY   473

PRIMARY PRODUCTIVITY

Primary productivity of an ocean area is defined as the amount of carbon fixed (assimilated) by plants in a given period of time. Rate of carbon fixation in a column of water, with a cross-section of 1 square meter, extending downward from the surface, is often used as a standard for productivity studies. Factors affecting productivity include: (1) the depth and stability of the photic zone; (2) the concentration of limiting nutrients and rate of resupply; (3) the rate of grazing by herbivores; and (4) the presence (or absence) of micronutrients (for example, vitamins) or growth-inhibiting substances in the water.

Net plant production during a given period is equal to the amount of organic carbon produced by photosynthesis minus the amount utilized by the plants themselves in respiration. A direct method for measuring this, developed in the 1920's, is to determine the amount of oxygen produced by the plants in seawater, since oxygen is given off by plants in direct proportion to the amount of organic carbon synthesized.

In this "oxygen-bottle" experiment, sealed bottles of seawater containing phytoplankton are lowered to varying depths below the surface in the presence of sunlight. After a known period, the oxygen content of the water in the bottles is measured and is compared with the amount present when the bottles were sealed. Near the top of the photic zone, oxygen content increases. The rate of oxygen increase diminishes with depth until a point of no net increase is reached. This is known as the *compensation depth*, meaning that at that point the oxygen produced by photosynthesis is exactly equal to the amount utilized in respiration. At greater depths, respiration exceeds photosynthesis, and no effective plant production can occur. Finally, no oxygen is produced at all and the plants eventually die. The concept of compensation depth, then, is used in reference to metabolism of an individual plant sample at a specific depth, which is determined by extent of light penetration.

Identical bottles wrapped in black material so that no light can enter (called "black bottles") may be used coincidentally in this experiment to measure oxygen depletion due solely to respiration. The amount of oxygen removed from the water in a black bottle during a given period is added to the amount of increase in a "light bottle" at the same depth during the same period of time. This figure is a measure of gross plant production in the "light bottle."

A method developed in the 1950's for measuring primary productivity uses a radioactive isotope, carbon-14. A given amount of CO_2 containing carbon-14 exclusively is added to a bottle containing a known amount of CO_2 made with stable carbon. The rate of assimilation of the carbon-14 CO_2 (i.e., rate of production) is computed by measuring the amount of beta-radiation emanating from the decay of the radioactive carbon in the phytoplankton at stated intervals. The greater the productivity, the more radioactive are the plankton owing to their incorporating carbon-14 in their tissues.

Most continental shelves are regions of high productivity. This is because they are shallow, well-mixed areas, with current patterns that hold nutrients in the coastal ocean, rather than permitting them to be drawn toward the deep ocean. Photosynthesis at mid- and high latitudes is inhibited during local winter, however, due to lack of sunlight. Plants can often grow in the top few centimeters, but they are constantly carried below the sunlit zone due to mixing of the water column by winter storms in the absence of a well stabilized surface zone. Thus there is no net plant production.

In the spring, increased sunlight causes the phytoplankton to grow and reproduce rapidly, provided the water is well supplied with nutrients. There can be no productivity increase, however, as long as plants are carried below the surface, causing net respiration in excess of production for the water column.

To understand the relationship between a stable surface layer and an increase in productivity of plants, we introduce the concept of *critical depth*. This is defined as the depth above which there is enough light present to support a bloom of phytoplankton. The critical depth depends on the penetration of sunlight below the surface; it is at a maximum in early summer and at a minimum in early winter at mid-latitudes. Until a thermocline is formed by warming at the surface (or until sunlight penetrates to the bottom of the water column), plants are transported below the critical depth, and no effective production can occur.

As light increases with the coming of spring (as early as February in Long Island Sound), phytoplankton begin to increase rapidly in numbers. This sudden growth spurt, typical of mid- and high latitudes and especially pronounced in subpolar regions, is known as a "*spring bloom.*"

At this time the compensation depth is very near the surface because sunlight is still relatively weak. The critical depth (always below the compensation depth because it represents a lower limit for the growing population as a whole) may occur at some point within the mixed surface layer. If that is the case, plant growth is retarded because part of the population is continually being mixed downward, below the critical depth. As sunlight becomes stronger, the critical depth may extend to the bottom in shallow areas. In deeper waters, a thermocline may form at or above the critical depth. In either case, the plant population expands rapidly as soon as the critical depth exceeds the depth of the mixed surface layer. This has been well documented in the North Sea as shown in Fig. 17–1.

Where large runoff of fresh water persists throughout the year, such as in parts of the Baltic Sea or in Chesapeake Bay, the low-salinity surface layer tends to be quite stable. This permits effective production of phytoplankton during the winter. Conversely, in the Bay of Fundy or Puget Sound, constant turbulence caused by strong tidal currents prevent forma-

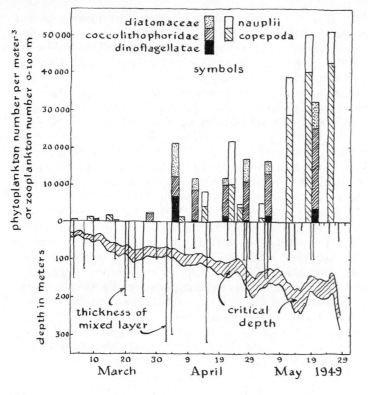

FIG. 17–1
Observations at a weather station (66°N, 2°E)
in the North Sea show that plankton do not
appear in large numbers until the critical
depth exceeds depth of the mixed layer.
Key: Dia.—diatoms; Coc.—cocolithophores;
Dif.—dinoflagellates; Nau.—nauplii; Cop.—
copepoda. (After Wimpenny, 1966)

tion of a stable surface layer, except in isolated inlets. The crop,
therefore, is never quite as rich as might be predicted from the
abundance of nutrients in the water. Here, too, constant stirring-
up of particles from the bottom creates turbidity and inhibits
penetration of sunlight.

With the beginning of active phytoplankton growth in
spring, generations of nannoplankton and diatoms reproduce in
a matter of hours. Often, the population is dominated by a
single species of diatom or dinoflagellate. As nutrients and
metabolites are used up, early-blooming forms may decline to
be succeeded by different species. Temperature is important
in determining species composition at any given time. In addition, by-products released to the water by one species may be
necessary metabolic requirements for succeeding populations.
Dinoflagellates often follow the spring and summer diatoms of
mid-latitudes because they require lower nutrient concentrations
and are able to utilize dissolved or particulate organic matter as
food.

Distribution of phytoplankton species and the order of their
succession varies widely throughout the world ocean, the result
of complex relations between weather, temperature, salinity,

stability of water masses, and the biological history of the ocean waters. In coastal waters, dominant populations often consist of relatively large phytoplanters, 100 microns or more in diameter. Especially in areas of upwelling, colonial species are common, their gelatinous masses forming clusters a few millimeters or even centimeters across. Fishes can feed effectively on plankton of this density, so that short, highly efficient food chains are established in exceptionally productive areas.

Offshore, in the relatively nutrient-poor open ocean, nannoplankton (from 1 to 25 microns in the largest dimension) replace larger plants as dominant species. Having a large ratio of surface to mass, and thus capable of utilizing nutrients in extremely dilute concentrations, these tiny plants account for a very large fraction of total carbon fixation in the world ocean.

Seasonal variation in productivity is most pronounced at subpolar latitudes, where the growing season is shortest, and becomes increasingly less marked as one moves toward the equator. Warm seas such as the Sargasso and the Mediterranean show small seasonal fluctuations, and the most productive season often falls between mid-winter and early spring. Off Bermuda, for instance, the upper layer of water is mixed nearly to the permanent thermocline (400 meters) during the stormy fall and winter season. In summer, however, pronounced thermal stratification creates maximal stability in surface waters, but productivity is low due to nutrient depletion.

In the tropics, seasonal effects are minimal and therefore not a major factor in distribution or productivity of phytoplankton. A high rate of photosynthesis is theoretically possible at all times, limited only by nutrient impoverishment, and plankton populations are more or less constant throughout the year. There is usually a permanent thermocline, through which water from deeper layers gradually rises into the sunlit zone. Coccolithophoridae replace diatoms as dominant species, due in part to the relative ease with which $CaCO_3$ precipitates in warm water (see discussion of skeletal minerals in Chapter 14).

GEOGRAPHICAL DISTRIBUTION OF PRODUCTIVITY

Primary productivity is highly variable in the ocean, differing one-hundredfold between the richest and the most impoverished areas. The open ocean, which includes roughly 90 percent of the marine environment, is relatively barren, with an average rate of carbon fixation estimated at about 50 grams of carbon per square meter per year ($gC/m^2/yr$). Production for the open ocean, calculated on the basis of 50 $gC/m^2/yr$[1] amounts to about 16 billion tons, as shown in Table 17–1, equivalent to 160 billion tons of phytoplankton. Average standing crop is probably about 1 percent of the total, or 1.6 billion tons wet weight. To imagine the density of material in such a crop, assume that all phytoplankton is in the top 10 meters; plants weighing about 0.5 gram would then be suspended in each cubic meter of seawater.

Table 17–1

GEOGRAPHICAL DISTRIBUTION OF
PRODUCTIVITY

477

DIVISION OF THE OCEAN INTO PROVINCES ACCORDING
TO LEVEL OF PRIMARY ORGANIC PRODUCTION*

Province	Percentage of ocean	Area (km^2)	Mean productivity (grams of dry carbon/m^2/yr)	Total productivity (10^9 tons carbon/yr)
Open ocean	90	326 × 10^6	50	16.3
Coastal zone†	9.9	36 × 10^6	100	3.6
Upwelling areas	0.1	3.6 × 10^5	300	0.1
Total	100.0		—	20.0

*After John H. Ryther, "Photosynthesis and Fish Production in the Sea," *Science*
166 (1969), 72–76.
†Includes offshore areas of high productivity.

Coastal areas are much more productive than this because
they are shallow. Nutrients there are readily recycled to the surface. In addition, the two-layered circulation of an estuary like
Chesapeake Bay tends to trap nutrients along the coast. River-
induced upwelling creates a net inward flow of deep water in
the estuary itself and in the nearby coastal ocean.

Nutrient and fresh-water runoff from the land also influence
productivity in certain coastal areas. In summer, rivers supply
few nutrients directly to the ocean because indigenous plankton
populations in the stream usually utilize them in transit. Some
dissolved and suspended organic debris is, however, river-
borne; and in winter, when river plankton decline, nutrients
are transported directly to the ocean. In coastal waters, and in
the open-ocean regions where divergences or other hydro-
graphic features bring deep water to the surface (as at the
equator in the Pacific and Atlantic Oceans), productivity probably averages about 100 gC/m^2/yr.

Where upwelling brings deep water to the surface, particularly off the subtropical western coasts of the continents and
around Antarctica, surface waters are enormously rich as indicated in Fig. 17–2. Primary productivity may range from 1
to more than 10 grams of carbon per square meter per day,
or probably in excess of 300 grams per year. The Peruvian
coast, for example, supports the world's largest fishing industry—7 to 8 million tons of anchovies per year that have fed
directly on the dense phytoplankton crop and associated zooplankton populations. Since coastal upwelling is dependent on
the constancy of prevailing winds, its effect in some areas is
quite sporadic and in others relatively dependable. It has been
estimated that upwelling normally occurs over about 0.1 percent of the world ocean (see Table 17–1).

In the vertical dimension, the gradient of marine productivity
is as dramatic as that of the horizontal, geographical dimension.
More than 80 percent (by weight) of living material in the ocean
is in the surface layer, to a depth of 200 meters. Another 16
percent is distributed through the next 2800 meters, with density decreasing generally downward. Only about 1 percent of
marine animals are known to live below 3000 meters.

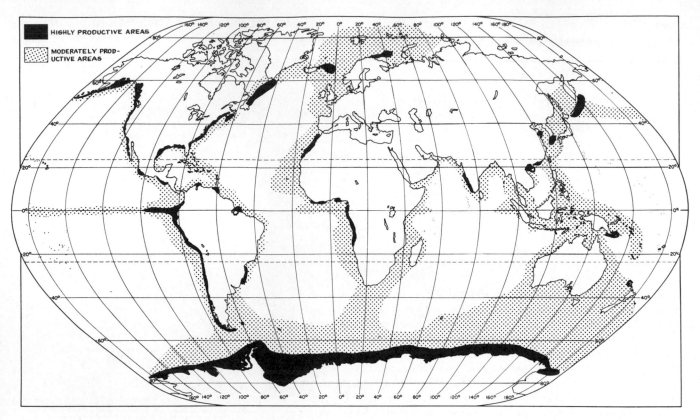

FIG. 17–2

Areas of high productivity of living matter in the world ocean. [After P. M. Fye, A. E. Maxwell, K. O. Emery, and B. H. Ketchum. "Ocean Science and Marine Resources," in *Uses of the Seas*, an American Assembly book, Prentice-Hall, Inc., Englewood Cliffs, N. J. (1968), pp. 17–68]

EFFECT OF GRAZING BY ZOOPLANKTON

The amount of phytoplankton present in an ocean area at a given moment is called the *standing crop*. This may be determined by filtering a known quantity of seawater to collect the algae, then extracting their chlorophyll with acetone (an organic solvent) and measuring the amount of light absorbed by the pigmented solution. For each phytoplankton species there is a known ratio of chlorophyll to carbon content in the cells; therefore, the carbon content of a single-species sample can be calculated by this method. The standing stock or *biomass* of a population or a biological community is often expressed as dry weight of carbon per cubic meter of water.

When phytoplankton are abundant, herbivores proliferate and feed voraciously, often greatly depleting the standing crop. The measurable biomass of phytoplankton in such an area may be quite low while gross production is in fact very high, because a large fraction is continually removed by grazing, as indicated in Fig. 17–3.

Surrounded by such plenty, herbivores may feed to such ex-

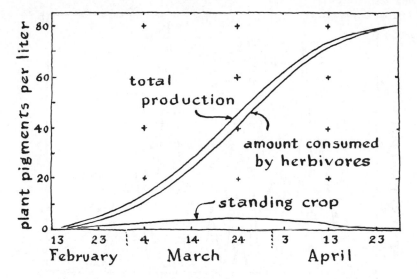

FIG. 17–3
Standing crop represents only a small part of total plant production. Data based on a study in the English Channel, showing that, of 85,000 plant pigment units/m^3 produced (according to predictions from PO_3^{4-} uptake), only 2500 units/m^3, or about 3 percent, were measured as a standing crop. Difference was assumed to be due to grazing. (After Raymont, 1963)

cess that much undigested plant matter is voided in the form of heavy fecal pellets. Some pellets are quickly decomposed by bacteria. The rest sink rapidly (at rates of 10 or more meters per day) and provide a rich food supply to detritus-feeders at greater depths. Decomposition of organic detritus within the photic zone returns nutrients to the water, permitting continued growth of phytoplankton. Phosphates dissolve readily from digested and undigested fecal matter, and about half of the nitrogen excreted by animals is in the form of ammonia (see the discussion of biogeochemical cycles in Chapter 14). In this way, significant amounts of limiting nutrients are returned to the ocean directly, without having to be first decomposed by bacteria.

As phytoplankton grow during a bloom, the abundance of nutrients in the water is reduced. Eventually, the lack of one or more limiting nutrients causes productivity to decline. Grazing by growing populations of herbivores quickly depletes the standing crop; thus the peak production of zooplankton in an area often coincides with a period of phytoplankton scarcity. In the Antarctic, increased numbers of broken diatom shells appear on the ocean floor following a bloom, indicating large-scale consumption by herbivores. Animal growth and reproduction is slow compared with that of plants, and food requirements are greatest during immature stages. Months may pass between the period of maximum grazing by herbivores and the point at which they attain full growth and are ready to spawn. With the appearance of the next generation, the biomass of zooplankton reaches its peak, as Fig. 17–4 indicates.

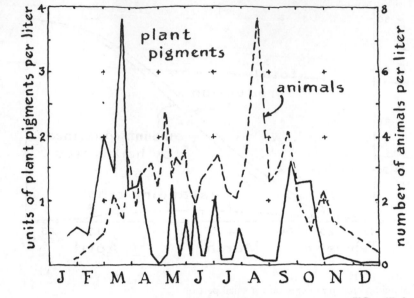

FIG. 17–4
High-density phytoplankton crop alternates with high density of zooplankton in a study made off Plymouth, England. Plants bloom in March and are rapidly consumed by zooplankton in April. Period of intensive feeding is followed by a peak in reproduction, so that high density of zooplankton occurs when phytoplankton density is low. Reproductive cycle brings peak zooplankton crop in August, but this population quickly diminishes for lack of food. Recycling of nutrients from animal excretions and decomposition enriches the water so that there is a late-summer plant bloom, at a time when zooplankton are few in number. Thus the size of the phytoplankton crop determines the size of the zooplankton crop, and vice versa. (After Raymont, 1963)

In the absence of an abundant food supply, zooplankton populations decline. Decomposition of organic detritus returns nutrients to the water. With the onset of autumn storms at mid-latitudes, nutrients from deeper water are mixed into surface layers. The resupply of nutrients while sunlight is still relatively strong commonly causes a second bloom during September or October. This is often followed by yet another growth spurt of herbivores. Figure 17–4 illustrates alternation between peak production of phytoplankton and zooplankton in the English Channel. At tropical latitudes, where plant populations are more constant throughout the year, there is correspondingly little variation in numbers of zooplankton.

To calculate the rate at which a phytoplankton population is experiencing growth or decline, the respiration rate and rate of depletion by grazing per unit of population are subtracted from the rate of photosynthesis. This figure, multiplied by the size of the population, gives the rate of change in the standing crop.

Distribution of herbivores is partly controlled by the "biological history" of the water, and by type as well as abundance of phytoplankton. Metabolic by-products of a species that has bloomed recently in a particular ocean area may be beneficial or even essential to a subsequently blooming species; conversely, a toxic residue may suppress the development of a succeeding population.

Another factor may be that certain animals can only ingest particles of a particular size or shape, and thus proliferate only when specific foods are available. For this reason, and also because food requirements of animals may vary throughout their life cycles, large populations of plants and animals may sometimes coexist. Dominant coastal herbivores include fish larvae, copepods, heterotrophic dinoflagellates, salps, and doliolids. Where plankton are tiny and sparse (as in the open ocean), radiolarians, foraminiferans, and tinntinids are more common, as well as larval forms of the smaller crustaceans. Tiny ciliate protozoans, a few microns across, are extremely abundant in open ocean waters, where they feed on the smallest of the nannoplankton.

ORGANIC GROWTH FACTORS

The exchange of organic trace materials among marine organisms offers a good example of how members of a community may depend on one another in ways other than directly as food. Certain photosynthetic algae and autotrophic bacteria require vitamins in trace amounts. Vitamin B-12 (cyanocobalamin) and related compounds, thiamin, and biotin are required by many marine algae. Other vitamins needed by bacteria and fungi are present in seawater.

Bacteria are the major producers of vitamin B-12 in the sea; certain algae may also synthesize vitamins in significant amounts. Most marine bacteria live on particles, and vitamins are readily absorbed on particle surfaces; thus muds are a rich source of vitamins in shallow water. Coastal waters contain much greater amounts than are found in the open ocean, where vitamin B-12 may be limiting in the growth of some algae.

Many marine bacteria require vitamins for growth, and they compete for them with algae and animals, some of which may also absorb these trace compounds directly from seawater. Consumption by bacteria may be quite important, since their rates of metabolism and cell division are higher than those of other consumers.

Besides vitamins, antibiotics and antibacterial substances are also released by certain organisms in the marine environment. Seaweeds and one-celled algae such as the brown flagellate, *Phaeocystis*, produce antibiotic substances that have an effect on organisms at higher levels on the food chain. Benthic invertebrates such as sponges, hydroids, and certain corals are also known to produce antibiotics of various kinds.

FOOD RESOURCES OF THE OCEAN

Much of man's interest in marine productivity centers on those species which he can effectively use as food. This includes animals and plants that man can raise himself or can capture efficiently, when their food value (particularly protein content) per unit weight is high enough to make the effort economically feasible.

Of the radiant energy that is changed into chemical energy by photosynthesis, it has been estimated that only about 10 percent on the average is finally converted into animal tissue by first-stage consumers. (This is referred to as a 10 percent *ecological efficiency*.) At each subsequent step in the food chain, another 90 percent of the energy ingested is lost through respiration and other metabolic activity, including much that goes into the water as dissolved organic matter. Typically, only about 10 percent of potential energy contained in the original food is passed along to the next consumer. Ecological efficiency varies according to the species involved and the age of the individual.

To describe the place that an organism occupies in a food chain, ecologists use the term *trophic level:* an herbivore represents the first (or lowest) trophic level, a primary carnivore the second, and so forth. Where productivity is low and primary producers are very small and sparsely distributed, as in most parts of the open ocean, food chains are usually long and extremely complex. It has been estimated that most of the commercially valuable open-ocean fishes taken in the 1960's represented the fifth trophic level, at least. In other words, these fishes are feeding at high trophic levels. Given the scarcity of reliable information on the diet of marine fishes, this is probably an oversimplification. For example, juveniles commonly feed at lower trophic levels than do adults of the same species.

Conversely, at every stage a certain number of individuals die and are consumed by bacteria rather than by animals at higher trophic levels. The bacteria may themselves be eaten by animals, but additional steps have been added to the food web in the process. In addition, within-level predation occurs at every stage; for instance, the larvae and young of plankton-eating fishes may be eaten in great numbers by mature individuals of other species of plankton-eaters. This also creates additional steps in the food web. In complex feeding relationships, there is increased likelihood of predation by species which are not normally eaten by food fishes (such as jellyfish), so that a dead end is reached as far as human resources are concerned.

In any case, fish protein is produced less abundantly in the open ocean than it is near continental margins, with exceptions as illustrated in Fig. 17–2. Among the commercially valuable animals that feed at high trophic levels are tunas, swordfish, seals, striped bass, sharks, sperm whales, and other toothed whales.

Loss of energy through many trophic levels is much reduced where food chains are short, as in areas of coastal upwelling, where fish can feed on plankton directly. Herring, sardines, shrimp, many flounders, haddock, cod, and baleen whales fall into this category. Loss of energy can be reduced even further and gain to the fisherman maximized through an understanding of the different ways in which organisms utilize food throughout their life cycles.

When a large amount of food is available to a young animal —for example, a fish—so that it does not need to expend a great deal of energy through hunting, it grows very rapidly. A large proportion of its food is converted into additional tissue; the ratio of growth to assimilation has been estimated at about 30 percent during this period of active growth. After it reaches approximately mature size, the food it eats will keep it alive and permit it to reproduce, but further growth will be slow and inefficient.

On land, this principle is applied in animal husbandry industries. Each successive crop of young animals (for example, chickens) is segregated throughout development, and as soon as a peak of rapid growth is attained, the crop is marketed. In fishing this is not so easy, for young fish and old are trapped together in the net. It is possible, however, to obtain an optimal sustainable yield from waters where fish populations live permanently, in the following manner.

A stock of fish is largest in a virgin, unfished state, when there is a large proportion of mature and older individuals. Overall normal growth of the young is balanced by death from disease and predation. As the resource is fished, the number of large-sized individuals is proportionally reduced—because fishing reduces the life expectancy. The average rate of growth for the population as a whole may increase, because more small fish are present. The stock will not decline as long as the catch is less than the difference between natural growth and loss by disease plus natural predation.

As fishing effort is increased, catches become larger, but at a decreasing rate. The largest catch that can be taken continuously, without impairing the capacity of the resource to renew itself, is known as the *maximum sustainable yield*. If this limit is exceeded, the catch will begin to decline even if fishing effort increases. Maximum economic yield, when the value of the catch is greatest per unit of cost, usually represents a smaller catch than maximum sustainable yield.

When a stock has been seriously depleted, profits decline until further fishing becomes economically unfeasible. This usually happens long before there is danger of extinction due to overfishing. Females typically lay tens or hundreds of thousands of eggs at a time, so that the number of fish surviving to catchable size is usually independent of the size of the standing stock.

When the largest sustainable yield has been attained, the stock usually has been reduced to about half its virgin abundance. Average sizes of the individuals caught will be smaller than in a virgin stock, but the rate of increase for the population is at a maximum. A fishery based on a relatively young, fast-growing population takes maximum advantage of the rapid and efficient growth of young animals.

During the late 1960's, the world catch of marine animals was about 55–60 million tons (fresh weight) annually (broken down by species in Table 17–2). Ninety percent of this was taken from coastal waters, where production was 80 times as great on a per-unit basis as in the open ocean. Nearly half of the total came from northern temperate waters, site of 43 percent of the world's continental shelves.

Predictions indicate that a catch of 100–200 million tons annually might be a realistic estimate for the year 2000, using presently developed equipment and methods. Underexploited stocks still exist, especially in the Southern Hemisphere, and systematic surveys still may uncover stocks previously unknown. The cultivation of animals that feed directly on plants, such as

Table 17–2

WORLD-OCEAN CATCH BY GROUPS OF SPECIES, 1967*
(excluding whales)

Species	Catch (millions of metric tons)
Herring, sardines, anchovies, etc.	19.7
Cod, hake, haddock, etc.	8.2
Ocean perch, bass, etc.	3.1
Mackerel, swordfish, etc.	2.7
Jack, mullet, etc.	2.0
Tuna, bonita, etc.	1.3
Flounder, halibut, sole, etc.	1.2
Salmon, trout	1.1
Sharks, rays	0.4
Crustaceans	1.4
Mollusks	3.1
Unsorted & unidentified	8.3

*From United Nations Food and Agriculture Organization, *Yearbook of Fishery Statistics, 1967,* Rome (1968).

bivalves, could be significantly expanded. Management of coastal and estuarine resources will be necessary to preserve these habitats for the many fishes and invertebrates (90 percent of the United States catch in the 1960's) whose lives are partly spent in nearshore waters. Present fisheries will have to be carefully managed to avoid overfishing and maintain maximal catches. Resources of less familiar animals could be developed, such as dogfish and oceanic squids.

To increase the ocean's total annual food yield past the estimated maximum of 100–200 million tons, it would probably be necessary to include species not widely utilized in the 1960's. New technology will be required to economically harvest and utilize the smaller organisms not now caught. Equipment for trawling at mid-levels might be developed to supplement the purse-seine (used at the surface) and bottom trawl that have been used throughout this century. Studies of fish behavior might lead to a better understanding of how and why schools move through the ocean, and perhaps to new methods of concentrating them so that they can be caught economically. Use of light, underwater sound, or electrical impulses might be based on studies of how fish themselves attract prey in the deep ocean.

By utilizing such small, abundant species as euphausids, lantern fishes, and stomiatoids, for which new processing and marketing techniques would have to be developed in order to make the product acceptable to consumers, a tenfold increase might be achieved over 1960's catches.

It has often been suggested that we harvest coastal plankton, for if concentrated, it would be high in protein. Difficulties arise, however, when we consider the necessary technology. The volumes of water that would have to be filtered might be prohibitive. Filtering the coastal ocean with a net fine enough to catch microscopic organisms is hard to imagine; a tough net would become clogged and a fine one torn. Spinning plankton out of the water with a centrifuge, another suggestion, is also an unreasonable idea—tremendous force is required to extract small particles from water by this method, and the cost would be high per unit yield. In addition, plankton spoils rapidly.

Fish, on the other hand, are rather efficient at making protein from planktonic raw material. Harvesting at low trophic levels would mean less food for high-level feeders; we cannot hope to take fish as well as plankton in quantity. Another factor is that the ratio of standing crop to annual production is extremely low for very small organisms. For phytoplankton, only about 1 percent of annual production is present in the ocean at any one time. It seems more economical to let fish, shellfish, or even whales collect and concentrate the plankton's protein for us.

It has also been suggested that we fertilize the ocean in order to increase primary productivity. This idea has met with various objections, chief among them again being the vast areas involved. Water movements would make selective enrichment difficult to control. Experiments in shallow, semienclosed areas have shown that fertilizers are quickly taken up by attached seaweeds and adsorbed on sediments, and then released rather slowly to the water. In some areas, artificial fertilization may, however, be feasible.

For meeting world food needs, fish is a promising source of protein but a limited source of calories. The 1968 world catch of slightly under 60 million tons represented about 60×10^{12} kilocalories of potential energy. Calorie requirements per year for the 3.5 billion people then living were estimated at 30×10^{14} kilocalories, 50 times the amount taken from the ocean.

For minimum nutrition, it has been estimated that 36 grams of "high-quality" animal protein per person are needed per day, or 43 million tons for the earth's population annually. After being processed to an acceptable human food product, 60 million tons of fish yields about 10 million tons of pure protein. Thus if the 3.5 billion people in the world had each eaten an equal share of fish protein, about 20 percent of world animal protein requirements could have been met with fish products in 1968.

Another aspect of the ocean's biological potential is in drug research. Toxins derived from many poisonous fish and invertebrates are known to have a variety of useful pharmacological properties. Some are antibiotic, antiviral, or fungicidal; some inhibit animal growth or rate of human heartbeat; many affect nerve function; and some are effective painkillers. Certain bivalves and echinoderms produce tumor-inhibiting chemical substances, some of which have been used in the study of cancer. Sterols from sponges, anticoagulants from clams, and insulin from sea stars are among the many other pharmaceuticals derived from marine invertebrates that have been used in medical research.

ESTUARIES AND FISH PRODUCTION

More than half of the fish and shellfish taken in United States coastal waters for sport and profit appear to be estuarine-dependent during at least part of their lives. Seven of the 10

FIG. 17–5
Food relationships along a stretch of beach and marsh. Marsh grass and algae are primary producers—animals eat the vegetation and each other in the complex web of matter and energy of a coastal area. [Redrawn from William H. Amos, *The Life of the Seashore*, McGraw-Hill (in cooperation with World Book Encyclopedia), New York) (1966), 123 pp.]

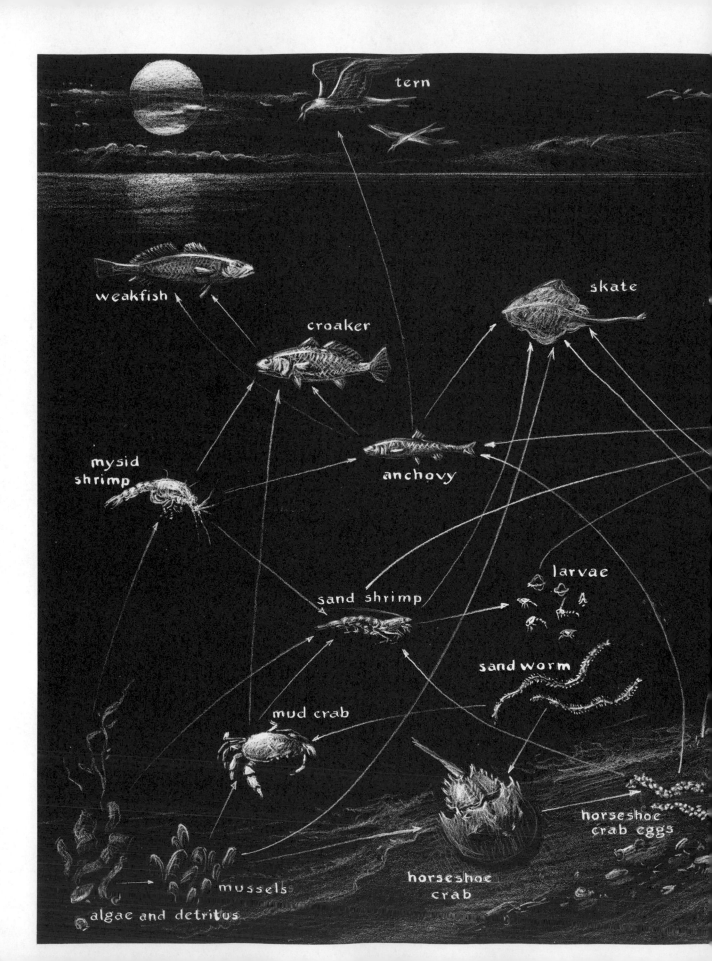

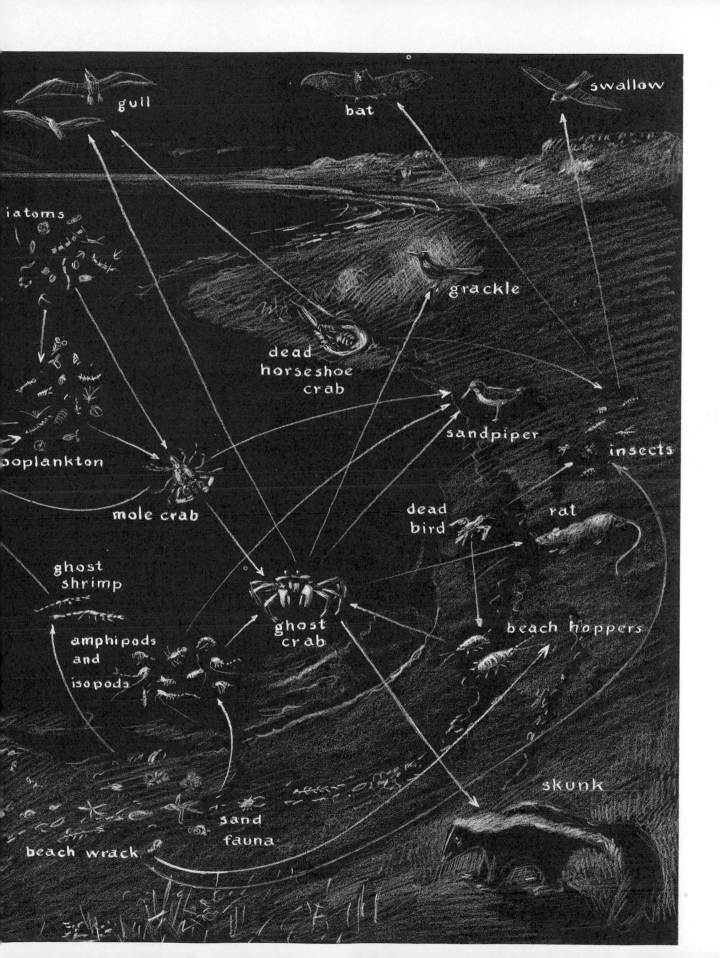

gull

bat

swallow

iatoms

grackle

dead
horseshoe
crab

sandpiper

insects

ooplankton

mole crab

dead
bird

rat

ghost
shrimp

beach hoppers

amphipods
and
isopods

ghost
crab

skunk

sand
fauna

beach wrack

most commercially valuable—shrimp, salmon, oysters, crabs, clams, menhaden, and flounders—are typically estuarine. Pacific salmon, shrimps from the South Atlantic and Gulf of Mexico, and the Atlantic menhadens together accounted for nearly half the total value of United States fish production in the 1960's.

Estuaries apparently serve as nursery areas for many species of coastal fish. Mineral and plant debris carried by rivers are retained by coastal and estuarine circulation systems. Abundant shallow-water vegetation provides protection for developing larvae and juveniles. Tidal flats, marshes, and shallow inlets are rich in nutrients and highly productive of planktonic and benthic plants. In Atlantic estuaries, the marsh grass *Spartina* together with dense growths of benthic and epiphytic algae may produce 250 tons of plant material per square kilometer (10 tons per acre) in a year. This is several times more productive than an average wheat crop.

Much of this food is consumed by insects, crabs, worms, snails, mussels, or bacteria within the marsh itself, as shown in Fig. 17–5, but a good deal is carried into the river by tides. Fishes, jellyfish, and benthic animals throughout the estuary and along the coast are nourished by the productivity of tidal marshes. Migratory and developing fish populations are major consumers. Food webs in shallow estuaries are based largely on microscopic benthic and epiphytic algae and on decaying vegetation. Thus the primary plant food is consumed directly by herbivorous fishes (such as mullet), crustaceans, and sedentary animal communities rather than by zooplankton, as is the typical pattern in open waters.

Patterns of spawning and development vary widely among estuarine-dependent salt-water fishes and shellfish. Off the Atlantic coast and the Gulf of Mexico, many coastal fishes and crustaceans spawn in the ocean, but their young soon move into estuaries. Examples are fluke (summer flounder), menhaden, king whiting, mullet, croakers, shrimp, and crabs.

Larvae often travel many tens of kilometers, cross or orbit with strong coastal currents, and enter small inlets where currents are quite strong. Larvae that swim at the surface (such as mullet) may come in on the tides. Bottom species such as shrimp, whose life cycle is illustrated in Fig. 17–6, are aided by inshore currents associated with estuarine circulation systems. Once inside, the larvae are distributed within the bays by currents; possibly some are sensitive to salinity gradients and swim toward less saline water. In general, lower salinities are preferred by fish larvae where waters are warmer. Tolerance for low salinities declines toward higher latitudes. North of Cape Cod, Massachusetts, for example, most fisheries depend on deep-water fishes, such as cod and haddock, which do not enter estuaries.

In the Northern Hemisphere in spring and summer, fish migrations are typically directed inshore and northward; in late summer and fall, the pattern reverses. The bottom-dwelling flatfish known as fluke, for example, winter offshore along the middle-Atlantic coast, sometimes as far as 150 kilometers from land. In spring they move toward the coast to feed, spending late spring and summer in nearshore waters. In early fall, fluke

FIG. 17–6

Life history of the Gulf Coast shrimp: (a) shrimp eggs; (b) nauplius larva; (c) late nauplius; (d) mysis; (e) postmysis; (f) juvenile; (g) adolescent; (h) adult.

start back toward their wintering grounds. They spawn during October and November, 15 to 100 kilometers offshore on the continental shelf, and surface currents carry the buoyant eggs (later the larvae) southward until early spring. They swim toward the coast and spend the summer in shallow estuarine waters; in autumn they re-enter the ocean. This type of breeding cycle, with variations, is common among Atlantic and Gulf coast species.

Some migratory fishes enter Atlantic estuaries to spawn. Striped bass, for instance, lay their eggs in the fresh waters of Chesapeake Bay and Albermarle Sound tributaries, and the young drift downstream into the estuary. At maturity (2 or more years) some of them enter coastal waters and migrate northward as far as New England during the summer. Winters are spent in deep holes or channels, and in spring they again swim upstream to spawn.

Permanent or semipermanent residents of mid-Atlantic estuaries include shallow-water species (such as silversides), filter-feeding fishes (river herrings, shad, alewife), marsh-dwellers with a preference for brackish water (killifish), members of eel-grass and oyster-bed communities (pipefish, gobies), sand-burrowers (cusk-eel), and species which often conceal themselves in vegetation along sandy banks (minnows).

Along the Pacific coast there are no mass offshore spawnings from which young fish swim into brackish waters, although juveniles may at times enter protected inshore areas. Many Pacific coastal fishes, such as salmon (illustrated in Fig. 17–7) certain smelts, lampreys, and sturgeon breed in fresh or slightly saline waters and later enter the sea. An extensive system of rivers and lakes for spawning and development, together with a large, rich offshore estuary to support adult populations, make the Pacific Northwest an ideal habitat for fishes with this type of life cycle.

FIG. 17–7
Chinook salmon (king salmon, spring salmon). Ranging from Southern California to northwest Alaska, chinook are especially prized by the Pacific Northwest's sport fishermen. They are common along the British Columbia coast. In spring and early summer, mature adults leave coastal waters to "run" up large rivers, where they spawn in fresh water and then die. The young may go to sea in the first year or remain for a year or more in the streams. Maximum length is about 1.5 meters. (Photograph courtesy Washington State Department of Fisheries)

Where man uses marine environments as waste-disposal areas, the net effect on marine communities may not necessarily be negative. We know that for most essential environmental factors, there is a minimum tolerable value and also a maximum—in between is a range of conditions some of which are likely to be optimal for various organisms. Problems arise from uncontrolled use of the ocean for waste disposal, without regard to the effect on local ecological systems.

Five major kinds of wastes are common in coastal areas:

1. bacterial contamination (human and animal wastes);
2. decomposable organic materials that deplete dissolved oxygen;
3. pesticides, herbicides, and toxic wastes from chemical manufacturing;
4. materials that act as fertilizers for some life forms at the expense of others;
5. inert materials that fill estuarine areas and smother benthic life forms.

Regardless of human activities, nutrients are carried to the coastal ocean by rivers. Soil erosion, plant decomposition, animal excretions, and nitrogen fixation by bacteria contribute varying amounts of nitrate and phosphate to rivers and ultimately to the ocean. This is one reason why coastal waters are so much more productive than open-ocean areas.

In the absence of excessive waste discharges, nutrients are steadily removed from coastal waters by planktonic and benthic plants. Grazing animals consume the plants at nearly the same rate as they are produced. Eventually, nutrients are recycled back to the dissolved state through respiration and decomposition processes.

Problems may arise, however, if the rate of nutrient supply suddenly begins to increase rapidly. Domestic, industrial, or agricultural waste-disposal operations have caused pronounced increases in nutrient levels in the waters of many estuaries. Species composition of plant communities may be drastically altered as a result. Populations of blue-green algae, for example, may flourish in areas where diatoms were formerly the dominant plants, because extreme nutrient enrichment is more favorable to the blue-greens. Local zooplankton, however, can eat only the native diatoms, and the blue-green algae sink uneaten to the bottom or accumulate in great rafts at the surface where it decomposes or washes ashore.

In such cases organic materials that would normally be incorporated into a grazing food chain are instead incorporated in sediment, and thus enter directly into a decomposition cycle. The grazing population starves, because its normal food supply has been choked out by blooms of inedible plants. Furthermore, bacterial decomposition in near-bottom waters and sediments depletes the deep-water oxygen supply faster than it can be replenished by bottom currents. This eventually leads to anaerobic respiration by bacteria who obtain oxygen through sulfate reduction.

Anaerobic respiration is slower and less efficient than its aerobic counterpart. Consequently, undecomposed organic material is permanently buried in sediments. Benthic organisms then die for lack of oxygen, and in some cases they may literally be smothered by the soft, easily eroded deposits. Anaerobic decomposition also produces many strongly odiferous gases, among them hydrogen sulfide (H_2S), and various organic compound with strong, unpleasant odors.

Habitat alteration caused by human activity has caused changes in many harbors and estuaries, but it is still relatively uncommon in the coastal ocean. It is important to realize, however, that when we permit marked alteration to occur in any natural habitat, a marked change in the structure of local ecosystems follows. Reduction in the number of species living in an area is a normal corollary: no amount of predation eliminates species so effectively as does a drastic change in habitat.

When stable compounds or toxic metals enter a food web, the possibility exists that they will be concentrated by organisms at each trophic level. Any substance not broken down or eliminated in respiration and not efficiently excreted tends to remain in the tissues of the organism that ingested it. Thus, at each successive trophic level, greatly increased amounts of the material are consumed.

Starting in the 1940's chlorinated hydrocarbons such as DDT were used to control undesirable insect populations in many parts of the world. A large fraction of what was originally released on land has been transported to the ocean. It is estimated that when crops are treated with pesticides, half of the sprayed material never reaches the crop but is transported to the atmosphere as vapor or associated with wind-borne particles.

Runoff from land is also an important source of chlorinated hydrocarbons in the ocean. Many of these compounds are derived from industrial products such as plasticizers and components of paints and rubber products. This latter type is comparatively resistant to degradation and much more soluble in fats than in water; as a result, the compounds become concentrated in algae and other marine organisms. Fats are conserved in food chains, whereas many skeletal materials eventually become incorporated in sediments.

Shrimp and other crustaceans are particularly sensitive to chlorinated hydrocarbons and may be killed outright, whereas fish populations at higher trophic levels, while showing no catastrophic effects, may yet contain such large amounts of DDT as to be unfit for human consumption. Sea birds such as gulls, ospreys, cormorants, and ducks are also second- or third-level consumers, and several species have suffered serious reproductive failures due to high concentrations of DDT in adult birds. Such effects on a worldwide scale may endanger entire species, rather than just a local population.

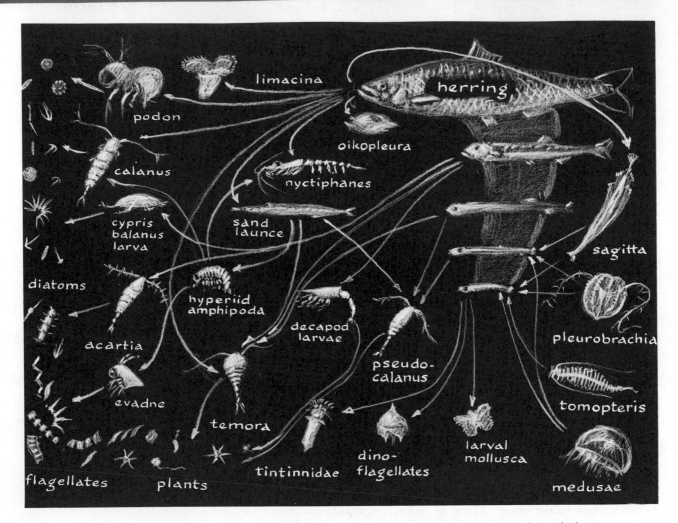

North Sea herring deposit their eggs in thick blankets which may cover 100 square meters or more of the continental shelf. Bottom-living fishes, especially haddock, consume great numbers of these eggs. Herring develop as tiny, wormlike creatures, curled around the yolk of their eggs. After a few weeks, at about 5 millimeters length, the transparent larvae break free of the egg membrane and rise toward the water surface. Their first food is phytoplankton, later nauplii, very small crustaceans, and young stages of copepods.

If suitable small food organisms have failed to develop in the area, or if predatory ctenophore or arrow-worm populations are unusually large, the herring may be eaten in great numbers. At a length of about 30 millimeters, they begin to eat some of the smaller adult copepods, and at about 40 millimeters they grow scales and begin to look like young fish instead of tiny eels. They form schools and travel toward shallow water, often passing into estuaries, now eating mainly estuarine crustaceans. After 6 months or so, the young herring disperse throughout the North Sea. When sexually mature, after a period of 3 to 5 years, they join schools of spawning fish, and the cycle is repeated.

The same species inhabits the Atlantic coast of Canada and the United States.

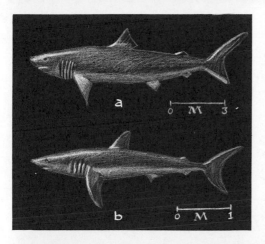

These giant fish commonly grow to 8 meters in length. They live in temperate seas throughout the world and travel open-mouthed, straining plankton from the water by means of a filtering device on the gills. Basking sharks travel singly or in groups, and spend the winter resting in deep water. The liver is a valuable source of oil. All sharks, skates, and rays have skeletons of cartilage rather than bone.

Mako (sharp-nosed mackerel shark)

Swiftest of all sharks, the mako feeds on schools of mackerel and herring, and on oceanic squids. Their average length is about 1.5 meters. When the Mako basks in the sun, the fin on its back may be seen above the surface. Its young are born alive, as with most sharks. The mako may be dangerous to man under certain circumstances, as when they have been stimulated by the smell of food and are in the mood to attack. They are commonly found in tropical and warm temperate Atlantic waters, in summer as far north as New England.

Artificial fishing reefs

Schools of migratory fish are sometimes attracted to a reef or sunken ship in the coastal ocean because the structure offers shelter from predators, particularly for juveniles. In addition, formation of benthic communities often provides a source of food. Along sections of the United States coast, heaps of quarry rock, construction debris, lengths of culvert pipe, and specially built concrete blocks with holes have been deposited as artificial fishing reefs. These have mainly been beneficial to parties of sport fishermen. Junk car bodies have also been tried, but they tend to break up after a few years and the metals thus released may have a deleterious effect on the environment.

Experiments with low- and high-profile fishing reefs have been sponsored by the Japanese government. The former, known as "constructed beach" type reefs, are best for raising seaweed as food and for attracting edible echinoderms. Grazing fishes also come to such an area. A reef of medium height with many entrance holes and large, hollow centers is most successful for raising crayfish. To accommodate large, roving predators, tall "fish apartment houses" of reinforced-concrete blocks or oil drums have met with success. Such schooling migrants as bluefish, pollock, and jacks are known to be attracted by tall structures (such as wrecked ships) along United States coasts.

The photographs here were taken at Artificial Shoals in Hawaii: (a) coral growth on a reef of junk car bodies; (b) a school of Chromis verator in a car body; (c) a reef made of damaged concrete pipe. (Photographs courtesy State of Hawaii, Department of Fish and Game)

Today's fishing methods are not very different from those used for centuries. Bottom-dwelling fish such as halibut are scooped up in a trawl net (a) that is dragged over the sea floor. The photograph shows a trawl catch before it is dumped on the deck of a boat of the National Marine Fisheries Service in the North Atlantic. Fish that feed in schools near the surface are often trapped or entangled in a vertically suspended net, or in a purse-seine as shown in (b) fishing for menhaden at Sealevel, North Carolina. Purse-seines are useful for catching densely schooling fishes in protected waters, but they are also used on the high seas in some parts of the world to catch tunas. The third common fishing method uses hook and line (c); here, fishermen land a 30-kilogram yellowfin tuna, using a three-pole, single-line rig in the Gulf of Guinea, off Africa. (Photographs courtesy National Marine Fisheries Service)

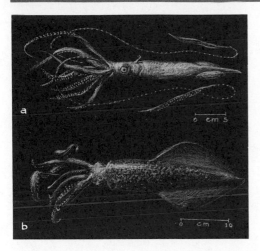

There are many species of these active, predatory cephalopods in the open ocean, ranging from luminous deep-sea species a few centimeters long to the giant squid that attains a length of over 16 meters. They live in all oceans, sometimes to depths of 3500 meters.

Famous for their agility, squids "jet-propel" themselves backward or forward, using the heavy "mantle" membrane to force water through their siphons. Speed is their principle defense against predatory birds, fish, eels, and sperm whales, and it is also an important asset in catching their own food. They often change color as a means of camouflage, and like other cephalopods can usually give off clouds of dark or luminous ink when alarmed.

Squids are fierce and fearless, with heavy, parrotlike beaks for biting and tearing their prey. They often seem to maim and kill other animals far in excess of immediate food needs. Studies show squids to be intelligent and capable of learning—they have quick reflexes and exert marvelous control over the actions of their ten snakelike arms, two extralong ones being equipped with flat, sucker-covered "hands."

Illustrated: are (a and c) two species of the deep-sea octopod Cirrotheuthis; (b) the common nearshore squid Loligo; two deep-water Atlantic squids —(d) Octopodotheuthis, and (e) Taonidium; and two mid-water species —(f) Callicuthis reversa, and (g) Histioteuthis bonnelliana. Histioteuthis has been captured at about 1000 meters depth, and Calliteuthis as deep as 1500 meters, but both are sometimes seen at the surface.

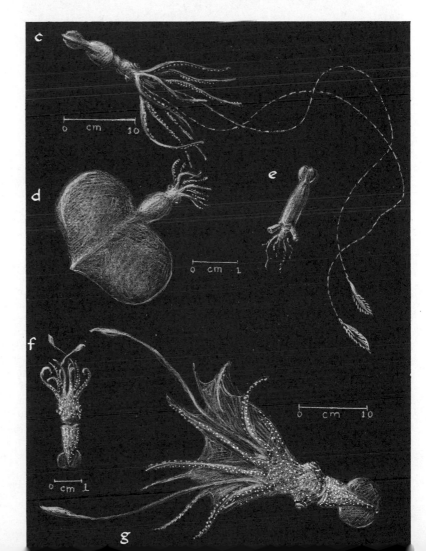

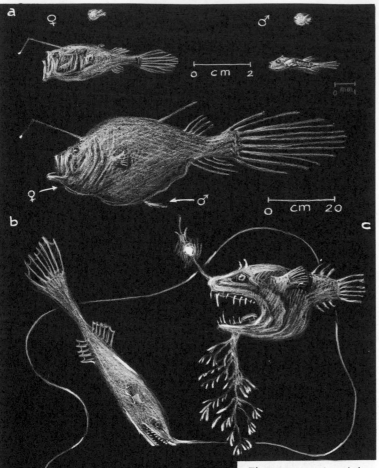

The many species of these curious fishes are most abundant at depths of 1500–2000 meters, where spaces are vast, unlit, and virtually without boundaries, and food is extremely scarce. In life cycle and in body structure, some angler fishes illustrate two remarkable adaptations for survival at great depths: a movable, lighted fishing lure which dangles in front of the huge mouth, and a foolproof means of guaranteeing mating—the male is parasitically attached to the female.

In the species Ceratius holboelli (a), the male swims freely until adolescence, at which time he is only a few millimeters long. Locating a female, he grips her belly skin with his teeth and grows to a length of several centimeters. Their fertilized eggs float to the surface, where the larvae hatch and feed on copepods until adolescence, when they return to deep water. Sexual maturity is not attained until after the fish have joined together permanently. Adult females are about 1 meter long.

Not all angler fishes have the "attached male" adaptation, but all females possess some sort of attractive light organ to lure prey. The female ceratioid angler Gigantactis (b), living at depths to 5000 meters, carries a maneuverable "fishing rod" many times the length of her body. Other species, such as Linophryne (c), have luminous "barbels" on their chins, or a light organ in their mouths, just behind the teeth.

A common adaptation at these depths is the ability to extend jaws and belly so that prey much larger than the predator can be swallowed.

Melanocetus johnsoni (d) is illustrated before and after having eaten a fish twice its own length.

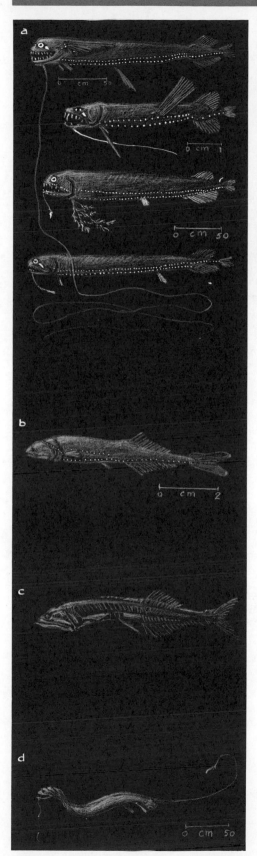

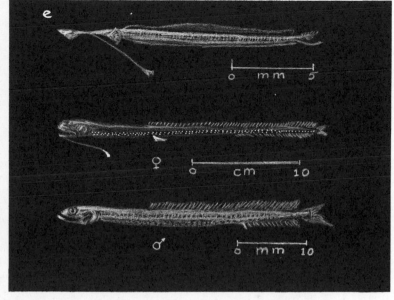

These slim, generally dark or silvery fishes range from a few centimeters to a few tens of centimeters long. Probably the most numerous fishes in the ocean, they are especially common at mid-depths of 500 to 2000 meters. Populations may not be dense in any one place, in the manner of coastal herring or anchovies, but they are common through a vast area of the ocean.

Melanostomiatoids (a) characteristically bear chin barbels, often associated with a luminous lure. These fleshy organs are sensitive to water movements, which cause the fish to snap in the direction of the slightest disturbance. Stomiatoids have big mouths and sharp, strong teeth, and many can swallow victims larger than themselves. Patterns of luminous dots along the body are common.

The mid-water Cyclothone (bristlemouths) are believed to be the largest single genus (group of species) among marine vertebrates. They are small (2 to 8 centimeters) and rapacious, with mouths that open nearly 180° to capture prey. Illustrated are (b) C. microdon and (c) C. pallida.

Idiacanthus niger (d), about 20 centimeters long, is shown after having swallowed a fish longer than its own gut, the prey being doubled up to fit inside the predator. Idiacanthus fasciola (e) is commonly found at depths of 900 to 1800 meters, in low-latitude waters. Females are black, reaching a length of 30 centimeters; males are pale and seldom exceed 5 centimeters in length, with lights on their cheeks, but lacking the characteristic barbel of the female. The larvae, whose eyes are on the ends of long stalks, drift at the surface while feeding on plankton. When they reach a length of about 5 centimeters, they sink to 900 meters or lower, and the eyestalks resorb to a normal position.

497

a

b

These flightless birds are completely adapted to life in the sea. They swim by "flapping" their long, flipper-like "wings." Darting rapidly through the water together, groups of penguins break the surface in unison for breath. Their principal enemy is the leopard seal (a).

Emperor (b) and Adélie penguins (c) breed on the shores of the Antarctic continent. The former has evolved a remarkable method of hatching chicks on fast ice: in the dark of midwinter it keeps the single egg warm on top of its feet, nestled in thick, downy folds on its legs.

Adélie penguins are migratory. Large flocks gather at rookeries in early summer, where they build nests of pebbles on dry ground or light snow. The eggs hatch in December; by February, the young can swim, at which time the flock moves off the continent onto floating pack ice, to feed on krill. Eggs and young are preyed upon by a species of gull known as the skua (d).

c

d

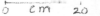

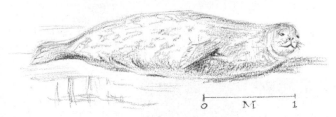

Weddell seals live in coastal waters off the Antarctic continent, where the sea surface is frozen for over half the year to a depth of a few meters. The seal lives on fish, which it hunts under the ice during the long Antarctic winter. In order to breathe, it must either find the widely dispersed cracks formed in the ice by wind or water movements, or ream out breathing holes through thin ice using its specially adapted front teeth.

These animals can remain underwater for up to 1 hour. During this time they may swim distances of several kilometers, traveling from one breathing hole to another. In search of food, they dive deeply, sometimes to 600 meters. When diving, their heartbeat and other metabolic functions slow to a fraction of the normal rate, and all blood circulation is diverted to the heart and brain. In this way, oxygen is conserved while the seal is not breathing.

An adult Weddell seal measures about 3 meters in length and may weigh a half-ton. The fur is dark, and splotched with lighter patches that coincide with the mottled appearance of light coming through the ice where the seal is swimming.

Baleen whales (whalebone whales)

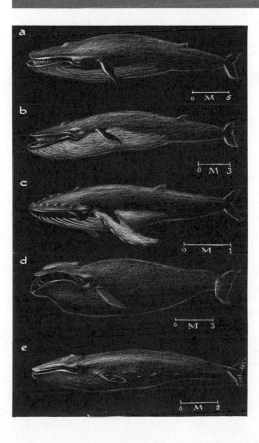

Whales are the largest animals that have ever lived. They are mammals, and thus breathe air and bear live young. A horny sieve which hangs in plates from the palate is the "whalebone"; it is used to strain copepods and krill from seawater. They have no teeth, and plankton or small fishes are their only food.

Blue whales (a) are the largest of all, reaching lengths of up to 31 meters. A 27-meter blue has been found to weigh 120 tons. Baleen species also include: (b) the finback, which attains a length of about 20 meters; (c) the humpback, about 15 meters at maturity; (d) right whales, 17 meters or more; and (e) the smaller California gray whale, about 12 meters.

Baleen whales live in all seas, but particularly in the Arctic and Antarctic, where during the short southern summer (with a peak in February), krill have been reported so thick that a large ship was "slowed to half speed by them." Not only whales but fish, penguins, seals, and huge populations of flying birds gorge themselves on the seasonally abundant food supply. The greatest numbers of baleen whales have been observed in the Antarctic when plankton are dense. During the rest of the year, they range widely throughout the oceans.

Sperm whales

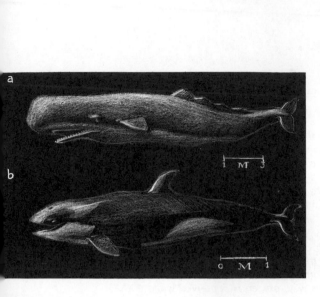

These are the largest of the toothed whales. Males average 15 meters, females are smaller. Their principal food is squid, which they hunt from the tropics to subarctic latitudes in all seas, diving to depths of 1000 meters or more and remaining below for as long as 1 hour. Giant squids up to 10 meters in length have been found intact in the stomachs of sperm whales. There is a reservoir of spermaceti oil in the animal's large, square head. Its function is unknown, although naturalists and scientists have speculated that it might be involved in the exchange of gases during long dives, or perhaps in sound production.

Before a dive, a whale fills its lungs with air; while it is underwater, most of the fresh gases are exchanged for carbon dioxide or other products of its metabolism. Large amounts of oxygen are stored in the myoglobin of its muscle tissue, as well as in blood. The heart beats very slowly while a whale (or other diving animal) is submerged, which helps to reduce oxygen demand. Small lungs and a flexible thoracic cavity permit compression without damage as the volume of air decreases with depth.

Sperm whales are sometimes attacked by packs of killer whales (b). These are not really whales at all, but members of the dolphin and porpoise family, most common in cold water. An adult sperm whale can usually resist such attacks, even if surprised while sleeping—probably the only animal in the ocean powerful enough to do so.

Sea turtles

Five kinds of sea turtles inhabit the warmer oceans of the world. They swim by working their big flippers up and down, instead of paddling in the manner of fresh-water varieties. Their eggs are laid in holes dug in the sand above high water. This is generally done at night, in late spring or summer, and repeated every 1 to 3 years.

Near the end of summer, young turtles hatch and head immediately for the sea, apparently directed by the quality of skylight over water. Females return to the same beach years later, to lay their own eggs. Several hundred may be deposited in a season.

Green turtles have been studied extensively in an effort to trace their long migrations between feeding and nesting grounds. Huge herds nesting at Tortuguero in Costa Rica may later be found feeding off the coasts of Cuba, Florida, Mexico, and Venezuela, up to 2500 kilometers away. Turtles hatched at Ascension Island in the mid-Atlantic probably feed 2350 kilometers away on beds of turtle-grass (a seed-bearing marine plant, Thalassia) that grow off the coast of Brazil. On the other hand, some herds feed within a few dozen kilometers of the nesting beach. The young are carnivorous and seem to remain at sea feeding on animal plankton until old enough to assume the adult herbivorous diet. (Photograph courtesy Marineland of Florida)

Tropical sea horse

The tropical sea horse is famous for its unusual breeding adaptation, in which the female deposits fertilized eggs in the abdomen of the male. There they incubate, nourished and supplied with oxygen from the father's bloodstream, before being hatched alive. Sea horses are weak swimmers; they remain stationary for long periods, the tail wrapped around a branch of weed or coral, snapping at food particles swept past by tidal currents.

A related species, the pipefish (15 to 30 centimeters long), lives among seaweeds and beds of eel-grass in United States Gulf Coast bay flats and Atlantic coast estuaries. They are rather eel-like in appearance, with a narrow fin along the back which vibrates to move the animals slowly through the water. Their food is plankton, carefully selected and then inhaled through the tubular mouth.

Eels

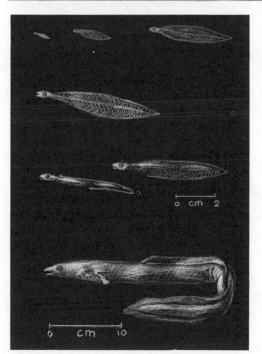

The anguillid "fresh-water" eels, 0.5 meter or longer, are well known to gourmets on both sides of the Atlantic. They live in rivers and lakes, but at sexual maturity (5–8 years old) begin a fantastic migration to the Sargasso Sea. After a journey of thousands of miles, and many months without food, they spawn at depths of 500 meters or thereabouts in the warm, salty water, and then they die.

The eggs float at the surface and soon hatch into tiny larvae that resemble little fish. Each begins the long journey to fresh water, according to the studies of the famous biologists Johannes Schmidt, the European eels taking 3 years to complete the journey and the American variety requiring only 1 year. There is a physical transformation at that time; the slender young eels ("elvers"), about 7 centimeters long, travel up the streams to live there until maturity.

This attractive animal is making a comeback along the west coast of the United States, after being nearly reduced to extinction during the eighteenth century by fur traders. Sea otters apparently have no enemies except killer whales, sea lions, and man. They themselves are regarded as a threat to the western abalone crop, since this huge, primitive gastropod (b) is one of the otter's favorite foods.

While sleeping, the sea otter floats on its back in the water, mothers cradling a cub in their arms. They can dive to 90 meters, seeking crustaceans, mollusks, sea urchins, or fish. They find these particularly in kelp beds, from the shore to distances up to 50 meters from land.

a

b

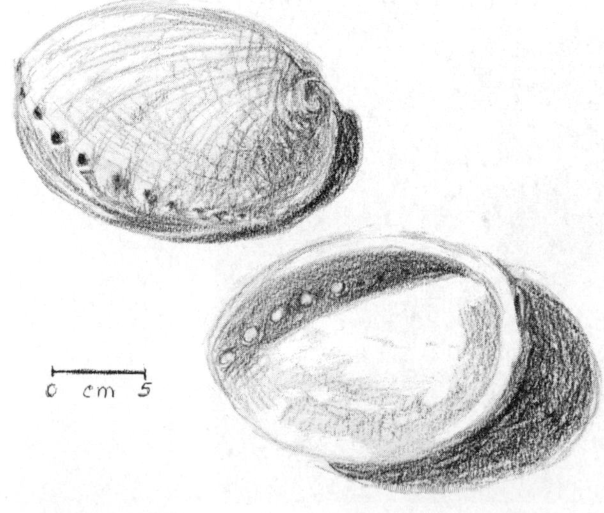

0   cm   5

502

Sargassum (sometimes mistakenly called "gulfweed" because of its supposed origins) is a particularly interesting brown alga because it is the only large nonattached seaweed. It is found in the south-central North Atlantic, where surface currents converge in an area of calm sea. There, it puts out shoots which break off as new plants. Patches of this algae form unique "floating islands"; they harbor communities of smaller plants and animals which can live nowhere else in the world. These include many kinds of specialized crabs, snails, copepods, worms, fish, sea horses, and small octopus, as well as the wonderful sargassum fish *itself in its* leafy-looking seaweed costume.

Patches of sargassum collect in windrows along lines of convergence formed by Langmuir cells and other circulation systems. Surface films, plankton, and particles of detritus are caught there and so become available to decomposers and consumers in the sargassum community. Large numbers of eggs and larvae of open-ocean fishes become trapped in the dense growth, which offers protection from predators as well as a rich supply of food particles. Some of these eggs and young fishes, representing an energy source that originated outside the sargassum community, are consumed within the community, thus augmenting the local nutrient supply.

Seven million tons of "gulfweed" is estimated to float in the approximately 5000 square kilometers of the Sargasso Sea. It provides oases of teeming life in an area where phytoplankton productivity is normally very low, due to the fact that organic material is trapped and recycled within the sargassum community itself.

503

SUMMARY OUTLINE *Productivity of ocean* is governed by oceanographic processes

Man competes with other organisms for the marine harvest

Nekton—large swimming organisms represent an important part of the harvest

Primary productivity—the amount of carbon fixed by plants in a given time period

> Net plant production—the carbon produced in photosynthesis minus that used in respiration; at the compensation depth these are equal
>
> Above the critical depth there is sufficient light to support a plankton bloom
>
> Phytoplankton bloom occurs when critical depth exceeds depth of the mixed surface layer
>
> Phytoplankton species bloom in sequence depending on physical and biochemical factors

Geographical distribution of productivity

> Primary productivity may differ by one-hundredfold between richest and poorest areas
>
> Open-ocean carbon fixation averages about 50 $gC/m^2/yr$
>
> Rich upwelling areas may produce as much as 10 $gC/m^2/day$ of carbon during periods of favorable conditions; upwelling occurs in about 1 percent of the world ocean
>
> 80 percent of living material is in the top 200 meters; only 1 percent below 3000 meters

Effect of grazing by zooplankton

> Standing crop of phytoplankton may be greatly depleted by grazing; therefore, biomass may be low although production is high
>
> Nutrients are returned directly to the water via dissolution of excreted matter
>
> Peak production of phytoplankton and zooplankton populations may alternate in a given area
>
> Rate of change in biomass of a phytoplankton crop is determined by rate of photosynthesis minus rate of respiration minus rate of grazing
>
> Biological history of the water affects species composition

Organic growth factors, such as vitamins and antibiotics, may favor or inhibit productivity in an area

Food resources of the ocean

> 90 percent of potential food energy is lost through transfer at each trophic level in the food chain
>
> Food chains tend to be long and complex in the open ocean
>
> Short food chains, as in the coastal ocean (especially in areas of upwelling), represent most efficient food production
>
> Largest sustainable yield found where there is a high proportion of young animals in a population
>
> 1960's world fish catch of 55–60 million tons per year might be increased to 100–200 million tons by the end of this century
>> Underexploited and (probably) unexploited stocks still exist
>> Cultivation of herbivores, management of coastal and estuarine resources and of present fisheries necessary
>
> Further increases in yield would require use of new types of fish, new technical developments, new marketing methods
>
> Neither harvesting plankton nor fertilizing the ocean seem practical
>
> World marine harvest could provide a significant proportion of world protein needs
>
> Drugs from the ocean a possibility

More than half of U.S. fish catch consists of estuarine-dependent species

Estuaries serve as nursery areas for many coastal fishes

Estuaries provide food and protection for larvae and juveniles

Many important food and game fishes breed in the ocean; young later enter estuaries

Many U.S. east and west coast fishes enter estuaries or rivers to spawn; as adults they enter the ocean

Several fishes are permanent residents of estuaries

Waste disposal in the coastal ocean

Controlled use of the ocean for certain waste disposal operations may be acceptable

Major kinds of wastes in coastal areas are:

Bacterial (human and animal)

Decomposable organic materials that deplete dissolved oxygen

Pesticides, herbicides, chemicals from manufacturing and agriculture

Fertilizers that upset ecological balance in the ocean

Inert materials that smother benthic life

Alteration of habitat is more dangerous to species survival than is predation

Many toxic substances are concentrated in marine food chains; effect of cumulation is greatest at higher trophic levels

SELECTED REFERENCES

HARDY, ALISTER. 1965. *The Open Sea: Its Natural History, Part 2: Fish and Fisheries*, one-vol. ed. Houghton Mifflin, Boston. 322 pp. A classic by a well-known British naturalist; nicely illustrated.

HOLT, S. J. 1969. "The Food Resources of the Ocean." *Scientific American*. 221(3) (Sept., 1969), 178–94. Survey of contemporary technology and outlook for the future of marine food resources.

MACKINTOSH, N. A. 1965. *The Stocks of Whales*. Fishing News (Books), London. 232 pp. Biology of whales and distribution of whale stocks as these relate to the whaling industry, past, present and future.

MARSHALL, N. B. 1958. *Aspects of Deep-sea Biology*. Hutchinson, London. 380 pp. Authoritative treatment of the deep-ocean environment and its ecology.

NATIONAL ACADEMY OF SCIENCES. 1971. *Radioactivity in The Marine Environment*. Washington D.C. 272 p. Comprehensive review of the behavior of radioactivity in the ocean and the controlling factors.

RAY, CARLETON, and ELGIN CIAMPI. 1956. *The Underwater Guide to Marine Life*. A. S. Barnes, New York. 337 pp. Classifies and briefly describes thousands of marine organisms, with line drawings.

RAYMONT, J. E. G. 1943. *Plankton and Productivity in the Oceans*. Macmillan, Pergamon Press, New York. 625 pp. Discusses productivity.

RYTHER, JOHN H. 1969. "Photosynthesis and Fish Production in the Sea." *Science*, 166:72–76. A summary of contemporary knowledge about productivity and ecological efficiency throughout the world ocean.

TAIT, R. V. 1968. *Elements of Marine Ecology: an Introductory Course*. Butterworths, London. 272 pp. Background survey for marine biology students.

WIMPENNY, R. S. 1966. *The Plankton of the Sea*. American Elsevier, New York. 426 p.

EXPONENTIAL NOTATION, THE METRIC SYSTEM, AND SELECTED CONVERSION FACTORS

appendix 1

EXPONENTIAL NOTATION

It is often necessary to use very large or very small numbers to describe the ocean or to make calculations about its processes. To simplify writing such numbers, scientists commonly indicate the number of zeros by *exponential notation*, indicating the powers of ten. Some examples:

1,000,000,000	= 10^9	(one billion)
1,000,000	= 10^6	(one million)
1,000	= 10^3	(one thousand)
100	= 10^2	(one hundred)
10	= 10^1	(ten)
1	= 10^0	(one)
0.1	= 10^{-1}	(one–tenth)
0.01	= 10^{-2}	(one–one-hundredth)
0.001	= 10^{-3}	(one–one-thousandth)
0.000 001	= 10^{-6}	(one–one-millionth)
0.000 000 001	= 10^{-9}	(one–one-billionth)

Multiplication: To multiply exponential numbers (powers of ten), the exponents are added. For example, $10 \times 100 = 1000$ which is written exponentially as $10^1 \times 10^2 = 10^3$

Division: To divide exponential numbers, the exponent of the divisor is subtracted from the exponent of the dividend. For example, $100/10 = 10$ or written exponentially as $10^2/10^1 = 10^1$

UNITS OF MEASURE

Length

1 *kilometer* (km) = 10^3 meters = 0.621 statute mile = 0.540 nautical mile

1 *meter* (m) = 10^2 centimeters = 39.4 inches = 3.28 feet = 1.09 yards = 0.547 fathom

1 *centimeter* (cm) = 10 millimeters = 0.394 inch = 10^4 microns

1 *micron* (μ) = 10^{-3} millimeter = 0.000394 inch

Area

1 *square centimeter* (cm^2) = 0.155 square inch
1 *square meter* (m^2) = 10.7 square feet
1 *square kilometer* (km^2) = 0.386 square statute mile = 0.292 square nautical mile

Volume

1 *cubic kilometer* (km^3) = 10^9 cubic meters = 10^{15} cubic centimeters = 0.24 cubic statute mile
1 *cubic meter* (m^3) = 10^6 cubic centimeters = 10^3 liters = 35.3 cubic feet = 264 U.S. gallons
1 *liter* (l) = 10^3 cubic centimeters = 1.06 quarts = 0.264 U.S. gallon
1 *cubic centimeter* (cm^3) = 0.061 cubic inch

Mass

1 *metric ton* = 10^6 grams = 2205 pounds
1 *kilogram* (kg) = 10^3 grams = 2.205 pounds
1 *gram* (g) = 0.035 ounce

Time

1 day = 8.64 × 10^4 seconds (mean solar day)
1 year = 8765.8 hours = 3.156 × 10^7 seconds (mean solar year)
1 aeon = 10^9 years (one billion years)

Speed

1 *knot* (nautical mile per hour) = 1.15 statute miles per hour = 0.51 meter per second
1 *meter per second* (m/sec) = 2.24 statute miles per hour = 1.94 knots
1 *centimeter per second* (cm/sec) = 1.97 feet per minute
= 0.033 feet per second

Temperature

Conversion formulas

$$°C = \frac{°F - 32}{1.8}$$

$$°F = (1.8 × °C) + 32$$

Conversion table

°C	°F
0	32
10	50
20	68
30	86
40	104
100	212

Energy

1 *gram-calorie* (cal) = $^1/_{860}$ watt-hour = $^1/_{252}$ British thermal unit (Btu)

GRAPHS, CHARTS, AND MAPS

appendix 2

In any scientific discipline, data gathered from many sources are usually at some point compiled and presented in graphic form. Oceanography is no exception, although the wide variety of material obtained from studying the ocean poses some special problems. In this section we shall review some of the common means of graphically presenting data—graphs, profiles, maps, and diagrams—in an effort to show the uses as well as the limitations of each technique.

GRAPHS

Of the various techniques used to portray scientific data, *graphs* are perhaps the most widely used. A graph permits general aspects of the relationship between two properties to be understood at a glance. Furthermore, values of various properties can be measured from a graph.

In constructing a graph, data are often first organized into tables. For example, if we are studying the increase in boiling point of seawater with salinity changes, we may arrange our data as follows:

Salinity ($^0/_{00}$)	Boiling point increase (°C)
4	0.06
12	0.19
20	0.31
28	0.44
36	0.57

From this table it is apparent that the boiling point of seawater increases with increased salinity. To see this more clearly, a graph can be drawn as in Fig. A2–1. Note that salinity is plotted on the x (horizontal) axis, with values increasing to the right. Boiling point increase, in degrees Celsius (°C), is plotted on the y (vertical) axis, with values increasing upward. We plot our experimental points, and then draw a line through these points.

Generally, a straight line is the best first estimate. In this case, it is a reasonably good approximation. For many other graphs, we may need to use more complicated curves; but even for complicated curves, a straight line is a reasonable estimate for small portions of the curve.

Note that the graph shows that the boiling point is raised as salinity increases. Also, you can estimate boiling point increases for salinities we did not study. For instance, for seawater with a salinity of $24^0/_{00}$, we can interpolate a boiling point increase of 0.375°C; for a salinity of $40^0/_{00}$, we extrapolate (or extend) the curve to indicate a boiling point increase of 0.63°C.

In reading a graph, always determine which property is plotted on each axis. Also, check both the scale intervals and the values at the origins. The appearance of the graph can be changed drastically by changing either; advertisers, for instance, often make use of graphs where the scales and origins are chosen to present their data in the most favorable light.

PROFILES

Profiles are used to show topography, either of ocean bottom or of land. An ocean-bottom profile can be considered as a vertical slice through the earth's surface. Such a profile can be drawn with no distortion—distances are equal vertically and horizontally. But imagine the problems involved in drawing a profile 10 centimeters long of the Atlantic Ocean between New York and London, a distance of 5500 kilometers, but only 3.4 kilometers deep at the deepest point. A pencil line would be too thick to portray accurately the maximum relief; thus such a profile conveys no useful information.

To get around this problem, profiles—including those in this book—are usually distorted. Profiles showing oceanic features are typically distorted by factors of several hundred or several thousand. This causes even gently rolling hills to look like impossibly rugged mountains. The effect of profile distortion can be seen rather dramatically when it is applied to the human profile, as in Fig. A2–2.

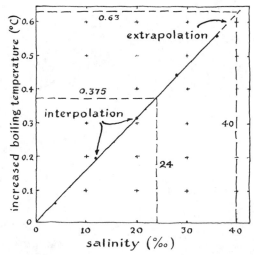

FIG. A2–1

Graph of the increase in temperature of boiling and water salinity. The dots represent the experimental data from the table.

FIG. A2–2

Distortions in a profile of a human face resulting from use of different standard exaggerations.

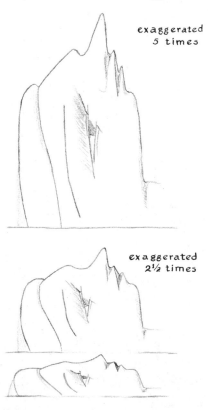

exaggerated 5 times

exaggerated $2\frac{1}{2}$ times

Various means have been used to portray land forms or topography; the most useful employ contours in contour maps. Fig. 2–10 is an example of a contour map. *Contour lines* connect points that are at equal elevations or at equal depths. Obviously, not all elevations (or depths) can be connected by contour lines. Only certain ones at selected intervals are shown; otherwise, the map would be solid black. The vertical interval represented by successive contour lines is called the *contour interval.*

To interpret a contour map, imagine the shoreline of a lake. The still-water surface is a horizontal plane, touching points of equal elevation along the shore. The shoreline is thus a contour line. If the water surface were controlled to fall by regular intervals, it would trace a series of contour lines, forming a contour map on the lake bottom (or hillside). Note that the shorelines formed at different lake levels do not cross one another; neither do contours on maps.

Contours reveal topography. For example, contours that are closed on a map (do not intersect at a boundary) indicate either a hill or a depression. To find out which, look to see if the elevation increases toward the closed contour(s). If so, you are looking at a hill. If the elevation decreases toward the closed contour, it is a depression. Contours around depressions are often marked by *hachures*—short lines on the contour pointing toward the depression. When a contour line crosses a valley or canyon, the contour line forms a V, pointing upstream.

Contours can also be used to depict properties other than elevation (or depth). For example, several maps in this book use contours to show distribution of properties such as surface-ocean temperatures (see Fig. 7–7, and 7–8) and salinities (see Fig. 7–14). We would consider them to be temperature (or salinity) hills and valleys. High temperature corresponds to a hill, low temperature to a valley. Contours may also be used to show the distribution of temperature or salinity in a vertical section of the ocean.

COORDINATES—LOCATIONS ON A MAP

A *coordinate* is a sort of address—a means of designating location. The most familiar coordinate system, found in many cities, is the network of regularly spaced, lettered or numbered streets, crossing one another usually at right angles, which enables us to locate a given address. This is an example of a *grid*. As soon as we determine how the streets and avenues are arranged, we can find Fifth Avenue and 42nd Street in New York, or 8th and K Streets in Washington, D.C., even though we may never have been in those cities before.

A printed grid is used on large-scale maps to designate locations of features. In making maps of relatively small areas, it is simplest to assume that the world is flat. This works for areas extending up to 100 miles from a starting point. For larger areas, the earth's curvature must be considered.

Since the earth is round, we must use *spherical coordinates*—a grid fitted to a sphere. A small town or even a state has rather definite starting points for a grid—the edges or the center. But a sphere has no edges or corners, so we must designate those points where our numbering system is to begin.

Distances north or south are easiest to deal with. We can easily identify the earth's geographic poles, where the axis of rotation intersects the earth's surface. Using these points, it is fairly easy to draw a line circling the earth and equally distant from the North and South Poles. This is the *equator*, which serves as our starting point to measure distances north and south. Going from the equator, the distance to the pole is divided into 90 equal parts (*degrees*). The series of grid lines which circle the earth and connect points that are the same distance from the nearest pole are known as *parallels of latitude*.

To see how this works, imagine the earth with a section cut out, as in Fig. A2–3. Now look at the angle formed by the line connecting any point of interest with the earth's center and the line from the earth's center to a point directly south of that point, on the equator. This angle is a measure of the distance between the chosen point and the equator. The North Pole has a latitude of 90°N, Seattle is approximately 47°N, and Rio de Janeiro, approximately 23°S.

Latitude was easily measured by early mariners. The angle between the pole star (Polaris) and the horizon provides a reasonably accurate measure of latitude. At the equator, the pole star is on the horizon (latitude 0°N). Midway to the North Pole (latitude 45°N), the pole star is 45° above the horizon. At the North Pole, Polaris is directly overhead. Although there is no star directly above the South Pole, the same principle holds, except that a correction would be necessary to allow for the displacement from the South Pole of the star used.

Measuring east–west distances on the earth poses the problem—where do we start? The answer has been to establish an arbitrary starting point—the *prime meridian*—and to indicate distances as east or west of that meridian. Several prime meridians have been used by different nations, but today the Greenwich prime meridian is most commonly used. It passes through the famous observatory at Greenwich (a suburb southeast of London).

Longitude—distance east or west of the prime meridian—is indicated on a map by north–south lines, connecting points with equal angular separation from the prime meridian. They are called *meridians of longitude*, and converge at the North and South Poles. Longitude in degrees is measured by the size of the angle between the prime meridian and the meridian of longitude passing through the given point, as in Fig. A2–3. Going eastward from the prime meridian, longitude increases until we reach the middle of the Pacific Ocean, when we come to the 180° meridian. Going westward from Greenwich, longitude also increases until we reach the 180° meridian, which represents the juncture between the Eastern and Western Hemispheres. Through much of the Pacific Ocean, the 180° meridian is also the location of the *international dateline*. This designation of the 180th meridian as the international dateline—where the "new

FIG.   A2–3
Latitude and longitude on the earth shown for Cape Hatteras, North Carolina.

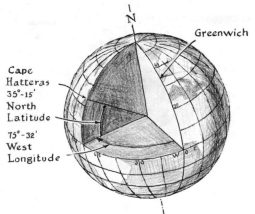

day" begins—is no accident. Its position in the midst of the Pacific avoids the problem of adjacent cities being one day apart in time. This also explains, in part, the choice of the Greenwich Meridian as the Prime Meridian.

Longitude and time are intimately related. The earth turns on its axis once every 24 hours. Since it takes the earth 24 hours to make one complete turn (360°), we calculate that the earth turns 15° per hour. We use this relationship to find our relative position east or west of the prime meridian.

Each meridian of longitude is a *great circle*. If we sliced through the earth along one of the meridians of longitude, our cut would go through the center of the earth. Of the parallels of latitude, only the equator is a great circle. All the other parallels are *small circles*, since a plane (or slice) passing through them would not go through the center of the earth. Great circles are favored routes for ships or aircraft because a *great circle route* is the shortest distance between two points on a globe.

To study time and longitude, let us begin at local noon on the prime meridian, when the sun is directly overhead. One hour later, the sun is directly over the meridian of a point 15° west of the prime meridian; two hours later, it is over a meridian 30° west of the prime meridian. Twelve hours later (midnight), it is over the 180th meridian and the new day begins.

If we have an accurate clock keeping "Greenwich time" (the time on the Prime Meridian), we can determine our approximate longitude from the time of local noon, when the sun is highest in the sky. Assume that our clock read 2:00 P.M. Greenwich time at local noon. The 2-hour difference indicates that our position is 30° from the prime meridian. Since local noon is later than Greenwich, we know that we are west of Greenwich and our longitude is therefore 30°W. Another example—if our local noon occurs at 9:30 A.M. Greenwich time, we are 2.5 hours × 15°/hour = 37.5° east of the prime meridian; our longitude is thus 37.5°E.

Degrees—like hours—are divided into 60 parts known as *minutes*. Each minute is further divided into 60 *seconds*. Consequently, in the last example we would give our position as 37°30'E.

To determine longitude, therefore, a ship need only have accurate time. With modern electronic communications, this poses no problems. For centuries, however, seafarers had no means of keeping accurate time at sea. Not until the 1760's, when the first practical chronometers—accurate clocks for use aboard ship—were designed, was it possible for most ships' pilots to determine longitude. Even the Greek astronomer Ptolemy (A.D. 90–168) made maps with relatively accurate positions north and south. But he overestimated the length of the Mediterranean by 50 percent, an error that was not corrected until 1700.

Each map in this book shows latitude and longitude (usually at 20° intervals) to indicate the positions of the map features (see Fig. 2–6). The parallels of latitude may also be used to determine approximate distance on a map. Each degree of latitude equals approximately 60 nautical miles (69 statute miles or 111 kilometers). Each minute of latitude is approximately 1 nautical mile, or 1.85 kilometers. At the equator, each minute of longitude is 1 nautical mile, but decreases so that 1 minute of longitude is only 0.5 nautical miles at 60° north or south latitude and vanishes at the poles.

MAPS AND MAP PROJECTIONS

A *map* is a flat representation of the earth's surface. Symbols are used to depict surface features. Since the earth is a sphere, making a flat map distorts the shape or size of surface features. The only distortion-free map is a globe, but a globe is not practical for the study of relatively small areas, so maps are used almost exclusively in science.

In making a map, we would like to make the final product as useful as possible. In general, we would like a map to preserve the following properties of the earth's surface:

Equal area—each area on the map should be proportional to the area of the earth's surface it represents.

Shape—the general outlines of a large area shown on a map should approximate as nearly as possible the shape of the region portrayed. A map which preserves shape is said to be *conformal*.

Distance—a perfect map would permit distance to be measured accurately between any two points anywhere on the map. Many common maps, such as the Mercator projection, do not accurately portray distances in a simple way.

Direction—ideally, it would be possible to measure directions accurately anywhere on a map.

No map has all these properties; only a globe preserves size, shape, direction, and distance.

A *map projection* takes the grid of latitude and longitude lines from a sphere and converts them into a grid on a flat surface. Sometimes, the resulting grid is a simple rectangular one where longitude and latitude lines intersect at right angles. In other projections, latitude and longitude lines are complex curves which intersect at various angles.

With the network formed from the grid lines, the map is drawn by plotting points in the appropriate spot on the new projection. In this way, the various types of maps are prepared. We shall consider only a few of the many map projections that have been developed to serve specific functions (listed in Table A2–1).

Without a doubt, the *Mercator projection* is most familiar. Parallels of latitude and meridians of longitude are all straight lines, and cross at right angles. The outline shape is a square or rectangle, as shown in Fig. A2–4).

In its simplest form, the Mercator projection can be visualized as being made by a light inside a translucent globe projecting latitude and longitude onto a cylinder surrounding the globe. Even through the cylinder used for the projection is curved, it is easily made into the flat map desired.

The common Mercator projection in our example is most accurate within 15° of the equator and least accurate at the poles.

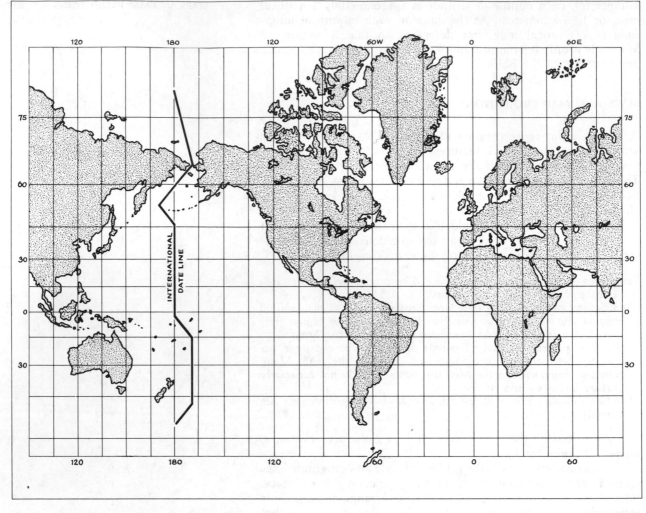

A Mercator projection. Compare the shape of
Greenland shown on this map with that
in Figs. A2–5 and A2–6 to see the distortion
at high latitudes.

Table A2–1

CHARACTERISTICS OF VARIOUS MAP PROJECTIONS

Name of projection (type)	Distinctive features	Desirable features (Undesirable features)	Uses
Mercator (cylindrical)	Horizontal parallels Vertical meridians	Compass directions are straight lines (Extreme distortion at high latitudes)	Navigation
Goode homolosine projection	Horizontal parallels Characteristic interruptions of outline	Equal area Little distortion of shape (Interruption of either continents or oceans)	Data presentation
Hoelzel's planisphere (modified	Horizontal parallels Characteristic nearly oval outline	Shapes easily recognizable (Moderate scale distortion at high latitudes)	Index map Data presentation

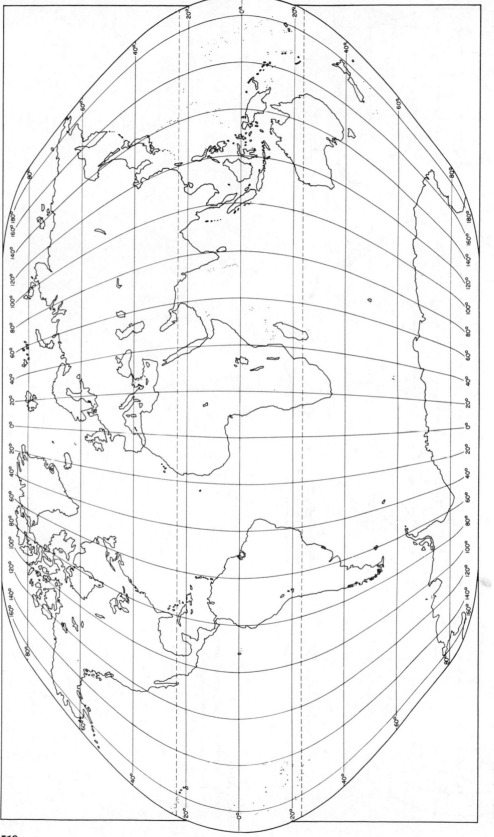

FIG. A2-5
A Hoelzel planisphere. Note that the meridians of longitude converge to a line shorter than the equator but still not a point.

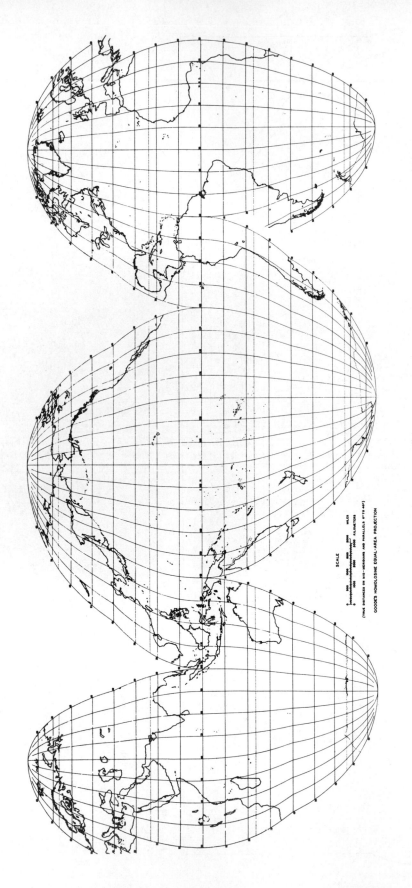

FIG. A2–6
Goode's homolosine projection interrupted to show the oceans to best advantage.

SCALE

500   1000   1500   2000   MILES
1000   2000   KILOMETERS

(TRUE DISTANCES ON MID-MERIDIANS AND PARALLELS 0°TO 40°)

GOODE'S HOMOLOSINE EQUAL-AREA PROJECTION

Although shapes are well preserved by this projection, area is distorted, especially near the poles. For example, South America is in reality 9 times the size of Greenland, but this is not obvious from common Mercator projections. The scale of a Mercator projection changes going away from the equator. If the reader uses the length of a degree of latitude as his scale, he can avoid serious error.

Another property of the Mercator projection useful to mariners is that a course of constant compass direction (a *rhumb line*) is a straight line on this projection. While a rhumb course is not a great circle, and thus not the shortest distance between any two points on the earth's surface, it is useful for navigation because a great circle course requires a constant changing of direction. A rhumb line is slightly longer but is easier to navigate.

For world maps, the Mercator projection has distinct limitations; but for relatively small areas, such as navigation charts, the Mercator projection is without equal. Nearly all navigation charts used at sea are Mercator projections.

For use in this book, a *Hoelzel planisphere* (shown in Fig. A2–5) is used. This modified cylindrical projection shows the continents well, permitting most of the world's coastal regions to be easily recognized. Like the Mercator, this projection distorts areas near the poles. Instead of converging to a point at the poles, the meridians converge to a line which is only a fraction of the length of the equator. Because the projection used in this book splits the Pacific Ocean in the middle, it is not overly convenient for showing properties in the ocean.

A special projection (the *interrupted homolosine*, shown in Fig. A2–6) was developed by J. P. Goode in 1923 to show the ocean basins without any interruptions. In addition, the projection shows areas equally. The continents are interrupted to show the ocean basins intact.

ALDRIDGE, B. G. 1968. *Mathematics for Physical Science.* Merrill Publishing Company, Columbus, Ohio. 137 pp. Elementary mathematical and data-plotting procedures.

COHEN, P. M. 1970. *Bathymetric Navigation and Charting.* United States Naval Institute, Annapolis, Md. 138 pp. Navigation through use of bathymetric data; nontechnical.

GREENHOOD, DAVID. 1964. *Mapping.* University of Chicago Press, Chicago. 289 pp. Elementary discussion of coordinates, maps, and map projections.

WILLIAMS, J. E., ed. 1963. *World Atlas,* 2nd ed. Prentice-Hall, Inc., Englewood Cliffs, N.J. 96, 41 pp. Includes graphical summary of different map projections.

SUGGESTED REFERENCES

GLOSSARY

appendix 3

*In part after B. B. Baker, W. R. Deebel, and R. D. Geisenderfer (eds.), *Glossary of Oceanographic Terms*, 2nd ed., U.S. Naval Oceanographic Office, Washington, D.C. (1966); and American Geological Institute, *Glossary of Geology and Related Sciences* and supplement, 2nd ed., Washington, D.C. (1960).

Abyssal–pertaining to the great depths of the ocean, generally below 3700 meters.

Advection—transfer of water, salt, or other properties by currents (mass movements).

Albedo—the ratio of radiation reflected by a body to the amount incident upon it, commonly expressed as a percentage.

Algal bloom—see *Plankton bloom*.

Algal ridge—the elevated margin of a windward reef built by actively growing calcareous algae.

Alternation of generations—mode of development, characteristic of many coelenterates, in which a sexually reproducing generation gives rise (by union of egg and sperm) to an asexually reproducing form, from which new individuals arise by budding or simple division of the "parent" animal.

Amphidromic point—the center of an amphidromic system; a nodal or no-tide point around which the crest of a standing wave rotates once in each tidal period.

Amphipods—elongate, usually compressed crustaceans, living throughout the marine environment.

Anaerobic—condition in which oxygen is excluded, with the result that organisms which depend on the presence of oxygen can not survive. Anaerobic bacteria can, however, live under these conditions.

Anion—a negatively charged atom or radical.

Anoxic—devoid of dissolved oxygen. See *Anaerobic*.

Antinode—that part of a standing wave where the vertical motion is greatest and the horizontal velocities are least.

Antipodal—located on opposite sides of the earth.

Aphotic zone–that portion of the ocean waters where light is insufficient for plants to carry on photosynthesis.

Aquaculture–the cultivation or propagation of water-dwelling organisms.

Archipelagic plain—a gently sloping sea floor with a generally smooth surface, particularly found among groups of islands or seamounts.

Archipelago—a sea or part of a sea studded with islands or island groups.

Arrow-worm—see *Chaetognath*.

Arthropods—animals with a segmented external skeleton of chitin or plates of calcium carbonate, and with jointed appendages—for example, a crab or an insect.

Asthenosphere—upper zone of the earth's mantle, extending from the base of the lithosphere to depths of about 250 kilometers beneath the continents and ocean basins; relatively weak.

Atoll—a ring-shaped organic reef that encloses a lagoon in which there is no pre-existing land and which is surrounded by the open sea. Low sand islands may occur on the reef.

Attenuation—the reduction in sound or light intensity caused by the absorption and scattering of sound or light energy in air or water.

Autotrophic nutrition—that process by which an organism manufactures its own food from inorganic compounds.

Backshore—that part of a beach which is usually dry, being reached only by the highest tides; by extension, a narrow strip of relatively flat coast bordering the sea.

Baleen (whalebone)—horny material growing down from the upper jaw of large plankton-feeding (baleen) whales, which forms a strainer or filtering organ consisting of numerous plates with fringed edges.

Bank—an elevation of the sea floor of a large area; a submerged plateau.

Bar—an offshore ridge or mound which is submerged (at least at high tide), especially at the mouth of a river or estuary.

Barrier beach—a bar parallel to the shore, whose crest rises above high water.

Barrier island—a detached portion of a barrier beach between two inlets.

Barrier reef—a reef that is separated from a land mass by a lagoon, usually connected to the sea through passes in the reef.

Basalt—a fine-grained igneous rock, black or greenish-black, rich in iron, magnesium, and calcium.

Basin—a depression of the sea floor more or less equidimensional in form and of variable extent.

Bathythermograph—a device for obtaining a record of temperature at various depths in the ocean, usually within a few hundred meters of the ocean surface.

Baymouth bar—a bar extending partially or entirely across the mouth of a bay.

Beach—seaward limit of the shore (limits are marked approximately by the highest and lowest water levels).

Beach cusp—one of a series of low mounds of beach material separated by troughs spaced at more-or-less regular intervals along the beach face.

Beach drift—see *Littoral drift*.

Beach face—this term is sometimes used to describe the highest portion of the foreshore, above the low-tide terrace and exposed only to the action of wave uprush, not to the rise and fall of the tide.

Bed load—see *Load*.

Benthic—that portion of the marine environment inhabited by marine organisms which live permanently in or on the bottom.

Benthos—bottom-dwelling marine organisms.

Berm—the low, nearly horizontal portion of a beach (backshore), having an abrupt fall and formed by the deposit of material by wave action. It marks the limit of ordinary high tides and waves.

Berm crest—the seaward limit of a berm.

Bight—a concavity in the coastline which forms a large, open bay.

Biogenous sediment—sediment consisting of at least 30 percent by volume particles derived from skeletal remains of organisms, which may contribute calcium carbonate, silica, or phosphate minerals to the sediment.

Biogeochemical cycles—the paths by which elements essential to life circulate from the nonliving environment to living organisms and back again to the environment.

Bioluminescence—the production of light without sensible heat by living organisms as a result of a chemical reaction either within certain cells or organs or in some form of secretion.

Biomass—the amount of living matter per unit of water surface or volume expressed in weight units.

Bird-foot delta—a delta formed by the outgrowth of pairs of natural levees, formed by the distributaries, making the digitate or bird-foot form.

Bivalves—mollusks, generally sessile or burrowing into soft sediment, rock, wood, or other materials. Individuals possess a hinged shell and a hatchet-shaped foot, which is sometimes used in digging. Clams, oysters, and mussels belong to this group.

Black bottle (dark bottle)—a container used in measuring respiratory activity of primary producers. The container is covered or coated to exclude light and thereby prevent photosynthetic activity.

Black mud (hydrogen sulfide mud)—a dark, fine-grained sediment formed in poorly aerated bays, lagoons, and fjords. This sediment usually contains large quantities of decaying organic matter and iron sulfides and may exude hydrogen sulfide gas.

Bloom—see *Plankton bloom*.

Blue-green algae—single-celled or simply filamentous plants in which the blue color is imparted by a water-soluble accessory pigment.

Bore—a very rapid rise of the tide in which the advancing water forms an abrupt front, sometimes of considerable height, occurring in certain shallow estuaries where the range of tide is large.

Bottom water—the water mass at the deepest part of the water column. It is the densest water that is permitted to occupy that position by the regional bottom topography.

Boundary currents—northward- or southward-directed surface currents which flow parallel and close to the continental margins. They are caused by deflection of the prevailing eastward- and westward-flowing currents by the continental land masses.

Brackish water—water in which salinity values range from approximately 0.05 to 17.00 parts per thousand.

Breaker—a wave breaking on the shore, over a reef, etc. Breakers may be roughly classified into four kinds, although the categories may overlap: *spilling breakers* break gradually over a considerable distance; *plunging breakers* tend to curl over and break with a crash; *surging breakers* peak up, but then instead of spilling or plunging they surge up on the beach face; *collapsing breakers* break in the middle or near the bottom of the wave, rather than at the top.

Breakwater—a structure, usually rock or concrete, protecting a shore area, harbor, anchorage, or basin from waves.

Brine—water containing a higher concentration of dissolved salt than that of the ordinary ocean. Brine can be produced by the evaporation or freezing of seawater.

Brown algae—greenish yellow to deep brown, filamentous to massively complex plants. They are most abundant in the cooler waters of the world.

Calcareous algae—marine plants which form a hard external covering of calcium compounds.

Calorie—the amount of heat required to raise the temperature of 1 gram of water by 1 °C (defined on basis of water's *specific heat*).

Canyon (submarine)—see *Submarine canyon*.

Capillary wave–a wave in which the primary restoring force is surface tension. A water wave in which the wave length is less than 1.7 centimeters is considered to be a capillary wave.

Cation—a positively charged ion, tending to be deposited at the cathode in electrolysis or in a battery.

Celsius temperature—temperatures based on a scale in which water freezes at 0° and boils at 100° (at standard atmospheric pressure); also called *centigrade temperature*.

Cephalopods—benthic or swimming mollusks possessing a large head, large eyes, and a circle of arms or tentacles around the mouth; the shell is external, internal, or absent, and an ink sac usually is present. They include squids and octopus.

Chaetognath—small, elongate, transparent, wormlike animals pelagic in all seas from the surface to great depths. They are abundant and may multiply rapidly into vast swarms. Some, especially *Sagitta*, have been identified as indicator species.

Change of state—a change in the physical form of a substance, as when a liquid changes to a solid due to cooling.

Chemo-autotrophic nutrition—that process by which an organism manufactures its food by using the energy derived from oxidizing organic matter.

Chitin—a nitrogenous carbohydrate derivative forming the skeletal substance in arthropods.

Chloride—an atom of chlorine in solution, bearing a single negative charge.

Chlorinity—a measure of the chloride content, by mass, of seawater (grams per kilogram, or per mille, of seawater).

Chlorophyll—a group of green pigments which occur chiefly in bodies called chloroplasts and which are active in photosynthesis.

Chordates—animals which possess a notochord (mid-dorsal cylindrical rod) and a series of paired gill slits—both of which features are present in the embryo only of air-breathing species; and a dorsal central nervous system. Representatives are the tunicates, fishes, and mammals.

Cilia—hairlike processes of cells, which beat rhythmically and cause locomotion of the cells or produce currents in water.

Clay—soil consisting of inorganic material, the grains of which have diameters smaller than 0.005 millimeter (or 5 microns).

Climate—the prevalent or characteristic meteorological conditions of a place or region—in contrast with *weather*, which is the state of the atmosphere at a particular time.

Coacervate—a colloidal solution, of a liquid phase, which appears as viscous drops instead of forming a continuous liquid phase.

Climax community—end-product of ecological succession in which there is a balance of production and consumption in the community and further change in species composition takes place only as a result of changes in environmental conditions or outside forces influencing the community.

Coast (coastal plain)—a strip of land of indefinite width (it may be several miles) that extends from the seashore inland to the first major change in terrain features.

Coastal currents—currents flowing roughly parallel to the shore and constituting a relatively uniform drift in the water just seaward of the surf zone. These currents may be caused by tides or winds, or associated with the distribution of mass in coastal waters.

Coastal ocean—shallow portion of the ocean (in general, overlying the continental shelf), where the circulation pattern is strongly influenced by the bordering land.

Cobble—a rock fragment between 64 and 256 millimeters in diameter, larger than a pebble and smaller than a boulder; usually rounded or otherwise abraded.

Coccolithophores—microscopic, often abundant planktonic algae, the cells of which are surrounded by an envelope on which numerous small calcareous discs or rings (coccoliths) are embedded.

Coelenterates—a large, diverse group of simple animals possessing two cell layers and a digestive cavity with only one opening. This opening is surrounded by tentacles containing stinging cells. Some are sessile, some pelagic (medusae), and some possess alternation of generations.

Coenzyme—a nonprotein molecule that works with an enzyme in catalyzing a reaction.

Compaction—the decrease in volume or thickness of a sediment under load through closer crowding of constituent particles and accompanied by decrease in porosity, increase in density, and squeezing out of water.

Compensation depth—the depth at which oxygen production by photosynthesis equals that consumed by plant respiration during a 24-hour period.

Conduction—transfer of energy through matter by internal particle or molecular motion without any external motion.

Conformal projection—a map projection in which the angles around any point are correctly represented—for example, a Mercator projection.

Conservative property—a property whose values do not change in the course of a particular, specified series of events or processes—for example, those properties of seawater, such as salinity, the concentrations of which are not affected by the presence or activity of living organisms, but which are affected by diffusion and currents.

Continental block—a large land mass rising abruptly from the deep ocean floor, including marginal regions that are shallowly submerged.

Continental climate—a climate characterized by cold winters and hot summers in areas where the prevailing winds have traveled across large land areas.

Continental crust—the thickened part of the crust forming continental blocks; typically, about 35 kilometers thick, consisting primarily of granitic rocks.

Continental margin—a zone separating the emergent continents from the deep-sea bottom; generally consists of the continental shelf, slope, and rise.

Continental rise—a gentle slope with a generally smooth surface, rising toward the foot of the continental slope.

Continental shelf—the sea floor adjacent to a continent, extending from the low-water line to the change-in-slope, usually about 180 meters depth, where continental shelf and continental slope join.

Continental slope—a declivity from the outer edge of a continental shelf, extending from the break in slope to the deep sea floor.

Contour line—a line on a chart connecting points of equal value above or below a reference value; can portray elevation, temperature, salinity or other values.

Convection—vertical circulation resulting from density differences within a fluid, resulting in transport and mixing of the properties of that fluid.

Convergence—area or zone where flow regimes come together or converge, usually resulting in sinking of surface water.

Copepods—minute shrimplike crustaceans, most species of which range between about 0.5 and 10 millimeters in length.

Coral—the hard, calcareous skeleton of various sessile, colonial coelenterate animals, or the stony solidified mass of a number of such skeletons; also the entire animal, a compound polyp which produces the skeleton.

Coral reef—a complex ecological association of bottom-living and attached calcareous, shelled marine invertebrates forming fringing reefs, barrier reefs, or atolls.

Coralline algae—red algae, having either a bushy or encrusting form and deposits of calcium carbonate either on the branches or as a crust on the substrate. Certain of them develop massive encrustations on coral reefs.

Core—a vertical, cylindrical sample of sediments from which the nature and stratification of the bottom may be determined; also, the central zone of the earth.

Coriolis effect—an apparent force acting on moving particles resulting from the earth's rotation. It causes moving particles to be deflected to the right in the Northern Hemisphere and to the left in the Southern Hemisphere; the deflection is proportional to the speed and latitude of the moving particle. The speed of the particle is unchanged by the apparent deflection.

Crinoids—a group of echinoderms, most of which are either permanently or when immature attached by a long stalk to the bottom; species without stalks either swim or creep slowly about. Crinoids (such as sea-lily, feather star) occur in shallow water as well as at great depths.

Critical depth—the depth above which occurs the net effective plant production in the water column as a whole.

Crust of the earth—the outer shell of the solid earth. Beneath the oceans, the outermost layer of crust is composed of unconsolidated sediment, the second layer of consolidated sediment and weathered lavas, and the inside layer probably of basaltic rocks. The crust varies in thickness from approximately 5 to 7 kilometers under the ocean basins to 35 kilometers under the continents. Its lower limit is defined to be the Mohorovicic discontinuity.

Crustaceans—arthropods which breathe by means of gills or similar structures. The body is commonly covered by a hard shell or crust. The group includes barnacles, crabs, shrimps, and lobsters.

Ctenophores—spherical, pear-shaped, or cylindrical animals of jellylike consistency ranging from less than 2 centimeters to about 1 meter in length. The outer surface of the body bears 8 rows of comblike structures.

Cumaceans—small (about 1 centimeter) shrimplike crustaceans, benthic or planktonic.

Current ellipse—a graphic representation of a rotary current in which the speed and direction of the current at different hours of the tide cycle are represented by vectors joined at one point. A line joining the extremeties of the radius vectors forms a curve roughly approximating an ellipse.

Current meter—any device for measuring and indicating speed or direction (often both) of flowing water.

Current rose—a graphic representation of currents for specified areas, utilizing arrows to show the direction toward which the prevailing current flows and the frequency (expressed as a percentage) of any given direction of flow.

Current tables—tables which give daily predictions of the times, speeds, and directions of the currents.

Cypris larva—the stage at which the young of barnacles attach to the substrate.

Daily (diurnal) inequality—the difference in heights and durations of the two successive high waters or of the two successive low waters of each day; also, the difference in speed and direction of the two flood currents or the two ebb currents of each day.

Daily (diurnal) tide—a tide having only one high water and one low water each tidal day.

Debris line—a line near the limit of storm-wave uprush marking the landward limit of debris desposits.

Decomposers—heterotrophic and chemo-autotrophic organisms (chiefly bacteria and fungi) which break down nonliving matter, absorb some of the decomposition products, and release compounds usable in primary production.

Deep scattering layer—a stratified population of organisms in ocean waters which causes scattering of sound as recorded on an echo sounder. Such layers may be from 50 to 200 meters thick. They occur less than 200 meters below the surface at night and several hundred meters below the surface during the day.

Deep-water waves—waves in water the depth of which is greater than one-half the average wave length.

Delta—an alluvial deposit, roughly triangular or digitate in shape, formed at the mouth of a stream of tidal inlet or a river.

Density—the mass per unit volume of a substance, usually expressed in grams per cubic centimeter. In centimeter-gram-second system, density is numerically equivalent to *specific gravity*.

Density current—the flow (caused by density differences or gravity) of one current through, under, or over another; it retains its unmixed identity because of density differences from the surrounding water.

Depth of no motion—the depth at which water is assumed to be motionless, used as a reference surface for computating geostrophic currents.

Detritus—any loose material produced directly from rock disintegration. *Organic detritus* consists of decomposition or disintegration products or dead organisms, including fecal material.

Diatom—microscopic phytoplankton organisms, possessing a wall of overlapping halves (valves) impregnated with silica.

Diffusion—transfer of material (e.g., salt) or of a property (e.g., temperature) by eddies or molecular movement. Diffusion causes spreading or scattering of matter under the influence of a concentration gradient, with movement from the stronger to the weaker solution.

Dinoflagellates—microscopic or minute organisms, which may possess characteristics of both plants (such as chlorophyll and cellulose plates) and animals (such as ingestion of food).

Discontinuity—an abrupt change in a property, such as salinity or temperature, at a line or surface.

Dispersion—the separation of a complex, surface gravity-wave disturbance into its component parts. Longer component parts of the wave travel faster than shorter ones and arrive at their destination first, as swell.

Disphotic zone—the dimly lighted zone extending from about 100 to 600 meters below the surface, or less, where no effective plant production can take place.

Distributary—an outflowing branch of a river, which occurs characteristically on a delta.

Diurnal—daily, especially pertaining to actions which are completed within approximately 24 hours and which recur every 24 hours.

Divergence—horizontal flow of water in different directions from a common center or zone, upwelling is a common example of divergence.

Doldrums—a nautical term describing a belt of light, variable winds near the equator; an area of low atmospheric pressure.

Drift bottle—a bottle released at sea for use in studying surface currents. It contains a card identifying the date and place of release and requesting the finder to return it, noting the date and place of recovery.

Drift current—slow-moving current, principally caused by wind, usually restricted to the waters above the pycnocline.

Drift net—fishing net suspended in the water vertically so that drifting or swimming animals may be trapped or entangled in the mesh.

Dynamic topography—the configuration formed by the geopotential difference (dynamic height) between a given surface and a reference surface, usually the layer of no motion. A contour map of dynamic "topography" may be used to estimate geostrophic currents.

Ebb current—the movement of a tidal current away from shore or down a tidal stream.

Echinoderms—principally benthic marine animals having calcareous plates with projecting spines forming a rigid or articulated skeleton or plates and spines embedded in the skin. They have radically symmetrical, usually five-rayed, body. They include the starfish, sea urchins, crinoids, and sea-cucumbers.

Echiuroids—unsegmented, burrowing marine worms.

Echo sounding—determination of the depth of water by measuring the time interval between emission of a sonic or ultrasonic signal and the return of its echo from the bottom. The instrument used for this purpose is called an *echo sounder*.

Ecological efficiency—ratio of the efficiency with which energy is transferred from one trophic level to the next.

Ecological niche—the position or functional status of an organism within the ecosystem of which it is a part.

Ecosystem—ecological unit including both organisms and the non-living environment, each influencing the properties of the other and both necessary for the maintenance of life as it exists on the earth.

Eddy—a current of air, water, or any fluid, often on the side of a main current; especially one moving in a circle, in extreme cases, a whirlpool.

Eddy viscosity—the turbulent transfer of momentum by eddies giving rise to an internal fluid friction, in a manner analogous to the action of molecular viscosity in laminar flow, but taking place on a much larger scale.

Eel-grass—a seed-bearing, grasslike marine plant which grows chiefly in sand or mud-sand bottoms and most abundantly in temperate water less than 10 meters deep.

Ekman spiral—a theoretical representation of the currents resulting from a steady wind blowing across an ocean having unlimited depth and extent and uniform viscosity. Specifically, it would cause the surface layer to drift on an angle of 45° to the right of the wind direction in the Northern Hemisphere; water at successive depths would drift in directions more to the right until, at some depth, the water would move in the direction opposite to the wind. Speed decreases with depth throughout the spiral. The net water transport is 90° to the right of the wind in the Northern Hemisphere.

Encrusting algae—see *Coralline algae, Red algae.*

Epifauna—animals which live at the water–substrate interface, either attached to the bottom or moving freely over it.

Epiphytes—plants which grow attached to other plants.

Equal area projection—a map projection in which equal areas on the earth's surface are represented by equal areas on the map.

Estuary—a semienclosed, tidal coastal body of saline water with free connection to the sea, commonly the lower end of a river.

Euphausids—shrimplike, planktonic crustaceans, widely distributed in oceanic and coastal waters, and especially in colder waters. They grow to 8 centimeters in length and many possess luminous organs.

Euphotic zone—see *Photic zone.*

Eutrophication—a process in the change or "aging" of a water body in which the nutrient level (phosphates, nitrates) increases and productivity increases, with accompanying depletion of dissolved oxygen in the bottom waters.

Fan—a gently sloping, fan-shaped feature normally located near the lower termination of a canyon.

Fault—a fracture or fracture zone in rock along which one side has been displaced relative to the other side.

Fault block—a rock body bounded on at least two opposite sides by faults. It is usually longer than it is wide; when it is depressed relative to the adjacent regions, it is called a *fault trough* or *rift valley.*

Fauna—the animal population of a particular location, region, or period.

Fermentation—a type of respiration in which partial oxidation of organic compounds takes place in the absence of free oxygen, for example the oxidation of glucose to alcohol by yeast.

Fetch—an area of the sea surface over which seas are generated by a wind having a constant direction and speed, also called *generating area;* also, the length of the fetch area, measured in the direction of the wind in which the seas are generated.

Fjord—a narrow, deep, steep-walled inlet of the ocean, formed either by the submergence of a mountainous coast or by entrance of the ocean into a deeply excavated glacial trough after the melting away of the glacier.

Flagellum—a whiplike bit (process) of protoplasm which provides locomotion for a motile cell.

Flocculation—process of aggregation into small lumps, especially with regard to soils and colloids.

Floe—sea ice, either as a single unbroken piece or as many individual pieces covering an area of water.

Flood current—the tidal current associated with the increase in the height of a tide. Flood currents generally set toward the shore, or in the direction of the tide progression; sometimes erroneously called *flood tide.*

Flora—the plant population of a particular location, region, or period.

Food web—the interrelated food relationships in an ecosystem, including its production, consumption, and decomposition, and the energy relationships among the organisms involved in the cycle.

Foraminifera—benthic or planktonic protozoans possessing shells, usually of calcium carbonate.

Forced wave—a wave generated and maintained by a continuous force, in contrast to a free wave that continues to exist after the generating force has ceased to act.

Forerunner—low, long-period swell which commonly precedes the main swell from a distant storm, especially a tropical cyclone.

Foreshore—see *Low-tide terrace.*

Foul—to attach to or come to lie on the surface of submerged man-made or introduced objects, usually in large numbers or amounts, as barnacles on the hull of a ship or silt on a stationary object.

Fracture zone—an extensive linear zone of unusually irregular topography of the ocean floor characterized by large seamounts, steep-sided or asymmetrical ridges, troughs, or long, steep slopes.

Free wave—any wave not acted upon by any external force except for the initial force that created it.

Fringing reef—a reef attached directly to the shore of an island or continental land mass. Its outer margin is submerged and often consists of algal limestone, coral rock, and living coral.

Fully developed sea—the maximum height to which ocean waves can be generated by a given wind force blowing over sufficient fetch, regardless of duration, as a result of all possible wave components in the spectrum being present with their maximum amount of spectral energy.

Gastropods—mollusks that possess a distinct head, generally with eyes and tentacles, and a broad, flat foot, and which are usually enclosed in a spiral shell.

Geostrophic current—a current resulting from the balance between gravitational forces and the Coriolis effect.

Giant squids—large cephalopods (length may be 15 meters or more) which inhabit mid-depths in oceanic regions but may come to the surface at night. They are eaten by sperm whales.

Gill—a delicate, thin-walled structure, often an extension of the body wall, used for the exchange of gases in a water environment, and sometimes for excretion of waste salts.

Glass-worm—see *Chaetognath*.

Glacial marine sediment—high-latitude, deep-ocean sediments that have been transported from the land to the ocean by means of glaciers or icebergs.

Glacier—a mass of land ice, formed by recrystalization of old, compacted snow, flowing slowly from an area of accumulation to an area of sublimation, melting, or evaporation, where snow or ice are removed from the glacier surface.

Grab—an instrument possessing jaws that seize a portion of the bottom for retrieval and study.

Graded bedding—a type of stratification in which each stratum displays a gradation in grain size from coarse below to fine above.

Gradient—the rate of decrease (or increase) of one quantity with respect to another—for example, the rate of decrease of temperature with depth.

Granite—a crystalline rock consisting essentially of alkali feldspar and quartz; in seismology, a rock in which the compressional-wave velocity varies approximately between 5.5 and 6.2 kilometers per second. *Granitic* is a textural term applied to coarse and medium-grained igneous rocks.

Gravel—loose material ranging in size from 2 to 256 millimeters.

Gravity wave—a wave whose velocity of propagation is controlled primarily by gravity. Water waves of length greater than 1.7 centimeters are considered gravity waves.

Great circle—the intersection of the surface of a sphere and a plane through its center—for example, meridians of longitude and the equator are great circles on the earth's surface.

Green algae—grass-green, single-celled, filamentous, membranous or branching plants in which the color, imparted by chlorophyll, is not masked by the accessory pigments. Green algae are cosmopolitan in the upper intertidal zone but most abundant in warmer waters.

Greenhouse effect—the result of the penetration of the atmosphere by comparatively short-wavelength solar radiation which is largely absorbed near and at the earth's surface, whereas the relatively long-wavelength radiation emitted by the earth is partially absorbed by water vapor, carbon dioxide, and dust in the atmosphere, thus warming the atmosphere.

Groin—a low artificial dam-like structure of durable material placed so that it extends seaward from the land; used to change littoral drift.

Group velocity—the velocity with which a wave group travels. In deep water, it is equal to one-half the individual wave velocity.

Gyre—a circular or spiral form, usually applied to semi-closed current systems.

Habitat grouping—an assemblage of organisms brought together by common environmental requirements. They do not necessarily depend on one another for survival.

Hadley cell—a semi-closed system of vertical motions in the earth's atmosphere. Warm moist air rises in equatorial regions, flows to mid-latitudes (30°N, 30°S), where it sinks and returns to the equatorial zone as the trade winds. Occurs near equator.

Half-life—the time required for the decay of one-half the atoms of a sample of a radioactive substance. Each radionuclide has an unique half-life, related to its rate of disintegration.

Halocline—water layer with large vertical changes in salinity.

Harmonic—a quantity whose frequency is an integral multiple of the frequency of a periodic quantity to which it is related.

Headland—a high, steep-faced point of land extending into the sea.

Heat budget—the accounting for the total amount of the sun's heat received on the earth during any one year as being exactly equal to the total amount which is lost from the earth by reflection and radiation into space.

Heat capacity—amount of heat required to raise the temperature of a substance by a given amount.

Heterotrophic nutrition—that process by which an organism utilizes only preformed organic compounds for its nutrition.

High carbonate island—an island similar to a low carbonate island; but elevated well above sea level.

High water—the upper limit of the surface water level reached by the rising tide; also called *high tide*.

Higher high water—the higher of the two high waters of any tidal day. The single high water occurring daily during periods when the tide is diurnal is considered to be a higher high water.

Higher low water—the higher of two low waters occurring during a tidal day.

Hurricane—a cyclonic storm, usually of tropical origin, covering an extensive area, and containing winds of 75 miles an hour or higher.

Hydraulic model—a scale model of a specific feature such as a harbor or estuary, in which water is made to move to reproduce currents and tides.

Hydrofoil—any surface, such as a wing or rudder, designed to obtain reaction upon it from the water through which it moves; also, a ship equipped with such planes to provide lift when the ship is propelled forward.

Hydrogen bond—the relatively weak bond formed between adjacent molecules in liquid water, resulting from the mutual atraction of hydrogen and nearby atoms.

Hydrogenous sediment—precipitates from solution in water.

Hydroid—the polyp form of coelenterate animals which exhibit alternation of generations. It is attached, often branching, and gives rise to the pelagic, medusa form by asexual budding.

Hydrologic cycle—the composite cycle of water exchange, including change of state and vertical and horizontal transport, of the interchange of water substance between earth, atmosphere, and ocean.

Hydrophilic—having the property of attracting water.

Hydrophobic—not attractive to water; unwettable.

Hydrosphere—the water portion of the earth, as distinguished from the solid part (called the *lithosphere*) and the gaseous outer envelope (called the *atmosphere*). It consists of liquid water in sedimentary rocks, rivers, lakes and oceans, as well as ice in sea ice and continental ice sheets.

Hypsometric graph—a representation of the respective elevations and depths of points on the earth's surface with reference to sea level.

Iceberg—a large mass of detached land ice floating in the sea or stranded in shallow water.

Ice shelf—a thick ice formation with a fairly level surface, formed along a polar coast and in shallow bays and inlets, where it is fastened to the shore and often reaches bottom.

Ice slush—an accumulation on the water surface of ice needles that are frozen together; it forms patches or a thin compact layer of a grayish or leaden color.

Igneous rock—formed by solidification of magma.

Indicator species—a species of marine plankton that is characteristic of a certain water mass to which it is restricted, so that, with proper precautions, its presence can be taken as an indication of the presence of water of that origin. Species of medusae, chaetognaths, euphausiids, pteropods, and tunicates, among others, have been shown to be indicator species.

Infauna—animals who live buried in soft substrate.

Insolation—in general, solar radiation received at the earth's surface; also, the rate at which direct solar radiation is incident upon a unit horizontal surface at any point on or above the surface of the earth.

Instability—a property of a system such that any disturbance grows larger, rather than diminishing, so that the system never returns to the original steady state; in oceanography, usually refers to the vertical displacement of a water parcel.

Interface—surface separating two substances of different properties (such as different densities, salinities, or temperatures)—for example, the air–sea interface or the water–sediment interface.

Internal wave—a wave that occurs within a fluid whose density changes with depth, either abruptly at a sharp surface of discontinuity (an interface) or gradually.

Interstitial water—water contained in the pore spaces between the grains in rocks and sediments.

Intertidal zone (littoral zone)—generally considered to be the zone between mean high-water and mean low-water levels.

Ionic bond—a linkage between two atoms, with a separation of electric charge on the two atoms; a linkage formed by the transfer or shift of electrons from one atom to another.

Iron—manganese nodules—see *manganese nodules*.

Island-arc system—a group of islands usually having a curving arch-like pattern, generally convex toward the open ocean, with a deep trench or trough on the convex side and usually enclosing a deep-sea basin on the concave side. Volcanoes usually associated.

Isopods—generally flattened marine crustaceans, mostly scavengers.

Isotherm—line connecting points of equal temperature.

Isotonic—equal in osmotic pressure.

Isotope—one of several nuclides having the same number of protons in their nuclei, and hence belonging to the same element, but differing in the number of neutrons and therefore in mass number or energy content; also, a radionuclide or a preparation of an element with special isotopic composition, used principally as an isotopic tracer.

Jellyfish—see *Medusa*.

Jet stream—high-altitude swift air current.

Jetty—in United States terminology, a structure built to influence tidal currents, to maintain channel depths, or protect the entrance to a harbor or river.

Juvenile water—water that enters for the first time into the hydrologic cycle; released at a slow rate from igneous rocks through volcanic activity.

Kelp—usually large, blade-shaped or vinelike brown alga.

Knoll—an elevation rising less than 1000 meters from the ocean floor and of limited extent across the summit.

Knot—a unit of speed equal to 1 nautical mile per hour, approximately 51 centimeters per second.

Lagoon—a shallow sound, pond, or lake generally separated from the open ocean.

Lamina—a sediment or sedimentary-rock layer less than 1 centimeter thick visually separable from the material above and below. *Lamination* refers to the alternation of such layers differing in grain size or composition.

Laminar flow—a flow in which the fluid moves smoothly in streamlines in parallel layers or sheets; a nonturbulent flow.

Langmuir circulation—cellular circulation, with alternate left- and right-hand helical vortices, having axes in the direction of the wind, which is set up in the surface layer of a water body by winds exceeding 7 knots.

Lantern-fishes (myctophids)—small oceanic fishes which normally live at depths between a few hundred and a few thousand meters depth and characteristically have numerous small light organs on the sides of the body. Many undergo diurnal vertical migration.

Latent heat—the heat released or absorbed per unit mass by a system undergoing a reversible change of state at a constant temperature and pressure.

Lava—fluid rock, issuing from a volcano or fissure in the earth's surface, or the same material solidified by cooling.

Layer of no motion—see *Depth of no motion*.

Layer 3—see *Crust of the earth*.

Levee—an embankment bordering one or both sides of a seachannel or delta distributary.

Light bottle—a container used for measuring photosynthetic activity.

Lithogenous sediment—sediment composed primarily of mineral grains transported into the oceans from the continents.

Lithosphere—the outer, solid portion of the earth; includes the crust of the earth.

Littoral—see *Intertidal zone*.

Littoral drift—sand moved parallel to the shore by wave and current action.

Load—the quantity of sediment transported by a current. It includes the *suspended load* of small particles which float in suspension distributed through the whole body of the current, and the bottom load or *bed load* of large particles which move along the bottom by rolling and sliding.

Longshore bar—see *Bar*.

Longshore current—a current located in the surf zone moving generally parallel to the shoreline; usually generated by waves breaking at an angle with the shoreline.

Low carbonate island—low sand and gravel islands associated with a reef or atoll structure, having a volcanic base covered at the top by calcareous material derived from the skeletons of encrusting organisms.

Low-tide terrace—the zone that lies between the ordinary high- and low-water marks and is daily traversed by the oscillating water line as the tides rise and fall. This area, together with the vertical scarp that often occurs at its upper limit, is sometimes called the *foreshore*, which ends at the highest point of normal wave uprush.

Low water—the lowest limit of the surface-water level reached by the lowering tide; also called *low-tide*.

Lower high water—the lower of two high tides occurring during a tidal day.

Lower low water—the lower of two low tides occurring during a tidal day.

Lunar tide—that part of the tide caused solely by the gravitational attraction of the moon, as distinguished from that part caused by the gravitational attraction of the sun.

Magma—mobile rock material, generated within the earth, capable of intrusion and extrusion, from which igneous rock solidifies.

Manganese nodules—concretionary lumps of manganese and iron, containing copper and nickel; found widely scattered on the ocean floor.

Mantle—the relatively plastic region between the crust and the core of the earth.

Map projection—a method of representing part or all of the surface of a sphere, such as the earth, on a plane surface.

Marginal sea—a semienclosed body of water adjacent to, widely open to, and connected with the ocean at the water surface but bounded at depth by submarine ridges—for example, the Yellow Sea.

Marine humus—partially or entirely undecomposed organic matter in ocean sediment.

Marine snow—particles of organic detritus and living forms. The downward drift of these particles and living forms, especially in dense concentration, appears similar to a snowfall when viewed by underwater investigators.

Maritime climate—climate characterized by relatively little seasonal change, warm moist winters, cool summers: result of prevailing winds blowing from ocean to land.

Marsh—an area of soft, wet land. Flat land periodically flooded by salt water is called a *salt marsh*.

Mean sea level—the average height of the surface of the sea for all stages of the tide over a 19-year period, usually determined from hourly readings of tidal height.

Mean tidal range—the difference in height between mean high water and mean low water, measured in feet or meters.

Medusa (jellyfish)—any of various free-swimming coelenterates having a disk- or bell-shaped body of jellylike consistency. Many have long tentacles with stinging cells.

Meiobenthos—animals in the size range 100 to 500 microns that live between the grains in nearshore sediments, including many kinds of small invertebrates and larger protozoans.

Meridian (of longitude)—a great circle passing through the North and South Poles. It connects points with an equal angular separation from the prime meridian.

Metabolite—substance necessary to survival and growth of an organism; this includes vitamins and other materials required only in trace amounts, as well as the nutrients required for synthesis of living tissue.

Metamorphic rocks—formed in the solid state, below the earth's surface, in response to changes of temperature, pressure or chemical environment.

Microbenthos—one-celled plants, animals, and bacteria living in or on the surface of bottom sediments.

Microcontinent—an island whose geophysical properties more closely resemble those of the continental blocks than those of the oceanic crust.

Mid-ocean ridge—a great median arch or sea-bottom swell extending the length of an ocean basin and roughly paralleling the continental margins.

Mid-ocean rise—see *Oceanic rise*.

Mixed tide—type of tide in which a diurnal wave produces large inequalities in heights and/or durations of successive high and/or low waters. This term applies to the tides intermediate to those predominantly semidaily and those predominantly daily.

Mohorovicic discontinuity—the sharp discontinuity in composition between the outer layer of the earth (crust) and the inner layer (mantle).

Molecular diffusion—see *Diffusion*.

Mole—formula weight of a substance in grams—for example, 1 mole of H_2O weighs 18 grams (oxygen = 16; 2 hydrogen = 2).

Molecular viscosity—an internal resistance to flow in liquids arising from the attractive interactions between molecules.

Monsoons—a name for seasonal winds (derived from the Arabic *mausim*, meaning "season") first applied to the winds over the Arabian Sea, which flow for 6 months from the northeast and the remaining 6 months from the southwest, but subsequently extended to similar winds in other parts of the world.

Mud—detrital material consisting mostly of silt and clay-sized particles (less than 0.06 millimeter) but often containing varying amounts of sand and/or organic materials. It is also a general term applied to any fine-grained sediment whose exact size classification has not been determined.

Mysids—elongate crustaceans which usually are transparent (or nearly so) and benthic or deep-living.

Nannoplankton—plankton whose length is less than 50 microns. Individuals will pass through most nets and usually are collected by centrifuging water samples.

Nansen bottle—a device used by oceanographers to obtain subsurface samples of seawater. The "bottle" is lowered by wire; its valves are open at both ends. It is then closed *in situ* by allowing a weight ("messenger") to slide down the wire and strike the reversing mechanism. This causes the bottle to turn upside down, closing the valves and reversing the thermometers. Usually a series of bottles is lowered, and then the reversal of each bottle releases another messenger to actuate the bottle beneath it.

Natural frequency—the characteristic frequency (number of vibrations or oscillations per unit time) of a body controlled by its physical characteristics (dimensions, density, etc.).

Nauplius—a limb-bearing early larval stage of many crustaceans.

Neap tide—lowest range of the tide, occurring near the times of the first and last quarter of the moon.

Nekton—those pelagic animals that are active swimmers, such as most of the adult squids, fishes, and marine mammals.

Net primary production—the total amount of organic matter produced by photosynthesis minus the amount consumed by the photosynthetic organisms in their own respiratory processes.

Nitrogen fixation—conversion of atmospheric nitrogen to oxides usable in primary food production.

Nodal line—a line in an oscillating area along which there is little or no rise and fall of the tide.

Nodal point—the no-tide point in an amphidromic region.

Node—that part of a standing wave where the vertical motion is least and the horizontal velocities are greatest.

Nonconservative property—a property whose values change in the course of a particular, specified series of events or processes—for example, those properties of seawater, such as nutrient or dissolved-oxygen concentrations, which are affected by biological or chemical processes.

Nuclide—a species of atom characterized by the constitution of its nucleus. The nuclear constitution is specified by the number of protons, number of neutrons, and energy content—or alternatively, by the atomic number, mass number, and atomic mass.

Nudibranchs (sea slugs)—gastropods in which the shell is entirely absent in the adult. The body bears projections which vary in color and complexity among the species.

Nutrient—in the ocean, any one of a number of inorganic or organic compounds or ions used primarily in the nutrition of primary producers. Nitrogen and phosphorous compounds are examples of *essential nutrients*.

Ocean basin—that part of the floor of the ocean that is more than about 2000 meters below sea level.

Oceanic crust—a mass of igneous material, approximately 5 kilometers thick, which lies under the ocean bottom.

Oceanic (mid-ocean) rise—general term for the discontinuous ocean-bottom province which rises above the deep ocean floor; area of crustal generation.

Ooze—a fine-grained deep-ocean sediment containing at least 30 percent (by volume) undissolved sand- or silt-sized, calcareous or siliceous skeletal remains of small marine organisms, the remainder being clay-sized material.

Open ocean—that part of the ocean which is seaward of the approximate edges of the continental shelves, usually more than 2 km deep.

Operculum—a flap covering an opening. In gastropods, it is the horny plate that closes the shell against dryness; in fishes, it is the bony or membranous flap covering the external openings of the gill slits.

Osmosis—passage of water across a somewhat porous membrane so as to equalize concentrations on both sides. *Osmotic pressure* is a measure of the tendency of water to pass through the membrane—in other words, it is the amount of pressure which must be applied to the more concentrated solution on one side so as to prevent the passage into it of the solvent through the semipermeable membrane from the more dilute side. *Osmoregulation* is the capacity to regulate the retention of water and its loss from the body, as is necessary when an organism is not in osmotic equilibrium with its environment.

Pack ice—a rough, solid mass of broken ice floes forming a heavy obstruction, preventing navigation.

Pancake ice—pieces of newly formed ice, usually approximately circular, about 30 centimeters across, and with raised rims caused by the striking together of the pieces as a result of wind and swell.

Pangaea—single continent thought to have existed prior to Permian times. Fragments of it split apart about 200 million years ago to form the present continents.

Parachute drogue—a current-measuring assembly consisting of a weighted parachute and an attached surface buoy. The parachute can be placed at any desired depth and the current speed and direction determined by tracking and timing of the surface buoy.

Partial tide—one of the harmonic components comprising the tide at any point. The periods of the partial tides are derived from various combinations of the angular velocities of earth, sun, and moon, relative to each other.

Patch reef—a term for all isolated coral growths in lagoons of barriers and atolls. They vary in extent from expanses measuring several kilometers across to coral pillars or even mushroom-shaped growths consisting of a single colony. Smaller representatives are called *coral heads* or *coral knolls.*

Patchiness of plankton—sudden changes in plankton density due to the occurrence of dense patches of organisms over large areas.

Pelagic—a primary division of the ocean which includes the whole mass of water. The division is made up of the coastal ocean (which includes the water shallower than 200 meters) and the open ocean (which includes that water deeper than 200 meters).

Pelagic deposits—deep-ocean sediments that have accumulated by the settling out of the ocean on a particle-by-particle basis.

Photic zone (euphotic zone)—the layer of a body of water which receives ample sunlight for the photosynthetic processes of plants; it is usually 80 meters deep, or more, depending on angle of incidence of sunlight, length of day, and cloudiness.

Photosynthesis—the manufacture of carbohydrate food from carbon dioxide and water in the presence of chlorophyll, by utilizing light energy and releasing oxygen.

Phytoplankton—the plant forms of plankton.

Planetary winds—see *Prevailing wind systems.*

Plankton—passively drifting or weakly swimming organisms.

Plankton bloom—an enormous concentration of plankton (usually phytoplankton) in an area, caused either by an explosive or gradual multiplication of organisms and usually producing a discoloration of the sea surface.

Pluteus—a free-swimming larva of the sea urchins and brittle stars.

Polychaetes—segmented marine worms, some of which are tube-worms; others are free-swimming worms.

Polyp—an individual, sessile coelenterate.

Prevailing wind systems (planetary winds)—large, relatively constant wind systems which result from the shape, inclination, revolution, and rotation of the planet. They are the northeast and southeast trade winds, the westerlies, and the polar easterlies.

Primary coastline—a coastline shaped primarily by marine forces, such as wave action, or by marine organisms.

Primary productivity—the amount of organic matter synthesized by organisms from inorganic substances in unit time in a unit volume of water or in a column of water of unit area cross-section and extending from the surface to the bottom; also called *gross primary production.*

Prime meridian—the meridian of longitude 0°, used as the reference for measurements of longitude, internationally recognized as the meridian of Greenwich, England.

Productivity—see *Primary productivity.*

Profile—a drawing showing a vertical section along a surveyed line.

Progressive wave—a wave which is manifested by the progressive movement of the wave form.

Protozoa—mostly microscopic, one-celled animals, constituting one of the largest populations in the ocean.

Pseudopod—an extension of protoplasm that has no permanent shape but can be projected or withdrawn by the animal, for capturing food or for locomotion; characteristic of some protozoans.

Pteropods—free-swimming gastropods in which the foot is modified into fins; both shelled and nonshelled forms exist. In some shallow areas, accumulated shells of these organisms form a type of bottom sediment called *pteropod ooze.*

Pycnocline—a vertical density gradient in some layer of a body of water, positive with respect to depth and appreciably greater than the gradients above and below it; also, a layer in which such a gradient occurs.

Radioisotope—any radioactive isotope of an element; also, a word loosely used as a synonym for *radionuclide*.

Radiolarians—single-celled planktonic protozoans possessing a skeleton of siliceous spicules and radiating threadlike pseudopodia.

Radionuclide—a synonym for *radioactive nuclide*.

Red algae—reddish, filamentous, membranous, encrusting or complexly branched plants in which the color is imparted by the predominance of a red pigment over the other pigments present. Some are included among the coralline algae.

Red clay—a more-or-less brown to red deep-sea deposit. It is the most finely divided clay material that is derived from the land and transported by ocean currents and winds, accumulating far from land and at great depths.

Red tide—a red or reddish-brown discoloration of surface waters most frequently in coastal regions, caused by concentrations of certain microscopic organisms, particularly dinoflagellates.

Reducing atmosphere—a region in which there is no free oxygen or other gas which can accept electrons. Oxidation, a reaction involving the removal of one or more electrons from an ion or an atom, does not readily occur in such an atmosphere.

Reef—an offshore consolidated rock hazard to navigation with a least depth of 20 meters or less.

Reef flat (of a coral reef)—a flat expanse of dead reef rock which is partly or entirely dry at low tide. Shallow pools, potholes, gullies, and patches of coral debris and sand are features of the reef flat. It is divisible into inner and outer portions.

Reef front—the upper seaward face of a reef, extending above the lowest point of abundant living coral and coralline algae to the reef edge. This zone commonly includes a shelf, bench, or terrace that slopes to 15–30 meters, as well as the living, wave-breaking face of the reef. The terrace is an eroded surface or is veneered with organic growth. The living reef front above the terrace in some places is smooth and steep; in other places it is cut up by grooves separated by ridges that together have been called *spur-and-groove systems*, forming comb-tooth patterns.

Reflection—the process whereby a surface or discontinuity turns back a portion of the incident radiation into the medium through which the radiation approached.

Refraction of water waves—the process by which the direction of a wave moving in shallow water at an angle to the contours is changed, causing the wave crest to bend toward alignment with the underwater contours; also, the bending of wave crests by currents.

Relict sediment—sediment having been deposited on the continental shelf by processes no longer active.

Replacement time—time required for a flow of material to replace the amount of that material originally present in a given area. Assuming a steady flow, replacement time can be calculated for any substance, such as salt or water.

Respiration—an oxidation–reduction process by which chemically bound energy in food is transformed into other kinds of energy upon which certain processes in all living cells are dependent.

Reversing thermometer—a mercury-in-glass thermometer that records temperature upon being inverted and thereafter retains its reading until returned to the first position.

Reversing tidal current—a tidal current that flows alternately in approximately opposite directions, with a period of slack water at each reversal of direction. Reversing currents usually occur in rivers and straits where the flow is restricted. When the flow is toward the shore, the current is flooding; when in the opposite direction, it is ebbing.

Rift valley—see *Fault block*.

Rill—a small groove, furrow, or channel made in mud or sand on a beach by tiny streams following an outflowing tide.

Rip—agitation of water caused by the meeting of currents or by a rapid current setting over an irregular bottom—for example, a *tide rip*.

Rip current—a strong current usually of short duration, flowing seaward from the shore. It usually appears as a visible band of agitated water and is the return movement of water piled up on the shore by incoming waves and wind.

Ripple—a wave controlled to a significant degree by both surface tension and gravity.

Rise—a long, broad elevation that rises gently and generally smoothly from the ocean bottom.

River-induced upwelling—the upward movement of deeper water which occurs when seawater mixes with fresh water from a river, becoming less dense, and moves toward the surface.

Rotary current, tidal—a tidally induced current which flows continually with the direction of flow, changing through all points of the compass during the tidal period. Rotary currents are usually found in the ocean where the direction of flow is not restricted by any barriers.

Sabellid—see *Tube-worm*.

Salinity—a measure of the quantity of dissolved salts in seawater. Formally defined as the total amount of dissolved solids in seawater in parts per thousand ($^0/_{00}$) by weight when all the carbonate has been converted to oxide, the bromide and iodide to chloride, and all organic matter is completely oxidized.

Salps—transparent pelagic tunicates. The body is more-or-less cylindrical and possesses conspicuous ringlike muscle bands, the contraction of which propels the animal through the water.

Salt marsh—see *Marsh*.

Salt-water wedge—an intrusion in a tidal estuary of seawater in the form of a wedge characterized by a pronounced increase in salinity from surface to bottom.

Sand—loose material which consists of grains ranging between 0.0625 and 2.0 millimeters in diameter.

Sargasso Sea—the region of the North Atlantic Ocean to the east and south of the Gulf Stream system. This is a region of convergence of the surface waters and is characterized by clear, warm water, a deep blue color, and large quantities of floating sargassum ("gulfweed").

Sargassum—a brown alga characterized by a bushy form, substantial "holdfast" (root-like structure) when attached, and a yellowish brown, greenish yellow, or orange color.

Scarp—an elongated and comparatively steep slope of the ocean floor, separating flat or gently sloping areas.

Scattering—the dispersion of light when a beam strikes very small particles suspended in air or water. In light scattering there is, theoretically, no loss of intensity but only a redirection of light.

School—a large number of one kind of fish or other aquatic animal swimming or feeding together.

Scour—the downward and sideward erosion of a sediment bed by wave or current action.

Scyphozoans—coelenterates in which the polyp or hydroid stage is minimized or insignificant and the medusoid stage is well developed. The true jellyfish belong to this group.

Sea—waves generated or sustained by winds within their fetch, as opposed to *swell*; also, a subdivision of an ocean.

Seabed drifter—a plastic float designed to drift in near-bottom currents; movements of the drifter indicate speed and direction of the current.

Sea breeze—a light wind blowing toward the land caused by unequal heating of land and water masses.

Seachannel—a long, narrow, U- or V-shaped shallow depression of ocean floor, usually occurring on a gently sloping plain or fan.

Seamount—an elevation rising 900 meters or more from the ocean bottom.

Sea state (state of the sea)—the numerical or written description of ocean-surface roughness.

Seawall—a man-made structure of rock or concrete built along a portion of coast to prevent wave erosion of the beach.

Second layer—see *Crust of the earth*.

Secondary coastline—coastline formed primarily by terrestrial forces rather than by wave action or other marine processes.

Sediment—particulate organic and inorganic matter which accumulates in a loose, unconsolidated form. It may be chemically precipitated from solution, secreted by organisms, or transported from land by air, ice, wind, or water and deposited.

Sedimentation—the process of breakup and separation of particles from the parent rock, their transportation, deposition, and consolidation into another rock.

Seiche—a standing-wave oscillation of an enclosed or semienclosed water body that continues, pendulum-fashion, after the cessation of the originating force, which may have been seismic, atmospheric, or wave-induced; also an oscillation of a fluid body in response to a disturbing force having the same frequency as the natural frequency of the fluid system.

Seismic—pertaining to, characteristic of, or produced by earthquakes or earth vibrations.

Seismic reflection—the measurements, and recording in wave form, of the travel time of acoustic energy reflected back to detectors from rock or sediment layers which have different elastic-wave velocities.

Seismic sea wave—see *Tsunami*.

Selective deposit-feeder—an animal which feeds by selecting or straining edible particles from bottom sediments.

Semidaily (semidiurnal) tide—a tide having a period or cycle of approximately one-half a tidal day; the predominating type of tide throughout the world is semidiurnal, with two high waters and two low waters each tidal day.

Semipermeable membrane—a membrane through which a solvent, but not certain dissolved or colloidal substances, may pass; used in osmotic-pressure determinations.

Sensible heat—the portion of energy exchanged between ocean and atmosphere which is utilized in changing the temperature of the medium into which it penetrates.

Sessile—permanently attached, by a base or stalk; not free to move about.

Set (current direction)—the direction toward which the current flows.

Shelf ice—see *Ice shelf*.

Shoreline—the boundary line between a body of water and the land at high tide (usually mean high water).

SIAL—a layer of rocks underlying all continents. The thickness of this layer is variously placed at 30 to 35 kilometers. The name derives from the principal ingredients, silica and aluminum. Its specific gravity is considered to be about 2.7.

Significant wave height (characteristic wave height)—the average height of the highest one-third of waves of a given wave group.

Sill—shallow portion of the ocean floor which partially restricts water flow; may be either at the mouth of an inlet, fjord, etc., or at the edge of an ocean basin—for example, the Bering Sill separates the Pacific and Arctic portions of the Atlantic Ocean.

Silt—particles between sands and clays in size.

SIMA—the basic outer shell of the earth; under the continents it underlies the SIAL, but under the Pacific Ocean it directly underlies the ocean water. Its specific gravity is about 3.3. It contains silica, with more magnesium, iron, and other heavy metals than occur in SIAL.

Sinking (downwelling)—a downward movement of surface water generally caused by converging currents or as a result of a water mass becoming more dense than the surrounding water.

Siphonophores—medusoid coelenterates, many of which are luminescent and some venomous. Some possess a gas-filled float. Some are colonial, so that polyp and medusoid individuals function as a single individual.

Sipunculids—wormlike marine animals, unsegmented, with the mouth surrounded by tentacles. The anterior (head) end can be withdrawn into the body. They are deposit-feeders.

Slack water—the state of a tidal current when its velocity is near zero, usually occurs when a reversing current changes its direction.

Slick—area of quiescent water surface, usually elongated. Slicks may form patches or weblike nets where ripple activity is greatly reduced.

Slump—the slippage or sliding of a mass of unconsolidated sediment down a submarine or subaqueous slope. Slumps occur frequently at the heads or along the sides of submarine canyons, triggered by any small or large earth shock the sediment usually moves as a unit mass initially, but often becomes a turbidity flow.

Slush—see *Ice slush*.

Solar tide—the (partial) tide caused solely by the tide-producing forces of the sun.

Sounding—the measurement of the depth of water beneath a ship.

Specific gravity—the ratio of the density of a substance relative to the density of pure water at 4°C; in the centimeter–gram–second system, *density* and *specific gravity* may be used interchangeably.

Specific heat—the quantity of heat required to raise the temperature of 1 gram of any substance by 1°C. The common unit is calories per gram per degree centigrade (cal/g/°C).

Sperm whale—see *Toothed whales*.

Spicules—crystals of newly formed sea ice; also, a minute, needlelike or multiradiate calcareous or siliceous body in sponges, radiolarians, some gastropods, and echinoderms.

Spit—a small point of land projecting into a body of water from the shore.

Spring bloom—the sudden proliferation of phytoplankton which occurs from the temperate zone to the poles whenever the critical depth (as determined by penetration of sunlight) exceeds the depth of the mixed, stable, surface layer (as determined by the establishment of a thermocline).

Spring tide—tide of increased range which occurs about every two weeks when the moon is new or full.

Spur-and-groove structure—see *Reef front*.

Stability—the resistance to overturning or mixing in the water column, resulting from the presence of a positive density gradient.

Stand of the tide—the interval at high or low water when there is no appreciable change in the height of the tide; its duration will depend on the range of the tide, being longer when the tidal range is small and shorter when the tidal range is large.

Standing crop—the biomass of a population present at any given moment.

Standing wave—a type of wave in which the surface of the water oscillates vertically between fixed points, called *nodes*, without progression. The points of maximum vertical rise and fall are called *antinodes*. At the nodes, the underlying water particles exhibit no vertical motion, but maximum horizontal motion.

Steady state—the absence of change with time.

Still-water level—the level that the sea surface would assume in the absence of wind waves; not to be confused with mean sea level or half-level tide.

Storm surge (storm wave, storm tide, tidal wave),—a rise or piling-up of water against shore, produced by wind-stress and atmospheric-pressure differences in a storm.

Stratosphere—that part of the earth's atmosphere between the troposphere and the upper layer (ionosphere).

Sublimation—the transition of the solid phase of certain substances into the gaseous—and vice versa—without passing through the liquid phase. Water possesses this property; thus ice can change directly into water vapor or water vapor into ice.

Submarine canyon—a V-shaped submarine depression of valley form with relatively steep slope and progressive deepening in a direction away from shore.

Subsurface current—a current usually flowing below the pycnocline, generally at slower speeds and frequently in a different direction from the currents near the surface.

Subtropical high—one of the semipermanent highs of the subtropical high pressure belt.

Succession, ecological—the orderly process of community change, whereby communities replace one another in sequence.

Surf—collective term for breakers; also, the wave activity in the area between the shoreline and the outermost limit of breakers.

Surf zone—the area between the outermost breaker and the limit of wave uprush.

Surface-active agent—substance, usually in solution, which can markedly change the surface or interfacial properties of the liquid fraction, even when present in trace amounts only.

Surface tension (surface energy, capillary forces, interfacial tension) —a phenomenon peculiar to the surface of liquids, caused by a strong attraction toward the interior of the liquid acting on the liquid molecules in or near the surface in such a way to reduce the surface area. An actual tension results, and is usually expressed in dynes per centimeter or ergs per square centimeter.

Surface zone (mixed zone)—water layer above the pycnocline where wind waves and convection cause mixing of the water, resulting in more or less uniform temperature and salinity with depth.

Surge—horizontal oscillation of water with comparatively short period accompanying a seiche (see also *storm surge*).

Surge channel—a transverse channel cutting the outer edge of a coral reef in which the water level fluctuates with wave or tidal action.

Swallow float—a tubular buoy, usually made of aluminum, which can be adjusted to remain at a selected density level to drift with the motion of that water mass. The float is tracked by shipboard listening devices, to determine current velocities.

Swash—the rush of water up onto the beach following the breaking of a wave.

Swell—ocean waves which have traveled out of their generating area.

Swimbladder—a membranous sac of gases, lying in the body cavity between the vertebral column and the alimentary tract of certain fishes. It serves a hydrostatic function in most fishes that possess it; in some, it participates in sound production.

Symbiosis—a relationship between two species in which one or both members are benefitted and neither is harmed.

T–S curve—see *Temperature–salinity diagram.*

Tablemount—flat-topped seamount; also called guyot.

Tabular iceberg—a flat-topped iceberg showing horizontal layers of compacted snow, usually calved from an ice-shelf formation.

Telemetry—the study and technique involved in measuring a quantity or quantities in place, transmitting this value to a station, and there interpreting, indicating, or recording the quantities.

Temperature–salinity diagram (T–S curve, T–S diagram)—the plot of temperature versus salinity data of a water column. The result is a diagram which identifies the water masses within the column and the column's stability, and allows an estimate of the accuracy of the temperature and salinity measurements.

Tethys—shallow sea which formerly separated what is now Africa from Asia. The Mediterranean, Black, and Caspian seas are remnants of this former seaway.

Thermocline—a vertical temperature gradient in some layer of a body of water, negative with respect to depth and appreciably greater than the gradients above and below it; also, a layer in which such a gradient occurs.

Thermohaline circulation—vertical circulation induced by surface cooling, which causes convective overturning and consequent mixing.

Tidal bulge (tidal crest)—a long-period wave associated with the tide-producing forces of the moon and sun; identified with the rising and falling of the tide. The trough located between the two tidal bulges present at any given time on the earth is known as the *tidal trough.*

Tidal constituent—one of the harmonic elements in a mathematical expression for the tide-generating force and in corresponding formulas for the tide or tidal current. Each constituent represents a periodic change or variation in the relative positions of the earth, moon, and sun.

Tidal current—the alternating horizontal movement of water associated with the rise and fall of the tide caused by the astronomical tide-producing forces.

Tidal day—the interval between two successive upper transits of the moon over a local meridian. The period of the *mean tidal day,* sometimes called a *lunar day,* is 24 hours, 50 minutes.

Tidal flats—marshy or muddy areas which are covered and uncovered by the rise and fall of the tide; also called *tidal marshes.*

Tidal period—the elapsed time between successive high or low waters.

Tidal range—the difference in height between consecutive high and low waters.

Tidal trough—see *Tidal bulge.*

Tide—the periodic rise and fall of the earth's ocean and atmosphere that results from tide-producing forces of moon and sun acting on the rotating earth.

Tide curve—a graphic presentation of the rise and fall of tide; time (in hours or days) is plotted against height of the tide.

Tide-producing forces—the slight local difference between the gravitational attraction of two astronomical bodies and the centrifugal force that holds them apart. Gravitational attraction predominates at the surface point nearest to the other body, while centrifugal "repulsion" predominates at the surface point farthest from the other body.

Tide pool—depression in a rock within the intertidal zone, alternately submerged and exposed with water remaining inside, by the rise and fall of the tide.

Tide rip—see *Rip*.

Tide tables—tables which give daily predictions, usually a year in advance, of the times and heights of the tide.

Tide wave—a long-period gravity wave that has its origin in the tide-producing force and which manifests itself in the rising and falling of the tide.

Tintinnids—microscopic planktonic protozoans which possess a tubular or vase-shaped outer shell.

Tombolo—an area of unconsolidated material, deposited by wave or current action, which connects a rock, island, etc., to the main shore or other body of land.

Toothed whales—this group includes dolphins, porpoises, killer whales, and sperm whales.

Trade winds—the wind system, occupying most of the tropics, which blows from the subtropical highs toward the equatorial trough; a major component of the general circulation of the atmosphere. The winds are northeasterly in the Northern Hemisphere and southeasterly in the Southern Hemisphere.

Transducer—a device that converts electrical energy to sound energy, or the reverse.

Trawl—a bag or funnel-shaped net for catching bottom fish by dragging along the bottom; also a large research net for catching zooplankton and fishes by towing in intermediate depths.

Trochophore—the free-swimming pelagic stage of some segmented worms and mollusks.

Trophic level—a successive stage of nourishment as represented by links of the food chain. Primary producers (phytoplankton) constitute the first trophic level, herbivorous zooplankton the second trophic level, and carnivorous organisms the third and higher trophic levels.

Tropics—equatorial region between Tropic of Cancer and Tropic of Capricorn; climate found in the belt close to the equator; characteristics are daily variation in temperature exceeding seasonal variation and generally high rainfall.

Tropopause—the upper limit of the troposphere.

Troposphere—that portion of the atmosphere next to the earth's surface in which temperature generally rapidly decreases with altitude, clouds form, and convection is active. At middle latitudes, the troposphere generally includes the first 10 to 12 kilometers above the earth's surface.

Tsunami (seismic sea wave)—a long-period sea wave produced by a submarine earthquake or volcanic eruption. It may travel unnoticed across the ocean for thousands of miles from its point of origin, and builds up to great heights over shallow water.

Tube-worm—any polychaete, chiefly the sabellids and related groups, that builds a calcareous or leathery tube on a submerged surface. Tube-worms are notable fouling organisms.

Tunicates—globular or cylindrical, often saclike animals, many of which are covered by a tough flexible material. Some are sessile, others are pelagic.

Turbidite—turbidity-current deposit characterized by both vertically and horizontally graded bedding.

Turbidity—reduced water clarity resulting from the presence of suspended matter. Water is considered turbid when its load of suspended matter is visibly conspicuous, but all waters contain some suspended matter and therefore are turbid to some degree.

Turbidity current—a gravity current resulting from a density increase by suspended material.

Turbulent flow—a flow characterized by irregular, random-velocity fluctuations.

Turtle-grass—a seed-bearing, grasslike marine plant which grows chiefly in sand or mud–sand bottoms and most abundantly in water less than 10 meters in depth. It serves as food for sea turtles in tropical and subtropical water.

Unselective deposit-feeders—animals which feed by swallowing large amounts of bottom sediment and assimilating food materials through the gut.

Upwelling—the process by which water rises from a lower to a higher depth, usually as a result of divergence and offshore currents.

Van der Waals forces—weak attractive forces between molecules which arise from interactions between the atomic nuclei of one molecule and the electrons of another molecule.

Varve—a sedimentary deposit, bed, or lamination deposited in one season.

Vector—a physical quantity which has magnitude (e.g., speed) and direction.

Veliger—the planktonic larval second stage of many gastropods.

Viscosity—internal resistance-to-flow property of fluids which enable them to support certain stresses and thus resist deformation for a finite time; arises from molecular and eddy effects.

Volcanic island—island formed by the top of a volcano or solidified volcanic material, rising above the sea surface.

Water budget—the accounting for the interchange of water substance between the earth, the atmosphere, and the ocean.

Water mass—a body of water usually identified by its T–S curve or chemical content.

Water parcel—water mass with a certain temperature and salinity, separated from surrounding waters by fairly sharp boundaries across which mixing occurs.

Wave—a disturbance which moves through or over the surface of the ocean (or earth); wave speed depends on the properties of the medium.

Wave age—the state of development of a wind-generated sea-surface wave, conveniently expressed by the ratio of wave speed to wind speed. Wind speed is usually measured at about 8 meters above still-water level.

Wave energy—the capacity of a wave to do work. In a deep-water wave, about half the energy is kinetic energy, associated with water movement, and about half is potential energy, associated with the elevation of water above the still-water level in the crest or its depression below still-water level in the trough.

Wave group—a series of waves in which the wave direction, wave length, and wave height vary only slightly.

Wave height—vertical distance between crest and preceding trough.

Wave length—horizontal distance between successive wave crests measured perpendicular to the crests.

Wave period—the time required for two successive wave crests to pass a fixed point such as a rock or an anchored buoy.

Wave spectrum—a concept used to describe by mathematical function the distribution of wave energy (square of wave height) with frequency (1/period). The square of the wave height is related to the potential energy of the sea surface so that the spectrum can also be called the *energy spectrum.*

Wave steepness—the ratio of the wave height to wave length.

Wave train—a series of waves from the same direction.

Wave velocity—speed at which the individual wave form advances; also, a vector quantity that specifies the speed and direction with which a wave travels through a medium.

Weathered—descriptive of ice or rock that has been destroyed or partially destroyed by thermal, chemical, and mechanical processes.

Wetland—see *Marsh.*

Whitecap—the white froth on crests of waves in a wind, caused by wind blowing the crest forward and over.

Wind-driven circulation—surface-current system driven by the force of the winds.

Wind mixing—mechanical stirring of water due to motion, induced by the surface wind.

"Wind tide"—vertical rise in the water level on the leeward side of a body of water caused by wind stress on the surface of the water.

Wind waves—waves formed and growing in height under the influence of wind; loosely, any wave generated by wind.

Windrows—rows of floating debris, aligned in the wind direction, formed on the surface of a lake or ocean by Langmuir circulation in the upper water layer.

Yellow substance—general term used to denote the pigmented products of organic decomposition which absorb blue and violent light in coastal waters, causing the water to be most transparent to yellow light.

Zonation—organization of a habitat into more-or-less parallel bands of distinctive plant and animal associations where conditions for survival are optimal.

Zone—a layer which encompasses some defined feature, structure, or property in the ocean.

Zooplankton—the animal forms of plankton.

MARINE FISHES

appendix 4

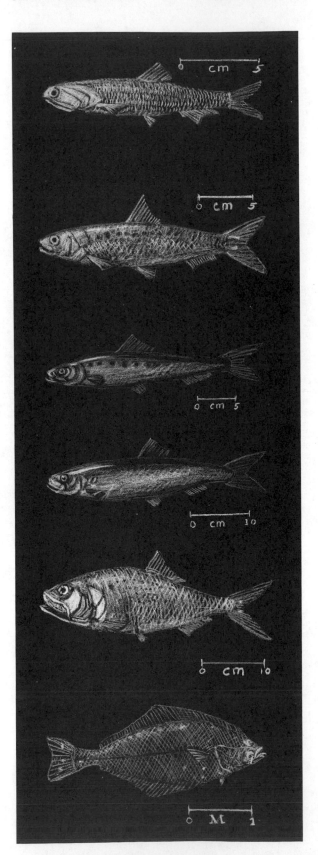

Anchovy (and related species)

Size: to about 14 cm; some larger

Distribution: abundant in warm seas, near shore, and in the open ocean

These valuable plankton-feeders are widely used as food and as fresh and frozen bait, for fish meal and oil. They are tolerant of a wide salinity range. Swimming in huge schools, they form a basic food for many larger fishes. Peruvian anchovettas yield the world's largest catch, about 8 million metric tons annually in the late 1960's, almost all of it converted to fish meal.

Pilchard

Size: over 20 cm

Distribution: coastal waters throughout the world—between isotherms 12° and 20°C

Like all plankton-feeders, this relative of the Pacific sardine and Atlantic herring has high oil content and is suitable for rendering into fish meal (for livestock feed) and industrial oils. Pilchard are of major commercial value off Australia and South Africa.

Sardine

Herring

Menhaden

Size: 30 to 45 cm

Distribution: Nova Scotia to Brazil

Used for fish meal, fertilizer, and industrial oil, these primary carnivores are a staple in the diet of many sea birds and fishes. Biological productivity of menhaden has been analyzed in an effort to predict fish yields per unit area.

Halibut

Size: to 3 m

Distribution: Arctic Ocean to New Jersey; Bering Sea to latitude of California; at lower latitudes it seeks deeper waters

This large, commercially valuable flatfish is an active predator. Adapted for lying on its side on the bottom, the sides are large and round or oval; both eyes are on the same side of the head, in this case the right.

Northern fluke (summer flounder)

Size: to 1 m or more

Distribution: Cade Cod to Cape Hatteras; related species in all but the coldest seas

This "flounder" is really a member of the halibut family. Its eyes are on the left side. The smaller "winter flounder" (Atlantic sole) is a true flounder, as are the European plaice and the turbot. Flatfishes sometimes bury themselves in the sand; only the eyes emerge. They eat large invertebrates and small fishes.

Winter flounder

European plaice

Turbot

Cod

Size: commonly to 0.5 m or more

Distribution: cold and temperate seas, chiefly in the Northern Hemisphere

This schooling carnivore lives in deep water over rocky or sandy bottoms. The eggs are planktonic and produced in millions to ensure survival. The smaller haddock is a relative, as are the North Atlantic pollock and whiting (hake). Cod catches are the third largest in volume on a worldwide basis.

Pollack

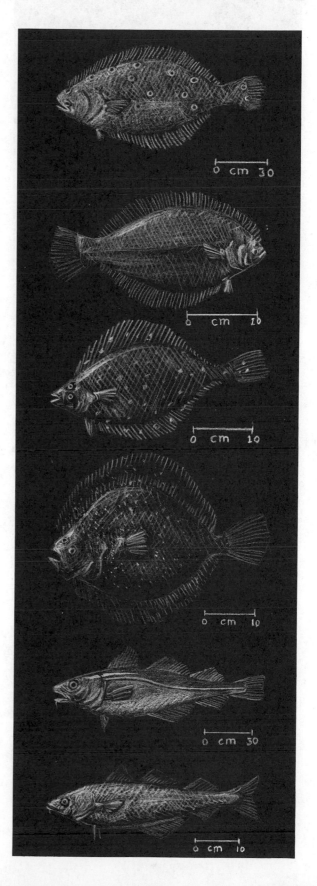

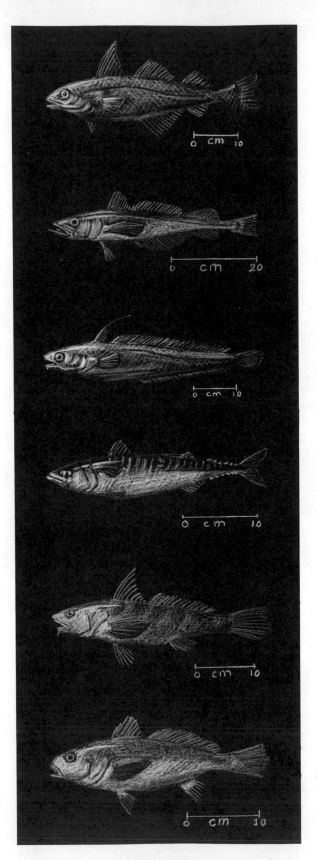

Haddock

Silver hake (whiting)

Size: to 0.5 m

Distribution: Newfoundland to the Caribbean, to 450 m depth

The hake family lives over continental shelves, preferring deeper water, in both hemispheres. It has a wide geographic and temperature range. The young eat planktonic crustaceans, adults prey on small fishes and squids. Like the rest of the cod family, and other less oily fishes, they can effectively be preserved by drying.

Squirrel hake

Atlantic mackerel

Size: averages over 0.3 m

Distribution: Labrador to Cape Hatteras, North Carolina; Norway to Spain

In the western Atlantic, mackerel migrate northward in schools along the coast in summer, southward offshore in winter. Preyed on by birds, squids, and other fishes, they are themselves omnivorous and will eat anything they can swallow. Members of this family are found in temperate and warm seas throughout the world and are widely caught for food.

Southern king whiting

Size: to about 40 cm

Distribution: New Jersey to Texas

The Gulf king whiting, found most abundantly from Chesapeake Bay to Texas, generally replaces the southern whiting in the Gulf of Mexico. The northern whiting ranges from Maine to Florida, and there is also a California relative, the corbina. They prefer sandy bottoms, feeding on shellfish and small fishes.

Atlantic croaker

Size: around 40 cm

Distribution: Massachusetts to Texas; common south of Chesapeake Bay

Most croakers make a drumming or grunting noise by creating a resonance within the swimbladder. Common in estuaries and coastal waters, they are related to the king whiting family and to the weakfish (sea trout), a southern estuarine native.

Skipjack tuna

Size: to over 1 m

Distribution: tropical and warm temperate seas

This economically important tuna of the tropics is related to the great bluefin sport fish (which may be up to 5 meters in length), and also to the white-meat albacore tuna and the yellowfin. These relatives of the mackerel are strong swimmers at the surface and at depth; they feed on squids, plankton, crabs, and small schooling fishes.

Jack crevalle (common jack)

Size: generally under 1 m

Distribution: Cape Cod to Brazil; Gulf of California southward

Related species occur around the world in temperate and tropical waters, often among coral reefs. A surface predator, it tends toward inshore waters. Related to pompano.

Pompano

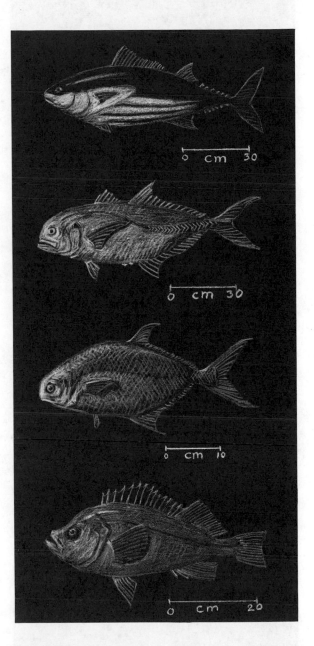

Ocean perch (rosefish, redfish)

Size: average 0.5 m or more

Distribution: North Atlantic, related species worldwide

Some of the many species in this family are referred to collectively as rockfishes because they prefer deeper waters. The flesh tends to be reddish and many groups have long spines, often poisonous in tropical members of the family. North Atlantic redfish and eastern Pacific rockfish are examples of major fisheries opened after World War II to exploit what used to be considered "trash" (noncommercial) fishes in meeting world protein needs.

LESSER-KNOWN FISHES OF POTENTIAL COMMERCIAL IMPORTANCE

Saury (garfish)

Size: about 20 cm depending on species

Distribution: lower California to Alaska and south off Japan to Asia, coastal and open-ocean species

These schooling fishes feed near the surface, at low trophic levels. Related species, such as the halfbeak, are some of the bullet-shaped "flying" fishes reported skittering along the surface in open waters. Pacific sauries have been fished commercially offshore, but to a limited extent.

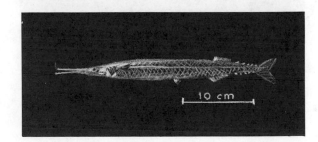

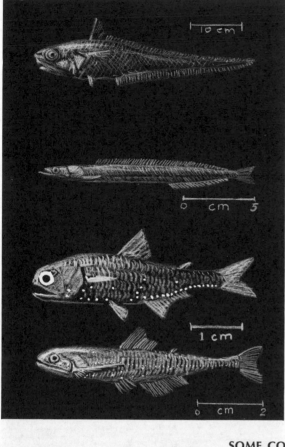

Grenadier ("rat-tail")

Size: around 30 cm or larger

Distribution: lower parts of continental slope worldwide—200–800 m

These are perhaps the most numerous of all fishes on the edges of the deep sea. They feed on benthic invertebrates, also on fishes and plankton at mid-depths. Like many deep-water organisms, they are fragile and eaesily damaged in fishing. The tapering "rat-tailed" body is characteristic of many deep-water species.

Sand lance (sand launce)

Size: to 15 cm

Distribution: northern seas

Sand lances swim in large schools near the shore. Their habit of diving into sand and sometimes remaining there even after the tide is out gives them their name. They are considered to be a tasty pan fish.

Lantern-fishes

Size: about 15 cm, depending on species; many are smaller

Distribution: throughout the world ocean, 500 to 200 m depth

Second in numbers only to the stomiatoids among deep-water fishes, there are at least 170 species of lantern fishes, in all but the very coldest seas. Great masses often migrate to the surface at night to feed. Light organs dotted along the sides and a bright light near the tail are characteristic. Although sunlight and even moonlight repel them, lantern fishes often experience a reversal of the normal reaction in the presence of a very bright light, so that they swim toward a searchlight on the water like moths.

SOME COMMON FISHES OF ATLANTIC AND GULF COAST ESTUARIES

Tidewater silverside

Size: to 7 cm

Distribution: Massachusetts to the Gulf of Mexico

This schooling, carnivorous fish is common in shallow, brackish waters. A famous California relative, the grunion, spawn on sandy shores during the highest tide of each summer month. Two weeks later, high tide washes the newly hatched larvae out to sea.

Killifish

Size: average 5 to 10 cm

Distribution: brackish marshes and inlets along all coasts

Active, swarming killies are hardy and omnivorous, tolerant of poor water quality and valued for their consumption of mosquito larvae. They like weedy places and tidepools.

Mullet

Size: average 30 cm or more

Distribution: cosmopolitan in warm seas

An excellent food fish, widely raised in the Orient and known in Polynesia as ama-ama. They grub in soft bottoms for food particles which they strain from the mud by a sieve in the throat.

Goby

Size: a few cm to under 0.5 m

Distribution: temperate and tropical zones—shallow water

Gobies are active, amusing little fishes—bold, friendly, and interesting to watch. They prop themselves up by their ventral fins in bays, mud flats, and tide pools, and some can leap from pool to pool with uncanny accuracy. Males select the rocky "nest" and guard the eggs. Gobies can tolerate wide ranges of water quality, temperature, and salinity.

Cusk-eel

Size: to 15 cm

Distribution: warm waters along shorelines

The cusk-eel is a sandy-bottom burrower, found in estuaries along the east and west coasts. It eats invertebrates and customarily enters a hiding place tail first.

American shad

Size: to 0.7 m, males smaller

Distribution: St. Lawrence River mouth, to Florida; introduced successfully on the west coast.

This member of the herring family lives in rivers and estuaries, feeding on plankton. It spawns upriver in spring. Both the fish and its eggs (roe) are considered a delicacy.

Bluefish

Size: to 1 m or more

Distribution: Cape Cod to Venezuela, and in most warm seas but not off Europe or the western U.S.

These fish travel in large, fast schools, preying voraciously on weakfish, herring, menhaden, or other small schooling fishes. A favorite of sportsmen and gourmets, they are usually caught in surf and in deeper water, although they occasionallly migrate into estuaries.

Common skate

Size: to 0.5 m

Distribution: Nova Scotia to Cape Hatteras; other species worldwide, especially in warmer waters

Small skates are usually found over sandy bottoms, to a depth of about 150 meters. They bury themselves in the sand by day, ambushing small fish, mollusks, and crustaceans and pinning them beneath their bodies. The huge side fins flap like a bird or gently undulate in swimming, and the jaws are on the lower surface of the head.

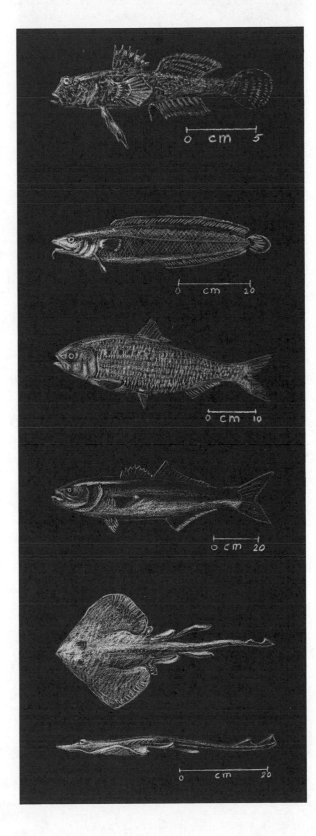

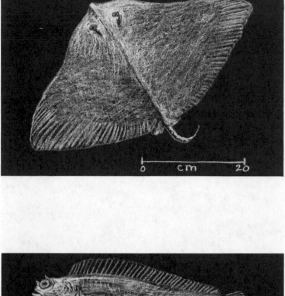

Butterfly ray (sand skate)

Size: average 0.6 m in length

Distribution: Massachusetts to Brazil; distribution of other species similar to the related sharks and skates

The poisonous spines at the base of a ray's tail can be a dangerous, painful weapon of defense. They are generally inactive, but are capable of considerable speed as swimmers, and when searching for crustaceans and mollusks they stir up the sand with their large side fins. Water is inhaled through two holes on top of the head.

SOME COMMON UNITED STATES WEST COAST FISHES

Sarcastic blenny

Size: to 20 cm

Distribution: Monterey Bay to San Diego—many species worldwide, including Arctic

Blennies tolerate a wide range of salinities and temperatures. Like gobies, they inhabit tide pools, reefs, rocks, and brackish inlets, and enjoy being out of water for short periods. Many are beautifully colored and patterned. The large mouth pictured here is unique to the sarcastic blenny. All members of the group "walk" on ventral fins and "nest" in caves, holes, and frequently tin cans.

Striped bass (rockfish)

Size: to 1 m

Distribution: St. Lawrence River mouth to Florida, originally; successfully introduced from California to the Columbia River.

Taken commercially in large numbers on the east coast, this is a popular game fish. It is often found in small or large schools near shores or at river mouths, feeding on whatever organisms are available.

Lamprey eel

Size: up to 1 m

Distribution: Pacific species from Alaska to Southern California, found in all temperate seas, they breed in fresh water

On the west coast, they prey mainly on salmon and the related steelhead trout. Their feeding method is to attach themselves to fishes, rasping away the skin and sucking the flesh; this attack is often fatal to soft-skinned species. Their invasion of the Great Lakes has threatened some of the food fishes there.

Pink salmon

Size: to 0.7 m

Distribution: northern California to northwestern Alaska; in winter throughout the Northern Pacific Ocean

Pink salmon spawn from late September to early November in lakes and streams. They migrate downstream as juveniles and spend some time in inshore waters; usually only 1 year is spent in the open ocean before they reach maturity at 2 years of age. Normally, each population "runs" only in alternate years.

Spiny dogfish

Size: 1 m or more

Distribution: North Pacific to central California

This cold-water shark is omnivorous and a nuisance to Pacific
coastal fisheries for its predation on small fish in schools.
It is found in cooler water throughout the world, moving to
deeper areas in hot weather.

Striped sea perch

Size: 20 to 35 cm

Distribution: Alaska to lower California

Distinctive bands of orange and green identify this common
resident of bays, wharves, and surf. It is a popular game fish. The
young are born alive.

Kelp greenling

Size: to 40 cm

Distribution: Central California to Alaska

This popular game fish is found off rocky coastal areas and
kelp beds. It primarily eats crustaceans.

Moray eel

Size: average 1 to 1.5 m

Distribution: tropical rocks and reefs throughout the world

These vicious, noctural predators live mainly on shrimp and
other crustaceans. They remain hidden in rocky caves during
daylight and will not attack unless provoked, when they
can be extremely dangerous. Morays are eaten by man in many
parts of the world.

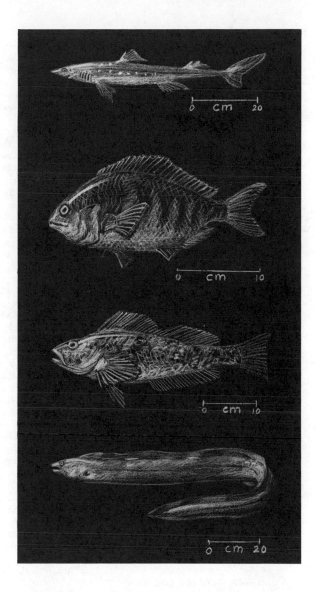

INDEX

Abalone, 502
Abyssal hills, 61
Abyssal plain, 13, 54, 61, 262
Acanthaster planci, 464
Adenosine phosphate, 100
Admiralty Inlet, 285, 316, 317, 318
Africa, 26, 30
  and Americas, 34
  and coastal upwelling, 127, 333
  and Eurasia, 30, 95, 96
  and fresh water discharge to the
      Indian Ocean, 37
  and rift valley in (*see* East African
      rift valley)
  and sediment discharge, 116
  and Walvis Ridge, 66, 90
Afsluitdijk, 386
Aging, rate of and metabolic rate,
      412
  rate of and size of individuals, 457
Airborne sediment (*see* Sediment
      transport, and wind)
Alaska, 36
  and Aleutian Islands, 68, 69
  and Arctic Ocean, 34
  continental shelf, 35, 342
  shoreline, 32, 342, 360
  and storms, 251
Alaska Current, 214, 219, 342, 350
Alaskan trench, 62
Albacore tuna, 554
Albatross (ship), 107
Albedo, 175, 176
Albermarle Sound, 298, 311, 489
Alberni Inlet, British Columbia, 304
Aleutian Basin, 68, 87
Aleutian Islands, 68, 69, 113, 231, 260
Aleutian Trench, 53, 68, 97
Alewife, 489
Alexander the Great, 268
Algae, attached (*see* Benthic algae;
      Seaweeds)
  blue-green, 102, 419, 420, 447, 454,
      490
  brown, 443, 447, 448 (*see also*
      Kelp; Sargassum)
  calcareous (*see* Calcareous algae)
  colonial, 99, 476
  coralline (*see* Coralline algae)
  and DDT, 491
  filamentous, 457, 463, 464
  as fouling organisms, 461
  green, 447, 448, 463
  nitrogen fixed by, 458
  red, 447, 448
  and vitamins, 481
Algal ridge, 75
Alpha radiation, 100
Alps, 340
Alternate generations, in
      coelenterates, 422
  in tunicates, 432
Alumina, 39
Aluminum, in rocks, 82
  in seawater, 155

Amazon River, 30, 35, 116, 117, 118,
      361, 363
Americas (*see also* North America;
      South America)
  as a continental land mass, 30, 87
  as split away from Africa and
      Eurasia, 34
  and young mountain ranges on
      west coasts, 32
Amino acids, non-biological synthesis,
      100, 101
  in organic matter, 100
  synthesis by bacteria, 458
  synthesis by plants, 396
  and utilization of by bacteria, 405
Ammonia, 458, 479, (*see also*
      Nitrogen cycle)
Ammonium ion, 396, 401, 402
Amoeba, 423, 457
Amphidrome systems, 280, 281, 291
Amphipods, 423, 467
Anaerobic respiration, 160, 161, 306,
      402, 458, 459, 491
Anchovy, 433, 477, 552
Andes Mountains, 43, 184
Angler fishes, 496
Angular velocity of the earth, 224
Antarctica, 4, 26, 30, 41
  as Atlantic Ocean boundary, 34
  and circulation of currents around,
      95, 206, 214, 216
  continental shelf, 41
  and crustal movements, 93
  encircled by mid-oceanic ridges, 57
  and icebergs, 205, 206
  and ice cover, 30, 118, 182, 205, 206,
      361
  as Indian Ocean boundary, 36
  as land mass antipodal to Arctic
      Ocean, 29, 30
  and sea ice, 201, 203, 222
  and sediment deposits near, 127
  and West Wind Drift, 235
Antarctic Bottom Water, 133, 195,
      201, 231, 235
Antarctic Circumpolar Water, 195
Antarctic Convergence, 214, 235, 436
Antarctic Deep Water, 436
Antarctic Divergence, 214, 228, 436,
      447
Antarctic Intermediate Water, 195,
      235
Antarctic Ocean, and barriers to
      water exchange, 66
  boundaries of, 31
  and migration of zooplankton, 436
  as a polar region, 207
  and sea ice, 201
  and seasonal abundance of
      plankton, 499
  and tides, 279
  and water masses, 195, 231, 233
Antibiotics, 481, 485
Anticoagulants, 485
Antinode, 263

Antipode, 29
Apatite, 109
Aphotic zone, 397
Appalachian Mountains, 89, 306
Aquaculture, 10, 12
Arabian Sea, 184, 186
Aragonite, 109, 114, 128, 159
Archidoris, 393
Archipelagic apron, 61
Archipelagic plain, 66
Arctic basin, sediment, 36
  topography, 36, 57, 58
Arctic communities, 42
Arctic Ocean, antipodal to
      Antarctica, 29
  and barriers to water exchange,
      66, 189, 231
  distribution of species in, 411
  and fresh water input, 30, 33, 35,
      36
  as part of the Atlantic, 34, 35
  as a polar region, 207
  and sea ice, 35–36, 201, 202, 204,
      213, 299
  and water budget, 33
  and water masses, 189, 231
Argon, 100, 160
Argon-40, 136
Argyropelacus gigas, 438
Arrow-worms, 425, 437, 470
Arthropods, 427–430
Artificial Shoals, Hawaii, 493
Ascension Island, 56, 500
Aseismic ridges, 40
Asia, as a barrier to water
      movements, 215
  as boundary of Indian Ocean, 36
  and coastal mountain ranges, 32
  and intersection with mid-oceanic
      ridges, 57
  and marginal ocean basins, 46
  and Monsoon winds, 218, 340
  and mountains formed by crustal
      deformation, 89
  as a source of sediment, 116, 117
Asterias rubens, 430
Asthenosphere, 85
Atlantic basin, dimensions, 34
  evolution of, 34, 93–96
  as a gulf, 31
  and sediment deposits, 34, 114, 115,
      123, 124, 127, 128, 131, 132
  topography of, 53–60, 61, 62, 66,
      124, 187
  and trenches, 54, 68
Atlantic Coast of the U. S., 4, 329
  and beaches, 368, 370
  and dunes, 374
  estuaries and lagoons, 297, 299
  and fall line, 306
  and Labrador Current, 217
  and oyster beds, 461
  and sea ice, 201
  and sediment transport by rivers,
      121

Atlantic Coast of the U.S. (*cont.*)
and shoreline, 342
and tides, 272
Atlantic Ocean, (*see also* North
Atlantic Ocean; South Atlantic
Ocean)
and barriers to water exchange,
66, 231
boundaries of, 26, 30, 34, 36–37
and convergences, 207
currents, 215, 219–220, 233–235
as a defense barrier, 6
and fresh water input, 33, 35
icebergs in, 206
islands in, 34–35
and isotherms, 179
and oxygen distribution, 402
salinity of, 33, 35, 184, 192, 194,
220, 231, 233–235
and sediment discharged by rivers,
116
and shorelines parallel, 34
surface and drainage areas and
average depth, 32
temperature of, 33, 35, 192, 193,
219, 233–235
and tides, 279–282
and water budget, 33
and water masses, 35, 195, 199, 231,
233–235
and zooplankton distribution, 434
Atlantis II, 5
Atmosphere, composition of, 26, 160,
176
and heat budget, 175, 177, 179
mass and abundance of, 27
origin of, 87, 97–100, 102
primitive, 99–100
and water budget, 182–184, 186,
188
Atmospheric circulation, 177–178, 184
and heat transport, 176, 177, 188,
191
seasonally variable, 340
Atolls, (*see also* Coral reefs)
distribution of, 37, 66, 71
formation and structure of, 72–74
Pacific Ocean, 18, 34, 72, 91
and submarine volcanoes, 60, 62,
91
Attached plants (see Benthic algae;
Epiphytes; Seaweeds)
Australia, 30, 34, 37, 72, 125, 186, 437,
551
as a barrier to water movements,
95
and river discharge of suspended
sediment, 116
Autotrophic nutrition, 101, 457, 458
Azores, 35, 56

Backradiation, 171, 175, 176, 177, 179,
191, 340
Bacteria, 12, 394
anaerobic, 160, 161, 306, 458, 491
autotrophic, 399–400, 458, 481
benthic, 305, 306
bioluminescence in, 398, 438
and carbon-dioxide cycle, 403
in the coastal ocean, 490

Bacteria (*cont.*)
as decomposers, 113, 160, 161, 393,
400, 401, 457, 458, 479, 490
in the deep ocean, 466–467
epiphytic, 454
in estuaries, 306, 488
in food chains, 400, 401, 423, 456,
457–458, 466–467, 482
fossil, 99
and fouling, 405, 461
and nitrogen cycle, 160, 161, 400,
401, 458
and organic detritus, 405, 406–407
and pressure, 396, 467
primitive, 102
and protein synthesis, 400, 458
in sediments, 451, 456, 457,
458–459, 490
sewage-associated, 349
spore formation in, 100
and vitamins, 305, 481
Baffin Bay, Canada, 35
Baffin Bay, Texas, 312, 313
Bahama Banks, 125, 159, 228
Balanus, 422
Baleen whales, 482, 499
Baltic Sea, 115, 203, 475
Barbados, 69
Barite, 109
Barnacles, 75, 395, 427
as fouling organisms, 405, 461
metamorphosis of, 422
in oyster beds, 460
zonation in, 447
Barrier beaches, formation of, 299–
300, 370, 371–372
U. S., 309, 342, 359, 387
Barrier islands, 362, 370
U. S., 310, 329, 342, 368, 377, 378
Barrier reefs, 72–73
Barrier spits, 371
Basalt, 39, 58, 82, 83, 92
Basking shark, 493
Bathymetry, 80
Bathythermograph, 169
Baymouth bars, 362, 371
Bay of Bengal, 337, 339, 363
Bay of Fundy, 272, 282, 305, 333, 475
Beach, profile of, 357
Beach cusps, 380
Beach drift (see Littoral drift)
Beach-dune complex, 372
Beaches, coral reef, 75
elevated, 359
erosion of, 369–370, 373, 374, 384
general features of, 373, 379
preservation of, 18, 383, 384
and sedimentation, 121, 360, 362,
369–372, 374–375, 377–378
study of, 13
submerged, 137, 341
and tides, 358, 360
U. S., 363, 370, 376, 377–378, 382,
384
variability of, 13, 360, 369
Virgin Islands, 369
Beach face, 358, 375
slope of, 356, 370
Beach grass, 373
Beach processes, 374–381

Beach sand, 111, 112, 369, 370, 373,
405, (*see also* Sand, on beaches)
Beach slope (*see* Beach face, slope of)
Beagle (ship), 72
Beaufort numbers, 428
Bed load, 120
Belgium, 291
Ben Franklin (submersible), 236
Benthic algae, 488 (*see also* Seaweeds)
Benthic colonies, settlement of, 445,
461
Benthic life, strategies of, 444
Benthic plants (*see* Seaweeds)
Benthos, collection of, 11, 13
estuarine, 305, 306
planktonic larvae of (*see* Planktonic
larvae, of benthic organisms)
Bering Sill, 30, 34, 231
Bering Strait, 34, 36
Berm, 374
Berm crest, 373
Bermuda, 9, 35, 65, 369, 476
Beroë, 425
Beta radiation, 100, 474
Bicarbonate ion, 157, 159, 161, 402
Biogenous sediment (*see* Sediment,
biogenous constituents)
Biological community (*see*
Community)
Biological concentration of elements
(*see* Concentration of elements)
Biological history of seawater, 476,
480
Biological oceanography, 11–13
Bioluminescence, 397–398, 438–439,
495, 496, 497, 555
Biomass, (*see also* Standing crop)
of deep-ocean benthos, 466
defined, 399
and energy flow, 456–457
and latitude, 412
and vertical distribution of, 477–478
Biotin, 481
Bird-foot delta, 368
Biscayne Bay, Florida, 253, 383
Bivalves, and cancer research, 485
cultivation of, 483
in the deep ocean, 467
deposit-feeding, 455
identified, 426
as meiobenthos, 456
metamorphosis of, 427
respiration of, 452
Black bottle experiment, 473
Black Sea, 46, 47, 95, 127, 402
Blake Plateau, 13, 115, 132
Bloom, 479 (*see also* Spring bloom)
causes of, 421
English channel, 401, 480
and patchiness of zooplankton, 437
Bluefin tuna, 554
Bluefish, 493, 557
Blue whale, 499
Body fluids, water content of, 407–408
Bolinopsis, 425
Bones, in sediment, 109, 114, 115, 128
Bonin Trench, 53, 68

Bony fishes, 407, 431

Bore (tidal bore), 305

Boron, 163

Boston, Mass., 5

Bottom currents (*see* Near-bottom currents)

Bottom topography, and barrier bars, 371–372
  bathymetric chart of, 53
  and distribution of species, 411
  and water movements, 66, 220, 231, 233, 279, 298, 304
  and waves, 255, 257, 258

Bottom trawl, 484

Boundary currents, 46, 218–220, 235, 332, 342

Bowditch, Nathaniel, 17, 20

Brahmaputra River, 30, 37, 116, 117, 118, 361, 363

Brazos River, Texas, 328

Breakers, 258, 357
  and coral reefs, 75, 77
  and energy released by, 247, 254, 255, 257
  and filter-feeding in sabellarian worms, 461
  formation of, 239, 243, 254–258, 259, 260, 377
  and longshore bars, 372, 374
  and water motion, 243, 256
  and waves, 240, 287

Breaker zone, 376

Brine, 145, 156, 201, 203, 204

Bristle-worm, 426

British Columbia, 298, 402

Brittle-stars, 430, 466, 467

Bromine, 165

Brown mud, 112, 129, 133

Bubbles, 100, 162, 164, 199, 406

Budding, of reef-building organisms, 462, 465

Buoys, 7, 212, 213, 284, 295

Burrow, of fiddler crab, 453

Burrows, in eel-grass communities, 454, 455
  in mud flats, 454–455

B-vitamins, 10, 421, 481

By-products of metabolism, and plankton bloom, 420–421, 475, 481

Cadmium, 403, 404

Cadophore, 432

Calanoid copepods, 412, 429

Calanus, 403, 492
  finmarchicus, 429

Calcareous algae, 65, 76, 125, 464

Calcareous material, and encrustation of coral reefs by coralline algae, 463

Calcareous plates, in sea lilies, 466

Calcareous sediment (*see* Sediment, calcareous constituents)

Calcareous shells, and carbon dioxide, 161
  of coccolithophores, 419
  of foraminiferans, 423

Calcareous shells (*cont.*)
  in sediment, 114, 128, 464

Calcareous skeleton, of corals, 465

Calcareous tubes, of worms, 460

Calcite, 109, 114

Calcium, in crustacean shells, 428

Calcium carbonate, capping extinct volcanos, 35, 74
  dissolution in seawater, 128
  precipitation in seawater, 157, 159, 399, 458, 462, 476
  and reef-building corals, 462–463
  in seawater, 118
  as skeletal material, 113, 399

Calcium ion, 157, 159, 408

Calcium phosphate, in bones, 114
  in sediments, 458

California beaches, 370, 448

California Current, 214, 218, 219, 221, 342, 350

California gray whale, 499

California-Oregon coast, and upwelling, 333–335

Callicuthis reversa, 495

Calocalanus, 429

Calorie, defined, 147

Cambrian period, 102

Camouflage, among reef-dwellers, 465

Campeche Bank, 62

Canada, 360

Canary Current, 214, 218

Canon Fjord, Ellesmere Island, 299

Cape Charles, 306, 307

Cape Cod, Mass., 9, 61, 172, 217, 333, 343, 488

Cape of Good Hope, 34, 37

Cape Hatteras, North Carolina, 219, 220, 333, 341, 343, 515

Cape Henry, 288, 306, 307

Cape Horn, 34

Cape Kennedy, Florida, 374

Cape May, New Jersey, 343, 377

Capillary waves, 247

Carbohydrate, as detritus, 400, 404, 407, 458
  and energy for protein synthesis, 400, 458
  non-biological synthesis, 100
  in organic matter, 100
  as potential energy, 400
  synthesized by plants, 102, 396

Carbon, organic (*see* Organic carbon production)

Carbonate ion, 157, 161, 402

Carbonates, in coral reefs, 69, 74, 75, 76
  in sediments (*see* Sediment, carbonate)
  solubility in seawater, 161

Carbon dioxide, in the atmosphere, 26, 100, 135, 176, 177
  and carbon-14 dating of sediments, 135
  and corals, 463
  and photosynthesis, 396, 402–403, 473
  and respiration, 161, 400, 403, 458, 500
  in seawater, 128, 159, 160
  solubility, 161

Carbon dioxide cycle, 161, 402–403

Carbon fixation, 473, 476–477

Carbon-14, and dating of sediment deposits, 135, 136
  and deep-water residence time, 233
  and primary productivity, 473–474

Carbonic acid, 402

Carbon monoxide, 100

Cardisoma, 454

Caribbean Sea, 30, 46, 62, 221, 272, 282, 342, 402

Carotenoid pigments, 397

Cartilagenous fishes, 431, 493

Cascadia Channel, 62

Caspian Sea, 87, 95

Celestite, 109

Cellulose, fermentation in deep ocean, 459

Centropages, 429

Cephalopods, 426–427, 438

Ceratius holboelli, 496

Ceylon, 31, 37

Chaetognatha, 425, 434

Chaetomorpha linum, 408

Chaetopterus, 455

Challenger Deep, 67

Challenger Deep-Sea Expedition, 3, 9, 21, 113, 114, 123

Channels, deep-sea, 62
  in deltas, 363–368
  in estuaries, 17, 314, 349, 383–384
  in marshes, 368, 382, 385, 386

Chemical energy, and photosynthesis, 393, 482

Chemical oceanography, 9, 11

Chemo-autotrophic bacteria, 458

Chesapeake Bay, 297, 298, 306–309, 341, 360
  cladocerans in, 430
  oysters in, 459, 461
  phytoplankton in, 475
  salinity distribution, 308, 309
  striped bass in, 489
  temperature distribution, 308, 309
  and tidal currents, 287–289, 307–308
  and tidal range, 304

Chile, 360

China, 34

Chitin, in blue-green algae, 419
  decomposition of, 459
  in radiolarians, 423

Chitinous bristles, of polychaetes, 426

Chitinous spikes, of chaetognaths, 425

Chloride ion, 144, 155, 156, 159

Chlorinated hydrocarbons, 491

Chlorine, in industrial discharges, 411

Chlorinity, 151–153

Chlorite, 109, 130, 132

Chlorophyll, 101, 396, 397, 400, 478

Chordates, 430–431

Chromis verator, 493

Chromium, 403

Chromium-51, 354

Chronometer, 25, 516

Chrysaora quinquecirrha, 424, 450

Cilia, in ctenophores, 425
  in oysters, 460
  in pteropods, 427
  in tintinnids, 424
  in trochophore larva, 426

Ciliate protozoans, 424, 456–457, 481
Cirrotheuthis, 495
Cladocerans, 430
Clams, 120, 268, 426
  anti-coagulants derived from, 485
  commercial value of, 488
  in estuaries, 408, 488
  sediment reworked by, 455
  variation in abundance of, 445, 446
Classification of marine organisms, 12
Clay, red (*see* Red clay)
Clay minerals, 109, 113, 120, 130–132
Clays, 113, 124
  and adsorption of surface active
    materials, 101, 405
  classified by size, 111
  and deposit-feeding, 452
  distribution and sources of, 112
  erosion of, 119, 370, 378
  settling velocity of, 119, 122
  transport of, 119, 122, 298
Cleaning symbiosis, 464–465
Climate, affected by the ocean, 6, 195,
    317, 334, 340
Climatic regions, 206–208
Climax community, 393, 461
Clio, 427
Clouds, 26, 34, 36, 176–177, 178, 184,
    185, 202
Clownfish, 465
Clyde Sea, 127
Coacervates, 101
Coal, 137
Coast, defined, 358
Coast Range, State of Washington,
    340
Coastal circulation, and boundary
    currents, 218
  Northeast Pacific coastal ocean,
    349–352, 353
  and recycling of nutrients, 47, 433,
    488
  and upwelling (*see* Upwelling)
  and waste retention, 349
  and winds, 216–217, 328, 331–333,
    346, 350–352
Coastal currents, 8, 216–217, 221,
    331–335
  New York Bight, 344, 348
  Northeast Pacific coastal ocean, 350
Coastal engineering, 8, 16–17, 18, 34,
    386
Coastal marshes (*see* Marshes)
Coastal ocean, and cladocerans in, 430
  as a climatic buffer, 340
  and conditions affecting plankton
    populations, 433, 476
  diversity of habitat in, 445
  general features of, 46–48, 328–
    329, 391
  management of, 349, 484
  productivity of, 305, 367, 445, 474–
    476, 477, 482, 483, 490
  and salinity distributions, 184, 186,
    219, 329–331, 335, 342 (*see also*
    Salinity distribution, New York
    Bight; Salinity distribution,
    Northeast Pacific coastal
    ocean)
  and temperature distributions, 219,

Coastal ocean (*cont.*)
    329–331, 332, 333–334, 339,
    342 (*see also* Temperature
    distribution, New York Bight;
    Temperature distribution,
    Northeast Pacific coastal
    ocean)
  and tides, 268, 331, 337
  turbidity of, 172, 310, 337, 338, 339,
    404
  upwelling (*see* Upwelling)
  vitamins in, 481
  wastes in, 11, 490–491
Coastal ocean regions of the U. S.,
    341, 342
Coastal plain, 38, 42, 62, 306, 341, 342
Coastal plain estuaries, 297, 306, 319
Coastal zooplankton, 433, 436
Coasts (coastlines), classification of,
    360–362
  of the world, 361
Cobalt, 115
Cobbles, 111, 370
Coccolithophoridae, 419, 420, 430,
    476
Coccoliths, 109, 112, 114, 134, 135,
    420
Coccosphaera, 420
Cockle, 452
Cod, 432, 482, 488, 553
Coelenterates, 422, 424–425, 434,
    460, 466, 468
Co-enzymes, 101
Cold-blooded organisms, 410–411
Collapsing breaker, 256, 258
Colloidal particles, 115, 405
Colonization (*see* Benthic colonies,
    settlement of)
Color, of marine organisms, 437–438,
    465, 466
  of ocean water, 172–173, 220
Colorado River, 117, 118, 123, 297,
    298, 305
Columbia River, 30, 62, 301, 352, 353,
    361
  effluent, 350–351, 352, 353
  sediment discharge, 121–122, 363,
    370
Columbia River estuary, 258, 350, 352
Comb-jellies, 425
Community, as an ecosystem, 392,
    460
  evolution of, 413, 461
  stability of, 445–446
Compass, magnetic, 91, 211, 521
Compensation depth, 473, 474
Competition, 413
  in coral reef communities, 75, 464
  and harvesting the ocean, 472
  in intertidal areas, 447
  and survival of oyster larvae, 460
  in trenches, 465
Composition of marine animals, 403
Condensation nuclei, 121, 163, 453
Congo River, 30, 35, 44, 116, 124
Connecticut River, 298, 314, 319
Conservative elements, 153
Continental air mass, 340
Continental blocks, antipodal to
    ocean basins, 28, 29

Continental blocks (*cont.*)
  boundaries of, 39, 41, 52, 66
  flooding of, 37, 38, 39, 46
  fracturing of, 95–96
  and marginal basins incorporated,
    45–46, 86
  movements of, 14, 15, 57, 87, 92,
    94–97, 98
  origin of, 87, 89–90, 97
  structure of, 39, 81, 87
  volume of, 28–29
Continental climate, 340
Continental crust, structure and
    properties of, 27, 39, 52, 81–
    82, 83
  and transitional area at continental
    margin, 86–87
  vertical movements of, 359–360
Continental drift (*see* Continental
    blocks, movements)
Continental ice sheets (*see* Glaciers)
Continental margin, 39, 40, 41–45, 84,
    132, 134, 220, 221
  and coastal ocean, 46–47
  and crust in transition at, 86–87
Continental rise, 39, 41, 42, 45, 54
  and sediment deposits, 44, 45, 130,
    132
Continental shelf, 41, 42, 47, 124, 346,
    351, 352
  Arctic Ocean, 35, 41, 125
  Atlantic Ocean, 34, 54, 342
  average water depth over, 328
  and coastal oceanic regions of the
    U. S., 342
  and coastal plain, 42, 43, 297
  erosion of, 42, 120, 341
  general features of, 41–43
  Indian Ocean, 37, 54
  Pacific Ocean, 32, 41, 54, 342
  remains of land organisms on, 38
  and sediment deposits, 41, 45, 121,
    122, 137, 362, 370
    biogenous, 125, 133
    clay minerals, 130
    glacial-marine, 125
    man-made, 116
    rate of accumulation, 127
    relict, 43, 121
    river-transported, 43, 132, 133
  and waste disposal, 349
Continental shelf break, 41, 42, 43,
    122, 328, 360
Continental slope, 39, 41–44, 47, 54
  and sediment deposits, 13, 43, 45,
    127, 134
Contours and contour maps, how to
    use, 514
Convergences, 207, 220, 221, 228, 337,
    339, 375, 406, 503
Coordinates, how to use, 514–515
Copepods, 395, 428, 429, 430, 492,
    496, 499
  composition of, 403
  as herbivores, 481
  as meiobenthos, 456
  polar species found in both

Copepods (cont.)
    hemispheres, 434
    in Sargassum community, 503
    and vertical migration of, 435
Copper, in industrial discharges, 411
    in manganese nodules, 115
Coralline algae, 69, 75, 463
Coral reefs, and continental margins,
        43, 44
    distribution of, 71, 125, 126, 133,
        341, 342, 361
    as ecosystems, 462–465
    Pacific Ocean, 18, 34, 72, 463, 464
    structure and formation of, 69–76,
        462–465
Corals, 448, 467, 481, 493
Corbina, 554
Core (of the earth), 26, 27, 91, 98
Core, sediment, 13, 14, 15, 82, 105
Core technique (water mass analysis),
        197–198
Coriolis effect, 223–225, 228
    and Chesapeake Bay, 308
    and tides, 278, 281
    and waves, 247, 263
Cormorant, 491
Corpus Christi, Texas, 312
Correlation of sedimentary deposits,
        134–135
Cosmic rays, 135
Cosmic spherules, 116
Cosmogenous sediment (see Sediment,
        cosmogenous constituents)
Crabs, 453, 454, 455, 554
    in eel-grass communities, 455
    in estuaries, 488
    in oyster beds, 460
    as predators, 444
    on rocky beaches, 448–449
    in Sargassum community, 503
Crassostria virginica, 459–460
Craten pilata, 450
Crayfish, 75, 450, 451, 493
Crest, of the tide, 279, 283
    of a wave (see Wave crest)
Crinoids, 466
Critical depth, 474, 475
Croaker, 201, 433, 488, 554
Crown of Thorns, 464
Crust (of the earth), (see also
        Continental crust; Oceanic
        crust)
    composition of, 26
    formation and destruction of, 14,
        87–88, 92
    origin of, 97–98
    structure and properties of, 50,
        81–86
    study of, 13, 81, 103
Crustaceans, 427–430
    and chlorinated hydrocarbons, 491
    deep-sea species, 434, 466
    in eel-grass community, 454, 455
    in estuaries, 488–489
    and evolution toward terrestrial
        habitat, 454

Crustaceans (cont.)
    larvae as open-ocean plankton, 435,
        481
    and light in the ocean, 438, 439
    in oyster beds, 461
    in tide pools, 448
Crustal deformation, 45, 86, 87, 89, 95
Crustal plates, movement of, 84, 87–
        91, 93
Crustal reworking, 98
Ctenophora, 425, 492
Cumaceans, 423
Current gyre, 215–216
Current meters, 7, 9, 212–213, 284,
        304
Current rose, 286–287
Current speed, and sediment
        transport, 113
Current table, 8
Currents, (see also Boundary currents;
        Coastal currents; Deep-ocean
        circulation; Geostrophic
        currents; Longshore currents;
        Near-bottom currents; Surface
        currents; Tidal currents and
        names of specific currents,
        e.g., California Current)
    charting of, 20, 212, 230
    deflected by land masses, 46, 179,
        215, 218, 219, 235
    and distribution of species, 411,
        418, 436, 455, 489
    and heat transport, 176–177, 179,
        188, 212, 218, 221
    Laguna Madre, 313–314
    and reef structures, 462, 465
    and ripple marks, 381
    and salinity distribution, 186
    and sediment transport, 113, 119–
        120
    and storms, 122
    study of, 7, 9, 212–213, 215, 236,
        284
    wave-generated, 374, 380
    wind-driven (see Winds, and
        surface currents)
Cusk-eel, 489, 557
Cuttlefish, 396
Cyanocobalamin, 481
Cyclothone microdon, 497
    pallida, 497
Cypris, 420, 492

Dabob Bay, Washington, 317, 319
Daily inequalities, 272, 277, 285, 286
Daily tides, 270, 271, 272, 278, 279,
        282
Darwin, Charles, 72–74, 413
Darwin Rise, 65, 71, 72
Dating of sedimentary deposits, 134–
        136
Davidson Current, 217, 227, 350
Davis Strait, 35
DDT, 11, 491
Dead reckoning, 20, 211
Death Valley, California, 146
Decomposer, defined, 393
Deep-drilling operations, 15, 62, 74,
        81, 82, 103

Deep-ocean benthos, 465–467
Deep-ocean circulation, 66, 120, 214,
        231–235, 402
Deep-scattering layer, 436
Deep-sea mud, 112, 114, 126, 127
Deep-water populations, characteris-
        tics of, 401, 434–435
    distribution of, 411, 434
Deep-water species, planktonic larvae
        of, 435, 466
    and use of in identifying upwelled
        water, 437
Deep-water waves, 242, 243, 244, 250,
        254–255
Deep zone, 191, 192
Delaware Bay, 333, 343, 360
Delaware River, 360
Deltas, 62, 121, 315, 360, 361, 362–
        368
Denmark, and tidal currents, 291
Density, and currents, 222, 227, 231–
        232, 304
    defined, 149
    as an environmental factor, 394–
        396
    and salinity, 149–150, 155, 188,
        190–192, 195, 215, 221, 222–
        223, 230, 231, 300–302, 306
    and sea surface topography, 230
    and state of water molecules, 145–
        146
    and temperature, 149–150, 188,
        190–192, 195–196, 208, 215,
        221, 222–223, 230, 231
    of water and anomalous density
        maximum, 149–150, 155
    of water compared with a "normal"
        liquid, 143
Density stratification, 127, 149–150,
        188–192, 198, 231
    in estuaries, 300–304
    and fresh-water discharge, 305,
        335, 475
    and internal waves, 261, 262
    Long Island Sound, 319–320
    New York Bight, 344–347
    in warm seas, 476
Denver, Colorado, 340
Deposit-feeding, 444
    of bivalves, 455
    in the deep ocean, 466–467
    of limpets, 447
    and sediment character, 451, 452,
        455, 458–459, 466
    selective, 451, 452
    unselective, 451, 452
Depth determination, 13, 38–39
Depth of no motion, 230
Desalinization, 165
Deserts, 206
Detergents, 164
Detrital food chain, 393, 457, 458, 490
Detritus, 393, 400, 411 (see also
        Organic detritus)
Dexiobranchaea, 427
Diamond Head, Hawaii, 362
Diamonds, 137
Diatomaceous mud, 112, 114, 126,
        127, 129, 133, 207

Diatoms, 419, 420, 492
  benthic, 457, 461
  composition of, 398
  division in, 419, 421
  in the English Channel, 401
  epiphytic, 454, 455
  and euphausids, 430
  and nutrient enrichment of
    estuaries, 490
  in polar regions, 434, 479
  shells in sediment, 112, 114, 127,
    128, 479
  sinking of, 420
  size of, 418
  and spring bloom, 475
  and tropical region, 476
  and upwelling, 420–421
Dikes, 8, 335, 382, 386
Dinoflagellates, 419, 420, 421, 430,
  470, 475
Diopatra, 455
Discoasters, 109, 134
Dispersal, of larvae, 393, 421
Dispersion, of waves, 250, 251
Disphotic zone, 397
Dissolved gases, 159–161, 162, 199,
  221, 233, 392, 393
Dissolved organic matter, 400, 401,
  404–407
  and adsorption by fine-grained
    particles, 451, 452
  as chemical energy lost from food
    chains, 482
  and light in the ocean, 172, 404
  and use by bacteria, 399–400, 458
  and use by plants, 399
Distephanus, 420
Distributaries (of a delta), 363–366,
  368
Ditylum, 420
Diurnal tides (*see* Daily tides)
Divergences, 127, 221, 228, 350
  and productivity, 401, 477
Diversity of species (*see* Species,
  number of)
Docks, 16, 17, 240
Dogfish, 484, 559
Doldrums, 207, 215
Doliolids, 431, 432, 481
Doliolum, 432
Dolphin, 201, 410, 500
Downwelling (sinking), 221, 227, 346,
  406, 421
Dredging, of channels, boat basins,
    etc., 17, 312, 314, 349, 362,
    367, 381, 383–386
  deep-ocean, 11, 14, 114
  New York Harbor, 8, 349, 393
  of tidal inlets, 359, 360, 378, 384
Drift bottles, 213, 344
Drift currents, 222, 225–226, 228
Drilling, deep-ocean, 14, 115
Drowned river valleys, 137, 296, 298,
    306, 308, 309, 310, 312 (*see also*
    Estuaries, submerged)
Drugs, derived from marine
    organisms, 485
Drumfish, 201
Ducks, 491

Dunes, formation of, 373–374
  Netherlands, 335, 373
  U. S., 121, 370–374
  and volcanic islands, 65, 75
Dutch Waddens, 382
Dynamical theory of the tide, 278–283
Dynamic topography, 230

Earth, age of, 97
  emission spectrum of, 171
  and interaction with moon in gen-
    eration of tides, 273–279
  photomosaic of, 182
  structure of, 26, 97
  temperature of, 146
  useful data about, 27
Earthquake waves, 14, 58, 80, 83–84,
  86
Earthquakes, 61, 90
  at crustal plate margins, 57, 87, 88,
    89
  deep-focus, 67, 83, 88–89
  distribution of, 57, 83, 85
  and seismic sea waves, 245, 260–261
  shallow-focus, 56, 58, 66, 88–89
  and turbidity currents, 123
East African rift valley, 55, 56, 57–58,
  96
Eastern boundary currents, 218–219,
    221, 333, 350
Eastern South Pacific Water, 195
East Pacific Rise, 53–54, 57, 59, 84,
    128
East River, New York, 319, 320
Ebb current, defined, 284
Echinocardium, 430
Echinocyamus, 430
Echinoderms, 430, 493
Echiuroid worms, 454
Echo sounding, 13, 14, 18, 39, 44, 62,
    63, 107, 142, 200
Ecological efficiency, 482
Ecological niche, 392, 393, 413, 460,
    461, 464
Ecological succession, 393, 461
Ecology, marine, 11
Ecosystem, 392–394, 404
  and alteration of habitats, 393, 491
  and reef communities, 459, 461
Ecuador, 31
Eddystone Lighthouse, 327
Eel, 495, 501
Eel-grass communities, 454, 455, 489,
    501
Ekman, Walfrid, 226
Ekman current meter, 213
Ekman spiral, 225, 226
Ekman transport, 226–227
Electrolytes, 156, 157
Electromagnetic spectrum, 177
Electron cloud (on water molecule),
    141, 143
Encrusting coral, 462
Endangered species, 491
Energy flow, in food web, 392–394,
    472
  and meiobenthos, 456–457
Energy transfer, and evolution of
    organisms, 101

England, 18, 46, 121
English Channel, and amphidrome
    system, 281
  nutrients in, 401
  plankton in, 479, 480
  and tidal currents, 285, 291
English coast, and storm surges, 335
  struck by long-period swell, 251
  and tidal currents, 291
  and tides, 268
Eniwetok Atoll, 74
Enzymes, 102
Ephyrae, 450
Epicenter, of an earthquake, 58
Epifauna, 445, 457
Epiphyte, 454, 455, 488
Equator, and Coriolis effect at, 223–
    224
  and heat budget, 178–179, 221
  and latitude determination of, 515
  and position relative to the sun,
    277
  relative amount of land and water
    at, 31
Equatorial Countercurrent, 208, 214,
    215, 217
Equatorial currents, 207, 215, 218,
    219, 235 (*see also* North Equa-
    torial Current; South Equa-
    torial Current)
Equatorial divergence, 401, 477
Equatorial region, 206, 207–208
Equilibrium concentration, 160
Equilibrium theory of the tide, 274–
    277, 278
Erosion, of coastal features, 255, 299,
    360–362, 363, 369–370, 374,
    378, 381, 384
  of coral reefs, 74–75
  and current speed, 113, 119
  and ocean bottom topography, 60
  and river-borne nutrients, 490
  and sedimentation, 98, 117, 119–
    120, 125, 129, 255, 361–362,
    363, 370, 378, 381–382, 384
  wave-induced (*see* Waves, and
    coastal erosion)
  worldwide distribution of, 118
Estuaries, defined, 296
  formation of, 296–298, 359
  management of, 484
  modification of, 312, 314–315, 383–
    384, 386
  North America, 297
  as nursery areas, 488–489
  oxygen in, 402
  productivity of, 305, 485, 487–488
  salinity distributions, 301, 302, 303,
    330, 384
  and sediment in, 8, 43, 120–121,
    297, 338, 362, 363, 366, 368,
    382, 383–384, 490
  submerged, 43, 297 (*see also*
    Drowned river valleys)
  and tides, 284, 291, 304–305, 336,
    337

Estuarine circulation, 300–305
  Columbia River, 352
  Laguna Madre, 314
  Long Island Sound, 319, 321, 322
  and migration of larvae, 488
  and nutrients, 305, 477, 488
  Puget Sound, 317
  and sediment transport, 120
  Strait of Juan de Fuca, 352
Estuarine conditions, and marine life,
    407–410
Estuarine-like circulation, coastal
    ocean, 344
  New York Bight, 344
  North Pacific Ocean, 350
Estuarine systems, study of, 387
Euchaeia, 429
Euphausia superba, life cycle of, 429
Euphausids, 428, 430, 435, 436, 484,
    498, 499
Euphotic zone (*see* Photic zone)
Eurasia, and Africa, 30, 95, 96
  and Americas, 34
  as antipodal to Pacific Ocean, 29
  as boundary of Arctic Ocean, 35
  climate of, 340
  and Indian subcontinent, 87
  and Mid-Atlantic Ridge, 57
  and Pangaea, 93, 95
Europe, 4, 26, 46, 97, 116, 178, 179,
    182
Eurytemora, 429
Evadne, 430, 492
Evaporation, and density, 221
  distribution of, 33, 182–184, 186,
    188, 207, 220, 314, 329–331
  and heat loss, 146–148, 172, 179,
    186, 188, 191
  in tide-pools, 448
Evolution, 14, 87, 99–100, 413, 414
  of bony fishes, 407
  of crustaceans toward terrestrial
    habitat, 454
Excretory mechanisms, 408
Extinction, of species, 413

Faeroe Islands, 66
Fall line, 306
Fats, in bacteria, 458
  and conservation of non-essential
    materials, 404, 491
  in copepods, 430
  in organic matter, 100
  in phytoplankton, 419
  synthesis of, 396
Faulting, defined, 43
Faults, at boundaries of crustal
    plates, 87, 89
  in continental blocks, 89
  and crustal movements, 61, 83
  and fracture zones, 54, 90
  and oceanic rises, 54–57, 89
  as sediment traps, 43, 59, 67
  in trenches, 67
  and volcanic provinces, 65
Feather-duster worms (*see* Sabellarian
    worms)

Fecal matter, of deposit feeders and
    resistance to decomposition,
    458–459
  and nutrient flow in food webs, 400
  as organic detritus, 404
Fecal pellets, and sedimentation, 120,
    122
  and turbidity reduced by, 452–453
  of undigested plant matter voided
    by zooplankters, 479
Feldspar, 109, 113
Fermentation, 396, 459
Fertilization of the ocean, 485
Fertilizers, discharged to coastal
    areas, 315, 490
Fetch, 248, 249
  North Atlantic, 253
  North Sea, 335
  Puget Sound, 319
  Strait of Georgia, 363
Fiddler-crab, 453
Fiji Islands, 33
Filter-feeding, 405, 407, 444
  of barnacle, 447
  of basking shark, 493
  of cockle, 452
  of copepod, 430
  of coral, 463
  in the deep ocean, 466
  of estuarine fishes, 489
  of euphausids, 428
  of *Oikopleura*, 430, 431
  in oyster communities, 460
  of sabellarian worms, 461
  and sedimentation, 120, 122, 455–
    456
  and sediment size, 451–452
  of *Urechis*, 454–455
  of whales, 430, 499
  of zooplankters, 422
Finback whale, 499
Fish, as protein, 5, 482, 485
  world catch, 5, 484, 485
Fish behavior, study of, 484
Fish eggs, as plankton, 431, 433
Fisheries, management of, 484
  North Atlantic Ocean, 488
  and prediction of North Sea
    productivity, 437
  and sustainable yield, 483
  temperate region, 207
  Texas, 315
  and upwelling, 334, 412, 477
Fishes, attracted to light, 439
  bioluminescence in, 438, 439
  and colonial phytoplankton eaten
    by, 476
  composition of, 403
  in coral reef communities, 75, 463–
    465
  and DDT, 491
  on deep-ocean floor after death,
    467
  deep-ocean species, 396, 466, 467
  and deep-scattering layer, 436
  in estuaries, 296, 305, 485, 487–489,
    501, 555–558
  flotation devices in, 396
  as high trophic level feeders, 482
  and marshes, 467, 488

Fishes (*cont.*)
  mid-level, planktonic larvae of,
    435, 496, 497
  migratory, 411, 472, 488–489, 493
  as nekton, 472
  and oyster beds, 461
  protective coating on, 408
  and skeletal debris in sediments,
    111, 112
  in subpolar regions, 412
  West Coast, 558–559
Fishing, methods of, 494
Fishing-grounds (*see* Fisheries)
Fishing reefs, 493
Fish kills, 314, 421
Fish larvae, in estuaries, 488
  as herbivores, 481
  as plankton, 431, 433, 481, 492
  in Sargassum weed, 503
Fish meal, 551, 552
Fjords, 42, 134, 296, 298–299, 303,
    306, 342, 360, 402
Flagellates, 421, 438
Flatworms, 460
FLIP, 6, 7
Flocculation, 120
Flod at London Brigge, 269
Floe ice, 204, 299
Flood current, defined, 284
Florida, 125, 219
Florida Bay, 383
Florida Current, 214, 219, 228
Flotation, adaptations for, 395–396,
    423
Flounder, 482, 488, 553
Fluke, 488, 553
Flying fishes, 554
Food chains, 405
  and bacteria in, 400
  and coastal upwelling, 482
  and conservation of trace elements,
    404
  in the deep ocean, 467
  and food web, 393
  and meiobenthos in, 456, 458
  and nutrient supply, 207, 490
  in the open ocean, 482
  and planktonic arthropods in, 427
  and skeletal materials not con-
    served in, 491
  and transfer of chemical energy,
    393, 482
  and tunicates as dead end in, 431
  and vitamins in, 481
Food requirements, and life cycle in
    animals, 434, 437, 479, 480, 482
Food resources, and the coastal
    ocean, 47
  and ecological efficiency, 481–482
  increasing the yield of, 484
Food webs, 11, 393, 397, 412, 461
  estuarine, 485, 486–488
  open ocean, 482
  and stable compounds conserved
    in, 491
Foraminifera, and calcium carbonate
    deposition, 60, 110, 404
  and correlation of sedimentary
    deposits, 135
  described, 423

Foraminifera (*cont.*)
  as herbivores, 481
  as meiobenthos, 456
  in shallow waters, 125
  shells in sediment, 109, 112, 114,
    135, 369
Foraminiferal mud, 112, 114, 126, 127,
    128, 129, 133
Forced waves, 246, 279
Foreshore, 370, 373, 374, 379
Fossil assemblages, and dating of
    sediments, 135
Fossils, 14, 97, 99, 102, 121, 136, 154
Fouling, 405, 411–412, 461
Fracture zone, 54, 58, 61, 69, 88, 89,
    90–91
Fram (ship), 22, 225
Franklin, Benjamin, 236
Fraser Delta, 127, 363
Fraser River, 30, 298, 316
Free-swimming larvae (*see* Planktonic
    larvae)
Free waves, 246, 279
Fresh-water discharge, 330–331, 337–
    338 (*see also* River discharge)
  held near the coast, 227, 332–333
  and upwelling inhibited by, 335
  and water budget, 33, 182–184, 186
Friesian Islands, Netherlands, 382
Fringing reefs, 72–73
Frontal systems, 178
Frustules, 207
Fully developed sea, 249
Funa Futi Atoll, 73
Fungi, and lichens, 447
  as meiobenthos, 456
  and vitamins, 481

Galveston Bay, 319, 329, 387
Galveston, Texas, 337
Gamma radiation, 100
Ganges-Brahmaputra Delta, 365
Ganges River, 30, 37, 116, 117, 118,
    361, 363
Garfish, 554
Gas, commercial production, 137
Gases dissolved in seawater (*see*
    Dissolved gases)
Gastropods, 426–427, 435
Genetic changes, and temperature,
    412
Geological oceanography, 13–15
Geostrophic currents, 215, 222, 227–
    230, 332
German coast, 291, 382
Gibbsite, 109, 131
Gigantactis, 496
Gill-chamber linings, function in
    semi-terrestrial crabs, 454
Gill-chambers, of fiddler-crabs, 453
Gill membranes, in mud-burrowing
    animals, 452
Gills, of basking shark, 493
  of *Diopatra*, 455
  of *Hemigrapsus*, 408
  of nudibranchs, 450
  and osmoregulation, 408
  of sabellarian worms, 461
Glacial-marine sediment, 111, 125,
    126, 132, 133 (*see also* Sediment

Glacial-marine sediment (*cont.*)
    transport, and glaciers)
Glacial moraine, 319, 361
Glacial period (*see* Ice Age)
Glaciers, 27, 205
  and changing sea level, 30, 37–38,
    42–43, 134, 297, 359–360
  distribution of, 203, 206, 361
  erosion caused by, 36, 42, 111, 117,
    299, 341, 361, 367
  fjords cut by, 42, 298, 299, 315, 360
  and water budget, 179, 182
Glass-worms (*see* Arrow-worms)
Glauconite, 136
Globigerina, 423
Glomar Challenger (ship), 14, 15, 103
Glucose, 393, 396
Gnathia, 428
Goby, 455, 461, 465, 489, 557, 558
Göteborg, Sweden, 107
Goethite, 109
Gold, 137
Goniaulax, 420
Goode, J.P., 521 (*see also* Homolosine
    projection)
Gough Island, 66
Graded bedding, 124
Grain size (*see* Sediment size)
Grand Banks, 123, 206, 219, 220, 231
Grand Canyon of the Colorado River,
    44
Granite, 39, 44, 82
Graphs, how to use, 512–513
Gravel, 112, 119, 137, 360, 361, 362,
    370, 378, 382
  Puget Sound, 315
Gravity, and rock density, 81
  and tides, 276, 278
  and waves, 243, 246, 261–262, 263
Gravity anomalies, 66, 67
Gravity corer, 14
Gravity waves, 247
Grazing, of phytoplankton crop, 419,
    421, 473, 478–481, 490
Great Barrier Reef, 71, 72, 125
Great circle, 516, 521
Great Lakes, 5, 263, 341, 368, 559
Greenhouse effect, 177
Greenland, 30, 35, 36, 518, 521
  continental nature of, 31, 34, 39
  and currents, 213
  and icebergs, 205, 206
  and land ice, 182, 205
  and oceanic ridges, 66, 189
  and origin of deep water masses,
    231, 401
  and rotation of crustal blocks, 87
Greenling, 559
Green turtle, 500
Greenwich time, 25, 516
Grenadier, 556
Gribble, 454
Groins, 360, 378, 384
Ground water, 182, 367
Group speed, 244, 247
Growth-inhibiting substances, 421,
    485
Growth rate, and standing crop, 445,
    480, 482–483
Grunion, 432, 556

Grunt (fish), 201
Gulf Coast of the U. S., estuaries
    and lagoons, 297, 299, 368
  and oyster beds, 460, 461
  shoreline of, 342
  and storm surges, 337
  and tides, 272
Gulf of Aden, 95, 96
Gulf of Alaska, 57, 68
Gulf of Aquaba, 46
Gulf of California, 30, 81, 298
  formation of, 93, 95
  and intersection of oceanic rise
    with continent, 46, 57
  and sediment deposits, 127, 134
  and tides, 282, 305
Gulf of Carioco, 402
Gulf of Maine, 42, 285, 341, 342, 343
Gulf of Martaban, 339
Gulf of Mexico, 312, 329, 341, 342
  and abyssal plain, 62
  and Baffin Bay, Texas, 312
  and fresh water input, 35, 298
  as a marginal ocean basin, 46
  and sediment deposits, 87, 121, 127
  and shrimp production, 488
  as a source of moisture, 340
  and spawning of marine organisms,
    488
  tides in, 270, 282, 313, 363
Gulf of Panama, 184
Gulf of St. Lawrence, 272
Gulf of Suez, 46
Gulf shrimp, 409, 488
Gulf Stream Drift Mission, 236
Gulf Stream system, 214, 219–220,
    342, 343, 344
  and absence of sediment on Blake
    Plateau, 132
  and North Atlantic current gyre,
    215, 219
  and sea surface slope, 228
  and study of, 220, 236
  as a western boundary current,
    218–220, 350
Gulf-weed (*see* Sargassum)
Gull, 491
Gymnodinium, 420

Habitat, 392, 421
  alteration of, 491
  diversity of, and species diversifica-
    tion, 445
Habitat grouping, 459
Habitats, intertidal, 446–451, 456
Haddock, 432, 482, 488, 492, 554
Hadley cell, 178
Hake, 553, 554
Halfbeak, 555
Half-life, of radionuclides, 135–136
Halibut, 433, 494, 552
Halobates, 430
Halocline, 192, 261
Harrison, John, 25
Hastigerina, 423
Hatchet-fishes, 438, 439
Havana, Cuba, 230

Hawaiian Islands, and beaches, 362, 369
    and seismic sea waves, 260, 261
    and volcanic coastal features, 361
    as a volcanic island chain, 33, 60, 65, 74, 90
Headland, defined, 371
Heat budget, 46, 170, 173–179, 235
    regional (local), 179, 191
    and salinity distribution, 188
Heat capacity of water, 142, 146, 177, 254, 331, 340
Heat flow, from the earth, 58, 60, 80–82, 84
Heat transport, global (see Heat budget)
    vertical, 222, 233
Helium, 160
Hematite, 109
Hemigrapsus, 408
Hepatitus, 460
Heptane, 143
Herbicides, 490
Herbivores, in the coastal ocean, 481
    as first trophic level, 482
Hermit crabs, 448–449
Herring, 423, 430, 431, 432, 482, 492, 552, 557
Hesperonoë, 454
Heterotrophic nutrition, 101, 102, 457
Hibernation, 410
High-carbonate islands, 33, 34, 74
Higher high water, 272
Higher low water, 272
High Plains of midcontinental U. S., 61–62
High pressure cell, 207
Himalaya Mountains, 89, 95, 340
Histioteuthis bonnelliana, 495
Hjort, Dr. Johan, 21
Hoelzel's planisphere, 518, 519, 521
Homolosine projection, 518, 520, 521
Homotrema rubrum, 369
Honolulu, Hawaii, tides, 271, 273
Hood Canal, 316, 317, 319
Hoover Dam, 123, 254
Housatonic River, Connecticut, 298, 320
Hover craft, 15
Hudson Bay, Canada, 46
Hudson River, New York, 44, 298, 304
    effluent, 345, 348
    estuary, 8, 319, 343
Hudson submarine canyon, 255, 257, 297
Humpback whale, 499
Hurricanes, 251, 253, 312, 335–337, 358, 377
Hwang-Ho River, China, 30, 116, 117, 118, 361
Hychoides dianthus, 461
Hydration atmosphere, 156–157
Hydraulic model, 8, 19, 387
Hydrocarbons, 100, 491
Hydrofoil 7, 15

Hydrogen acceptor, 100
Hydrogen atom, in water molecule, 143, 145
Hydrogen bond, 141, 143, 145, 147–149, 163
Hydrogen chloride, 100
Hydrogen ion, 161
Hydrogen sulfide, 161, 306, 402, 458
Hydroids, 408, 422, 424, 448, 449–450, 454, 461, 481
Hydrological cycle, 183 (see also Water budget)
Hydrolysis, 405
Hydrophilic compounds, 164
Hydrophobic compounds, 100, 164, 407
Hydrosphere, 26
Hydrostatic pressure (see Pressure)
Hydroxides, in marine sediments, 109
Hyperiidae, 423, 492
Hypsographic curve, 28

Ice, (see also Floe ice, Glaciers, Pack ice, Sea ice, Shelf ice)
    crystal structure, 144–145
    density of, 145, 146, 149
    effect of heat on, 147–150
Ice Age, (see also Pleistocene glacial period)
    and glacial-marine sediments, 132
    and ice cover during, 201, 203
    and populations, 413, 465
    and sea level, 30, 37
Icebergs, 18, 125, 203, 205, 206, 207, 299
Iceland, 18, 35, 56, 66, 81
Ideal waves, 241, 244
Idiacanthus fasciola, 497
    niger, 497
Igneous rocks, 82, 98, 112
Illite, 109, 130, 131, 132
Incoming solar radiation (see Insolation)
India, 37
    and coastal upwelling, 334
    and Eurasia, 87, 95
    and monsoon circulation, 340
    and mountains formed by crustal deformation, 89
    as a source of sediment, 117
Indian basin, dimensions, 37
    formation, 94, 95
    as a gulf, 31
    sediment deposits, 37, 112, 124, 131–132
    subsidence, 72
    topography, 37, 39, 57–58, 61, 62, 124
    trenches, 54, 68
Indian Central Water, 195
Indian Equatorial water, 195
Indian Ocean, 30, 36–37, 96
    and barriers to water exchange, 95
    boundaries, 30, 34, 36–37
    and coral reefs, 69
    and currents, 215, 217–218
    and fresh water input, 30, 33, 37
    and oxygen distribution, 402
    salinity of, 33, 192, 194

Indian Ocean (cont.)
    and sediment discharged by rivers, 116
    surface and drainage areas and average depth, 32
    temperature of, 33, 192, 193
    and tides, 268, 282
    and water budget, 33
    and waves, 252, 253
    and zooplankton distribution, 434
Indicator species, 437
Individuals, number of, in coral reef communities, 463–464
    and ecological succession, 393
    and temperature, 412–414
Indonesia, 31, 37, 46, 69, 70, 95, 125
Indus River, 30, 37, 117, 118, 268, 361
Industrialization, and estuarine systems, 312, 315
Industrial waste, 116, 490
Infauna, of coastal sediments, 451–455
    and destruction of varves, 134
    in marine food cycle, 457
Infra-red radiation, 171, 172, 175, 177
Insects, 405, 427, 488, 491
Insolation, and distribution of, 174–175, 179, 192, 206–207, 397
    and heat budget, 146, 159, 172, 173–179, 212
Insulin, 485
Internal waves, 122, 261–262, 301
International Airport, San Francisco, 385
International Date Line, 515–516, 518
International Ice Patrol, 18, 205
Intertidal zonation, 446–451
Intertidal zone, colonization of, 461
    and deposit-feeders, 452
    and diversity of habitat, 445
    and meiobenthos, 456
    and oysters in, 461
Iodine, 163
Ionic bond, 143, 144
Ion pairs, 157
Iron, in earth's core, 97
    as a mineral resource, 137
    and red tides, 421
    in rocks, 39, 82, 91
    in seawater, 65, 99, 102, 114, 348
    in sedimentary deposits, 99, 102, 115, 116, 121, 122, 458
Iron-manganese nodules (see Manganese nodules)
Irrawaddy River, Burma, 30, 117, 118, 339
Isias, 429
Island arcs, 40, 46
    and destruction of crust at, 87, 88
Island arc-trench complexes, 15, 40, 43, 66–69, 70, 83, 87–88
Islands, (see also Volcanic islands)
    Atlantic Ocean, 34
    Indian Ocean, 37
    and mid-oceanic ridge, 35, 58
    Pacific Ocean, 33–34, 72
Isohalines, 331, 347, 350
Isopods, 428, 454
Isotherms, 179, 193, 207, 347

Jack, 493, 555
James River, Virginia, 298, 306, 307
Japan, 12, 33, 69, 121, 261, 493
Japan Trench, 53, 68
Java Trench, 53, 68, 69, 70, 123
Jeanette (ship), 213
Jellyfish, 418, 422, 424
    and bioluminescence, 438
    in Chesapeake Bay, 424
    and dead end in food chains, 482, 492
    density of, 436
    and tidal marshes, 488
Jet stream, 121, 132, 221
Jetty, 16, 240, 312, 314, 359, 360, 367
John F. Kennedy International Airport, New York, 385
Julius Caesar, 268

Kamchatka Peninsula, U.S.S.R., 69, 121
Kaolinite, 109, 130, 131
Kelp, 502
Kermadec Trench, 68
Key West, Florida, 230, 368
Kilauea, Hawaii, 90
Killer whale, 500, 502
Killifish, 461, 488, 556
King whiting (see Whiting)
Kovachi, 64
Krill (see Euphausids)
Krypton, 98, 160
Kullenberg corer, 107
Kure Island, Hawaii, 91
Kurile-Kamchatka Trench, 53, 68
Kuroshio Current, 7, 214, 216–219

Labrador Current, 18, 206, 214, 217, 342
Lagoons, and calcium carbonate, 159
    and coral reefs, 34, 72–76, 462, 464
    development of, 299–300, 359, 370
    as estuarine systems, 296, 319
    and sea ice, 331
    and sediment in, 75–76, 370, 378
    submerged, 43, 297, 342
    U. S., 299, 309–315, 342, 368, 370, 377–378
LaGuardia Airport, New York, 385
Laguna Madre, Texas, 297, 298, 312–315
Lakes, and deltas, 366
    and lagoons, 300
    and Langmuir circulation, 221
    and oxygen, 150
    as sediment traps, 367–368
    and standing waves, 262–263
    and surface films, 406
    and turbidity currents, 123
Lamination (beach feature), 378–380
Lamont-Doherty Geological Observatory, Columbia University, 103
Lamprey eel, 489, 558
Langmuir, Irving, 221
Langmuir circulation, 221, 406, 436, 437, 503
Lantern-fishes, 439, 484, 556
Larvae, non-pelagic (see Non-pelagic larvae)

Larvae (*cont.*)
    planktonic (see Planktonic larvae)
Latent heat, 148, 177
    of evaporation, 148
    of melting, 148
Latitude, parallels of, 515–517
Lava, volcanic, 55, 63, 65, 66, 88, 91, 98, 361, 369
Layered bedding, 123, 124, 134, 306
Layer 3 (see Oceanic crust, structure and properties)
Lead, in organisms, 403
Lefroyella, 467
Lena River, U.S.S.R., 36, 57, 361
Leopard seal, 498
Lepidocrocite, 109
Leptogorgia, 463
Levee, 44, 62, 367
Lichen, 447
Life, origin of, 79, 99–102
Lighthouse Point, Florida, 373
Light in the ocean, (see also Bioluminescence)
    and animals attracted by, 439, 484, 555
    and distribution of seaweeds, 447
    extinction of, 172–173, 397
    and migration, 397, 406, 435, 436, 500
    and photosynthesis, 172, 496–497
    and physiological processes, 397
    and protective coloration, 438
Limestone, 35, 65, 74, 463 (see also Calcium carbonate)
Limpet, 447
Linophryne, 496
Lipids, 404
Lithogenous sediment (see Sediment, lithogenous constituents)
Lithosphere, 81, 83, 96
Lithothamnion ridge, 75
Littoral current, 374
Littoral drift, 375, 376, 377–378
Lobate delta, 364
Lobster, 395, 427, 450, 451
Logan International Airport, Boston, 385
Log-line and glass, 211
Loligo, 495
London Bridge, 269
Long Island, New York, 255, 320, 343, 344, 348, 359, 361, 377
    coastal features of, 363, 368, 370, 372, 374, 381
Long Island Sound, 9, 297, 298, 319–322
    natural period of, 290
    net circulation in, 320
    and tidal currents, 284, 289–291, 319, 321–322
Longitude, meridians of, 515–517, 519, 521
    and time, 25, 516
Longshore bars, 370, 372, 374
Longshore currents, 121, 255, 312, 362, 363, 371–372, 374, 375, 376, 380
Longshore troughs, 372
Los Angeles, California, 5

Los Angeles, California (*cont.*)
    and tides, 273
Louisiana, 62
Low-carbonate islands, 33–34, 37, 75
Lower high water, 272
Lower low water, 272
Low tide terrace, 373, 378
Luidia, 430
Lunar cycle, 269, 271, 277
Lunar partial tide, 281
Lunar tide, 277

Mackerel, 433, 554
Macrobenthos, defined, 456
Madagascar, 30, 31, 37, 39, 57
Magma, 58, 61, 81, 87, 88, 90, 98
Magnesium, commercial production of, 165
    in rocks, 39, 82
Magnesium ion, 156–157, 159
    in *Hemigrapsus*, 408
Magnesium sulfate, 156, 199
Magnetic field of the earth, 14, 15, 91
    reversals of, 91–92, 93
Magnetic stripes, in ocean basin, 91–92
Maine, coast of, 443, 447, 448
Mako, 493
Mammals, extinct, 38
Manganese, in seawater, 65, 114
    in sedimentary deposits, 111, 115
Manganese nodules, 13, 14, 111, 114–115, 126, 132, 133, 137
Mangrove swamps, 133, 342, 382–383
Manhattan Island, 385
Man-made sediment, 116
Man-o'-war fish, 424
Mantle (of squid), 495
Mantle (of the earth), and alteration of rocks in, 58, 84
    composition and properties of, 26, 27, 83
    and continental blocks segregated from, 87
    formation of, 98
    and lithosphere, 81
    movements in, 61, 65, 89, 96
    and re-assimilation of crust, 88
    and speed of earthquake waves in, 83, 84, 86
Maps and map projections, 517–521
Marginal ocean basins, 45–46, 62
    and sediment deposits in, 45, 86–87, 97, 108, 127
    as transition between continental and oceanic crust, 45, 86–87
Marginal seas, Atlantic Ocean, 34, 35, 46
    Indian Ocean, 37
    Pacific Ocean, 46, 184, 282
    and river discharge, 45
    salinity of, 46, 184
    seasonal variability of, 46
    and tides, 282
Marianas Trench, 53, 67, 68
Marine animals, composition of, 403

Marine geophysics, 14, 15
Marine humus, 457, 458
Marine "snow," 406
Maritime climate, 317, 340
Mars, 15
Marshall Islands, 74
Marshes, dredging of, 384, 385, 386
    erosion of, 386
    filling of, 384–386
    H₂S in, 161
    and productivity of, 367, 382, 453,
        488
    sedimentation of, 360, 366–367, 370,
        381–383
    temperature of, 331
    and tides, 331, 381–382, 453
    U. S., 363, 368, 372, 377, 381
Marsh grass, 382, 486–487, 488
    and sediment stabilized by, 453
Massachusetts Bay, 258
Mastodons' teeth, in sediments, 121
Mathematical models, 8, 19, 386
Mauna Loa, Hawaii, 90
Maury, Matthew Fontaine, 7, 17, 20,
    212
Maximum sustainable yield, 483
Mean tidal range, 272
Mediterranean region, 340
Mediterranean Sea, and convergence
        of continental blocks, 89, 95
    and crustal sinking, 359
    as a marginal ocean basin, 46
    and productivity, 476
    salinity of, 35, 186, 330
    and tides, 268, 282
Mediterranean Water, 195, 198, 233
Medusa, 422, 424, 425, 450 (see also
    Jellyfish)
Meiobenthos, 456–458
Mekong River, Viet Nam, 30, 117,
    118, 361
Melanocetus johnsoni, 496
Melanostomiatoid fishes, 497
Menhaden, 488, 494, 552, 557
Mercator projection, 517, 518
Mesopotamia, 363
Messenger, 7
Metabolic rate, and salinity, 408
    and temperature, 401, 410, 412, 435
    and vertical migration, 436
    of Weddell Seal, 499
Metabolism, primitive, 101–102
Metamorphic rocks, 98, 112
Metamorphosis of larvae, 421
    and bivalves, 427
    and euphausids, 429
    and gastropods, 426
    postponement of, 445
    and segmented worms, 425–426
meteorites, 115
Metridia, 429
Mexico, 96
Miami, Florida, 228
Mica, 109, 113, 131, 379
Microbenthos, 456–457
Microcalanus, 429

Microcontinent, 31, 33, 37, 39, 58, 91
Microfossils (see Fossils)
Micro-nutrients, 421, 473 (see also
    Organic growth factors)
Microorganisms, 135, 456–458, 460
    (see also Bacteria)
Mid-Atlantic Ridge, and abyssal hills,
    61
    as a barrier to east-west transport,
        58
    and crustal movements, 89, 95
    and crustal structure, 84
    as a high-relief oceanic rise, 53–56
    and magnetic survey of, 92
    and rift valley, 55, 56, 58
    and sediment deposits on, 55, 82–
        83, 128
    and volcanic rocks, 83
Middle America Trench, 68
Middle Atlantic Bight, 333
Mid-Indian Ridge, 53, 57, 128, 131
Mid-oceanic ridge system, 13, 14, 15,
    36, 52–59, 103   (see also
    Oceanic rises)
Midway Island, Hawaiian Islands, 74,
    76, 91
Migration, of Adélie penguins, 498
    of eels, 501
    of fishes, 411, 472, 488–491, 493
    of lobsters, 450
    seasonal, 411, 436
    of sea turtles, 500
    vertical (see Vertical migration)
    of zooplankters, 435–437
Mineral resources, 47, 137, 165
Minnow, 489
Minot's Lighthouse, Massachusetts,
    258
Mississippi Cone, 62
Mississippi Delta, 121, 297, 363, 364,
    366, 367
Mississippi River, 30, 35, 62, 118, 121,
    301, 341, 361
    sediment load of, 363
Mixed tides, 271, 272, 279, 282, 286,
    316
Mixing, bottom-associated, 47, 303,
    317
    in the coastal ocean, 47, 190, 328,
        331, 332, 338–339, 350
    in estuaries, 301–303
    and indicator species, 437
    Long Island Sound, 320–322
    Puget Sound, 316–317
    and wind (see Winds, and mixing
        of surface waters)
Mixing of water masses, 195, 196–
    197, 198
Mohorovicic discontinuity, 83, 85, 92
Mojave Desert, 186
Mole (defined), 144
Molecular diffusion, and phytoplank-
    ton, 420
Molgula, 405
Mollusca, described, 426–427
Mollusks, and carbonate contributed
        to coral reefs, 69, 464
    protective coatings on, 408
    and reworking of sediments, 455

Molten rock (see Igneous rock)
Moncton, New Brunswick, 282
Monsoon circulation, 206, 217–218,
    222, 340
Monsoon currents, 215, 217, 218, 223
Montmorillonite, 109, 130, 131
Moon, 4, 15, 91, 97, 146
    and tides, 246, 268, 270, 272, 273–
        277, 278, 279 (see also Lunar
        partial tide; Lunar tide)
Moraine (see Glacial moraine)
Moray eel, 465, 559
Morocco, 251, 334
Mountain building, 45, 67, 86, 89, 90,
    95
    Pacific margins, 32, 297–298, 359
Mount Everest, 27
Mud, brown (see Brown mud)
Mud flats, 454–455
Muds, 12, 111, 130, 360, 456
    and adsorption of organic matter,
        100, 451, 452
    distribution and sources of, 112,
        129
    and richness of populations in, 456
    subsurface, 458
    and vitamins adsorbed on, 481
Mullet, 407, 488, 556
Municipal wastes, New York Bight,
    348
    and oyster colonies, 460
    as a source of sediment, 116
Murray, Sir John, 21, 114, 116
Mussels, in the deep ocean, 467
    element enrichment in, 404
    in estuaries, 408
    as fouling organisms, 461
    in marshes, 453, 488
    in oyster beds, 460
    on rocky beaches, 447, 450
    and sedimentation, 120
    in tide pools, 407
Mutations, 412, 413
Mysids, 423
Mysis larval stage, 429, 488

Nannoplankton, 418
    in the open ocean, 424, 435, 476,
        481
    as sediment contributors, 134
    and spring bloom, 475
Nansen, Fridtjof, 22, 225–226
Nansen bottle, 7, 11, 190
Nantucket Shoals, 285, 286, 333, 341,
    348
Nantucket Sound, 285
Narrows, New York Harbor, 284, 285
National Airport, Washington, D.C.,
    385
Natural gas, 5, 137
Natural selection, 101, 413
Nauplius, 422, 428, 429, 488, 492
Naval architecture, 15
Navigation, and Mercator projection,
    518, 521
Navigational instruments, 25, 211
Neap tide, 272, 277, 278, 285
Near bottom currents, 7, 212, 213–233
    and filter-feeding, 452, 455–456

Near-bottom currents (*cont.*)
  Northeast Pacific coastal ocean, 351, 352
  and oxygen transport, 401, 491
  and sediment transport, 108, 123, 354
Near-surface currents, measurement of, 213
Nekton, 12, 422, 457, 472–473
Neon, 98
Neon goby, 465
Netherlands, 8, 16, 335, 341, 373
Netherlands coast, tidal currents, 291
Net plant production (*see* Primary productivity)
Neuse River, North Carolina, 298, 310
Neutrally stable density system, 188–189, 191
Newark Bay, N.J., 385
Newark International Airport, 385
New England, 18, 42, 493
  beaches, 370
  glaciated coastline, 41, 299, 341
  hurricanes, 358
  marshes, 381
New England seamounts, 61
Newfoundland, 123, 220
New Guinea, 33
New Jersey coast, beach development and sediment movement, 370, 377–378
  and groins, 384
  and Hudson River effluent, 345
  and wave refraction, 255
  and wind-driven currents, 348
New Orleans, 224
Newton, Sir Isaac, 273, 276–278
New York Bay, 16
New York Bight, 343–349
New York City, 5, 385
New York-Europe shipping lanes, 206
New York Harbor, 8, 173, 298, 319, 393
  contamination problems, 348–349
  and fresh-water effluent, 346
  and tidal currents, 284, 285, 290
  and tides, 270, 273, 284, 304
  and wave refraction, 257
New York metropolitan area, 343
New Zealand, 37
Nickel, 97, 115, 411
Niger Delta, 361, 368
Niger River, 30
Nile Delta, 363
Nile River, 30, 46, 361
Nimbus-4, 202
Nitrate ion, 127, 160, 161, 399, 401
Nitrates, anaerobic decomposition of, 402
  concentration in the ocean, 400
  depletion in surface waters, 400
  in the English Channel, 401
  in estuaries, 305
  in polar regions, 434
  and protein synthesis, 399–400
  river-borne, 490
Nitrite ion, 400
Nitrogen, in the atmosphere, 26, 100, 176

Nitrogen (*cont.*)
  as a by-product of metabolism, 402, 479
  fixation by bacteria, 458, 490
  solubility, 160
Nitrogen cycle, 399–401
Node (of a standing wave), 262–263
Nonconservative elements, 152, 153
Non-living organic matter (*see* Organic detritus)
Non-pelagic larvae, distribution of, and temperature, 434, 435
  and stability of communities, 445–446
Normandy coast, 282
North America, 26, 30, 34, 35
  and coastal ocean, 35, 46, 341
  and earliest evidence of solid crust, 97
  and East Pacific Rise, 57
  and estuaries, 297
  and fresh-water drainage, 35
  and glaciers, 179
  and Gulf Stream System, 220
  and marginal seas, 46
  and river discharge of suspended sediment, 116
  and sediments on continental margin, 115, 121
  and tides, 272
  and transition from continental to oceanic crust, 86
  and West Wind Drift, 350
North American basin, 60
North Atlantic basin, and sediment deposits, 125
North Atlantic Central Water, 195
North Atlantic Current, 215, 219
North Atlantic Deep Water, 195, 231, 235, 402
North Atlantic Ocean, and barriers to water exchange, 66, 189
  and bottom topography, 56, 59
  as center for movement of crustal blocks, 87
  and convergence of surface currents, 503
  and deep currents, 231
  and fresh-water input, 30, 35
  and icebergs, 206
  magnetic survey of, 92
  productivity of, 127
  and tides, 268, 270, 280–281
  and water masses, 198, 199, 231, 233
  and waves, 251, 252, 253
North Atlantic subtropical gyre, 215, 219
Northeast Pacific coastal ocean, 342, 349–354
Northern Equatorial Current, 214, 215, 217, 218, 219, 235
Northern Hemisphere, and Coriolis effect, 224, 225
  and currents, 30, 215, 218, 219, 226, 228, 230, 235
  distribution of land and water, 29, 30, 31
  and Ekman spiral, 226
  and heat transfer, 177

Northern Hemisphere (*cont.*)
  and tidal currents, 285
  and tides, 280, 281
  and winds, 207
  and zooplankton distribution, 434
North Pacific Basin, and sediment deposits, 127
North Pacific Current, 214, 215
North Pacific Ocean, and fresh-water input, 30
  and halocline, 192
  salinity of, 192
  and water masses, 233
  and waves, 252, 253
North Polar Water, 195
North Pole, 34, 179, 223, 515
North Sea, 30, 46, 285, 374, 382, 386
  and amphidrome system in, 281, 291
  and plankton, 430, 437, 474, 475
  and storm surges 335–336, 345
  and tidal currents, 290–291
Norway, 202, 298, 360, 402, 445
Nuclear power stations, 411
Nuclear reactor, 10, 353
Nuclear weapons, atmospheric testing of, 121, 354
Nucleic acid, 100
Nudibranchs, 403, 450, 451
Numerical taxonomy, 12
Nutrients, 10, 11, 420
  absorbed by marine algae, 448
  and blooms, 421, 475, 479–480
  in coastal waters, 433, 474, 477, 490
  concentration of, and nannoplankton, 476
  and productivity, 473
  in coral reef ecosystems, 462, 463
  depletion in surface waters, 154, 218, 317, 352, 398–399, 400, 401, 475, 476, 479
  in the English Channel, 401, 480
  in estuaries, 305, 488, 490
  and flow in food chains, 393, 399, 400, 479
  and protein synthesis, 400, 458
  re-cycled to surface waters, 47, 127, 154, 207, 208, 219, 227, 305, 338, 350, 352, 394, 401, 420, 433, 477, 480
  river-borne, 305, 477, 490
Nutritional requirements, for man, 485

Obelia, 422, 424, 437
Ocean, origins, 87, 97–99
Ocean basin, age, 39, 92, 135
  antipodal to continental blocks, 28, 29
  bathymetric chart of, 53
  boundaries, 39, 45, 52, 66, 85, 86
  evolution of, 79, 93–97
  general features, 31, 59–61
  and heat flow from, 82
  layered structure, 15, 82, 83 (*see also* Oceanic crust)

Ocean basin (*cont.*)
  as a province, 39, 40, 52, 54
  study of, 13, 14, 15, 80, 81, 82
  volume of, 28–29, 32, 37
Ocean bottom, erosion of, 120
  study of, 14–15, 38–39, 80–81
  topography (*see* Bottom
      topography)
Ocean engineering, 15–17
Oceanic air mass, 340
Oceanic crust, formation (*see* Crust
    [of the earth], formation and
    destruction)
  origins (*see* Crust [of the earth],
      origins)
  and reversal of earth's magnetic
      field, 92
  structure and properties of, 27, 39,
      52, 60, 81–82, 83, 84, 85
  study of, 14, 15, 80
  subsidence, 65, 67
  and transitional areas near con-
      tinental margins, 45, 86–87
Oceanic front, 220
Oceanic rises, (*see also* Aseismic
    ridges; Mid-oceanic ridge
    system)
  and crust overlying, 58
  and earthquakes, 55, 57, 61, 84, 89
  and formation of new crust at, 87–
      88, 89, 92, 97, 103
  and fracture zones, 61, 69
  and heat flow from, 58, 81, 82, 84
  high relief, 54
  inactive, 65
  and intersection with continental
      blocks, 46, 57
  location of, 88
  low relief, 54, 55
  as a province, 39, 40, 52, 54
  rifting of continents over, 95–96
  and sediment deposits, 58–59, 82–
      83, 103, 122–123
  and volcanic ridges, 66
  and water movements, 66, 189
Oceanography, defined, 4
Octant, 25
Octopod, 495
Octopodotheuthis, 495
Octopus, 201, 426, 450–451, 503
Ocypode, 454
Oikopleura, 430, 431, 492
Oil (*see* Petroleum)
Oolite, 159
Ooze, 114, 466
Opal, 109
Open ocean, 207
  average water depth of, 328
  circulation of, 218
  and food chains in, 482
  and plankton populations, 433–435,
      476
  productivity of, 476, 477
Operculum, 427
Ophiothrix, 430
Orbital motion, wave-induced, 210

Orbital motion (*cont.*)
      244, 283 (*see also* Water
      parcels, movement of)
Oregon, 57, 59, 122
Oregon coast, 342, 350, 371
  and upwelling, 335, 353
Organic carbon decomposition, 458
Organic carbon production, 335, 473
      (*see also* Carbon fixation)
Organic compounds, non-biological
      synthesis of, 99–101, 102
  surface activity of, 100–101, 405–
      406
Organic detritus, composition of, 404
  decomposition of, 400, 458, 467,
      480, 490
  in the deep ocean, 467
  and eel-grass, 454
  and food chains, 405, 456–459, 461,
      464, 479
  as food for marine animals, 423,
      434, 447, 450, 453, 455, 503
  and Langmuir circulation, 436, 503
  and oxygen utilization, 122, 401,
      452, 490, 491
  and protein synthesis, 400, 458
  river-borne, 477
  in sediments, 445, 453, 456–459, 491
Organic growth factors, 481 (*see also*
      Micronutrients)
Organic matter, chemical constituents
      of, 100
Organic molecules, adsorbed on sedi-
      ments, 101
Oscillatoria, 420
Osmoregulation, 408
Osmosis, 407
Osmotic gradient, in tide pools, 407
Osmotic pressure, 156, 407, 408
Osprey, 491
Oxides, in sediments, 109
Oxygen, in the atmosphere, 26, 99–
      100, 102, 176
  biological cycle, 401–402
  and conservation of in deep diving,
      499, 500
  and decomposition of organic
      detritus, 122, 401, 490, 491
  depletion in subsurface waters, 134,
      160, 161, 227, 306, 402, 491
    New York Harbor, 349
    Puget Sound, 317
  distribution, 161, 192, 219, 401–402
  free, 99, 101–102, 112, 396, 397
  and iron, 99, 122
  in lakes, 150
  and manganese nodules, 114
  photosynthetic production, 396,
      397, 473
  Puget Sound, 317, 318
  and respiration, 399, 401, 402, 452,
      456, 458, 473
  in sediments, 122, 306, 452, 456,
      458–459
  solubility, 160
  in tide pools, 448
  as a tracer for water masses, 215
  uptake, 410, 412
  in water molecule, 143, 145
Oxygen acceptors, 102

Oxygen-bottle experiment, 473
Oxygen-minimum layer, 134, 402
Oyster bed communities, 460–461,
      489
Oyster drill snail, 409, 461
Oysters, in Chesapeake Bay, 459, 461
  copper in, 412
  culturing of, 12
  element enrichment in, 404
  in estuaries, 408, 409, 488
  life-cycle of, 427, 459, 460
  mangrove, 382
  and preference for oyster-shell
      substrate, 445, 459
  and reef-building, 460, 461–462
  and sedimentation, 120
Ozone, 100, 102, 176

Pacific basin, antipodal to Eurasia, 29
  and continental margins, 32, 43,
      44, 84, 299, 328
  dimensions of, 31–32
  evolution of, 95, 97
  as a gulf, 31
  and sediment deposits, 32, 59, 111,
      115, 123, 124, 127, 131, 132,
      135
  subsidence, 72, 74
  topography of, 57, 58, 61, 65, 91,
      124
Pacific Coast of the U. S., 122, 258,
      328
  beaches and dunes, 121–122, 299,
      370, 374, 376, 380
  climate, 340
  and coastal ocean circulation, 349–
      352
  and estuarine systems, 297, 298
    as nursery areas, 489
  shoreline, 342
  and tides, 272
Pacific Equatorial Water, 195
Pacific Northwest, coastal features,
      298, 299, 342
  coastal ocean region (*see* Northeast
      Pacific coastal ocean)
Pacific Ocean, (*see also* North Pacific
      Ocean; South Pacific Ocean)
  and barriers to water exchange,
      46, 95, 231
  boundaries, 30, 34, 37
  and coral reefs, 18, 69
  and currents, 213
  as a defense barrier, 6
  and fresh water input, 32, 33, 116
  and icebergs, 206
  islands in, 33–34, 72
  isotherms, 179
  and military operations, 18
  oxygen distribution, 402
  salinity, 32, 33, 184, 192, 194
  and seismic sea waves, 260–261
  surface and drainage areas and
      average depth, 32
  temperature, 32, 33, 192, 193
  and tides, 271, 282
  trenches, 54, 68
  and water budget, 33
  and water masses, 231
  and waves, 251, 253, 254

Pacific Subarctic Water, 195
Pack ice, 22, 35, 202, 204, 205, 207
Padre Island, Texas, 312
Pago Pago, 361
Pakhoi, China, tides, 273
Palmer Peninsula, 34
Pamlico River, North Carolina, 298, 310, 311
Pamlico Sound, North Carolina, 297, 298, 309–312, 313, 318
Pancake ice, 204
Pangaea, 93, 94
Paracalanus, 429
Parachute drogue, 213
Parapod, 455
Parrot-fish, 464
Partial tide, 279, 281 (see also Lunar partial tide)
Patchiness, of deep-ocean benthos, 465
of plankton, 437
Patch reefs, 76
Pea crab (see *Scleroplax*)
Peat, 367
Pebbles, 111, 370
Pelagic sediment deposits, 124, 127
Penguins, 498, 499
Pensacola, Florida, tides, 271
Peptides, 405
Perch (ocean perch), 555 (see also Sea perch)
Period, of a basin, 280, 282
Persian Gulf, 37, 46, 186, 331
Peru, 32
Peru-Chile Trench, 43, 68, 83, 85
Peru Current, 214, 216
Peruvian coast, fisheries, 477
Pesticides, 315, 460, 490, 491
Petitcodiac River, New Brunswick, 271, 305
Petroleum, 5, 11, 17, 62, 137, 412
Pettersson, Hans, 107
pH, of seawater, 403
Phaeocystis, 481
Philadelphia, Pennsylvania, 5
Philippine Trench, 53, 68, 123
Phillipsite, 132
Phoenecians, 8
Phosphates, depletion in surface waters, 399–400
dissolved from fecal matter, 479
in the English Channel, 401
in estuaries, 305
"hoarded" by phytoplankton, 401
as a mineral resource, 137
in the North Sea, 437
and photosynthesis, 400
in polar regions, 434
and protein synthesis, 399–400
in Puget Sound, 318
river-borne, 490
in seawater, 113, 399, 400
in sediments, 109, 458 (see also Sediment, phosphatic constituents)
as skeletal materials, 113, 128
Phosphorus, in organic acids, 101
Phosphorus cycle, 398, 400
Photic zone, and coral reefs, 463
depth of, 391, 397, 401

Photic zone (cont.)
and detritus layer near the bottom of, 400
as food source for deep-water organisms, 397, 466
and mixed zone extending below, 401, 474, 475
and nutrients recycled within, 479
and nutrients transported to, 401, 421
oxygen produced in, 401
and productivity, 473
Photon, 397
Photosynthesis, and attached plants, 444
and CO_2 utilized in, 396, 403
and dissolved oxygen, 160–161
evolution of, 99, 101–102
and food web, 12, 393, 396–397, 482
and insolation, 172, 191, 334, 396–397
and oxygen produced during, 396, 401
and phosphate concentrations, 400
and primary productivity, 473, 474
and production of biogenous sediment, 127
Puget Sound, 317
and temperature, 410
Photosynthetic zone (see Photic zone)
Physalia, 424, 425
Physical oceanography, 7–9
Phytoplankton, characteristics of, 419, 420
in the coastal ocean, 305, 474–476, 490
colonial, 476
composition of, 399
and dissolved organic matter, 172
distribution of, 419–421, 437, 475–476
in the English Channel, 479, 480
in estuaries, 305
flagellated, 419, 420, 492
in food webs, 393, 456, 492
as fouling organisms, 405
and "hoarding" of phosphates, 401
and Langmuir circulation, 406
in marshes, 488
in the North Pacific Ocean, 350
in the Northeast Pacific coastal ocean, 352
and nutrient utilization, 400–401
in Peruvian coastal waters, 477
and primary productivity, 473–474
and productivity of the western side of ocean basins, 218
protein in, 418
in Puget Sound, 319
and small size of, 420
and standing crop of, 478, 485
in subpolar regions, 207
and symbiotic associations, 423
and upwelling, 219, 227, 334
and zinc-65, 354
Phytoplankton bloom (see Bloom)
Piccard, Jacques, 236
Pilchard, 552
"Pillow" lavas, 63
Pipe-fish, 489, 501

Plaice, 432, 553
Planetary circulation, 488
Plankton, classification of, 418
collection of, 10, 12, 418
composition of, 398, 399
and "deep-scattering layer", 436
defined, 418
harvesting of, 484
in organic detritus, 404, 406
in rivers, 477
Planktonic fishes, deep-water species, 432
Planktonic larvae, of benthic organisms, 421, 433, 434–435, 445
and community instability, 445–446
of deep-water species, 435, 466, 496, 497
of echinoderms, 430
of fishes (see Fish larvae)
of gastropods, 426, 434, 435
of mollusks, 426, 427, 435
of nekton, 422
of polychaetes, 426
Plankton net, 10, 417, 418
Plant debris, in marshes, 367, 488
in organic detritus, 400, 404
in sediments, 124, 382, 467
Plants, attached (see Seaweeds)
evolution of, 102
fossil, 99
marine, 113 (see also Phytoplankton)
and sediment stabilization, 117, 120, 371, 374, 378, 382, 453
(see also Beach grass; Marsh grass)
Platelets, 114
Pleistocene glacial period, 43, 74, 203, 297, 361 (see also Ice Age)
Pleurobrachia, 425, 492
Pliny the Elder, 268
Plunging breaker, 256, 257, 258, 354
Pluteus, 430
Plutonium, 353
Podon, 430, 492
Poisons, 421
Polaris, and determination of latitude, 515
Polar molecule, 143, 164
Polar regions, 206
and Atlantic Ocean, 34, 35
and currents, 215
defined, 207
and epifauna of, 445
and heat budget, 218
and icebergs, 206
and insolation, 175
and populations in, 410, 413, 434
and precipitation, 184
and sea ice, 201
and sediments in, 117, 132
and sound channel in, 201
and temperatures of, 410
and thermohaline circulation, 231
Polar species, breeding cycles of, 434, 436
found in both hemispheres, 434,

Polar species (*cont.*)
    469
    lack of diversity among, 412, 433
    metamorphosis of, 434
Polder, 8
Pollen, 135
Pollock, 493, 553
Polychaetes, 426, 455
Polyp, function in *Physalia*, 424, 425
    structure and function in coral,
        462, 463
    structure and relation to medusa,
        422, 425, 450
Polysaccharides, 405
Pompano, 555
Poorly-sorted sediment, defined, 111
Population, of U. S. coastal areas, 5
Porpoise, 410
Port Adelaide, Australia, tides, 273
Portuguese man-o'-war (*see Physalia*)
Potassium, radioactive decay of, 81,
    100, 136
Potassium chloride, 156, 408
Potassium-40, 136
Potassium ion, 156, 157, 408
Potomac River, Maryland, 289, 298,
    304, 306, 307
Precipitation, distribution of, 33, 182–
    184, 186, 188, 207, 220, 314,
    330, 340, 350
    and heat gain, 179, 186, 188
Predation, among benthic animals,
    444
    and food web, 482
    and standing crop in fish, 483
    study of, 484
Pressure, in the deep ocean, 161–162
    and marine life, 396, 467
    and sound, 199
    and viscosity, 161–162
Prevailing winds, and surface cur-
        rents, 222, 228
    and upwelling, 219, 333, 477
    and West Wind Drift, 350
Priapus, 469
Primary carnivore, 482
Primary coast, 361
Primary consumer, 393, 481
Primary production, 458, 482, 486–
    487, 488
Primary productivity, and critical
        depth, 474
    estimated for ocean areas, 476–478
    measurement of, 12, 473–474
    methods of increasing, 485
    Northeast Pacific coastal ocean, 352
Prime Meridian, 515, 516
Producer, defined, 393
Productivity, of the coastal ocean,
        305, 367, 445, 474–476, 477,
        482, 483, 490
    of coral reef communities, 463
    distribution of, 207, 208, 218, 476–
        478
    and distribution of biogenous sedi-
        ment, 127, 133
    of estuaries, 305–306

Productivity (*cont.*)
    of the open ocean, 477, 482
    prediction of in the North Sea, 437
    of Puget Sound, 317, 319
    seasonal variation in, 474–476
    and seaweeds, 445
    and stability of surface waters,
        306, 335, 474–475
    and thermocline, 334, 474
    and upwelling, 127, 219, 334–335,
        401, 412, 477
Profile, how to use, 513
    of the ocean bottom, 14–15, 39
Progressive waves, 241–244, 262, 263
    and tides as, 278, 280, 283, 284,
        287–289, 304
Protective coloration, 438
Protein, in bacteria, 458
    and conservation of trace elements,
        404
    in copepods, 430
    as detritus, 405, 407
    in fishes, 5, 482, 485
    and nutritional requirements, 485
    in organic matter, 100
    in plankton, 418, 484
    synthesis of, 396, 400, 458
Proto-co-enzymes, 101
Protoplasm, 99, 396
Protozoans, benthic, 461
    epiphytic, 454
    planktonic, 422–424, 481
    in sediments, 456, 457
Provinces, of the ocean bottom (*see*
    Ocean bottom)
Provincetown, Mass., 374
Psammechinus, 430
Pseudocalanus, 429, 492
Pseudopod, 423
Psychropotes, 466
Pteropods, 427
    shells in sediment, 109, 112, 114,
        129, 427
Ptolomy, astronomer, 516
Puerto Rico, 85
Puerto Rico Trench, 53, 68, 83, 85,
    123
Puget Sound, 9, 42, 297, 298, 315–318,
    341
Purse-seine, 484, 494
Pycnocline, 191–192
    and areas of convergence, 222, 228
    distribution of, 193, 194
    and estuarine circulation, 300
    and internal waves, 261, 301
    and rise of bottom waters, 233
    and surface currents, 220, 226
    in subtropical regions, 207
    and wind effect on depth of, 218,
        227, 332
Pyrolusite, 109

Quartz, 109, 111, 369, 379
Quebec, 304
Quinalt River, Washington, 339

The Race, Long Island Sound, 289,
    320, 322
Radiant energy, and photosynthesis,
    393, 394, 396–397, 481

Radiant energy (*cont.*)
    and physiological processes, 397
Radioactive decay, and crustal move-
    ments, 87
Radioactive wastes, 353–354
Radioactivity, 11
Radiolaria, 423, 481
    shells in sediment, 110, 112, 114,
        128, 135
Radiolarian mud, 114, 126, 129, 133
Radiometric dating, 97, 131, 135–136
Radionuclides, in the coastal ocean,
    354
    and dating of sediments, 135–136
Rainfall (*see* Precipitation)
Rainshadow, 340
Rappahannock River, Virginia, 306,
    307
Rare gases, 160, 161
Raritan Bay, New York-New Jersey,
    298
Raritan River, New Jersey, 298, 343
Rat-tail, 556
Ray, 493, 558
Recent geologic period, rate of sedi-
    mentation, 62
Receptivity, 413
Redbeard sponge, 463
Red clay (red mud), 114, 129, 161
Redfish, 555
Red Sea, 30, 46
    formation of, 93, 95, 96
    and intersection of oceanic rise
        with continental block, 46, 57
    salinity of, 37, 186, 330
    temperature of, 331
Red Sea Water, 195
"Red tide," 421
Reducing atmosphere, 100, 101
Reef-building, 459, 461, 462–463
Reef flat, 75, 77
Reef front, 74–75
Reefs, and environment modified by,
    462
Reflection seismology, 13, 80
Refraction seismology, 80, 81
Relict sediments, 43, 121, 133
Reproduction, failure of, and en-
        dangered species, 491
    rate of, and size of individual, 457
Residence time, 155, 163, 233
    and biological processes, 113
    of elements in seawater, 158–159
    of water in coastal ocean, 332–333,
        347–348
    of water in Long Island Sound, 322
    of water in Pamlico Sound, 312
Respiration, and energy release, 396–
        397, 456, 457, 482
    and mud-burrowing animals, 452
    and oxygen utilization, 401, 452,
        458, 473
    and pH of seawater, 403
    of plants, 473, 474, 480
    and recycling of materials, 490, 491
Reversed magnetism, 91
Reversing thermometer, 7, 9, 189, 190
Reversing tidal currents, 283–285,
    289, 291, 307, 337
Rhincalanus, 429

Rhine Delta, 363, 382
Rhine River, 16, 361
Ribbon reef, 34
Rift valley, 55, 58, 96
Right whale, 499
Rill, 378
Rio Grande, Texas, 312, 313
Rio Grande Delta, 363
Rip (see Tide rip)
Rip current, 376–377
Ripple mark, 381
Ripples, 240, 247
River discharge (see also Fresh water
        discharge)
    to Chesapeake Bay, 298, 306
    and coastal currents, 332–333, 348
    and estuarine circulation, 300–304,
        337, 477
    to Laguna Madre, 298, 314
    to Long Island Sound, 298, 319–
        320, 322
    to marginal ocean basins, 45
    to New York Bight, 343–346
    to Northeast Pacific coastal ocean,
        342, 350–351
    and oysters killed by, 408
    to Pamlico Sound, 298, 312
    and productivity, 305, 475, 477, 490
    to Puget Sound, 298, 315, 317, 319
    of salt to the ocean, 154, 155
    and sediment transport (see Sedi-
        ment transport, and rivers)
    and stable surface layer in coastal
        ocean, 305, 335, 475
    and tidal currents, 285, 291, 338
    and tides, 337
    to U. S. coastal ocean regions, 342
    and water masses formed by, 337–
        338
River drainage basins, 30, 116, 117,
        297, 298, 305, 341
River herring, 489
Roaring Forties, 251
Rockfishes, 433, 555
Rockweed, 448
Rocky beaches, and marine life, 443,
        446–451
Romanche Trench, 53, 56, 66, 68, 69,
        231
Rosefish, 555
Rotary tidal currents, 285–287, 291
Rotation of the earth, and currents,
        218, 222, 223, 224, 225, 228
    and tides, 276
Royal Society of London, 3

Saanich Inlet, British Columbia, 118
Sabellarian worms, 461
Sacramento River, California, 298
Sagitta, 425, 437, 470 (see also
        Arrow-worms)
Sahara Desert, 125, 186
Saint John, New Brunswick, 282
Saint Lawrence River, 30, 35, 297,
        304, 341, 368
Salinity, changing, and distribution of
        species, 408–409, 433, 445
    and physiological responses to,
        314, 315
    and tide-pool organisms, 448

Salinity (cont.)
    and density, 149–150, 155, 188,
        190–192, 195, 215, 221, 222–223,
        230, 231, 300–302, 306
    gradient, and distribution of
        species, 409–410, 433, 461, 488
    measurement of, 11, 150–153, 154,
        159, 188
    and osmotic pressure of body
        fluids, 407–408
    of seawater, 154, 155
    and sound, 199
    and temperature of initial freezing,
        150
    as a tracer for water masses, 9,
        153, 170, 195–198, 215, 231,
        233, 235
    of world ocean, 195
Salinity distribution, Atlantic Ocean,
        192, 194, 195, 198, 220, 231,
        233–235
    Chesapeake Bay, 308, 309
    coastal ocean (see Coastal ocean,
        salinity distributions)
    and coral reefs, 462
    in estuaries (see Estuaries, salinity
        distributions)
    Indian Ocean, 192, 194
    Laguna Madre, 314
    Long Island Sound, 320–321, 322
    New York Bight, 343–347, 348
    Northeast Pacific coastal ocean,
        350–352
    open ocean basins, 186
    Pacific Ocean, 192, 194
    Pamlico Sound, 310–312
    Puget Sound, 317, 318
    regional, 184, 186, 188, 206–207,
        230–231, 233, 335
    Strait of Gibraltar, 198
    surface ocean, 187
    vertical, 190–198, 223, 230–231, 233,
        235
Salinometer, 153
Salmon, 407, 432, 488, 489, 558
Salp, 431, 432, 437, 481
Salt, commercial production of, 137,
        165
Salt dome, 62
Salt marshes (see Marshes)
Salt wedge, 300, 301–302, 337
Sampling techniques, 7, 11, 12, 13, 14
San Andreas fault, California, 57
Sand, 111, 113, 137
    on beaches, 75, 360, 369–370, 371,
        374, 375, 377–378, 379, 380, 384
        (see also Beach sand)
    carbonate, 62, 75, 76, 369
    in deep-ocean sediments, 123
    distribution and sources of, 112
    erosion of, 119, 361, 362, 378, 384
    and filter-feeders, 451, 452
    in graded bedding, 124
    and intertidal populations, 456
    and marshes, 382
    Puget Sound, 315
    in relict sediments, 121
    settling velocity of, 119
    size of, 111

Salinity distribution (cont.)
    transport of, 119, 121–122, 124, 358
    in worm tubes, 461
Sand bars, 362, 383
Sand dunes (see Dunes)
Sand eel, 431
Sand-gravel islands, 72–73, 75
San Diego, California, 5
    and tides, 271
Sand lance, 556
Sandy Hook, N.J., 377, 384
San Francisco, California, tempera-
        ture range in, 340
San Francisco Bay, 258, 297, 298, 312
San Francisco earthquake, 1906, 57
San Francisco Lightship, and tidal
        currents, 283
Sardina pilchardus, 433
Sardine, 482, 552
Sargasso Sea, 207, 219–220, 476, 501,
        503
Sargassum, 207, 221, 503
Sargassum fish, 503
Satellites, 4, 97, 171, 202, 212
Saturation concentration, 160
Saury, 555
Scallop, 404, 455
Scandinavia, 202
Scarp, 373, 374, 379
Scattering of light, 173
Scavenging, 444, 450, 467 (see also
        Deposit-feeding)
Schmidt, Johannes, 501
Schooling of fishes, 18, 439, 465, 484
Scleroplax, 454, 455, 461
Scotch Cap, Alaska, 261
Scotland, 189, 335
Scotoplanes, 466
Scripps Institution of Oceanography,
        103
Sea, 244, 245, 247–249 (see also Fully
        developed sea)
Sea anemones, 449
    as predators, 444, 465
    on rocky beaches, 447, 448, 449
    and symbiotic relationships, 449,
        465
    in tide pools, 407, 448
Seabed drifters, 212, 351
Sea cucumbers, deep-ocean, 466, 467
    as deposit feeders, 451, 466
    as echinoderms, 430
    on rocky beaches, 449
Sea-floor spreading, 87–97, 98, 103
Sea gooseberries (see Ctenophora)
Sea-hair, 447
Seahorse, 501
Sea ice, Arctic, 35–36, 201, 202, 204,
        213, 299
    Chesapeake Bay, 308
    coastal ocean, 331, 344
    distribution of, 118, 201, 203
    drift of, 22, 225–226
    formation of, 201–204
    oceanwide influence, 170, 201
    and sediment deposition, 125
Sea lettuces, 448

Sea level changes, 37–38 (see also Glaciers, and changing sea level)
  and atoll formation, 72–74
  and continental margins, 30, 42, 44, 47, 121, 134, 297, 360
  and shorelines, 38, 358, 359, 360
  and storm surges, 334, 335, 336
Sea lily, 430, 466
Sea lion, 410, 502
Seal, 201, 482, 499 (see also Leopard seal; Weddell seal)
Seamount, 61, 465
Sea of Okhotsk, 30, 46, 233
Sea otter, 502
Sea perch, 559
Sea salt, in the atmosphere, 162, 163, 164
  constituents of, 9, 98, 151–153, 157
  exclusion from ice structure, 145, 156, 203
  origin of, 154
  residence time of, 154–155
Sea squirts (see Tunicates)
Sea star, 448
  as coral predator, 463, 464
  as an echinoderm, 430
  insulin derived from, 485
  larvae of, 434
  as oyster predator, 409, 461
  on rocky beaches, 448
  and skin breathing, 452
Sea surface topography, 228–230
Seattle, Washington, 5, 316, 515
  and tides, 271, 273
Sea turtle, 500
Sea urchin, 434, 444, 449
Seawalls, 16, 360, 378, 384
Sea-walnuts (see Ctenophora)
Seawater, acoustical properties of, 200
  biological properties of, 142, 152, 153
  composition of, 98–99, 118, 151–154, 157, 158–159, 161, 398, 403
  conductivity of, 11, 153, 159
  factors controlling density of, 149–150
  pH of, 403
  residence time of elements in, 158–159
  and temperature of initial freezing, 156, 201, 331
Seaweeds, (see also Benthic algae)
  as analogous to terrestrial plants, 420
  cultivation of, 493
  fertilizers absorbed by, 485
  as fouling organisms, 405
  limited by depth of photic zone, 444, 445
  and nutrients in the coastal ocean, 490
  on rocky beaches, 443, 447–448, 450
  as shelter for marine organisms, 448, 450, 488, 501

Seaweeds (cont.)
  in tide pools, 407, 448
Secchi disk, 173
Secondary coast, 360–361
Secondary consumer, 393
Sediment, age determination, 134–136
  in the atmosphere, 121
  biogenous constituents, 109–111, 113–114, 115, 117, 125, 128–129, 134–135 (see also Sediment, calcareous constituents; Sediment, carbonate; Sediment, phosphatic constituents; Sediment, siliceous constituents)
    destruction of 113–114, 118, 128, 154, 160
  biological reworking of, 134, 455, 456
  calcareous constituents, 109, 112, 118, 127, 128, 129, 458, 464
  carbonate, 109, 114, 118, 137
    ancient, 98
    and bacterial activity, 458
    and coral reefs, 75
    destruction of, 128, 133, 160
    distribution of, 59, 128, 129, 133
    and volcanic islands, 65, 74
  chemistry of, and bacterial activity, 458–459
  classification of, 109, 111
  and coastal infauna, 451–456
  on continents, 82, 89, 306
  cosmogenous constituents, 112, 115–116
  deep-ocean, age determination, 136
    and benthic life, 465, 466
    constituents of, 112, 114, 125–134, 160
    distribution of, 58–59, 61, 66, 67, 82, 103, 108, 122–123, 125–134
    exposed on land, 69, 70
    and organic detritus, 122, 402
    and oxygen, 122, 402
    rate of accumulation, 127
    study of, 13, 14, 15, 80, 107
    volume of, 45, 60
  and fertilizers adsorbed on, 485
  glacial-marine (see Glacial-marine sediment)
  hydrogenous constituents, 109, 112, 114–115
  intertidal (see Intertidal zone)
  and layered structure of ocean basin, 82, 83
  lithogenous constituents, 109, 112–113, 116, 117, 118, 125, 127, 128–129, 132–134
  man-made (see Man-made sediment)
  and nutrients incorporated in, 400
  oldest known, 98
  phosphatic constituents, 109, 118–119, 128
  and Puget Sound, 315, 319
  siliceous constituents, 109, 110, 112, 114, 118, 120, 128, 129, 130
  stabilized by plants, 117, 120, 371, 374, 378, 382, 453
  suspended, 337 (see also headings under Sediment transport)

Sediment (cont.)
  and light in the ocean, 172
  and reef-building animals, 69
  residence time of, 122
  trace elements in, 404
  and transition between continent and ocean basin, 45, 85, 86
  and water retained in, 26, 182
Sedimentary rocks, 26, 39, 44, 82, 306
  organisms contained in, 90
  and water budget, 26, 28, 182
Sedimentation, of beaches, 121, 369–372, 374–375, 378
  of continental rises, 44–45
  of continental shelves, 41, 43, 116, 122, 125, 132, 133
  of continental slopes, 43
  of the deep-ocean floor, 60, 61, 115, 120, 124–134
  of deltas, 62, 121, 361, 363, 364, 366–368
  of estuaries, 43, 120–121, 297, 362, 363, 368
  and evolution of ocean basins, 95
  and filter-feeding animals, 120, 122, 455–456
  of marginal ocean basins, 45, 97, 108, 116, 127
  of marshes, 360, 366–367, 370, 381–382
Sediment size, and beach slope, 370
  classification by, 109, 111
  and distribution of coastal infauna, 451–452, 455
  sediment distribution and sources related to, 112
  and transport, 113, 119–122, 124, 130–131
Sediment transport, 119–121
  in the coastal ocean, 341, 363, 368
  and continental margin topography, 134
  and current speed, 113, 119–120
  and glaciers, 36, 112, 125, 131, 298, 361
  and hydraulic models of, 387
  longshore movements, 121, 299, 319, 362, 363, 368, 370–372, 374, 375, 377–378, 383–384
  and rivers, 36, 45, 113, 116, 117, 118, 120–121, 125, 131–132, 153, 297, 338, 342, 362–363, 368, 370, 381, 383
  and sea ice, 125, 132
  and tidal currents, 378, 381–382
  and turbidity currents, 123–124
  and waves, 299, 358, 362, 363, 369–370, 372, 375, 381
  and wind, 112, 117, 121, 125, 131, 132, 153, 370, 373, 379
Seiche (see Standing waves)
Seismic profiling, 107
Seismic reflection (see Reflection seismology)
Seismic refraction (see Refraction seismology)
Seismic sea waves, 245, 246, 254, 260–261
Seismic waves, 15
Seismology, defined, 80

Semi-daily tides, 270, 271, 272, 276, 277, 278, 279, 282, 285
Semi-permeable membrane, 407
Semi-terrestrial life, 454
Sensible heat, 148, 177, 179
Set (of a current), 212
Settling velocity, 113, 119
Sewage, 11, 348, 386, 460
Sexual maturity, and temperature, 412
Shad, 407, 489, 557
Shallow-water waves, 242, 244, 254–261, 279, 283
Sharks, 493
    and absence of young from plankton, 432
    on deep-ocean floor after death, 467
    as high trophic level feeders, 482
    relation to bony fishes, 407
    teeth, in sediments, 114, 115
Shelf ice, 203, 205
Shells, in sediment, 109, 113–114, 121, 124, 127, 128, 137, 161, 369, 378
Shetland Islands, 66
Shore, defined, 358
Shoreline, 37, 41, 42, 342, 358–359, 361
Shrimp, 427
    and chlorinated hydrocarbons, 491
    in eel-grass community, 454, 455
    in estuaries, 488
    life history of, 488
    as low trophic level feeders, 482
    and release of luminous ink by, 439
SIAL, 82
Siberia, 34, 35, 36, 57, 68, 213, 233
Sierra Nevada, 96
Significant waves, 252
Sigsbee Abyssal Plain, 62
Silica, 117, 399
Silicate minerals, 109, 120, 132
Silicate rocks, 39, 82, 85, 87, 97, 98, 112, 369
Silicates, in seawater, 113, 118, 401
Siliceous sediment (see Sediment, siliceous constituents)
Siliceous shells, of diatoms, 114, 128, 398, 419
    of radiolarians, 114, 128, 423
    of silicoflagellates, 114, 419
Siliceous skeletons, dissolution of, 128
    in sponges, 114
Silicoflagellates, 112, 114, 419, 420
Sill, 298, 303, 315, 317
Silt, 111, 113
    and adsorption of surface-active materials, 405
    bacteria in, 458
    calcareous, and coral reefs, 464
    classified by size, 111
    and deposit-feeding, 452
    distribution and sources of, 112
    erosion of, 119, 370, 378
    in glacial-marine sediment, 111, 125
    in graded bedding, 124
    in Laguna Madre, 315
    in marshes, 382

Silt (cont.)
    and oyster beds, 459–460
    in relict sediments, 121
    settling velocity of, 119
    transport of, 119, 121, 122
Silver, 403
Silverside, 489, 556
SIMA, 82, 88
Sine wave, 241, 244, 245
Sinking, (see also Downwelling)
    of marine organisms, 395–396, 419
Siphonophores, 424–425
Sipunculids, 452
Size, and biomass, 457, 485
    and temperature, 412
Skagit River, Washington, 298, 315, 316
Skate, 432, 493, 557, 558
Skeletal materials, 400, 491
Skipjack, 555
Skua, 498
Slack water, defined, 284
Slicks (see Surface films)
Slugs, 426
Slumping, of sediment, 43, 45, 124
Slush, 203
Smelt, 432, 489
Snails, 393, 426 (see also Gastropods)
    as deposit-feeders, 405, 444
    in eel-grass community, 455
    in marshes, 488
    in oyster beds, 461
    on rocky beaches, 447, 448
    in Sargassum community, 503
Snapper, 201
Snohomish River, Washington, 298, 316
Soap, 164
Sodium chloride, 144, 154, 155, 156, 204
Sodium ion, 144, 155–157, 159
    in deep-sea squid, 396
    in Hemigrapsus, 408
Sodium sulfate, 204
SOFAR, 201
Solar tide, 277
SONAR, 18, 199
Sound, generated by marine animals, 201, 557
    and study of ocean basin (see Reflection seismology; Refraction seismology)
    transmitted in seawater, 39, 142, 198–201
    transmitted in sediment, 15, 80
    and use of to attract fish, 484
Sounding, 38 (see also Echo sounding)
South Africa, 37, 184, 551
South America, 26, 30, 521
    as a barrier to water movements, 219, 235
    coastal relief of, 43
    and coastal upwelling, 127, 333
    connected with volcanic ridge, 90
    and marginal seas, 46
    and river discharge of suspended sediment, 116
South Atlantic Basin, 90
South Atlantic Central Water, 195

South Atlantic Ocean, barriers to water exchange, 233
    and fresh water input, 30
    and storm waves, 251, 253
    and tides, 279
    and water masses, 66, 184, 195, 231, 235
Southern California, 42, 44, 115
Southern Equatorial Current, 214–217, 219
Southern Hemisphere, and Coriolis effect, 224
    and currents, 30, 215, 216, 218, 228, 235
    distribution of land and water, 29, 30, 31, 36
    earliest evidence of solid crust, 97
    and Ekman spiral, 226
    and oceanic heat transport, 177
    and precipitation, 186
    and sediment transported by winds, 117
    and underexploited fish stocks in, 483
    and winds, 207
Southern Ocean (see Antarctic Ocean)
South Island, New Zealand, 29
South Pacific Ocean, and fresh water input, 30
    and islands, 33
    and waves, 252
South Pole, 179, 216, 515
South Sandwich Trench, 53, 68, 123
Southwest Monsoon Current, 217, 218
Spartina, 488
Spatangus, 430
Species, number of, and coastal regions, 433–434, 445
    and coral reef communities, 463–464
    and deep-ocean benthos, 465–466
    and ecological succession, 393
    and evolution, 412–413
    and habitat alteration, 491
    and open ocean regions, 433–434
    and salinity, 409
    and temperature, 412–414, 433
Specific gravity, 397
Specific volume, 150
Sperm whale, 482, 495, 500
Spicules, of sea ice, 203
    of sponges, 110, 112, 114
Spilling breaker, 256, 257, 258
Spirialis, 427
Spit, 362
Sponges, 75, 408, 463
    antibiotics produced by, 481
    deep-ocean species, 466, 467
    in oyster beds, 460
    spicules (see Spicules, of sponges)
    sterols derived from, 485
Spores, 394, 408
Spring bloom, 474–475
Spring tide, 272, 273, 277, 278, 282, 381
Spur and groove structure, 75

Squid, 426, 495, 500, 554
  deep-sea, 396
  as a food resource, 484
  giant, 13, 495, 500
  and light in the ocean, 439
Stable density system, 188, 192 (see
    also Density stratification)
Standing crop (see also Biomass)
  and effect of grazing on, 478–480
  and growth rate, 445, 480, 482–483
  measurement of, 478, 480
  and open ocean populations, 476
  and size of individuals, 457, 485
Standing waves, 262–263
  in estuarine systems, 304, 314
  and storm surges, 335
Standing wave tide, 278, 280–282, 291
Starfish (see Sea star)
Steady state (of world ocean), 47,
    117, 154
Steelhead trout, 558
Sterols, 485
Stokes' law, 119
Stomiatoid fishes, 484, 497, 556
Storms, New Jersey coast, 377
  Puget Sound, 319
  and shoreline features, 75, 314, 358,
    360, 369, 372, 373
  temperate region, 207
  and tides, 291
Storm surges, 335–337
  caused by winds, 225, 328
  and sea level changes, 334, 335,
    336, 359–360
Storm waves, 239, 240, 247, 249–253,
    377
Strait of Dover, 285, 291, 335
Strait of Georgia, Canada, 42, 297,
    298, 315, 316, 363, 437
Strait of Gibraltar, 35, 198, 282
Strait of Juan de Fuca, 42, 297, 298,
    315–317, 318, 350, 352
Stratification (see Density stratifica-
    tion)
Stratosphere, 102, 125, 177
Striped bass, 482, 489, 558
Sturgeon, life history of, 489
Subantarctic Water, 195
Subarctic communities, 412
Subarctic species (see Subpolar
    species)
Submarine canyons, 42, 43–45, 62,
    124, 255, 297
Submarine volcanoes (see Volcanoes,
    submarine)
Submersibles, 7, 15, 236, 406
Subpolar region, 184, 206, 207
  and biomass of organisms, 412
  and current gyres, 215–216
  productivity of, 207, 476
  and sediment transport, 132
  temperatures of, 410
Subpolar species, breeding cycles of,
    411, 434
  lack of diversity among, 412
Subsidence, of deltas, 361, 366–367
  of the ocean bottom, 72

Subsurface currents, 7, 30, 230, 233–
    235
  coastal ocean, 337
  Long Island Sound, 321, 322
  New York Bight, 348
  study of, 195, 213, 215
  and zooplankton migration, 435–
    436
Subtropical convergence, 31, 207, 214,
    228
Subtropical region, 183–184, 192, 206,
    207, 215, 228, 433
  and beaches, 369
  and current gyres, 215, 219
  productivity of, 207
  temperature of, 410
Subtropical species, 411, 413, 434
Sulfate ion, 156, 157, 159
Sulfates, 109, 161, 306, 402, 458, 491
Sulfer, as a mineral resource, 137
Sulfer dioxide, 100
Sun, emission spectrum of, 171
  and tides, 246, 268, 277, 278
Sunlight (see Insolation; Radiant
    energy)
Sunlit zone (see Photic zone)
Supersaturation, 160
Surf, 256, 258, 447
Surface activity, 164, 405
Surface currents, 214, 215–230, 233–
    235
  convergence of, 220–221, 228, 406,
    503
  depth of, 192, 220
  east-west trend of, 195
  mapping of, 7, 18, 20
  and ocean bottom topography, 220
  and sea ice, 204, 225–226
  and sediment transport, 120, 354
  Southern Hemisphere, 31
  and zooplankton migration, 435–
    436
Surface films ("slicks"), 164, 406
  and Langmuir circulation, 406, 503
Surface layer, stability of (see
    Density stratification)
Surface tension, 143, 163–164, 246,
    430
Surface zone, 191, 192, 193–194
  as a climatic buffer, 195
  and daily temperature cycle in the
    coastal ocean, 330
  productivity of (see Photic zone)
Surf perches, 430
Surf zone, 254, 363, 369, 374
  colonial sabellarians in, 461
Surge channel, 75
Surging breaker, 256, 258
Suspended load (see Sediment trans-
    port, and rivers)
Suspended organic matter, 404–406,
    451, 452
Suspended sediment (see Sediment,
    suspended)
Susquehanna River, 298, 306, 307, 360
Suwanee River, Florida, 368
Swallow, John, 213
Swallow float, 213
Swash, 375, 378
Swedish Deep-Sea Expedition, 1947–

Swedish Deep-Sea Expedition (cont.)
    48, 107
Swell, 247, 250–251, 257, 374
Swim bladder, 296, 436, 557
Swordfish, 482
Symbiotic relationship, of hermit-
    crab and sea anemone, 449
  of radiolarians and phytoplankton,
    423
  among reef-dwelling animals, 463–
    465
Synoptic oceanography, 171

Tabular iceberg, 205
Tacoma, Washington, 5
Tacoma Narrows, 316, 317
Tampa Bay, Florida, 312
Taonidium, 495
Tasmania, 37
Tees River, U.K., 409
Teeth, in sediment, 109, 114, 115, 121
Telemetrics, 9
Tellina, 452
Temora, 429, 470, 492
Temperate populations, 410, 411, 413
Temperate region, 206, 207, 412
  productivity of, 207
  temperature of, 410
Temperature, and breeding cycles,
    411, 434
  and change of state in water, 146–
    148
  and convergences, 406
  and density, 149–150, 188, 190–192,
    195–196, 208, 215, 221, 222–
    223, 230, 231
  and distribution of species, 410–
    411, 412, 433, 434–435, 475
  and flotation, 396
  and larval development, 421, 434–
    435
  measurement of, 169, 188, 189, 220,
    339
  and metabolic rate, 401, 410, 412,
    435
  and sound, 199
  and structure of water, 146–149
  as a tracer for water masses, 9,
    170, 195–198, 215, 231, 233,
    235, 339
  of world ocean, 195
Temperature distribution, Atlantic
    Ocean, 192, 193, 198, 219–220,
    231–235
  Chesapeake Bay, 308, 309
  coastal ocean (see Coastal ocean,
    temperature distribution)
  earth's surface, 182
  Indian Ocean, 192, 193
  Laguna Madre, 314
  Lateral, 190
  Long Island Sound, 320, 321, 322
  New York Bight, 343–347, 348
  Northeast Pacific coastal ocean,
    339
  Pacific Ocean, 192, 193
  Pamlico Sound, 311
  Puget Sound, 317, 318
  regional, 206–207, 230–231, 433
  Strait of Gibraltar, 198

Temperature distribution (*cont.*)
surface ocean, 177, 179–182, 195, 206–207
vertical, 179, 190–198, 223, 230–231, 233–235
Temperature of initial freezing, 156, 188, 201, 231
Temperature of maximum density, 149–150, 155, 188
Temperature tolerance, 314, 410–411, 413, 437, 448, 466
Temporary plankton (*see* Planktonic larvae)
Tentacles, of cephalopods, 427, 451
of coelenterates, 424–425, 429, 448, 450, 451, 463
of ctenophora, 425
of sipunculid worm, 452
Ten Thousand Islands, Florida, 383
Tethys, 94, 95, 97
Texas, 62, 312
Thalassia, 500
Thames River, Connecticut, 172, 320
Themisto, 423
Thermocline, 192, 201
coastal ocean, 331, 334
and internal waves, 261
and productivity, 334, 401, 474, 476
Thermohaline circulation, 223, 230–231, 233
Thiamin, 481
Thomson, Sir Wyville, 3, 21
Thorium, 81, 97, 136
Tidal bore, 282
Tidal bulge, 275, 276–277, 279
Tidal constituents, 278, 279, 282
Tidal currents, 213
and beach processes, 359, 370, 374
Chesapeake Bay, 287–289, 308
and coastal engineering, 8, 17
and deltas, 315, 362, 363
and erosion, 120, 363, 372
in estuaries, 301–303, 304, 336, 383
and filter-feeding organisms, 451–452
and idealized tide, 283
in lagoons, 299, 372, 387
Laguna Madre, 314
Long Island Sound, 284, 289–291, 319, 321–322
in marshes, 381–382
and mixing, 47, 301–303, 338, 350
and mud flats, 454
Nantucket Shoals Lightship, 285–286
New Jersey coast, 378
New York Harbor, 284, 285, 289
Northeast Pacific coastal ocean, 350
North Sea, 290–291
and phytoplankton in the Bay of Fundy, 475
prediction of, 269, 284, 291
Puget Sound, 315–317
reversing (*see* Reversing tidal currents)
rotary (*see* Rotary tidal currents)
and San Francisco Lightship, 287
and seahorse, 501
and sediment transport, 120

Tidal currents (*cont.*)
and shipping, 260, 268, 285
strength of, 285
study of, 287
Tidal curve, 269, 271, 279
Tidal day, defined, 270
Tidal flats, 120, 378, 382, 488, 501
Tidal inlets, formation, 372
Galveston Bay, 387
Laguna Madre, 312, 314
Long Island, 359, 363
New Jersey coast, 377, 378
Pamlico Sound, 309–310, 312
productivity of, 488
sedimentation of, 300, 359, 378, 384
Tidal marshes (*see* Marshes)
Tidal period, defined, 270
Tidal range, 272, 277, 279, 282
Columbia River, 363
and delta formation, 363
determination of, 269
Hudson River, 304
Laguna Madre, 313
and longshore bars, 372
and marsh formation, 381
Pamlico Sound, 310
Puget Sound, 316
and tidal currents, 285
Tidal rhythm, and microorganisms in intertidal sediments, 456
Tidal scouring, 447
Tide gauge, 269, 270, 276
Tide-generating forces, 274, 275–278, 282–283
Tide pools, and algae, 448
and coral reefs, 75
and rocky beaches, 448
and salinity variation, 407
and tube-worms, 461
Tide rip, 337–338
Tides, and beaches, 358
in estuaries, 301–302, 304–305, 337–338
and estuarine organisms, 488
hydraulic models, 387
and marshes, 331, 381–382, 453
New Jersey coast, 377
prediction of, 269, 270, 272, 277, 279, 283, 291
Puget Sound, 316
and salinity gradients, 409–410
and shoreline processes, 358, 360, 362
and storm surges, 335
and tide pools, 407
and water movements, 246
as waves, 246, 254, 262, 278–282, 283–284, 287, 304
Tide station, 269
Tide table, 8, 269, 291
Tierra del Fuego, 298
Tigris-Euphrates delta, 363
Tillamook Rock, Oregon, 258
Time, and longitude, 25, 515–516
Timor, Indonesia, 69, 70
Tin, 137
Tintinnids, 424, 481, 492
Titanic (ship), 18
Toegeria, 467
Tombolos, 371

Tomopteris, 426, 492
Tonga Trench, 53, 68, 123
Toothed whales, 482, 500 (*see also* Sperm whales)
Tortuguero, 500
Trace elements, 403–404
Trade wind coasts, 375
Trade winds, 132, 178, 206, 207, 215, 219, 222
Transform fault, 54, 89
Trawling, 484, 494
Trenches, 13, 15, 54, 66–67, 68, 124, 161 (*see also* Island arc-trench complexes)
animal populations, 411
associated with continental block margins, 40, 43, 58, 66, 83, 86, 97
destruction of crust at, 87, 88, 89, 95, 97, 98
distribution of, 53, 88
and heat flow from, 82
infauna of, 465
sediment deposits in, 85, 123
Trinity Shoals Lighthouse, 327
Tristan de Cunha, 35, 56, 66, 90
Trochoid curve, 241
Trochophore, 426
Trophic level, 393, 482, 491
Tropical communities, diversity of species in, 412–414
Tropical region, 179, 184, 207, 218, 433
beaches, 369
constancy of conditions, 413
populations, 410
productivity of, 476
and sediments in, 117, 132, 133
temperature of, 410
Tropical species, breeding cycles of, 434
and diversity of, 413–414
and tracing of water masses, 437
Tropopause, 177
Troposphere, 102, 177
Trough, of the tide, 279, 283
of a wave (*see* Wave trough)
Troy, New York, 304
Trunk Bay, Virgin Islands, 369
T-S diagram, 196–198
Tsunami (*see* Seismic sea waves)
Tuamoto Archipelago, 72
Tube-worms, in eel-grass communities, 454, 455
in oyster beds, 460
and reef-building, 461, 462
on rocky beaches, 447
Tufts Abyssal Plain, 62
Tuna, 482, 494, 555
Tunicates, 430–431, 432, 455
Turbidites, 123, 124
Turbidity, coastal ocean (*see* Coastal ocean, turbidity)
and distribution of species, 433, 451–452, 455
and fecal pellets, 452
Laguna Madre, 314

Turbidity (*cont.*)
  and productivity inhibited by, 475
  Puget Sound, 319
Turbidity currents (turbidity flows), 123–125
  and absence of biogenous sediment near continents, 129
  and barriers to the flow of, 59
  and burial of bottom features, 60
  and formation of continental rises, 45
  and plant debris in the deep ocean, 124, 467
  and river-transported sediment in the deep ocean, 132
  and submarine canyons, 44
Turbot, 553
Turbulence (turbulent flow), 119, 120, 221, 226, 301, 374
  and deep-ocean benthos, 465
  and recycling of phytoplankton, 420, 474
Turtle-grass, 500
Typhoons, 253
Typhus, transmission by oysters, 460

Ultra-violet radiation, 100, 101, 102, 177
Umbellula, 468
Underwater houses, 17
Underwater photography, 173
University of Miami, Institute of Marine Sciences, 103
Upwelling, 227
  and the coastal ocean, 328, 333, 334–335
  and deep-water species, 437
  and diatoms, 420
  in lakes, 221
  and land climate, 219, 334
  Northeast Pacific coastal ocean, 317, 352, 353
  and productivity, 127, 219, 334–335, 401, 412, 476, 477, 482
Uranium, 81, 97, 136
Urbanization, and estuarine systems, 312, 315
Urban wastes (*see* Municipal wastes)
Urea, 408
Urechis caupo, 454
U. S. Army Corps of Engineers, 8
U. S. Coast Guard, 205
U. S. Federal Water Quality Administration, 490
U. S. S. Ramapo, 253

Vancouver, British Columbia, 5
Van der Waals bonds, 143–144
Vapor pressure, 156
Varves, 134, 306
Vegetation (*see* Plants)
Veliger, 427
Venerable Bede, 268
Verhoyansk, U. S. S. R., 146
Vertebrates, 431
Vertical migration, 397, 418, 435–436, 438

Virginia, 201, 217
Viruses, 394
Viscosity, 143, 156, 161–162
Vitamin B-12, 10, 305, 481
Vitamins, and productivity, 473
  synthesis of 305, 397, 481
Volcanic ash, 61, 121, 125, 136
Volcanic glass, 63, 131
Volcanic island groups, 61, 65, 66
Volcanic islands, 62–63
  Atlantic Ocean, 35, 56
  and atolls, 73, 91
  formation of, 33, 64, 90
  Indian Ocean, 37
  Pacific Ocean, 33, 61, 68
Volcanic provinces, 65
Volcanic ridges, 40, 46, 54, 66, 90–91
Volcanic rock, and archipelagic plains, 65
  as base of atoll, 74
  formed from lava, 63
  in Layer 2 of ocean basin, 83, 85
  as lithogenous sediment, 111, 112, 125, 126
  and magnetic field reversal, 91
  on Mid-Atlantic Ridge, 83
  as nucleus for Fe-Mn nodules, 115
  on slopes of submarine volcanoes, 63
  and subsidence of ocean basins, 65
  and transport by sea-floor spreading, 90
  transverse to oceanic rises, 66
Volcanoes, Atlantic basin, 35, 54, 56, 90, 125
  as coastal features, 361
  at crustal plate margins, 57, 66, 87, 88, 89
  distribution of, 57
  eroded by waves, 33, 63, 64, 65, 74, 90
  and "hot spots", 90
  Indian basin, 54, 125
  Pacific basin, 33, 54, 62, 64, 65
  submarine, 33, 60, 61, 62–65, 73–74, 90
  in trench and island arc areas, 69, 84–85
  in young ocean areas, 96
Vulcanism (volcanic activity), 63, 66, 90, 114

Waikiki Beach, Hawaii, 362
Wake, of a storm, 336
Walvis Ridge, 66, 90, 233
The Wash, England, 291
Washington coast, and upwelling, 335
Washington, D. C., 5, 304, 340
Washington-Oregon coast, 217, 342, 351, 352
Washington (State), 59, 122
  University, 103
Waste chemicals, 348, 353, 460, 490
Waste heat, 308, 348, 411
Wastes, in the coastal ocean, 349, 490–491
  dilution and dispersal, 19, 47, 349, 354
  in estuaries, 296, 383, 387
  in harbors, 384

Wastes (*cont.*)
  human, 404, 490
  Laguna Madre, 315
  and marine life, 11, 315, 349, 353–354, 384, 403–404, 490–491
  in marshes, 385, 386
  metabolic, 393
  New York Bight, 343, 349
  nutrient-rich, 305
  regulation and control of, 349
  Washington-Oregon coast, 353
Water, boiling point, 143, 146, 148
  commercial production from sea-water, 165
  compared with chemically similar compounds, 143, 147
  as a constituent of organic matter, 100
  density (*see* Density)
  distribution on the earth, 26, 28, 179, 182–183
  freezing point, 143, 146, 147, 148–149, 156
  juvenile, 98
  melting point (*see* Water, freezing point)
  solvent properties, 143, 155
  viscosity (*see* Viscosity)
Water budget, 33, 155, 170, 179, 182–184, 186, 188, 191 (*see also* Hydrological cycle)
Water masses, formation of, 9, 194–197, 208, 231, 232, 233, 235, 303
  identified by plankton populations, 437
  mixing of (see Mixing of water masses)
  movement of, 153, 170, 192, 195–198, 208, 213, 223, 230–231, 233, 235
  vertically stratified, 188–191, 194, 231
Water parcels (water particles), movement of, 221, 225, 230, 242–244, 420
Water spiders, 164
Water structure, in deep-sea organisms, 396
  molecular, 141, 143–144, 146
  molecular aggregates, 146, 147, 148–149, 150, 156, 162, 163, 164
Water vapor, 26, 176
  in atmospheric circulation, 177, 184, 188
  origin of, 87, 100
  physical properties of, 144, 148, 149
  and water budget, 184, 186, 188
Wave age, 250
Wave approach, 255, 374, 375, 376
  and beach cusps, 380
Wave crest, 239, 241–243, 248–249, 250, 255, 257, 260
Wave energy, changed to heat energy, 247, 254
  and wave frequency, 245
  and wave height, 247, 248–250, 251, 253, 257
  and wave period, 246, 249, 253–254

Wave energy (cont.)
  and wave refraction, 254–255, 256
  and wave speed, 244
  and wind direction, 253
  and wind energy, 247–249
Wave erosion, of coastal features, 33, 63, 75, 120, 122, 255, 360, 362, 363
Wave frequency, 241, 244, 245
Wave height, 241, 243, 244, 252, 254–255
  and beach cusps, 380
  and fetch of the wind, 249
  of internal waves, 262
  and water depth, 255
  and wave energy, 247, 248–250, 251, 253, 257
  and wind speed, 248, 253
Wavelength, of ocean waves, 241, 243, 244
  of tides, 279
  and water depth, 254–255, 260, 377
  and wave speed, 250–251
Wave period, and restoring forces, 246
  of standing waves, 263, 281
  and water depth, 254, 260
  and wave energy, 246, 249, 252–254
  and wave speed, 241, 243, 250–251
Wave reflection, 258
Wave refraction, 255, 256, 257, 263, 374
Waves, (see also Breakers; Deep-water waves)
    Shallow-water waves; Storm waves)
  and beach processes, 358, 369–380
  components of, 244
  and delta formation, 315
  disturbing forces, 244–247
  and intertidal organisms, 447, 456, 461, 462
  and longshore currents, 374, 376
  and mixing, 190–191, 301, 338, 350
  and motion of water particles, 242, 243, 244, 256, 262, 375, 377
  and reflection of light, 173
  restoring forces, 244–246
  and sea ice, 204
  and sediment transport, 299, 358, 362, 363, 369–370, 372, 375, 381
  and surface slicks, 406
Wave spectrum, 244, 252
Wave speed, 241, 243, 244, 253, 254
  of internal waves, 261–262
  and water depth, 254
  and wavelength, 250–251
  and wind speed, 250–251
Wave steepness, 241, 242, 249, 250, 255
Wave tanks, 243, 244
Wave train, 243, 244, 255, 279
Wave trough, 239, 241, 243
Weakfish, 557
Weather, 177, 178, 185
Weathering, 112, 117, 120, 122, 131,

Weathering (cont.)
  154
Weddell Sea, 231
Weddell seal, 499
Weight in the ocean, 394–396
Well-sorted sediment, 111
West Coast of the U. S. (see Pacific Coast of the U. S.)
Westerly winds, 178, 207, 222
Western boundary currents, 218, 219, 228
Western North Pacific Water, 195
West Indies, 69
West Point, New York, 304
West Wind Drift, 207, 214–219, 235, 350
Wetlands, 382
Whales, (see also Baleen whales; Sperm whales; Toothed whales)
  and copepods, 430
  on deep-ocean floor after death, 467
  earbones in sediment, 111, 112, 114, 115
  early records of, 18
  filter-feeding in, 430
  as mammals, 410
  as nekton, 472
  size of, 12, 394
  sounds produced by, 201
  in subpolar regions, 207
Whaling grounds, 412
Whip coral, 463
Whiting, 488, 553, 554
Wick Bay, Scotland, 258
Windrows, 406, 503
Winds, (see also Atmosphere)
  Chesapeake Bay, 308
  and coastal circulation, 216–217, 328, 331–333, 346, 350–352
  and coastal ocean salinity, 329–330
  and coastal ocean temperature, 329
  and Coriolis effect, 223
  and heat transport, 176, 178
  Laguna Madre, 313–314, 318
  and mixing of surface waters, 47, 190–191, 221, 328, 401
    New York Bight, 346–347
    Northeast Pacific coastal ocean, 351
  Pamlico Sound, 310, 311, 318
  Puget Sound, 318
  and sea ice, 204, 499
  seasonably variable, 47, 215–218, 222, 246
  and sediment transport, 117, 121, 125, 131, 132, 153, 370, 373, 379, 381
  and storm surges, 335, 359
  and surface currents, 207, 208, 212, 215–218, 221, 222, 225, 226, 235, 328, 346–347, 348
  and surface slicks, 405–406
  and tidal currents, 285, 291
  and upwelling, 219, 227, 328, 333,

Winds (cont.)
  352
  and wave generation, 225, 246–250, 251, 252, 304, 382
Wisconsin glacial period, 62
Within-level predation, 482
Woods Hole, Massachusetts, 455
Woods Hole Oceanographic Institution, 5, 103
World catch, 5, 483, 484, 485
World ocean, physiographic provinces, 54
  salinity of, 33, 195
  surface and drainage area and average depth of, 32
  temperature of, 33, 195
  volume, density, mass and abundance of, 27
Worms, 393 (see also Tube-worms)
  as deposit-feeders, 451, 452, 454–455, 466
  in the deep ocean, 466, 467, 469
  in marshes, 488
  as meiobenthos, 456
  in oyster beds, 460
  planktonic, 425–426
  as predators, 444, 469
  and salinity changes, 408
  in Sargassum community, 503
  segmented, 426
  in tide pools, 407, 448

Xenon, 98, 160
X-ray techniques, 113, 130

Yangtze River, China, 116, 117, 118, 361
Yellowfin tuna, 554
Yellow River (see Hwang Ho River)
"Yellow substance," 172
Yenisei River, U. S. S. R., 36
Yoldia limatula, 455–456
York River, Virginia, 306, 307
Yucatan Peninsula, 125

Zeolites, 109, 132, 133
Zinc, 403
Zinc-65, and marine organisms, 354
Zones, intertidal (see Intertidal zonation)
Zooplankton, 207, 334
  composition of, 399
  at convergences, 406
  distribution of, 421, 433–435, 437
  English Channel, 480
  in estuaries, 305, 490
  and grazing of phytoplankton crop, 421, 478–481, 490
  migration of, 435–437
  Peruvian coast, 477
  and zinc-65, 354
Zooxanthellae, 462–463
Zuider Zee, Netherlands, 16, 386

OCEAN SURFACE CURRENTS IN FEBRUARY-MARCH

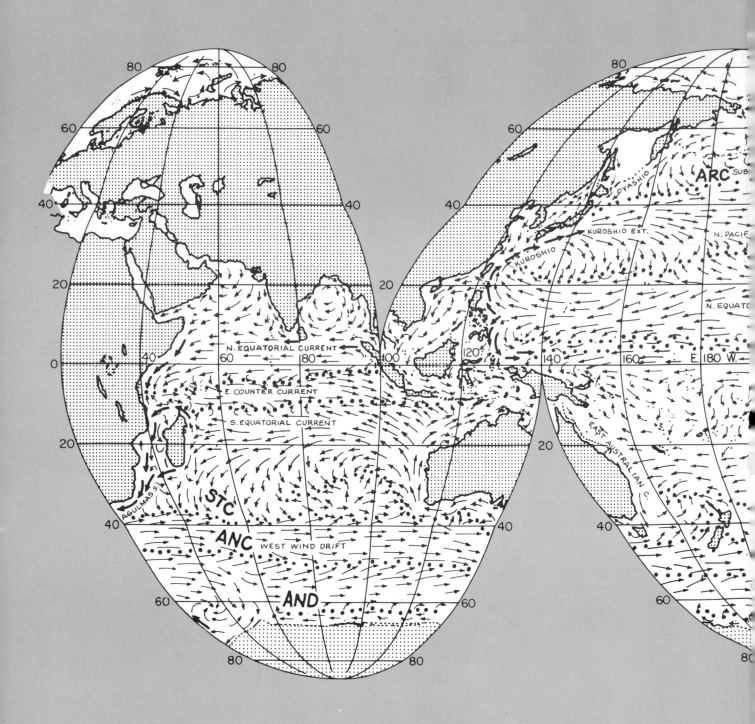

ARC – Arctic convergence   ANC – Antarctic convergence

STC – Subtropical convergence   AND – Antarctic divergence